What Oxygen Users Are Saying About Ed McCabe© and Oxygen Therapies!

"Take note. The interest this book will unleash is noteworthy. Both lay and professional persons who secretly will not admit their ignorance in this emerging field of study will use it as a guide."

Mona Harrison, MD, Former Assistant Dean, Boston University School of Medicine

"Your name is well-known in Russia particularly among the specialists in the field of Ozone Therapy."

Natalia Berdnikova,
Manager Foreign Business, Medozons, Nizhny Novgorod, Russia

"Ed's information really ignited my brain."

Viktor Goncharov, ND, Ph.D., New York

"I now literally thank God for giving us Ed McCabe. It is through his courage and conviction that my life was saved. Thank you, Ed."

Chris Savage, Australia

"May God continue to bless you Ed as you have blessed us with your compassionate service. Humankind is the better for your being."

Jerry Bergemeyer, Arizona

"Thank you Mr. McCabe. You're a gift from a higher entity."

Linda Dale, Florida

"Ed, I applaud your courage to press on despite all that you've had to come against. God will never forget your sacrifices for him to help his children."

Kathleen Swan, Oregon

"Ed, God bless you my man. I'll always owe you more than I will ever be able to repay. I'm heartfelt about the tribulations you've had to endure over helping people like me. My deepest appreciation and respect."

Bob Carlisle, Massachusetts

"Thanks Ed, you're a God damn legend and are brave. I am sending out your message daily to hundreds in my presentations. Thanks Ed, you saved my brothers life."

J.F., Australia

"I thank you for your stand and crusade for our health in spite of the dangers you have had to endure. I would say your stance would describe one of my favorite sayings, 'One man with courage makes a majority.'"

M.G., New Hampshire

"Ed, it was a lucky day for us when you found your way into our lives. One rarely has the opportunity to meet such wonderful and caring people."

Al Sands, Florida

"I am doing my best to promote your work and the concept of Oxygen Therapy. The work is so important and many more people must know about this. Keep up the fantastic work. God Bless."

John Smith, England

"I think you are a very brave and courageous person and your stand against the very malicious authorities deserves the admiration and support of all thinking people."

Nancy Adams, Oregon

"We all owe a tremendous debt of gratitude to Ed McCabe for researching and exposing the range of Oxygen Therapies available, many of which are within the reach of anyone regardless of location or resources. Ed has, at great personal sacrifice, continued to teach and to research, and we owe him a lot."

M. B., Florida

"Since early times the chosen ones have referred to thunder as the 'voice of God' and the O_3 produced in such thunder storms as the 'Breath of God.' The proper spelling for the Supreme Eternal Purifier of all living things is actually GO3D. And Mr. Ed McCabe is today's messenger. So, arise, purify thyself and restore the earth to all its OXYGENATED brilliance."

Zero$_3$ Man, California

"I became aware of Ozone and Oxygen Therapies some 13 years ago when listening to a Radio broadcast (KPFK) FM out of L.A. and I was amazed at the information I was hearing about how cancer, hepatitis, HIV, and most of the other life-threatening diseases of mankind were all 'anaerobic' and could be effectively killed with ozone (A higher form of oxygen) with little to NO side effects to the human body.

I had just become a journalist and an investigative reporter for Paragon Cable Television and later to become TIME WARNER Cable television all over the U.S. I was so excited and amazed by the information that I was hearing I could not sleep. I had to find out more about mind-blowing ozone and oxygen killing all of these diseases. I thought to myself, this could be the biggest story of the decade if not the century. I started investigating and heard about Ed McCabe and his book 'Oxygen Therapies.' I read the book and was completely blown away.

I attended the very next Cancer Control Society health convention in California and I was coming around a blind corner and ran right into Mr. Ed McCabe. I attended his lecture and found him to be very precise, very knowledgeable, and very credible in his presentation.

It was very hard for me at the time to find others in the United States that were using or talking on this subject other than Ed McCabe. Ed was engaged in something that almost no one was doing at this time, lecturing, investigating, researching, and reporting on ozone all over the world—and especially in America where almost no one knew anything about it. I had interviewed and talked to hundreds of doctors in the U.S. that didn't know anything about the medical use of ozone that actually started in America in 1900 with Nikola Tesla.

Ed continues to this day to be one of the leading forces in bringing knowledge and awareness of Ozone and Oxygen Therapies to the world and especially here in the United States.

Ed, keep the flame alive and burning. Humanity and all those suffering unnecessary deaths from disease need you. We thank you soooooooooooooo much for never stopping your mission."

Keith Ranch, producer, reporter, and journalist, California.

FLOOD
YOUR
BODY
WITH

OXYGEN

Happy Oxygen!

Ed McC

FLOOD
YOUR
BODY
WITH

OXYGEN

THERAPY FOR OUR POLLUTED WORLD

By

"Mr. Oxygen™" Ed McCabe©

∞ Energy Publications ∞

The previous book in the Oxygen Awareness Series Oxygen Therapies, A New Way of Approaching Disease, first edition 1988, sold more than 250,000 copies, mostly by word of mouth.

Book design: Ed McCabe©
Cover design: George Foster of Foster and Foster
Editing: Ron Kenner of RKedit - www.rkedit.com
Editing: Leeda McCabe

Contributors

Dr. Mona Harrison, Dr. Martin Dayton, Dr. George Freibott, Dr. Fereydoon Batmanghelidj, Dr. Larry Lytle, Dr. Saul Pressman, Dr. Oldrich Capek, Dr. Charles Farr, Dr. Darryl Wolfe, Dr. Frank Shallenberger, Dr. James Boyce, Leeda McCabe, Udo Erasmus, Sam Biser, Ron Logan, Keith Ranch, Nancy Adams, Jim Brown, Ken Hughes Basil Wainwright, Norman Cowell, Mark Lester, Richard Kroll, and Rev. Mary Seid.

Catalog suggestion

McCabe, Ed
    Flood Your Body With Oxygen: Therapy For Our Polluted World / Ed McCabe
    © 2003. – 6th edition.

ISBN: 0-9620527-2-8(pbk)

Health–Practical Solutions
Disease-Practical Solutions
AIDS–Practical Solutions
Cancer–Practical Solutions
Naturopathy–Treatments
Medicine–Complementary Alternatives
Pets-Practical Solutions
Veterinary–Complementary Alternatives
Animal Husbandry–Practical Solutions
Gardening–Practical Solutions
Farming–Practical Solutions

DEDICATION

To My Continually Amazing Wife
Leeda

In Service With and For

The Infinite Ocean of Love and Mercy

Acknowledgments

My sincere heartfelt thanks to:

John Taggart for running the fulfillment and distribution during the first book, so the people could get what they needed and also for his years of help.

George Freibott, Wally Grotz, Charlie Farr, Richard Wilhelm, Duncan Roads, Bruce Smeaton, Hugh Sangster, Sherill and James Sellman, Marjie Adleman, Keith Ranch, Faye Joseph, Howard Griswold, Bill Galkowski, Sandy Copeland, Carl Vollmer, Arleim Steiner, The Garrods, Carol Kuria, Claudine Menta, Ruth Hutchins, Kris Galli, Norman Colwell, John Smith, Mark Lester, Everett Hale, Lorin Dyrr, Guy Roberts, Jim McMahon, Jeremy Reiss, Tom Corbett, Phil Ratte, Peter Christian, Rick Kroll, Jim Brown, Leeda McCabe, Carol Gunther, and others–for their help when it was needed. Abraham Wilson, Paul Twitchell, John Dalton, Janelle Wardell, Harold Klemp, Marienne DuBois, Clint Miller and Jack O'Brien for their sage counsel.

All those who raised funds, and those who donated at just the right time.

All those who generously gave me rides, food and shelter, or worked the booth as needed during all those years on the road.

All the doctors, naturopaths, nurses, healers, media people, buyers, meeting promoters and administrators who wisely listened instead of turning away.

All the local pioneers and opinion leaders who presented me to their flocks.

And all the people who stood up at my lectures and told everyone else that they had to really listen closely to what I was saying, because the information had already worked for them in a miraculous way.

All the original American oxy-heroes whose accomplishments went unsung, people such as—Nickola Tesla, F. M. Eugene Blass, Otto Warburg, William F. Koch, Carl Edward Rosenow, William Turska, and Robert Mayer. Their efforts were followed by the modern works of Charlotte Gerson, George Freibott, Julius LaRaus, Andrija Puharich, F. Sweet, M.S. Kao, S. Lee, W. Hagar, Richard Wilhelm, Walter Grotz, Charles H. Farr, Lee Devries, Migdalia Arnan, Phil Seifer, Renate Viebahn-Haensler, Rip Rice, Ray Evers, Terrance McGrath, Lucas Boeve, Horst Kief, Richard Ribner, Frank Sedarnee, Erwin Dorsch, John Taggart, John Shilling, Bill Fry, Bud Curtis, Michael Carpendale, Tom Vallentine, Eustace Mullins, Michael B. Schacter, Ron Hoffman, Robert Atkins, Rathna Alwa, Kenneth Wagner, Wells, Latino, Galvachin and Bernard Poiesz, Alexander Preuss, Michael Shannon, James Boyce, Basil Wainwright, Julian Whittaker, John Waldron, James Caplan, Frank Shallenberger, Kurt Donsbach, Joseph Passero, Saul Pressman, Brad Hunter, John Pittman, Berkley Bedell, Tom Harkin, Geoff Rogers, Gary Null.

Others included—Tom Kopko, Morton Walker, Dan Wiliamson, Cameron Tapp, Norm McVae, Jane Heimlich, Lorraine Day, Dale Pond, Bob Willner, Jim Karnstedt, Norm Ralston, Roger Clemente, Hugh Sangster, Bruce Smeaton, Duncan Roads, Sid Safon, Bill Bump, Jerry Decker, Larry Thatcher, Donna Andrew, Skoshi Farr, Ken Bock, Howard Robbins, Carlos Jimenez, Bruce Hedendal, Mark Konlee, Christian Coffinet, Dan Chittock, Keith Ranch, Steve Kurzweil, Tom Corbett, Murray Susser, Roy Kupsinel, Johnathan Wright, Kirk Morgan, John Parks Towbridge, Bernard Bleem, Bernie Kirshbaum, Richard and Ginger Neubauer, William Campbell Douglass, Martin Dayton, William Faber, Ravi Devgan, Robert Rowen, Matthew Morton, Leonard Haimes, Seth J. Baum, Darrell Wolfe, Marcel Wolfe, Maurice Hathaway, Dave Sterling, Adam Locke, Abe Chaplan, Ziegfried Rilling, Betsy Russell Manning, Nathanial Altman, Bob Beck, Ross Turlog, Norm Colwell, Roger Doucette, Burton Goldberg, Jeff Harrison, Darell Stoffels, Bernd Friedlander, Eddie Barr, James Julian, Waves Forrest, Daniel Bangs, Larry Duffy, Jessie Partridge, Tim Gunns, Bob King, Thom Dean, Susan Lark, A J McDonald, Robert Strecker, Mike Davis, Hector DeLafuente, Roy Tuckman, Paul Crosswhite, Frank Sontag, Alex Duarte, Scott Witt, Jim Brown, Ken and Mardy Theifault, Norman Fritz, Lorraine Rosenthal, Andrew Pincon, Buck McCabe, Hector deLafuente, Tom Brown, Den Rasplica, Gary Gordon, Tom Vallentine, Ken Mouw, Dean Stonier, Alwyne Pillsworth, Joel Wallach, Elmer Heinrich, Clay Lewis, Joan Priestly, Leo and Diego, Peter McGratten, Peter Aykroyd, Jacob Swilling, Deepak Chopra, Richard Murray, Karen Calabrese, Steve Atkins, H.E. (Helfried Eric) Satori, Hans Neiper, Moe Lepenven, Bill Sailing, Alex Nahow, Claudia and Frank McGruer, Jennifer Bolen, Christine Beaudoin, George Klabin, Bill Lyons, Steve Stein, Shaun S. Pierson, Chris Shaw, Mary Lou Brady, Lance Manning, the members of The International Bio-Oxidative Medicine Association, The International Oxidative Medical Association, Medizone, The International Ozone Association, The American Naturopathic Association, and all the families, teams and other kind friends who personally helped me whom I have unfortunately failed to mention by name, especially the Australian, Kiwi, and European oxy-helpers.

All I can do is say that a big oxy-thank you goes out to all of you. This group of souls deserves the center stage for the entire world to see. Each made some positive contribution to either my personal mission or to the Oxygen Therapies being forwarded here. We owe all of these people a debt of gratitude.

Disclaimer, Intent, and Rights Reserved

I am the author of the best selling 1988 book, *Oxygen Therapies A New Way Of Approaching Disease*. As the first and only book of it's kind, it detailed a whole new way of approaching and treating disease safely and effectively without causing any poisonous side effects. Incredibly, some people saw this as somehow threatening and did their best to prevent the spread and disclosure of these facts to you.

The main reason I am writing this disclaimer is that I do not know how you will use this information. Some of those among us have questionable realities. Will you use this information to help yourself and others, or engage in the opposite? I do not take responsibility for your choices. Choose wisely.

A major aim in my life is to educate every household on the planet about the tremendous value of proper oxygenation. So you will know my findings, I publish what I discover as often as I am able. As a journalist, I research medicine and cutting edge healing while promoting oxygen awareness, health, and environmental consciousness. I simply report all the facts to you so that YOU can make fully informed decisions. As an analyst, I also editorialize to help you put everything in perspective.

My publications are frank discussions of healthcare and environmental issues. I'm not giving or intending to give, anyone medical or veterinary advice, so please don't think you're getting that here, you're not. I'm not advising you to do anything. I'm simply telling you what I have found out, as is my right, and as necessitated by your right to be fully informed. 'Mr. Oxygen™' is my trademarked name. This book and the writings, information, and personal intellectual creations within it are federal and common law copyright©2002 by Ed McCabe©1967 All Rights Reserved. Licensed to Energy Publications.™1988

Our right to liberty reaches far beyond obtaining permission from someone else in order to live freely. Our right to liberty encompasses our right to health and also the very important right of being fully informed. This means FULL disclosure of all health solutions available for your consideration at all times. Only then can you choose wisely.

Nothing stated herein is a diagnosis, treatment, or cure for anything, and it is not intended to diagnose, treat, or claim to cure. I also do not claim, or intend to claim that the information herein will affect the function or structure of any humans, animals or vegetables, no matter how convincing the evidence I present. The way you use the information means individual results will always vary. I will say this repeatedly to remind you. I leave it all up to you and the healers, doctors, and politicians. You must prove or disprove any therapy's effectiveness in each and every individual case.

I do not diagnose or prescribe, and everything I say in this book is preceded with, "In my opinion." But rather than my irritating you by saying this phrase over and over, please apply it mentally throughout the book to everything I say. Thank You.

Table of Contents

The Beginning

Part One: The Problems

Part Two: The Solutions

Part Three: The Evidence

Part Four: The Politics

Part Five: The Resources

Dr. Mona Harrison, MD
Former Assistant Dean—BU School of Medicine

"Ed McCabe is a breed of journalist from the 'old school' of investigation. His quest for knowledge and truth has been exhaustive–even when it has potentially meant personal harm. His realizations that water and air are not just *water and air* have spurred his efforts forward.

"The story of oxygen and its primal role in water and health enhancement are the book's key ingredients. Not only is oxygen a key to the air we breathe, it is equally important in the water we drink and otherwise use. In his treatise, Ed McCabe takes one well into this new century with an understanding of the applications of water and Oxygen Therapies.

"Both lay and professional persons who secretly will not admit their ignorance in this emerging field of study will use it as a guide.

"It is only here in North America that the significance of this science is just awakening. The rest of the Western Scientific World caught on a long time ago.

"Take note. The interest this book will unleash is noteworthy."

Mona Harrison, MD

SIGNIFICANT POSITIONS HELD By Dr. Mona Harrison
Assistant Dean–Boston University School of Medicine
Director–Adolescent Medicine, Boston City Hospital
Chief Medical Officer–D.C. General Hospital–Directed Trauma Center Emergency Services, Outpatient Services and Employee Health

Medical Director-Rehab Inc., a multifaceted physical rehabilitation agency

EDUCATION
MD University of Maryland School of Medicine
Residency, Family Medicine, University of Maryland Hospital
Residency, Pediatrics, Boston City Hospital, Boston University
In-Service Fellowship:
Adolescent Medicines Boston Children's Hospital, Harvard University
Neonatology Research Fellowship
Boston University Hospital, Boston University

Dr. Martin Dayton, MD

"Ed McCabe, affectionately referred to as 'Mr. Oxygen' by his admirers, has literally educated millions around the world on the health benefits of oxygen. I know from personal experience the validity of his teaching, for as a physician I have used oxygen in its various forms in my practice for overcoming disease, restoring health and improving human performance since the 1980s. In the 1980s, the few existing doctors treating with one or more forms of oxygen often did not know that colleagues using other forms existed. I had come to personally know Mr. Ed McCabe as a meticulous investigator on a mission to tell the world about all the known therapies in this relatively unknown field at that time.

"Today, Ed McCabe is a celebrated authority on Oxygen Therapies. He is an acclaimed author, avid researcher, and energetic humanitarian in the field. In order to share his discoveries with the world, he wrote numerous stimulating articles in a gifted style that allows for large amounts of pragmatic knowledge to be effortlessly assimilated. Thanks to the knowledge disseminated by Mr. McCabe, public awareness of the therapies has remarkably increased and the medical community using Oxygen Therapies has grown in both wisdom and population.

"The driving force behind Mr. McCabe's popularity, more so than his talents in communication, is the enormous value of the knowledge he offers for living in the 21st century. He teaches how oxygen related therapies, such as ozone, peroxide, hyperbaric and photoluminescence, may be applied to complement or replace various in-vogue medical and health practices, yielding superior outcomes with less cost and undesirable side effects. The huge list of conditions that are prevented and treated by Oxygen Therapies range from fleeting minor colds to chronic wounds that do not heal to dreaded life taking diseases such as cancer. Mr. McCabe, the scientist, teaches why oxygen related methods work so well and how to apply them.

"He provides us knowledge on how we can all take advantage of exciting, scientifically backed health related opportunities available today which still remain largely unknown to the public and the medical profession. His book, *Flood Your Body with Oxygen, Therapy for Our Polluted World* is a must read and must share! Thank you 'Mr. Oxygen.'"

Martin Dayton D.O., MD, MD (h), C.C.N., F.A.A.F.P.

Dr. George Freibott, ND

"Wally Grotz, who with the late Father Richard Wilhelm refined decades of Hydrogen Peroxide materials into an oxidative protocol for the general public, first referred Ed McCabe to I.A.O.T. (International Association for Oxygen Therapy). Being involved in 'Oxygen Therapies' for over a century, the Institute, known formerly in Germany as the *Institut fur Sauerstoff Heilvahfahren* and in America, as the *Eastern American Association for Oxygen Therapy*, had its roots in the 'air therapies' of German Naturopathy. If our counsel was at all of worth, perhaps it was to excite Ed to examine just how broad a field the oxidative sciences really are. Many times he zigzagged around the globe compiling international archives and spreading the truth about Oxygen Therapies. He wrestled with new scientific concepts and long-standing social apathy. Ed, then a worthy student, is now a knowledgeable teacher. An honored friend, Ed has proven to be a solid investigator and reporter, an animated educator, and a steadfast truth seeker.

"Ed has risked his all many times in the hopes that the common good would prosper. As a layman, he tackled the very bastions of medicine. Addressing governmental hierarchies, Ed rallied for freedom of choice. Sometimes single-handedly, Ed appealed to corporate cartels with less than popular empirical evidence.

"Ed, (more popularly known as 'Mr. Oxygen') has rattled the 'dry bones' of our society. He has taught us about 'Oxygen Therapies' but as importantly, he has shown us how to live our lives with style and with spiritual tenacity. In closing his first book, he opens us to infinite possibilities. His conclusion? 'We all have a lot to discover.'"

George A. Freibott IV, ND, MD (h), DNB.
President and Lifelong Trustee
American Naturopathic Association
January, 2002, President of the American Naturopathic Association

The
Beginning

The Ed McCabe Story

How a Regular Guy Became "Mr. Oxygen™"

Ed McCabe, "Mr. Oxygen" is a best selling author, investigative journalist, analyst, writer, lecturer, and consumer advocate in the innovative oxygen and health area. The moniker "Mr. Oxygen" started as a gag, a play on Ed's expertise in his favorite subject combined with the name of Ed's favorite childhood TV program, "Mr. Wizard." Many a cold snowy New England Saturday morning in the '50s found young Ed glued to the TV while being fascinated by the science show's constant parade of new mind-expanding concepts. His constant experiments with

what his father called 'tinkering' with everything mechanical and electrical in the house caused his mother many a worry and clean up session. He was often hospitalized with pneumonia as a child due to being born prematurely and consequently not having strong lungs. He says God was trying to make sure he got the message to be compassionate for the sick later on. When not playing neighborhood sports many of his after school sessions were spent in the local library's science and history sections.

New Zealand and Australia Oxy-movement volunteers scheduled rounds of lectures, TV, and radio 'talkback' venues for Ed, and took turns delivering him safely and well fed to each event.

As an adult, Ed lectured worldwide, and has been the honored recipient of several awards. He is the first and only one to create a modern mass public awareness of the existence and benefits of Oxygen and Ozone Therapies.

One milestone in his career was when Mr. McCabe, along with Senator Tom Harkin and former Congressman Bedell organized and brought former AIDS patients and their doctors to the top levels of the National Institute of Health. The patients had become healthy by using Oxygen Therapy. Mr. McCabe's international expertise, recognition, popularity, and experience have also enabled him to appear on U.S. network television.

Mr. McCabe holds a degree in Educational Media from the University of Massachusetts. He is a leading international author, lecturer, consultant, and promoter of Oxygen and Ozone Therapies. His ongoing involvement with advanced healing modalities encompasses a span of over 25 years. As a journalist, he solely focused upon studying Oxygen Therapies during 12 years of intensive investigation, research, experimentation, interviews,

On location in San Francisco.

and travel. He has personally interviewed thousands of Oxygen Therapy users and suppliers. As a result, he is recognized and acclaimed as a top international expert on the subject of using oxygen and ozone for health and environmental improvement.

Without a major publisher, his best-selling, and now cult classic, 1988 book *Oxygen Therapies, A New Way of Approaching Disease,* sold more than 250,000 copies through word of mouth. His was the first book to detail every known therapy that used special forms of oxygen to eliminate disease and restore health by ridding the body of toxins and microbes, while simultaneously boosting the immune system.

As a direct result of Mr. McCabe's writings, his audio and video tape publications, and his extensive worldwide appearances and lecturing, millions have now learned that the proper use of oxygen supplementation and therapies are of prime importance in order to stay healthy, to optimize performance, and to successfully treat disease. "After all," he teaches, "your immune system runs on oxygen, and disease causing microbes, like most of the bacteria and viruses, can't live in it."

In addition to *Oxygen Therapies*, Mr. McCabe has written a syndicated newspaper and Internet column, 'Ask Mr. Oxygen', and numerous national magazine articles. He has been published in 'AIDS Patient Care', 'Health Freedom News', 'Explore!', 'East West Journal', 'New Perspectives', 'What Medicine', 'Nexus' and 'Well Being Journal' magazines, and his image and his work were featured on the cover of 'Health Consciousness' and 'New Times' magazines.

After his appearances on more than 1,500 radio and television programs and speaking platforms in the U.S., England, Scotland, Australia, Canada, Mexico, Holland, and New Zealand, he became internationally recognized as a prominent and captivating lecturer. A *Maury Povich* TV talk show was devoted to his emerging Oxygen Therapy work, and featured Mr. McCabe and a no-longer-sick-with-AIDS oxygen/ozone user as central guests. The Bio-Oxidative Medicine Foundation–an international Oxygen Therapy physician's group–honored him with its prestigious 'Special Recognition,' and 'Distinguished Speaker Awards.' Mr. McCabe is unique in his distinction as the only person who has ever interviewed thousands of Oxygen Therapy using patients and hundreds of doctors worldwide. He did this while simultaneously investigating the Oxygen Therapy research and treatment centers and at the same time publishing and lecturing on his findings.

Although several Oxygen Therapies have been quietly in use for more than one hundred years prior to Mr. McCabe's body of work, the general public was unaware of them. His undertakings also earned him popular usage of his title of 'Mr. Oxygen.' The numbers of professional and lay adherents to the therapies continues to grow rapidly owing to his promotion of their simple effectiveness.

U.S. manufacturers and the professional organizations surrounding the Oxygen Therapy concepts are now flourishing, in large part, owing to Mr. McCabe's years of relentless lecturing and purposely focusing the public's attention on the efficacy of the therapies. His promoting created a demand for them, which then naturally induced people into becoming suppliers to fill the need. The knowledge that Mr. McCabe gathered, created and then taught us is the very foundation of our modern public understanding of how the therapies work and why we should employ them.

This foundation, or 'informed group consciousness' that we all now stand upon, which he repeatedly laid down, is so large, and so much a part of the now 'common knowledge' of Oxygen Therapy, that we scarcely notice that it did not exist before his pioneering work began.

The Start

His work started with a vision, a beautiful, joyous inner offer to do this work. He accepted life's offer, and two weeks later, after forgetting about the vision, he mentally decided to write the book. He reasoned that he knew more about this subject than anyone he knew. Too many people were getting sick, and the oxygen-using doctors were all embroiled in disputes that scared away patients. Someone needed to step forward.

About a month later, while writing the book, he remembered and realized what the vision meant, and there he was, already beginning to fulfill the contract. He states that from then on, it was like 'grabbing the tail of a comet.'

In order to reach all the people, Mr. McCabe meticulously planned how to bring the awareness and presence of these wonderful therapies out of obscurity and into the modern mainstream. During earlier experiences, life had prepared him specifically for this daunting task with 20 years of spiritual training running concurrently with 10 years of marketing reality training as a Fortune 300-corporation representative.

Just before the first U.S. Oxy-TV appearance!

After product placement, the main emphasis of his corporate experience was upon coordinating; distributor sales forces, advertising, promotion, shipping and distribution, the very keys necessary for first creating public demand and then getting his book placed everywhere for distribution. He states the success of the book was due to expending 10 percent of his energy writing, followed by 90 percent salesmanship. The salesmanship was necessary in order to get radio, TV, and other media and venue leaders to let an unknown with an unknown concept be seen and allowed to do the lecturing.

He concurrently had to get major distributors to buy and stock an unknown first time author so the people attending his lectures and listening to him on the radio and TV could have universal access to purchasing his book. He did all this single-handedly. There was no staff. The phenomenal success of his book including selling so many copies *without a major publisher* is the proof of his pudding.

As part of his how to 'increase the public awareness' plan, he carefully created the original 'Oxygen Therapy' name itself, and titled his first book after it. He wanted to always be driving home the main point by concept repetition, familiarizing the people with the name and solidifying the idea in the public awareness. He realized this was necessary if the therapies ever were to achieve prominence and the full respect they deserved here in America.

He took all the scattered unknown single oxygen-related therapies, and their practitioners, who were mostly in competition and personally fighting with each other, and harmoniously combined them all under the OXYGEN THERAPY name, as though under a big umbrella in the rain. This was accomplished by attending and speaking at medical meetings and visiting clinics and teaching the practitioners what their therapies had in common, while at the same time vigorously promoting the practitioners and their treatments and wonderful results to the public.

Drug medicine was fully entrenched and stomping on alternatives in 1988, and upstarts to the status quo were routinely being jailed. No big health food chains existed. He was the only public representative of an unknown group of related therapies, and without funding or corporate sponsorships. He had no titles or professional letters other than 'BA' after his name. Neither did he have a university behind him or a successful track record in the field. Once he started the work and realized the huge number of

obstacles facing his mission, he could have easily made the case that he didn't stand a chance, and taken the easy road and gone home. Instead, he hit the trail with vigor.

To give his subject respectability, and in contrast to most of the other lecturers who were appearing at the many diverse people's events he attended, he made sure that he always wore suits at his lectures (even though he

Ed often lectured with Dr. Joel Wallach, originator of *Dead Doctors Don't Lie.*

prefers not to). He worked at only giving presentations that combined a tamed passion with a lively professionalism.

He knew he needed to break through the clutter of all the other items competing for people's attention by using the right amount of passion in the delivery, but the expressed emotion must always be restrained enough to give the message professional respectability, and therefore continued acceptance. A delicate balance of opposites had to be maintained. It was

hard for him to suppress his own personal feelings of outrage at the continued unnecessary suffering going on, for he knew solutions existed, yet people were dying daily and just so much 'politics as usual' was actually responsible for killing them.

Fortunately, everywhere he went contacts emerged and metaphorical entrance doors magically opened the moment he arrived, and the doors behind seemed to slam shut, sealing him in protection. He had the continual assistance of many people from all over the world who also sensed the same calling to help with oxygen. They came from everywhere and filled every need. It was like a huge unseen hand brought everyone together into what became an oxy-movement charged with freeing health care from the restrictions placed upon it by various egos.

Especially during his European, Australian and New Zealand tours, every day different volunteers would coordinate and tell him the itinerary and drive him everywhere. From the time he got off the plane alone, a different stranger was always meeting him daily. Each would bring him to

new radio interviews, and then on to a 2-to-4 hour lecture that night, the after lecture restaurant meeting, and later to a new home or motel. The next morning he would be dropped off in another city somewhere further along the route, and met by another stranger oxy-enthusiast or healer, and this cycle was repeated for weeks. Along the way there, and everywhere else he went, he survived years of people showing up in his motel rooms, calling all day and night, and everyone wanting to go to dinner with him because they wanted more knowledge of the therapies.

He appeared at every venue requested, from back-rooms to studios and big halls. None of the sincere oxy-knowledge seekers were turned away; and few of them realized that someone or some other group probably had just left him before they got there. They naturally also expected Ed to be fresh, and in top cheery form and able to also instantly give them hours of intense individualized attention. He said it was extremely demanding but he loved every second of it. His oxygen supplementation definitely enabled him to have enough energy to keep up this intense non-stop demanding pace.

Due to his many activities, he is considered the modern 'father' of public awareness of the field by the original medics, patients, customers, suppliers, and manufacturers that now make up our modern field of Oxygen Therapy. This is his legacy.

Over the past 12 years, 47 distributors in 8 countries, including the largest U.S. health book distributors, sold Mr. McCabe's publications, tapes, and videos.

In 1998 the Oxygen Therapy community was shocked to learn of Ed McCabe's sudden imprisonment for 18 months. His house was raided and his computer, new book, and research files illegally stolen through the use of fictions and faulty warrants that were mysteriously 'sealed' so no one could prove their defects. In an attempt to ruin his credibility and make it hard for him to have any legal recourse later, Ed was sent to live with criminally insane prisoners in a Federal mental hospital for one month of special punishment thinly veiled as an 'evaluation.' This device was used against him without any foundation in reality—solely because he dared to question in court what they were illegally doing to him. He was shuffled through 11 other prisons. For a total of 18 days he was—from morning to night—locked in leg, arm and waist chains. He was also–despite not breaking any regulations to cause it—illegally locked alone in solitary confinement and cut off from all human contact on five separate occasions for a total of two months. Four of them while still presumed innocent before going to trial.

A Federal Binghampton, NY Judge, a Syracuse NY Magistrate, an Asst. U.S. Attorney, and two IRS agents were actors in a conspiracy to imprison Ed. There was no valid Grand Jury vote. The defense attorney forced on the case against Ed's will secretly met with the Magistrate and they planned how to imprison Ed, and then hatched their plot in court. The court transcript has lies in it. One of the Agents falsely swore as to Ed's identity. The Magistrate ignored Ed's demand for a 'show cause' hearing and falsely claimed jurisdiction where there was none. The Judge ignored all the illegalities and refused to produce the Res upon demand for public inspection. Mispersonation and a multitude of frauds and fictions were used. They allege they convicted Ed, and have records allegedly proving it; but all their paperwork was concocted using the ruse of actually convicting a fictitious person with an assumed fictitious name (similar sounding to his own) for non-liable and non-existent tax violations. Then they put Ed in prison.

While out on probation with some freedoms restored, he and his wife Leeda tirelessly compiled the stored highlights from his previous lectures, articles, and unpublished writings into the basis of this book, and together they prepared it for this publication against many odds.

Mr. McCabe continues consulting with doctors, manufacturers, and the media. He writes, lectures, appears as a media guest, and his advice is sought by noted personalities, the rich, poor and powerless, and the public worldwide.

Rick Kroll, President, Oxygen America
www.oxygenamerica.com

The Wholistic One Page Version
WHY Oxygen and Ozone Work So Well

You, me, them, it, and all the bodies, animals, and plants have spent Eons evolving while surrounded by, and swimming in, a sea of oxygen which is itself swimming in a sea of magnetic/gravitic particles of sunlight energy. Oxygen stores the sun's energy so that all life can feed off of it. Even anaerobes (life forms that do not use oxygen) require oxygen using life forms to live in or around, because oxygen carries the energy of life itself. Prana, Chi, Ki, and Bions are all names for the Life Force stored in Oxygen. 'Pneuma' as in air, is ancient text for 'Soul.' Ozone, from the Greek, is a form of oxygen so important that the translation means, 'The Breath of God.'

We all know this oxygen stuff is important. It is just as important right now to animals and plants.

If something this important to us is taken away, everything in life goes downhill fast. If it is *s l o w l y* and effectively taken away by insidious and ever encroaching soups of greed-caused-pollution, we call the ensuing responses *plagues*, *chronic diseases*, *illness*, and *poor animal and crop yields*. This is where we are now, with the pretty but low yield nutritionally empty crops, the poor quality animals, and the chronic diseases with their skyrocketing incidences.

The whole solution is to put back the missing oxygen. Back into the environment, by removal of oxygen-robbing pollution combined with reforestation, and back into the human and animal bodies through supplementation and delivery systems specializing in active forms of oxygen and minerals. Problems solved.

Solving the Human Health and Disease Riddle

The most common cause of human death worldwide is waterborne disease. The reality is that these diseases carried in water are caused by anaerobic microorganisms. Anaerobic microorganisms cannot live in active forms of oxygen. Water borne disease causing microbes and parasites can be completely eliminated by putting something containing active forms of oxygen, perhaps something as simple as common hydrogen peroxide, in all drinking water. Individual results may vary, but, in my humble opinion, there would be another added benefit to properly oxygenating all water; namely the elimination of most, if not all, of the diseases already present in the bodies of the people drinking the water. This can be accomplished easily and inexpensively.

To rid the inhabitants—in all the impoverished and richest nations of the world—of most disease would be simple. What we need to do is continually put a harmless weak solution (65 parts per million) of food grade hydrogen peroxide by itself, or in combination with ozone, or even just ozone by itself, into all water before it is used for drinking, or washing, or cooking.

Peroxide distribution in the African slums.

Doing this would eventually kill almost all water-borne diseases inside and outside the body. In low-tech environments, peroxide is cheap and sold in bulk, and can easily be added to the water. In Industrialized nations both peroxide and/or ozone can be metered into water supplies. In this book *Flood Your Body With Oxygen, Therapy For Our Polluted World*, I report detailed solutions to the major causes of human physical suffering. I hope you will apply them.

Solving the Animal Health and Disease Riddle

The most common cause of animal disease worldwide is **anaerobic microbes.** Animals (our pets and those used in our food supply) can get sick just as we do. They can also be a source of disease and parasite transmission for us, especially when a diseased animal ends up as food on our tables.

Because anaerobic microbes cause almost all animal diseases, it has already been proven that we can eliminate most animal diseases simply and inexpensively with available chemical supplies and low technology solutions.

Courtesy- small flowing dog.

In low-tech environments needing the liquid active oxygen forms available from peroxide and ozone, they can simply be poured or bubbled into local water supplies. In industrialized nations, both can be metered and injected into the municipal water supplies.

Through oxygen, I have seen animals given up for dead because of disease (or whose immune systems were 'burnt out' from antibiotics and other over-medication) returned to complete vibrant health. I witnessed that through the use of a local sodium chlorite oxygenating formula, several barns full of chickens in New Zealand were cured of coccidiosis, and that stables of Kiwi horses were also cured of encephalitis. I interviewed knowledgeable animal owners all over the world who had been successfully giving their pets and animal's oxygen supplements so that these animals wouldn't get sick in the first place.

Individual results may vary, but animal disease-causing anaerobic microbes can't live in active oxygen either, just like the ones that bother humans. In this book I explain detailed products and solutions for the major causes of the physical suffering of animals and of the degradation of our food supply. I hope you will spread this knowledge far and wide and apply these solutions wherever and whenever you can for the betterment of us all.

Solving the Plant Health and Disease Riddle

The most common cause of crop disease worldwide is a lack of **oxygen** in the soil. Plants (houseplants, crops, and our food, vegetables and fruits) only live because they have aerobic (oxygen-loving) bacteria clinging to their roots. These bacteria act like their stomachs. They pre-digest the soils around them in order to get the soil minerals into a softened form enabling them to be sent up the stalk or trunk and absorbed.

You can watch this happen with a potted plant or a tree in your yard. As it grows and gets bigger and bigger, the soil level drops lower and lower. That's where the plant comes from, the soil that's been used up to create it.

Just as in the case of people and animals, the minerals are combined and used by the universal Life Force, or Mother Nature, to build the plant bodies. It is indisputable that all human, animal, and plant bodies are built strong by having lots of minerals, trace minerals, and oxygen that combine and allow them to have good immunity.

With plants, you can watch it happen quickly. The more oxygen we give them, the more these root-clinging bacteria proliferate. We can visually watch how more oxygen added to the soil equals more minerals being taken up into the oxygenated plants and the bigger and more disease resistant these plants become. Knowledgeable farmers, hothouse managers, and other commercial growers feed oxygen supplements to plants by oxygenating their watering systems. Their plants invariably commonly grow larger and faster than non-oxygenated ones. They also stay disease free, often to the amazement of county agricultural agents whose skepticism they have to overcome when their animals test cleaner than any others.

In this book you will find solutions to the major causes of food shortages and crop and seedling loss. I hope you will apply them.

Isn't it obvious that life itself runs on oxygen?

Ed McCabe Homepage: www.oxygenhealth.com
Book orders: www.floodyourbodywithoxygen.com

Introduction

I will now spend a few hundred pages repeating in more detail the same message I have just given you. I hope to further convince you how simple and powerful our oxygen truths are when humans use their free will for the betterment of all—and then apply these truths correctly in this world.

Problem Solving

Our problems are seldom solely caused by whichever 'them' you blame. 'They' are always very few indeed, and no matter how powerful someone is, none of our leaders are above us. They all have children in schools, plays, and sports and they all have wives, or lovers and anxieties. They all shop and eat at restaurants. For these reasons they are all accessible to opinions, and no one wants to be unpopular. Remember that there is a solution for everything, and that everyone answers to someone else.

Politicians have to keep jumping in front of public opinion as they pose at leading the crowd. The people around them can deliver the good messages to those standing in the way of implementation. Our problems invariably stem from our own inaction. If ego, greed, money or politics seem to stand in the way of a solution, the people in the relevant group can change the situation by organizing, grouping together, and convincing the people closest to the decision makers to advise them in better ways.

So even the big problem solving comes back to you and me. Now that you are holding this book stuffed with right-out-in-the-open-secret-knowledge in your hand, at this very moment you are personally capable of solving many of the world's major human, animal, and crop problems. What will you do with this information?

Our Mutual Voyage

When you started reading this book, we formed a link and began a mutual voyage into the fascinating world of Oxygen Therapies. My intent is to educate you and hopefully have an effect that might ultimately improve your well being, and perhaps the well being of your family and friends. Partners can only work effectively together if they trust and know one another. I would like you to get to know me through my history and work, and to develop confidence in my many years of global experience gained. I want you to trust that I am here to help you.

First, "What makes Ed McCabe© such an authority on the Oxygen Therapy subject?" Good question. The short answer is that I have personally collected a *huge* amount of information only available through thousands of interviews of people with years of hands-on-experience. Information that is not available in schools or anywhere else. And why am I known as "Mr. Oxygen?" In short, because I was the first to make the modern mass public aware of this subject and everyone in the field has learned something directly from my work, although it often came disguised as it filtered through people that had studied under me.

I will answer in more detail as we go along, drawing from real life experiences to illustrate and validate what indeed makes me an expert in my field, and why you can look to all the Oxygen Therapies with confidence when used correctly.

How My Work Was Accomplished

I publicly began my own voyage into the world of Oxygen Therapies when AIDS came along in the '80s. I didn't have it, but I watched the media tell us that it was spread through sex and that most if not all of the world was eventually facing death from this terrible disease—and that there was NO cure! Naturally it concerned me. I was compelled to seek solutions for myself and my friends and family and for you, my yet unmet brothers and sisters.

Asking the right questions in the first place determines your journey, so I carefully formulated the best questions I could think of. I started with three questions:

- ✓ *Why hasn't the riddle of disease been solved?*
- ✓ *Out of everything, exactly which therapies have the best performance records and are also capable of treating the most diseases—what's the 'best of the best?'*
- ✓ *What did these 'best of the best' therapies all have in common?*

I knew the answers I sought would be found within this ballpark.

Once I formulated the right questions, the search was on. Years later I ended up publishing my *Crown Jewels of Health* in various places to announce the results of my research. It turned out that at the end of the day it all comes back to the old Naturopathic adage: "The closer we are to Nature, the healthier we are. The further away from her we get, the more

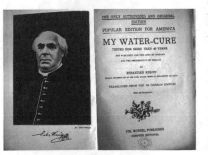

disease we have." Father Sebastian Kneipp, a famous naturopath of his time was onto this in 1846.

Ah, yes, we ask. But how do we get back to her and still not have to give up our houses, cars, or businesses, and not have to start living in trees? And, we also may realize, Nature isn't really as healthy as she once was. Man's voracious appetite to rape Mother Earth for profit has destroyed her once pristine condition. So how can we get back to what may no longer be around for us to get back to?

The Search Was On

I became deeply inquisitive. After years of looking at all the known healing methods I could find, there was only one real logical choice. The answers were found within the very therapies that had the greatest amount of proven safety and effectiveness as their historical documented track records proved, and yet were still obscured in modern America. You may have already guessed that I of course am referring to Oxygen Therapies.

As my fascination with the subject of Oxygen Therapies developed, my thirst for gaining more knowledge about them grew at a rapid pace. I had discovered a wonder of magnanimous proportions, almost unknown within modern society, and these therapies were based on Nature's infinite evolved wisdom, not just a soup of chemicals thrown together.

Oxygen Therapy has always been around in the U.S. and elsewhere. As I interviewed first a few, then a handful, and then hundreds, and later on, thousands of oxygen/ozone users and suppliers, my mental and computer database grew, and more and more people began to ask me about the effectiveness of my discoveries. Before long, I was in demand for the knowledge I had collected, and many were asking for guidance through this huge remarkable secret that had lain dormant since the turn of the century. I spent weekend after weekend all over the globe, and crisscrossing the U.S., speaking in city after city, endlessly, without making a profit, (all profit was eaten up by travel expenses) and solely because I believed in what I was doing. No one had previously achieved this goal of oxy-education of the masses.

Purposeful Positioning

The reason the Oxygen Therapies concepts have proliferated in the public eye was not pure chance. It resulted from inspiration, pre-planning, and long hours of exhaustive work by many people. I was constantly positioning their image, safeguarding it, fighting for it at every turn, and taking knowing advantage of every opportunity. But, of course, you probably wouldn't know this as a new observer simply reaping the modern benefits of this once pioneer campaign. I mention it because this has to go on the historical record.

The very phrase *Oxygen Therapies* was coined by me in 1987 as a way to classify everybody and everything 'active oxygen related' under an umbrella concept. In the process I taught everyone I met who the past GIANTS in Oxygen Therapy were; and I was careful to always give my teachers full credit. Now their names are common knowledge in the field;

and after all, we *are* standing on their shoulders. The unsung oxy-heroes often spent their lives undertaking painstaking lab and clinical research without much recognition or reward. Occasionally, the recognition would come after they died.

When I first started, the Oxygen Therapies were living under a public relations rock as far as the general public knew, and that included the alternative therapy community.

My Contribution

I will tell you from my perspective what I did, though I do not intend to brag about my accomplishments. As a reporter, I feel bound to relate the facts to you as I experienced them.

I am the only person to have traveled around the world, and back and forth across America for 12 years, telling the public, doctors, officials, and the media about Oxygen Therapies in every possible venue you could think of. This book is the culmination and summation of everything I experienced and have learned since then.

After spending a long time doing the preliminary research, and just before I started writing it all down, I realized that I knew more about a broad spectrum of Oxygen Therapy information than just about any other lay person. And no one had ever written a book on the whole subject. At that time there were a few experts but only within their own specialized areas of the various Oxygen Therapies. My personal distinction was that no one else had investigated this whole subject on a worldwide basis, while interviewing thousands of people and many hundreds of doctors from a journalist's point of view, while also publishing the results. I kept saying: "This is just amazing; the world needs to hear about this. Tell us more."

Filling the Need

My primary obstacles to telling everyone, apart from the medical politics discussed later, were economic and logistical. Since mysteriously no major publisher would publish the successfully selling book, and I had two professional agents chasing the major publishers for me, I had to do it myself. Vindication came later when the book became a bestseller without them. That is impossible, by the way, without having a major publisher behind you. I sold far more books than many books that do make it to the NY lists. At that time, the publishing houses retained conservative 'old-method' doctors on staff as advisors but their fears and lack of knowledge still couldn't keep the book from the people. Remember that this was the late 1980s, long before all of our now taken-for-granted easy access to alternatives, and alternative doctors, and before the rise of all the mega health food stores.

The more lectures I gave, the more United Parcel Service would come to my door. Finally, I got overwhelmed boxing and addressing the cases all by myself. The Universe provided the first solution to the problem in the form of Charlie Pixley, and for a while he and his wife did the fulfillment work. Later on, Johnny Taggart was forming *The Family Health News*, an oxygen-related supplements catalog/news company, so I eventually switched fulfillment over to his operation.

Johnny and I had become friends in the infancy days of modern American ozone, back when I went to Miami in 1989 and investigated Lucas Boeve's nefarious Miami 'underground cowboy' ozone treatment house operation. This was early on, long before Basil Wainwright had moved to Miami and became a serious ozone player as well. At the time, my book was just out and I was looking for more experiences and facts and interviews.

Ozone Goes into Hiding

Ozone and Oxygen Therapy has been around in the U.S. since the late 1800s.

Nikola Tesla was successfully experimenting on disease and rejuvenation with ozone in the late 1800s and early 1900s. Oxygen and Ozone Therapy's previous heyday in the '20s and '30s included many limited circulation books published on the subject of its use, with all sorts of devices being manufactured, and Oxygen/Ozone Therapy using hospitals, sanitariums and clinics springing up and doing very well.

They would have still been succeeding to this day. According to author, historian, and researcher Eustace Mullins, they were all eventually put out of business by the newly conquering large commercial drug monopolies. The big railroad barons had quickly become oil barons too, and their syndicates needed more places to put their money. Simply continuing their then current business practices, they bought politicians and endeavored to crush any methods competing with their patented petrochemical drug therapies.

Along the way, through their man Abraham Flexner, they set up old 'Doc' William Simmons, a doctor with *two* fake medical degrees who never practiced medicine, as their journal editor in 1898. Eventually they succeeded in fraudulently discrediting anyone outside their inner club who was not allowed to pay to advertise in their new AMA trade journal. Flexner was John D. Rockefeller's 'stool pigeon' in setting up the takeover of the entire medical school industry by the Carnegie Foundation, which was a Rockefeller Foundation subsidary at that time. He produced *The Flexner Report*, and this changed the medical schools of the United States from homeopathic and naturopathic medicine to allopathic medicine—which was a German school of medicine which depended on the heavy use of drugs, radical surgery, and long hospital stays. That's the established system today, allopathic medicine.

That is why only drugs, radiation, and surgery, and eventually only one class of doctor gained political legitimacy in mainstream America. Along came the World Wars. With the need for quickly stamping out actors for the battlefield, all doctors were quickly processed through medical schools that had dropped about four years of diagnosis training requirements from their curriculums. All new doctors were only taught allopathic drug therapy.

During the war, the Allied bombers in Europe destroyed all of the German ozone clinics and left the building housing the world's largest drug company, IG Farben the only one standing. Talk about blowing away your competition! IG Farben industries, a chemical cartel which, prior to conviction and being disbanded for war crimes (during the Nüremburg Trials) after World War II, was the *fourth* largest company in the world.

Since World War II, the multinational commercial interests and their high priced and side-effect-riddled therapies were taken for granted by 'everyone' as the only way to go. Many laws were written to legitimize this position, resulting in much suffering and many lives being lost unnecessarily, along with many legitimate healers unnecessarily jailed. So today's so-called 'alternatives' to drug therapies, such as Oxygen and Ozone Therapies, are actually working traditional therapies previously forced into hiding for fear of persecution.

By the 1950s, medical doctors became rich and god-like, and people worshipping (by shopping) at their altars learned the short-sided consequences of personality worship when they went to their gods for treatment. Their will to live was sapped by multiple drug side effects, and they unnecessarily passed away drugged up, in pain, and lying in bed with tubes sticking into them. All their money had been drained away, thus feeding the beast that tortured them. Too many of these failures swung the public opinion pendulum back the other way and now, with many states passing 'health freedom of choice' laws, we are seeing more choices becoming available. The one word I hear most often from people describing their unhappy encounters is *arrogance.*

The Re-emergence of Ozone in Modern Times

As far as I can tell, I have generally traced the re-emergence of modern U.S. Ozone Therapy—at the level where the average person looking real hard could find it again—to a series of events during the early '80s.

The background to this story shows Oxygen Therapies being used in many countries, especially in Germany. Here in the U.S., in February, 1982 Lee Devries, a Dutch-American Lutheran from Michigan, was studying a system of medical codes he maintained were hidden in Scriptures; he was working with a Poughkeepsie General Hospital pathologist, Dr. Migdalia Arnan, to research his findings.

Dr. Migdalia discovered that there were two types of cancer, environmental (from irritants) and genetic (budding under proper conditions). His research also revealed that T-cells produced ozone, an active form of oxygen, so if we put ozone in the bodies of cancer patients we would be following Nature's laws and removing cancer.

Phil Seifer attended one of Lee Devries' lectures and became his student. He put human cancer cells on microscope slides and filmed them. The cells would lose their red color and turn green and then die, all within 11 minutes.

Lee and Migdalia had explained that the waste product of photosynthesis in plants was oxygen and that if you fed any organism its own waste products it would die. The cancer cells were actually plant type cells growing in humans. The cells would start to grow under the microscope's light, and produce so much oxygen that they would bleach out the red color and then destroy themselves with their own oxygen wastes trapped under the slide. In the body, the oxygen was not trapped as under the slide, so the cells kept growing.

Phil once asked the Universe which path to take. He pulled over into a rest area to meditate, to decide whether to continue with his film career or to help Lee Devries. He closed his eyes, and in a few minutes heard laughter, and opened his eyes to witness a busload of pale, mustard skinned, young children suffering with cancer alighting from a bus and heading for refreshments. They were out for a day in the park after having chemotherapy. The waking dream had answered his question for him.

He started building air ozone machines. Others were successfully using and selling air ozonators at that time, but were not very well known. One of his first customers was the Gerson Clinic in California. They didn't necessarily believe the theories, but they did want to do something about the polluted local air.

The hospice patients started feeling better from just breathing the fresh ozone air. This was verified by Dr. Arnan and also Dr. Bob Mayer from Miami Children's Hospital who had been using ozone since WWII.

Dr. Mayer asked Phil Seifer to call a certain Lucas Boeve, who was using ozone

Charlotte Gerson and Ed.

commercially in Puerto Rico, to help spread the word about ozone. When Phil told Lucas what he found out about cancer, Lucas jumped at the chance to partner with him. Lucas wanted to use ozone to help his family members suffering from cancer.

The media was starting to mention AIDS, but it seems that not until movie stars came down with it was it taken more seriously. The Gerson people now wanted not just ozone air purifiers, but medical ozone machines as well to use for doing blood work. Phil started importing them for use in Mexico.

The success stories quickly spread, starting with a Mr. Green from New York City who had a tracheotomy lung cancer and a tumor the size of a tennis ball. Mr. Seifer brought him to Dr. Ribner's office along with Mr. Boeve, Dr. Horst Kief, and Biozone President Dr. Ribner from Germany. Dr. Ribner injected him over a two hour period; then, after a rest, Mr. Green said he felt better than ever! They now knew why ozone had been used so widely in Germany since the First World War. Mr. Green's wife was a nurse and when she picked him up she was astonished with joy.

Dr. Ribner introduced another cancer patient to Phil and Lucas, a Mr. Terrance McGrath. He was also given ozone, and immediately joined Phil and Lucas in forming Medizone, the first modern U.S. ozone medical equipment sales company.

Dr. Ribner (Biozone), Dr. Ron Hoffman, Dr. Migdalia Arnan, Dr. Robert Atkins (Atkins Diet), Dr. Horst Kief, Dr. Ray Evers, Dr. Kurt Donsbach, Dr. Rogers, Dr. Robert Carpendale, and others were all helping promote ozone and the results produced from using it.

Ed & ANA Pres. Dr. George Freibott ND.

In 1983 the Gerson Institute published a *Progress Report* by Norman Fritz. The report tells of the Gerson Institute receiving their first medical ozone generator, which was cheerfully supplied by Dr. George Freibott, ND, and how they used it on a legally blind man with DDT, dioxin, and mercury poisoning. The ozone restored most of his normal sight in less than 15 minutes. They were suitably convinced.

The *Progress Report* related how earlier in the year, during May, Dr. George Freibott invited the Gerson people to attend the International Ozone Association's Washington, DC meeting. During the meeting, European physicians related how for 40 years they had been getting good results using ozone against all sorts of diseases. They proved that ozone and hydrogen peroxide are non-toxic antiseptics. Used inside and outside the body they destroy nearly every virus, bacterium, and parasite.

Mr. Seifer told me that in 1985 he received a call from Dr. Johnson, the personal doctor to President Reagan, asking about ozone for cancer. He proceeded to tell Dr. Johnson all about ozone. Our former president secretly went to Dr. Hans Neiper's clinic in Germany and had his cancer cured by ozone, while at the same time U.S. doctors using ozone were being persecuted by the Food and Drug Administration. That is why German ozone doctor, Hans Neiper would always say, "I can't divulge all the names of the heads of the U.S. government, and the heads of your cancer institutes (officially disparaging of ozone), and the names of their friends, who have come for ozone treatment in my clinic…. but President Reagan is a very nice man."

Dr. Robert Mayer & Ed.

Miami Dr. Robert Mayer had used ozone injections and rectal insufflations in Miami hospitals for more than 40 years, since the Second World War, after learning about it from a German prisoner of war he was treating. The Ellis Island prisoner had complained to Dr. Mayer that he was using ineffective methods on him, and that he should "use ozone instead, like they do back home in Germany."

Dr. Mayer was later commonly curing spinal meningitis by injecting ozone into the spinal fluid of children suffering from it at Miami Children's Hospital. The hospital later told him to stop using it—"Because no one else is."

In the early '90s, I videotaped Dr. Mayer about his life full of experiences using ozone daily since the German prisoner woke him up in the '40s. I also received ozone injections from him at his office in Miami. He told me he had safely injected hundreds of children over 40 years without incident, and had many cures of 'incurable' diseases.

I also spent a whole day in Mist, Oregon, video interviewing Dr. William Turska, another senior ozone therapist, an ND, and he also reported that he had been direct IV injecting and curing thousands with ozone over the course of his career safely and without incident. These two brave pioneers are warmly remembered posthumously for their pioneering work in our field.

Ozone Cowboys!

As I noted earlier, Dr. Mayer encouraged Phil Seifer to bring Lucas Boeve up from Puerto Rico. Lucas was patenting ozone use in water purification and while in Puerto Rico did human studies on volunteers who had cancer or who had AIDS and were willing to try using active oxygen. Lucas told me the following at his Key Biscayne Spa in 1989—just after my first book came out.

"We had tremendous results with treating AIDS patients in particular. We didn't have our research completely done on cancer at that time, so we didn't push that as much as AIDS. Again we came back to petition the American Red Cross and the National Institute of Health, showing them the studies we had done in Puerto Rico. We were told that nothing could be presented to the government without being in the form of double blind studies. Being a small engineering company, we decided that rather than spend time trying to find eight or nine million dollars to do these studies, we would approach alternative thinking doctors who recognize the importance of oxygen. We began our campaign that way.

'We went to the 20 doctors in this country who were using ozone and wanted to get a few years of good clinical studies to give the government proving there were no problems and good results, in the hopes that they then might give us permission to use ozone on a large scale basis. Unfortunately the FDA found out what we were doing and went to the doctors and confiscated all their ozone machines. This was about three years ago (1985).

'The reason they gave at the time was that the German machine we were using back then, in daily use in medicine in Germany for the past 50 years, had a piece of literature that came with it stating particular diseases ozone was effective in curing.'

Lucas and Phil formed an alliance, and after treating Terrance McGrath with ozone, and getting rid of his Prostate cancer, they formed the first modern U.S. medical ozone company, Medizone. But, as is too often the case, they soon fell into fighting amongst themselves and split up. Terrance assumed control, but years later, Medizone's stockholders displaced him from the CEO position and moved on without him. Medizone currently trades on the U.S. market as 'MZEI.'

Before he was run out of town for practicing medicine without a license, Lucas was renting a nice house in Key Biscayne, Florida, a quiet upscale neighborhood, outside Miami. He gave direct IV ozone treatments and sold machines while dreaming of super big deals. After all, he reasoned, wasn't he watching IV delivered active Oxygen/Ozone Therapies

Site of Key Biscayne 'cowboy clinic.'

curing cancer, arthritis, and other diseases right before his eyes in his own house? Surely the world would soon flock to him with mega investment dollars.

All of this was happening right in plain sight under the noses of the local authorities, in fact, in their backyard neighborhood. Many shenanigans went on there during American ozone's underground 'cowboy' days. Lives and dreamed of fortunes were at stake, and it proved to be an explosive mixture.

The First Modern U.S. 'Experiment' Using Direct IV Ozone

Lucas was to be kind, 'erratic' and he claimed former military Colonel Status with CIA employment. He was out to use breakthrough science facts about ozone to cure cancer and get rich doing it.

His approach was simple. Before a new patient was even brought to their room, he or she was required to take all of their prescription drugs and put them in the trash. Then he would stuff em' full of ozone and organic food.

According to Lucas, his client list included ozone secretly saving Rupert Murdoch, the Fox Network media mogul. Fox was about to trash ozone on a TV expose program, so I called them from Lucas's clinic and informed the producers that contrary to their about-to-be broadcast smear on ozone, the owner of their network was successfully using Ozone Therapy. The top producer was incredulous when I told him. The segment never aired. I also tried unsuccessfully to get General Electric to clean up their PCB environmental problems with Lucas' ozone based bio-remedial equipment.

German national Erwin Dorsch was working there, building super high tech and very effective $35,000 reclining clamshell ozone water/steam cabinets called 'Rejuvenators' and later 'Aquacizers' in the garage, and Johnny Taggart was independently at the same time building and pitching ozone water sterilizers for home use. A few of the Aquacizers are still in existence. We were all injecting ourselves with ozone. Erwin helped Lucas procure a Haensler ozone dealership and lots of translated clinical ozone studies from Germany. Lucas then sold the 42 mcg/ml^3 ozone generating German Haensler machines for $15,000 at first, then $7,000, then $4,000.

These generator sales coupled with high clinic fees allowed his operation's rapid expansion, including the financing of his continual supply of imported Warsteiner beer. Lucas drank cases of it constantly, but claimed it did not affect him as he kept injecting himself with ozone to counteract it. He married a woman there, and later it was discovered that she and her later to appear lesbian true lover were working him as a *mark* so they could take over the soon-to-be-rich operation.

Problems with Non Regulated Clinics

Despite all the rampant human dysfunction surrounding the therapies in the cowboy ozone medical underground, the therapies still worked. One of the practical and political problems faced by the alternative clinics back then was that the clinic would only get patients already sent home for dead. The establishment's used up burned-out medical castoffs. This meant the clinic was often getting patients washed up on its doorstep who had no hope. Amazingly, despite this handicap, many recovered due to ozone's miraculous effectiveness.

Since individual results may vary, others with problems died because there was nothing left to repair after having had too much surgery, radiation or chemotherapy poisons. Then the mishap would be used by the establishment press to further denigrate the new therapy. I always used to ask: To be fair, why not print how many people died in your local hospital this week? Count them up, and put these failures on the front page as well with a similar damnation of current medicine like you just did to ozone.

No editor ever dared to venture into such realms of honesty, but I hope you get the point. Why are so many mainstream failures ignored, and a new therapy economically threatening to the status quo vilified for an occasional mistake or someone dying who was likely to die anyway?

Some criticism was doubtless deserved, but when used openly and correctly the science of Ozone Therapy itself has been proven safe, sound, and effective for more than 50 years. Due to the enforced illegalization of this then budding therapy in the U.S., the renegade underground operators were often found wanting.

There were some wild stories. On one occasion, a very sick ozone patient had passed out from incorrect application of the therapy while at the Key Biscayne location. Lucas, afraid of being caught with the goods in his clinic, drove the patient back to his residence unconscious, propped him against the door, pressed the bell, and ran away! As I said, outrageous shenanigans. Of course, when the local press got wind of this kind of stuff it inflamed them. In case you're wondering, the propped up patient naturally recovered without further incident, once given enough time for his body to recover from his way too rapid detox, and scratching his head upon waking up and wondering why he was at his home.

Because there were no rules or oversights, the first U.S. underground ozone clinic moved the practical application of the still unknown (in the U.S.) Ozone Therapies forward, since without any restrictive regulations they were free to try anything right away. The only checks and balances were history and common sense. The underground clinic's direct IV ozone philosophy easily eclipsed the old methods, proving the German autohemotherapy methods too slow, expensive, and outmoded by comparison. In this way, the clinic was producing many secret cures.

The undocumented case histories proved that the combinations of twice a day direct IV ozone delivery, ozonated water saunas, all natural diets, and adjusting it all according to the real time results showing up in darkfield microscopes was highly successful. But when publicly examined, the operation was completely outside the law and the lack of rules inflamed the press and authorities. Unfortunately this polarized the local press against medical ozone which is an otherwise valuable medical therapy used daily by thousands of physicians outside of the U.S. That part of the story was never told to the people.

Mr. Boeve's political downfall came when the ex-wife of the local newspaper mogul ripped him off, and for good measure ratted him out to her former hubby who quickly unleashed his media might. Of course, it was instant end game front-page news, 'Quack Clinic Plunders Patients.' Like the renegade Colonel Kurtz operating outside the law with a native army just over the Cambodian border in the film 'Apocalypse Now,' the methods employed, though highly effective, were unacceptable.

The authorities showed up and inspected the clinic. Lucas said he would comply with whatever they needed, and they went away. The agents had to think about all that they had seen going on in this quiet suburban neighborhood home. They called back, and Lucas promised to immediately come downtown to speak with them. Once his car was loaded with everything he could gather, he drove directly to Miami airport and disappeared.

The few guests/patients in the clinic didn't know what to do. Lucas had simply disappeared without a word, they were mostly sick, and they couldn't get the ozone anywhere else. So, they just decided to stay there as long as the equipment worked and there was oxygen in the tanks. They hoped for the best while desperately continuing to treat themselves.

In a few days Lucas' 'wife's' lesbian lover showed up with a gun, a truck, and a gang of thugs. Still dreaming of riches, they stole all the remaining equipment. One of the sick guests tried to stop them from taking the ozone generators and the Aquacizer that were keeping him alive, but she pulled out the gun and threatened him. Word on the street was that the equipment ended up on the west coast of Florida.

They say Lucas fled the U.S. with the authorities hot on his trail, and ended up in the Dominican Republic. The last we heard from Tom Corbett, a New York ozone technician hired as a consultant by many Caribbean ozone clinics, Lucas had surrounded himself and his new clinic with shotgun toting guards on horseback, and the local Dominican authorities were being paid to look the other way. He was faxing out threats and innuendoes to his imagined enemies and it was reported he finally ended up sick, and then dead, from sleeping with too many of his AIDS patients. He was constantly re-infecting himself with new infections at a rate faster than he could drink to forget about, or inject himself out of with ozone. Speculation still exists over whether or not he faked his death, but I spoke with a friend of mine who saw him very thin and sick.

Erwin Dorsch and Jimmy Brown partnered for a while and then went on to form their own separate ozone companies. I have watched ozone spread widely by this first partner and the break up phenomenon, which always creates and then leaves the world with more ozone truth spreading companies after the breakup pattern cycles.

My First Ozone IV, and the Importance of Detoxification

After my first book *Oxygen Therapies, A New way of Approaching Disease* came out; I received my first IV ozone treatments in Key Biscayne, and despite having done peroxide cleansing for a long time, that's where I learned what *real* detoxification with the big boy, ozone, feels like. After a few days of twice daily increasing IV injections, my blood became black and almost stagnant due to all the displaced-by-ozone built-up inner sludge thrown into the solution. I got so tired from the too rapid detox that I went swimming but could barely move my arms.

I firmly learned that no matter what you think, you *do not* know what real detoxification is unless you do a few weeks of daily ozone IVs yourself. Only then will you realize that all previous fasts and detoxification efforts, no matter how well intentioned, are mere grains of sand on oxygen's beach by comparison to an increasing quantity of daily ozone IV shots.

I proved all this to myself by being one of the first people in the U.S. looking at my own blood, and the blood of others daily in a darkfield microscope in 1989. Dr. Frank Sedarnee was supplying the cowboy clinic with his prototype darkfield microscopes, which later went on to become so common. I was easily able to track blood chemistry reactions to the massive courses of the ozone shots being given. In time, I daily watched black diseased patient blood routinely being turned bright cherry red and sterile under the influence of ozone. That's the true precursor to health, clean blood.

Year's later; network TV did a story on one lady who went to Lucas' Dominican clinic, and got a little bit of ozone but unfortunately then went home and died. You would think from the way the TV story was edited that ozone was no good, and people were being fleeced by clinics using it. But, if you knew what to look for, as I did, you would know right away that the TV story was a scam.

I could see plainly—in the color TV video shot of the woman injecting herself back home—that her blood in the tubing was still *black in color*. Compare the color of the two drops of blood on the back cover of this book. Which color blood would you rather have circulating inside you and your organs all day long? People (especially politicians and TV producers for some reason) get a little knowledge about ozone and pontificate like they know what they're talking about. Usually they don't know, but because they don't know enough to recognize what they don't know, we have problems.

Despite this poor woman thinking so, and claiming to the TV producers that she was using ozone correctly, her blood color plainly revealed right in the film that she was not properly injecting enough ozone often enough to keep it cherry red. For someone in her late-stage cancerous condition that was a fatal mistake. Of course, due to typical newbie outsider thinking that any minuscule use of ozone must be the same as the far different continual proper use of enough ozone, the TV show erroneously blamed the *ozone* as being ineffective and responsible for killing her.

Those who aren't familiar with our field hear the name 'Ozone Therapy' and erroneously think that some flawed minuscule use is the same as all possible proper uses. They then wonder why some minor dabbling in its use doesn't give them the spectacular results they had previously heard on the street that ozone was capable of.

Roy (Tuckman) of Hollywood was the late night all night Public Radio talk show host and almost permanent fixture on KPFK. I was on his popular '*Something's Happening*' show many times.

The show is on southern California's biggest Public Radio station *KPFK,* and for several years they used my book as an incentive during their fundraising. Roy stood up for me when a certain manager tried to have my

Ed with *Dynamite Radio for Night People*
Legend, Roy of Hollywood.

truthful viewpoint silenced, and although his show was threatened, he backed me up and kept the truth flowing in Southern California.

Roy's partner is a nurse, and she told me of one poor man who ended up in her hospital emergency room. He had poisoned himself running behind busses in L.A. while sucking up diesel exhaust and thinking he was giving himself Ozone Therapy. The poor fellow kept being told on the nightly news that ozone was smog, and incorrectly equated the two. Here's just one more reason, as I've suggested previously, some of those among us have questionable realities.

In actuality, 'modern' medicine would love to have ozone's record of effectiveness and safety. By comparison, count all those who die in hospitals every day after receiving the 'best of standardized care.'

Too often I sadly watched diseased people begin to get well from Ozone Therapy after the ozone had cleaned them out a little, but their insurance wouldn't cover any more therapy, or they had run out of their own money, so they went home unable to continue treatment. Inevitably, in time, the disease would again reestablish itself because the patient never cleaned out enough of the garbage and microbes to get over the hump.

As for myself, I paid a heavy price personally, financially, and physically during my continual 10 year expenditure of energy traveling the planet on behalf of the public good and Oxygen Therapies. However, at the end of being burned out by it all, I still had the personal satisfaction of knowing I had been instrumental in saving many lives. It was especially rewarding when the winners would boldly stand up in my lectures and tell the others to stop doubting and listen to me.

Was it worth all the toil and sacrifice? Yes, because my efforts championed the cause for those who needed it, and for those who couldn't do it. Although I faced impossible odds, including deeply entrenched alternative medicine ego-kingdoms and societal and well-heeled corporate obfuscation of Oxygen/Ozone Therapy's truth, I could not sit and do nothing. I could not be complicit in hiding the truth I knew would help so many. The Oxygen/Ozone Therapies, when used correctly, are probably the most outstanding and successful therapies ever discovered.

The Seed Began to Grow

Since 1988, when I first published the classic *Oxygen Therapies, A New Way of Approaching Disease*, I have witnessed a whole new industry built upon Oxygen Therapies come to life. When I started researching and lecturing on this, the availability of the therapies in America was slim to none. As I went from town to town, more and more people joined in. They got treated, then requested more treatments and/or started their own

oxygen businesses. Others started incorporating the oxy-knowledge in their own lectures and clinics, and still others started publishing books and articles and websites on the subject, thereby giving me the sincerest form of flattery.

A whole new industry evolved around oxygen and ozone device sales, supplements, products, technicians, training, treatments, literature, web sites, and mailing lists. Once impossible to find, or at least *extremely* rare and difficult to access, and without any common public knowledge of it left, or any available practitioners around, Oxygen Therapy websites, physicians, naturopaths, supply companies, and home treatments are now commonplace. Collectively these businesses and their healing activities add up to a large industry employing a great many people and selling millions of dollars worth of products and services each year, and it's all based upon Oxygen Therapies! If the Oxygen Therapies did not work, this industry would not exist. It's heartwarming to know that my continued efforts as a modern day pioneer expounding the benefits of oxygen were not in vain.

My First Book

My first book *Oxygen Therapies*, now an oxy-cult classic and best seller in the United States, was sold all over the world. I've lectured in Australia, New Zealand, Holland, Scotland, England, Mexico, Canada, and the Caribbean, and my tapes took the message everywhere. After 13 years in print, people continue to seek this book out. The book was a first of its kind, unique in laying down so many foundations for our U.S. Oxygen Therapy industries.

Worldwide distribution of more than 250,000 books without a major publisher is supposed to be almost impossible to achieve, but I did it with much work and help from my oxy-friends. Consider this, if there isn't at least half a grain of truth in what I am saying about how good this stuff is, how could this first book have been such a long-legged best seller without having a public relations gang behind it.

In my early days, the Oxygen Therapies were too advanced, and too unknown, so usually, few medical school graduates who evolved into any position of authority would dare hold themselves up to possible ridicule for using unknown modalities back then. It was people, just plain folks who were the key to the spread of my information. These people bought my publications and gave them to their friends, and this is now going on into its 14th year. I'm told books are supposed to run out of popularity in two years. The first *Oxygen Therapies* book is now out of print after selling like water in the desert.

Why did mine have 'legs?' I was told it was because people found the book, tried the products or therapies, and later excitedly told their friends about the successful results. Grateful users kept buying cases of the book simply to hand out as presents. Others who 'got' the oxy-message gave the books away at lectures they were doing to help me spread the awareness.

People were thankful that I publicly came out with this one important idea that so many brilliant minds had overlooked for so many years, the idea that most disease is not able to thrive in active oxygen. It was like watching their minds being freed one by one from some mental prison in the past. This popular support along with my constant lecturing and radio shows was the main reason why my old *Oxygen Therapies* book sold as well as it did.

My first book had grammatical mistakes in it. Some words are misspelled, and it's printed in a poor choice of typeface. I had to get it out quickly because I had large orders for the book from many people who were sick. I decided in 1988 that I just couldn't work on it anymore. I simply had to put it out even though it was unfinished. After hitting the road, I was in such demand that I never had time to go back and fix it. Now that it has become a cult classic, we can lovingly reflect upon its imperfections.

The Pioneer Role

Being a pioneer isn't easy. When the going got tough, I tried to remind myself, psychiatry and chiropractic treatment were once labeled as quackery, and many a health pioneer before me has faced obstacles, ridicule, bitter opposition, and death. For example, and it's hard to believe, but a surgeon washing his hands before he operated was once regarded as disgraceful quackery. The most popular surgeons were always covered in blood, and their unhygienic display was advertising their popularity. Hey, if you're covered in blood you must be busy, and the best. Unfortunately, the cleaner and therefore more hygienic surgeons were thought unpopular, and not worthy of giving your business to! Such is the power of the weight of public opinion. When I found the doors to academia locked to our oxy-philosophy, I purposely set out to direct public opinion for the good of everyone.

As noted, I approached my task with a determination born of knowing a single powerful truth, namely: disease can't live in active oxygen. The truth is what got me through and prevented my knuckling under due to all the unfortunate experts saying I was wrong simply because, "If it was any good, we'd all already be doing it." This type of lazy thinking is just a restatement of that old phrase once sincerely pronounced: "There is nothing left to learn."

As I told others so many times, "Well, I've tried it on myself to see if it would work. I don't get colds anymore, and it got rid of the arthritis that I had, and somehow it also got rid of the chronic bone spurring in my spine." When I first tried Oxygen Therapies, my immune system was shot. I didn't have any specific disease, but I was no longer in my twenties enjoying youth's seemingly bulletproof immunities. I was in my thirties, and getting really sick every few months with the flu, or a fever, and amazingly, all that went away when I started oxygenating. I haven't been sick with recurring fevers in years. I've told all my friends about oxygen, and the ones who tried it correctly have reported the same kinds of results.

My Job

As a journalist and news analyst I specialize in all the various Oxygen Therapies. I continually research the history and applications of the Oxygen Therapies best suited to helping the most people, and I also educate. Unlike most authors, I do not write opinions gathered from oxygen textbooks or other mere literature that I have reviewed and pasted together. Unlike some Oxygen Therapy books riding on the popularity of my original work, my books and publications come from the real life experiences of people that I meet.

I have personally witnessed with my own eyes the diseases of the people I interviewed, and then watched these diseases go away. I traveled widely to meet and spend time with Oxygen Therapy doctors and healers who were working tirelessly in the field, trying to help people and, at the same time, I was educating them. Thus I've personally interviewed thousands of people who have used Oxygen Therapies, and hundreds of doctors also using them, and visited scores of clinics. To my knowledge these individuals were, and still are, using Oxygen Therapies on a daily basis. If nothing else, we can see that if that many people are using it worldwide they can't all be crazy, they can't all be wrong, and they can't all just be running on a belief factor. There must be something more to it.

My Results

Since 1988, things have really moved forward in the world of Oxygen Therapies. There are many more practitioners and advanced and updated products, with new ones added almost daily. Much more information is available, and laws are being changed to allow widespread access to them. Public demand for more knowledge about the Oxygen Therapies has been overwhelming. Once an obscure therapy, my lectures and activities have fueled creation of a new demand. Oxygen Therapy is now regarded as an essential part of most alternative treatments.

Now, more than ever, the world needs to be alerted to what I've long been saying about the shortage of fresh available oxygen in our bodies and on our planet, and how our task is to do something about it. We must assert ourselves to solve the rampant pollution and forest clear cutting that is causing our planet earth to run out of clean, life-giving oxygen.

The Reason for This New Book

There is so much more to tell you since the first one! My main purpose, my demographic aim, is to enable as many people as possible to benefit from Oxygen Therapies in as many countries as possible. Since my first book, things have advanced in leaps and bounds. *Flood Your Body With Oxygen* is a people's book for ordinary people. It is written in layman's terms so that it can easily be understood, and it's written especially for sick people who have no time for complicated technical information. However, due to the many professionals that approach me concerning the various Oxygen Therapies, I have included some technical information for their use.

The Approach to Take

There are only two diseases. The one that a person has and the one that he or she can eventually get. Which category are you in? We live in a very sick world, and diseases and all those little things that just won't go away of every type imaginable exist all over the globe waiting for you. People are still in the medical dark ages searching for relief from the common cold, let alone the most horrific of new diseases. Where are the cures? Are there solutions? Hint: everyone I ever met who correctly used and maintained these therapies also got a nice little side benefit. They all stopped having common colds. In some cases 15 or more years have gone by and it still has not returned.

Within the traditional or the alternative field, there are many therapies and countless books containing conflicting advice as to which route is the best for you to take for health. If you are not familiar with Oxygen Therapies, the approach for you to take is to remain open-minded and carefully conduct your own research. Whether or not you end up agreeing with us is your choice, but take the time to consider what I reveal to you. Remember, actions speak louder than words. Please don't be one of those all too-common 'experts' who forms an opinion without thorough study and personal *experiences*.

Since they have caused no harm to millions of users, *why not* try Oxygen Therapies after you have examined the evidence for yourself? While you consider that, ponder on this overwhelming basic fact: At birth, your *first* encounter with life outside the womb was to draw a deep breath of oxygen into your lungs. This is the whole point, the life-giving point, the life-generating source point. My friend, the health answer is indeed right under your nose.

This whole Oxygen Therapy concept is actually very simple. In this book I make it very easy for you to follow along, to try it out, and then to be able to explain it to others if you wish.

It's as simple as 1-2-3. Problems. Solutions. Evidence.

Part One

The
Problems

Chapter 1

Life and Health on This Planet

Everybody reading this is a bag full of dirty fluid. Please excuse the plain speaking to drive home my point, but you really are a bag full of dirty water. Two-thirds or more of you is fluids. You're supposed to be an elastic carbon based sack filled with clean oxygen and water and bright red blood, but you've become a bag full of dirty fluid. Because I share this environment with you I'm also a bag full of dirty fluid, but perhaps less so. I live on this planet just like everybody else, but the difference is that many others and I have taken steps to remove some of the garbage and put more oxygen in its place. Because of this daily ritual most of us haven't been sick in a long time other than from purposeful detoxification. We shouldn't all be bags full of dirty fluid, and our planet earth's fluids, our water, shouldn't be dirty either, yet the sober fact remains that we have been severely compromised. How did we get that way?

City People

On a trip to California, I looked out the plane window and saw mountaintops sticking out of and up above the huge dark smog clouds hanging over the coastline. The city was completely engulfed in a carpet of smog. I sighed, and thought to myself, great, here I am staring right into the problem. I'm flying in to do another round of health lectures and media appearances, trying to help people living in this bowl of toxins to get healthy. The irony is that I've now got to descend into this toxic soup city and swim in it just like they do. Like a fish in the sea surrounded on all sides, I'm going to have to swim in this pollution for four or five days—yuck. I was thankful that at least I get to leave later.

I had arrived early so I went to the movies. As I walked out of the theater I crossed the street and sat down on a bench. There I was in Los Angeles, staying at Peter Aykroyd's house, and I'm watching all the Hollywood stars and starlets going by in their sleek cars. I was sitting on the corner of Sunset and Laurel Canyon Boulevard without having done any real exertion, and my only effort was walking across the street.

Isn't this interesting. Here I am a tourist in Hollywood sitting down and feeling very relaxed, but noticing that I am having to gasp at the end of each breath. I wondered, "What's wrong with my body?" What was wrong was that there was no oxygen in that five-way intersection crowded with cars under a low-level atmospheric inversion layer. All there was for breathing was a little oxygen and a lot of smog.

Most people live in a downtown metropolitan area where you have the smokestacks, the car and truck exhausts, the buses, the diesel trains, and the planes all spewing their angry fumes. These people are accustomed to being coated inside and out with pollution; and so, just like old smokers, they hardly notice how low their health has sunken. They use stimulants like hip caffeine-laced drinks and mood-altering alcohol, cigarettes, and drugs to try and get an edge of energy—but there is no cheating Nature. The Piper will be paid.

All my contemporaries who partied hard in their youth and never got off that path are getting older now, and they are in various rough shapes and getting sicker. Some are mad I don't age as fast. Their bodies' filters, the kidneys and liver, were overtaxed for too long and are starting to fail, and it's becoming noticeable, especially in their eyes and skin. If you ask them, they're all singing the same tune, "I need more energy!"

I just had the image in my mind of the old stereotype of young people gaily dancing around and around to the music, and abusing their bodies with various substances. Off to the side stood Ol' Scratch, the metaphorical Mr. Bad himself, chuckling, and wringing his hands and saying to himself, "That's it, my pretties, have your fun, and keep putting that stuff into your body temples. I'll be back for you later—you won't be laughing then!"

Do you live near cars? Every time someone brakes to a stop, everybody breathes asbestos particles scraped off of the brake linings. Every time you drive you put unburnt and burnt petroleum hydrocarbons in our breathing air. Everyone living in pollution and those consuming that which they don't really need are seen by some as a rich source of healthcare and/or addictive substance sales profits. They figure that if someone is stupid enough to live like that, or to use that stuff, they deserve what they get and are just waiting to be plucked at the inevitable harvest. My advice to the city people? You don't need that much action. If possible go somewhere clean to live while you still can, and phone it in.

Message to the party monsters: You can start drinking oxygenated water instead, and gradually work at fitting in and feeling good all by yourself without putting anything stupid in your hands. Don't you realize you were tricked into some really perverted message if you presently can't simply happily exist every day without some crutch? We are not that irreparable. Lots of people will help get you clean, but *you* have to *ask*.

By the way, I have watched a good many country people pour plenty of things into their bodies every day as well, so the 'geographic cure' won't change you completely; it will only change where you're standing.

The Air Oxygen Content

A friend of mine measured the oxygen content in the area near Gary, Indiana, where they have a bunch of steel mills, huge blast furnaces, and toxicity in the air. You drive by on the highway and this place really stinks! He measured the oxygen there as only 9-to-11 percent. We have shortness of breath when the oxygen level drops into the teens, and below 7 percent oxygen we cease to live anymore. Allowing for pollution in the cities, our society as a whole has allowed so much pollution to accumulate, and so much of the environment to be destroyed, that our available oxygen commonly drops below 21 percent in the air, depending upon location sampled.

So we are right now trying to run these physical vehicles, our bodies, on insufficient oxygen. Think of your body as a simple machine magically infused with and animated by the Life Force. This physical machine we walk around in was designed to exist here on the planet within an atmospheric sea full of high-level fresh oxygen, perhaps up to 38 percent or 50 percent oxygen! Our ancient ancestors were living in a much cleaner highly oxygenated environment. That's when your genetic body was created by them, and now you've got less than half, probably a third, of the oxygen your body was designed to run on. This lack of cellular oxygen is ultimately where all your health troubles come from.

There is definitely much less fresh, untainted (not bound in pollution) energetic (high atomic charge) oxygen available to us, and the air we breathe today is reported to have only about 21 percent oxygen. The other 79 percent is mostly nitrogen, which is an inert gas, plus trace elements and noble gasses.

Oxygen gives us life, and equally as important, it's what Nature uses to clean out us humans and everything else. Is it any wonder—since we don't have enough oxygen for our body machines to efficiently and smoothly run on—that our bodies are now bags full of dirty water? Much of our atmospheric oxygen is bound with pollution. So much so, our Life Force is crying out for good clean oxygen and water to carry it through using all the pain and discomfort and low energy warning signals and chronic illness it can muster enough energy to send you.

A peer-reviewed journal in the United States, *Science News*, Volume 132, published November 7, 1987, stated—"Air bubbles trapped in 80 million year old amber are giving scientists an unprecedented opportunity to sample and analyze the atmosphere from the earth's Cretaceous period when dinosaurs roamed the planet. And the preliminary results are suggesting that these creatures breathed an air far different from the atmosphere we breathe today."

"We were able to recalculate the original concentration of oxygen" says Robert A. Berner of Yale University who has analyzed the samples along with Gary P. Landis of the U.S. Geological Survey in Denver. "The oxygen level appeared to be around 30 percent more (of the atmosphere) as opposed to 21 percent today."

Scientists tell us our air oxygen today is around 21 percent, not the higher 30 percent plus quoted above. We all go about our daily business slightly oxygen starved, subtly huffing and puffing at the cellular level, and filling up with our daily load of pollution from the food, water, and the environment. Everybody in smog zones is coughing with the fumes in the air, and holding his or her hands up to their mouth due to the stench of the pollution—yet we just blindly accept all this as normal. It isn't.

We do not even know the amount of oxygen present in the atmosphere of the earth. We can test for oxygen percentages anywhere (this is easy); however, the percentages are different at different altitudes and different latitudes. This makes an estimate more difficult. What makes it REALLY difficult is that the differences at different latitudes and at different altitudes and polluted loads are not consistent everywhere. In other words, the air oxygen content is different EVERYWHERE.

Our Atmosphere

One study using sediment abundance data, along with assumed rapid recycling of sediments to stabilize oxygen, shows a pronounced and extended rise in atmospheric oxygen over the period of 375-to-275 million years ago, spanning the Carboniferous and Permian periods. What could have brought this about? The modeling indicates that increased oxygen production caused by increased burial of organic carbon is the chief suspect. This increased burial is attributed to the rise and spread of large woody vascular plants on the continents, beginning about 375 million years ago. The plants supplied a new source of organic matter to be buried on land and carried to the oceans via rivers. This 'new' carbon was added to the amount already being buried in the oceans, thus increasing the total global burial flux—Robert A. Berner:—*Atmospheric oxygen over Phanerozoic time*. (Proc. Natl. Acad. Sci., U.S., 28 Sep 99, 96:10955) QY.

As I write this, I am in Florida, stuck right by the interstate freeway and downwind from the filthy Fort Lauderdale power plant. Although the cooler season has started, I often still have to keep the windows closed and use the air conditioner simply to filter the air; because when I have the windows open all night, upon waking, my throat is congested after filling up with pollution during sleep. Burning at night so that no one will easily see the dark smoke that comes out of the smokestacks is very deceptive. I use ozone air purification and oxygen supplementation and ozonated showers to combat the toxic effects of what other people have come to accept as normal. Due to this pollution overload, if I did not constantly use oxygen to clean myself out I would have had a much shorter and severely compromised life experience, and it would probably be filled with lots of chronic disease.

Natural Oxidation Environmental Cleanup

We can clean up all the contaminated toxic dump sites! All the big news stories you have seen where they're stymied are wrong. I know a company with a proven on-site hydrocarbon, oil, gas, coal tar, mgp, PCB, dry cleaning solvent, and other contaminated soils eliminating bio-remediation technique. They come in, do their technology thing, and in a couple of months the soils are all clean. It's completely safe. No trucking the problem somewhere else, no big expense, no being tied up in endless litigation and enforcement actions. But nobody knows about it.

The Natural World Life Cycle and Oxygen

You are breathing oxygen right now. But where does the oxygen come from? Our planet earth consists of trees, water, the oceans, and the aquifers. The trees have fruit on them, and the animals come along and eat the fruit from the trees, grasses, and other foods. Man comes along and eats the fruit and the animals. The oxygen we breathe is produced mainly from two places. The plankton in the ocean and the new growth in the forests. The sun's light strikes the plankton, and when they're struck with light they produce O_2, oxygen. The sun's light stimulates the ground plants as well and they create O_2 as a waste product of photosynthesis. That's where most of our oxygen comes from. It's not, as many presume, that we have a fixed load of oxygen for our planet and it's going to be here forever. No. Life is a living thing, constantly being created and transformed. We participate in a live real-time oxygen generating system. This oxygen mixes into the planetary atmosphere and becomes a part of what we are now breathing, eating, and drinking.

Our Eventual Collective Deaths

As I have lectured so often, the reason we are in this alarmingly, and rapidly worsening disease and plague situation—with mad cow disease, West Nile virus mosquitoes, drug resistant TB, arthritis, AIDS, anthrax, and cancer, and on and on—is because we are losing our planetary oxygen. The result is that everyone with a body will have more disease, more often, for longer and longer as the availability of Nature's natural curative and detoxifying oxygen disappears from the scene.

Good Bye, Children

Today we face the existing health problems created by ego, greed, and a lack of holistic leadership that began with the dawn of industry. Amazingly, *Science News*, on July 14th, 2001, Vol. 160, No. 2, page 24, quietly announced the projected date of the death of our species!

Why was this little tidbit about the demise of humanity buried in a brief mention? Were they afraid of 'offending' someone? Isn't it important for us to be aware of this? Shouldn't we know we are allowing our children's deaths by our apathy and inaction? As I said, we receive most of our oxygen from the rainforests and the ocean plankton. *Science News* announced, "The Amazon rain forest could disappear much more rapidly than expected. . . . Logging and burning for agriculture currently claim about 1 percent of the Amazon rain forest per year. . . . After farmers and loggers cut and burn broad swaths of rain forest, more precipitation runs out of the area via the rivers. This leaves less moisture to return to the atmosphere—and that means, in turn, less rain. As a result, forest areas that people have cleared don't grow back as quickly. . . . The Amazon rain forest could disappear sometime between 2020 and 2030." Stop and think a minute, how old will you and your children be in about 20 years? And remember, waiting until 'then' will be too late.

O_2, O_3, H_2O_2, H_2O Cycle, Nature's Ozone and Peroxide

Oxygen, (atomic symbol 'O') is the major part of the continuous cycle of life here. That's why we can't keep cutting the trees and polluting the ocean. They are the source of life itself. Because of the clear cutting of the ancient rainforests cited above, new African dust storms are blowing strange bacteria, fungi, and molds worldwide into places they should not be—where there are no natural predators to keep them in balance—thereby disrupting ancient ecosystems worldwide.

The chemical symbol for the two atoms making up the oxygen molecule is O+O or O_2. Once created in Nature, the sun turns the oxygen into ozone, O_3. The O_2 oxygen generated on the planet rises up and up into the heights of the atmosphere, pushed there by atmospheric winds until the sun's light (at 185-to-254 nanometer wavelengths, in the ultraviolet range) energetically strikes these O_2 oxygen molecules and splits them apart. Most of this starts happening in the upper atmospheric layer about 6 kilometers or 3.7 miles above the earth. Stratospheric ozone is found in a broad band, generally extending from about 15-to-35 kilometers (9-to-22 miles) above the earth.

Ozone is created when three O_2 molecules are split up, or disassociated, by the sun's light and then they immediately recombine into two O_3 molecules. At first the oxygen O_2 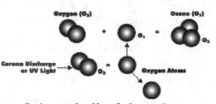 molecules are broken apart into lots of two O_1's and all of these hungry-to-combine-with-something O_1 atoms quickly attach to each other. They reassemble and form long chains of oxygen molecules, O+O+O = O_3, O+O+O+O = O_4, then O_5, O_6, O_7, O_8, etc., all the way up to O_{21} and beyond.

These polyatomic (many atom) forms of oxygen are all called 'ozone.' For simplicity's sake we simply say 'ozone' or use the symbol O_3, to roughly include all these different types of ozone. However, there are significant differences between the various ozone forms, especially when we talk about their use in healing. O_3, O_4, and O_8 seem especially significant.

O_3 ozone is so energetic it constantly wants to become stable O_2 oxygen by giving up O_1. That's what makes it useful in healing; all those singlet oxygen forms are constantly being created in a cascade of O_1 releases. The O_1's are negatively charged, and they are the heroes that perform all of the healing work and do all of the free radical scavenging by combining rapidly with the positively charged garbage in our bodies.

This is all a little confusing at first, but all you need to know is that way up high in the region of the ozone layer the sun turns fresh oxygen into ozone, and this creates the ozone layer. If we were to compress the entire atmospheric ozone region it would only be paper thin.

The Ozone Layer

Ozone creation is a transitory phenomenon; it happens momentarily, and continually repeats so long as the sun is shining. Usually no one realizes that the ozone layer is a paper-thin boundary layer. It's produced constantly and is part of the Grand Circle of Life. The trees and plankton make the oxygen. The oxygen comes up, the sun turns the top of the oxygen layer into ozone, and then the ozone falls to earth because it's heavier than air.

The trees and plankton make more oxygen. More oxygen comes up to replace the falling-down ozone, and the new replacement oxygen also gets turned into ozone. This is a living biosphere. The ozone layer is not some stationary gas band. It's not a fixed quantity layer, and there is no fixed oxygen pool below it.

This is normally how the oxygen we breathe is constantly produced, cleaned, and recycled. Ozone is so full of the sun's clean energy that we use it in healing to remove impurities. In the same way, and with the same purifying action, we use the sun and ozone when airing out an old mattress or sleeping bag under the sunshine pouring into our backyards. Sunlight bouncing off water and snow increases Nature's ozone production because the bounced sun rays get another chance to strike the atmospheric oxygen. Breathing elevated natural ozone levels is one of the reasons we unconsciously feel energized during swimming and snow sports.

Natural Hydrogen Peroxide

Now what about another natural Oxygen Therapy used in healing, hydrogen peroxide? There is always water in the air. Remember that the falling newly created O_3 ozone in the atmosphere quickly wants to become O_2 again, and does so by giving up O_1. The O_2 oxygen we breathe is very, very stable. O_3, however, is very unstable; that's why it quickly wants to turn into stable O_2. To do that, it must naturally give up the O_1. So O_1 is easily created by Nature's cycles.

As the falling O_3 ozone passes through moisture, or clouds in the air, it gives up O_1, which immediately combines with the water leaving stable O_2 oxygen. The water, H_2O, then has this O_1 combined with it; and that turns it into H_2O_2, or $H_{2+}O_1$ water plus O_1 oxygen. This is what we call hydrogen peroxide, just like the stuff in the drugstore that we put on cuts, except that Nature's peroxide is milder.

Hydrogen peroxide is found everywhere in Nature because it falls to earth in the form of rainwater. The normal rainwater, the clouds, the fog, and the morning mists spread around the atmospheric water carrying a small amount of hydrogen peroxide. Rain is formed high up in the atmosphere, within a highly ozone/oxygen rich atmosphere. The low concentration peroxide formed up there then goes into our groundwater, our aquifers, and our drinking water. The roots of plants and trees reach down into the ground water and take up the peroxide, and Nature continues this cycle by delivering the cleansing, oxygenating peroxide to us.

Men and Women get the peroxide when they come along and eat the fruit and vegetables from the trees and plants that are all 'drinking' the peroxide and depositing it into their fruits. Man breathes oxygen from the air. He also breathes trace amounts, approximately .05-to-1 part per million, of ozone in the air. We also absorb oxygen through our pores.

These are the fundamental ways that we get oxygen from our ecosystem. We also eat animals, which have also been eating the peroxide containing fruit, vegetables, and grasses. In this way, the oxygen-ozone-peroxide is transferred from the air into water into the earth into the plants into the animals into man's food chain for the benefit of the human body so that it can stay healthy and thus provide a reproducing vehicle for the human consciousness.

The Importance of Ozone in Nature

No doubt you've heard of the good ozone layer above the planet? The 'good' ozone layer above the planet is supposed to be shielding us from the harmful effects of ultraviolet rays. That's why noble but misapplied laws are passed attempting to protect the ozone layer.

Here's how it works, and what few understand. I have stood at many podiums saying this for many years, in contradistinction to media-created public opinion. The sun's ultraviolet light at 185-to-254 nanometers is used up because it's absorbed into the reaction which breaks apart the O_2 oxygen into the O_1's which quickly reassemble as O_3's. Via this phenomenon, we have the sun's UV light energy actually transferred into the newly created long chain ozone molecules. That high-energy gradient is what allows oxygen to maintain itself despite being unstable as singlets or as ozone. The harmful UV light energy is mostly used up in the oxygen to ozone conversion process. This is what the public and media inaccurately call the ozone 'shielding' us from the sun's harmful rays.

It's not that the ozone is some sort of wall protecting us; it's simply that the UV is used up to create the ozone. So the truth is actually a little different from what you've been taught.

As I explained, the O_3 is heavier than air and as the O_3 falls back to earth it will go through rain clouds and moisture, and the water picks up O_1. Most ozone is so energetic that it only lasts for about 20-to-30 minutes (its half-life) and probably longer in Nature. If we are considering one of those previously mentioned long chain O_{17}, or O_{21} forms of ozone existing beyond O_3, you'll find that maybe traces of it stick around for a couple of days in the upper atmosphere. That's where there is little else for it to react with, but generally we say 20-to-30 minutes for ozone before it gives up O_1 and breaks down into the stable two oxygen form again, turning back into what we call oxygen.

The heavier-than-air ozone is falling to earth. If a waterfall is tumbling over rocks, or if waves are crashing on the seashore, the mechanical agitation stretches the water surface area reducing the water's surface tension. Because the surface tension of the water is lessened. the water now easily absorbs oxygen from the air. Because the water molecules are stretched apart they are no longer so tightly packed; so there's plenty of room for oxygen in-between the molecules of water. This is how Nature naturally cleans up our water; by first oxygenating it, and then the oxygen oxidizes the pollutants, dead things and microbes.

Water absorbs oxygen and ozone from the atmosphere. That's why if you ever went swimming downstream from a waterfall, you always felt really, really good and invigorated. At waterfalls everybody swimming and breathing in the mist usually laughs and talks loudly, and that's because they're absorbing the oxygen and Life Force from the water in the form of ozone and oxygen and hydrogen peroxide. The water itself has absorbed oxygen and now passes it right through the swimmer's skin and into their blood, naturally making them feel more alive. Another one of Nature's natural oxygen delivery systems at work.

The peroxide falls to earth, the ozone is circulating in the air, the oxygen is circulating in the air, and it rains, so the oxygenated rain falls to the ground and gets down into the aquifers. When we draw water from wells, we pick up this oxygen, and peroxide, H_2O_2. The roots of the plants and the trees and all the vegetables and grasses rooting down into the earth start taking up the peroxide and depositing it into the trees, fruits, and vegetables we eat. Then the plants put fresh oxygen into the air. This completes the cycle of environmental oxygenation.

Have you ever noticed that when you go out and water your lawn in the summer it grows an average amount, but when it rains it really grows much faster, and so much better and greener? It really comes alive. What's the difference? When measured for natural structure and vitality, the water coming through the pipes down from the local water 'purification' operation comes up as 'dead' water—full of chemicals that have used up the oxygen. We're giving this unnatural water to living organisms, which survive, but don't really like it that much. Give the plants pure natural water with Nature's oxygen and peroxide in it and they grow like they have never grown before. Just like humans do when given a completely natural environment with plenty of oxygen in the food, air, and water.

The Importance of Oxygen Combining with Hydrogen

The balance between Hydrogen and Oxygen is what keeps us alive. Hydrogen (H) is the most basic physical substance having only a single electron and proton. Everything you see is held together by it. It is the first element, and the most abundant one in all of creation. Hydrogen holds its chemical arms out and bonds with just about everything, and strings molecules together into different compounds. Our body is put together this way. Water is put together this way. The things we eat are put together this way. Hydrocarbons are put together this way. Chemicals, PCBs, and all pollution are held together by hydrogen bonding. Over-hydrogenation leads to excessive structuring and stiffness.

Oxygen comes to our rescue because it readily combines with hydrogen. When oxygen combines with hydrogen, we call this process oxidation caused de-hydrogenation (pulling out the hydrogen). We commonly use oxygenating compounds such as ozone or peroxide in healing and environmental clean-up because they generate active singlet oxygen, O_1. This is of prime importance because lots and lots of active oxygen can break apart the hydrogen bonds holding together pollution and disease while leaving healthy cells alone.

Oxygen and the Environmental Solutions

As I mentioned, right now there are oxidation based environmental clean-up solutions that can clean up any toxic spill (gasoline, oil, PCBs, dry cleaning solvents, coal tar, etc.) on site for far less cost and permanently. They aren't being used much because everyone is feeding on the billions of available environmental dollars with endless testing, monitoring, and writing reports year after year. If it stays this way then 20 years down the road not much will be changed, most of the toxic plumes will still be there, and along the way everyone looked properly 'concerned' and stayed busy in politically correct fashion. There is scant motivation to actually clean it all up in the real world. As long as the polluters and the agencies can have 'meetings' and 'look busy' why spend the money to actually clean it up, and why end your lucrative consulting and monitoring business?

Under the influence of singlet oxygen, all manner of pollution is oxidized, reduced to its elemental structures and burnt up. The hydrogen bonds collapse and the remaining base elements are washed away.

Nature has devised a perfect solution for cleaning away all of man's harmful artificially constructed pollution; active oxygen. And the beauty of it all is that normal healthy living human cells are immune from this de-hydrogenating action. We'll get to that later.

Chapter 2

The Problem of Anoxia
Your Lack of Oxygen

Go out into your driveway, pop open the hood of the car and put something half way over the opening of your engine air intake to block the air from getting in. Then try to drive around like you normally would.

Mechanics understand that a machine is designed to run at a certain air to fuel ratio. This body we walk around in is a machine. It's designed to run well with a certain level of oxygen present. Like any engine, if you deny the body machine half the oxygen it is designed to use, is it going to run as well? Is it going to run clean? Is it going to run efficiently? Is it going to run without creating pollution because it efficiently burns the fuel cleanly? No. It's going to produce and have very little energy, and it will merely chug erratically. It's going to smoke, it's going to belch, and it's going to emit noxious odors and fluids. That's you I'm describing, and the covered air intake is your lack of oxygen in your fluids and cells owing to a lack of it in our food, water, and air.

As you get older, your imperfections grow worse as time and life here pile up more and more wastes; and in come the microbes to live on and in the waste. We call the result disease. When you aren't burning your food/fuel mix cleanly because of low oxygen, you get so full of toxicity and so full of garbage that there's little oxygen left in the cells. Just as a machine must mechanically break down, your body fails. And then you have to go crawling to the doctor with a bag of problems that gets bigger and bigger with each passing year.

Anoxia literally means: 'A lack of oxygen.' Society has created a world full of anoxia, and that is the problem we are facing on an international scale. Through industrialization and pollution based on ego and greed, we have destroyed the natural balance of Nature. We've lowered the amount of fresh oxygen normally found in our environment. We need this oxygen back to help us flourish and maintain good health.

Did you ever do that trick where you take the vacuum cleaner and put a porous cloth over the intake opening, and then go vacuum the bed sheets with it? You can easily do this to show someone all of the dead skin cells we are continually sloughing off. There is always a pile of filthy dark and gray powder on the cloth.

That's equivalent to the condition that your lungs are constantly forced into while you breathe during your residence here. Oh, and if you're reading this while basking in the sun and surf of some vacation isle—no you didn't get away from it. You only made it less. There is no exemption on this planet, not even for those above it or inside it. That means everyone is personally responsible to pitch in and fix it so that we can all survive.

Every day each lung in each individual is breathing in an aggregate-heaping tablespoon of dirty particulate matter. Day after day, month after month, year after year it's piling up, piling up, and piling up. Wherever you go or wherever you live, your lungs are being bombarded and overloaded with dirty particulate matter seeping into the blood and the cells. Without sufficient oxygen, how and when will you clean your blood and cells out?

Movement and exercise helps clean the tissues, but consumerism has led the majority of cultures into a sedentary lifestyle surrounded with creature comforts. You don't even have to get up to turn the TV on or off anymore.

You just sit there and go zap with the remote. You drink your beer, smoke your cigarettes, snack on hydrogenated fast foods, and you're still sitting there on that couch.

Granted, it could be that you're one of many so freaked out because the day has been so hard on you and your nerves and cells are so oxygen depleted that you just can't stand any more activity, so all you can do is sit down. It's this revolving cycle where you get sicker and sicker and more and more tired; then, before you know it, you're gasping for—what? Oxygen. You're dead tired, beat all the time, and totally washed out; wasted, with little energy to move.

This is what you're locked into if you just continue the downward slide by going along using shortcuts to get along. You have no 'get up and go' cuz it got up and went, and you're always feeling tired. So, you lean on your doctor as the health authority as if you were taking a car to a mechanic. But your doctor has only been taught a certain way of using medicine owing to the politics involved in his training, and the accepted political situation has institutionalized its errors, so even if he wants to help, he has limitations. By all means please go to your doctor. Do that, but ask yourself, what doesn't he know, and how am I going to help solve my own health problem, how am I going to live, how am I going to eat and drink, how am I going to breathe, and how am I going to oxygenate my body?

How Pollution Affects Our Bodies

A single human body is made up of as many as one hundred trillion cells. Most cells in the body are floating cells. What do I mean by that? Picture a man weighing about 150 pounds. One hundred pounds (or two-thirds) of him is water. Our body is 66 percent water. So what you look at in the mirror is around 100 plus pounds of water and the rest is solid matter. That majority weight of 100 pounds of dirty, filthy, stinking, bacteria, virus, and parasite laden toxic liquid mess of an elasticized carbon sack is stumbling through life and calling itself Man. The only thing missing is an abundance of common sense. The writer Mark Twain was great on common sense: He wrote, *"Be careful of reading health books, you may die of a misprint."*

Common sense in our case is fixing those 100 pounds of dirty water that you are carrying around before it fixes you for good. Every day we all go through the process of drinking water. If you're normally drinking city water, or ever drink commercial beverages, or eat food prepared in city water, welcome to chemical city. If you're fortunate enough to have water from a well, you still don't know what happened upstream from the well where the farmer sprayed the crops.

And while we're at it, let's not forget how the farmer put a little tag in the cow's ear so the could be identified. Did you know the little numbered ear tag is a drug delivery system? And do you realize the animal feed has antibiotics and hormones and medicine in it? The cattle are nicely injected even from birth, and they've got all that stuff in their cells and fluids that does not belong there, and it definitely does not belong inside you. But you drink the water and eat the cow. What has that got to do with me, you ask? What it's got to do with you, is the fact that what took place next door or thousands of miles away is all piling up inside of you, and has been piling up inside you for a long time.

Polluted Food, Air, and Water Filling Us Up

Now what about our food—the food that is so essential to life and keeping our bodies well maintained. Every day, with any intake of food, air, or water in this microscopically and macroscopically polluted world, we're filling up our 100 pounds of fluid with things that don't belong there.

Our personal 100 pounds of water is being filled up, filled up, filled up, filled up, day after day, breath after breath, mouthful after mouthful, beer after beer, cigarette after cigarette, greasy food morsel after greasy food morsel, irradiated bite after irradiated bite killing the enzymes, and it just piles up, day after day, week after week, month after month, year after year, generation after generation. It keeps right on piling up, and even if you get bored of reading about this concept, it's *still* piling up and piling up and then, all of a sudden BANG, or maybe with a slower *whimper*, something's not right. Out goes your energy, your bones hurt, you can't remember stuff, and you have a list of problems longer than a freeway.

Let's take a minute to think about one of the parts of the body—our bones, for example. The bones are roughly 21 percent and, just like with every other cell, the water that the bone cells are floating in is getting filthier day after day, particle after particle, breath after breath, puff after puff, and injection after injection. Here it all is, filling up, filling up, filling up and it doesn't stop. Have another slice of super hydrogenated margarine along with the additives in your beer while it keeps on piling up. And as it piles up, all the cells, all the organs are floating in this piling-up pollution.

You're getting increasingly tired and stiffer, and your face is starting to get wrinkled, and you've got this twitch, and there's a fungus on your toes, and your girlfriend's got this rash, and you yell, "Help, what's wrong with me? Quick, fix me up Doc, give me a pill." You rush to the pharmacist and they offer you the pill and down it goes. You take the pill because you're supposed to, even though it's another unnatural substance to add to the pile; but you take it anyway, and fast, because it's going to fix you up real well. It turns out, however, that the pill has some kind of side effect; so it tricks your inner sensors and short circuits a brain cell or neuron and you bloat up or get more side effects that make you think you need more pills. Or maybe you're new to the world of symptom chasing and you go, "Wow, I feel better. I feel pretty good. Watch out, world, I'm back on track!" Meanwhile you never cleaned anything out. It's all *still* silently piling up, piling up, and piling up in the cells. Did all that repetition about the piling up get you to FEEL how relentlessly the pollution is piling up even as you read this? It's still piling up and piling....

Obviously I have belabored the problem to get any overly polluted brains to learn through repetition. The number one problem is your cells being forced to float in this accumulated liquid dirty mess.

Undifferentiated embryonic cells are being created in this mess, and then oxygen and the Life Force molds them and they specialize and aggregate replacing our organs. These cells don't have walls; they have semi-permeable membranes; cellular membranes. 'Membrane' is a way to describe the fact that things go through them, in and out. Our polluted fluids flow around our bodies in the blood and lymph constantly, in and out of everything. If the fluids are full of garbage and continually passing through more garbage, and the fluids flow in and out of the permeable cell walls, in and out of the cells, what's going to happen to our friends the cells that are making up our organs? We could be talking about your heart, your liver, your spleen, your brain, your colon, or your big toe. We could be talking about the follicles of your hair, the cornea in your eye, macular degeneration, and leading up to your immune system, your digestive systems, your reproductive organs, your ability to sleep at night, your moods, your brain chemicals, your hormones.

Do you really think that your body's inner systems are going to work correctly and stay in balance if they're filled with, and floating in, garbage? What cell or organ wouldn't have a hard time living and functioning under the strain of all this garbage?

If I was a cell and I was trying to listen to orders from the nervous system, or from the other cells to do something, how would I react while swimming in the dirty soup? Imagine that a cell in our bone marrow gets an order to produce white blood cells. He gets busy inside a bone and he's making white blood cells and more white blood cells and more white blood cells. Meanwhile, another cell that also can barely see or hear its way through all these filthy polluted fluids somehow receives its orders via a garbled electrical or chemical or telepathic message. The message is garbled because it's coming through all the garbage in the body, and it says, "Tell that bone cell to stop making white blood cells. Tell him to stop making white blood cells!"

But because it's floating in the polluted fluids, the cell doesn't hear very well; so it complains and says, "What? I can't hear you!" "Make more blood cells? No, no, change that, say again, what? Make more? O.K., more white blood cells, here they come boys!" Of course never receiving correct instructions and also having its own self-confusing interior pollution to contend with, the cell might do just the opposite of what the original nervous system message was asking for.

Scientists would say at this point that the cellular signaling pathways are not functioning properly. When any cell is so imbibed and imbued with toxicity that the cellular self-regulatory function is damaged, it doesn't work anymore. It *can't* work anymore, so it's going to do something dysfunctional. And if it happens to be a cell in the bone marrow making white blood cells and it just keeps making them, and making them, and making them, despite everything else saying, "Stop! Shut it off, shut it off!" how do you think your body is going to react?

You don't feel so good and your friend comes along and you both head for the health food store to try all the latest herbs and powders, and you manage to get some relief for a while, but it doesn't last long, because it's all *still* piling up, piling up, piling up....

You go back to the doctor and say, "I don't feel good again, the pill must have stopped working, and I just can't shake this thing off." He says, "Well son, I'm afraid you've got leukemia. Too many white blood cells. We're going to have to give you a few more pills. And we're going to have to inject this new stuff into you. But don't worry. It's the best there is, and we're going to have a good chance of straightening all this out. By the way, you know, that organ there is just about shot. We're going to have to cut it out. But don't worry, we've got this new pulsed magnetic laser saw and we're going to go inside this bit and cut this part out because we now have the latest technology you know. And for good measure, we'll hit that spot with radiation. We'll make it glow in the dark. Yeah, that'll work real well." Folks, what is wrong with this picture?

This illogic piled upon illogic is what many are dealing with. We know that throwing more fantastic futuristic marvels into the soup and cutting parts out is not going to clean out the problem. More ingredients in the soup, no matter how hip they are, what the latest name for them is, or how expensive the ingredient in question is, or how famous your doctor is, or how your chosen institute is one of the top centers in the world, or how rare anything is, or how many thousands of people are doing it this week, or whether it's legal or illegal, is not going to solve the problem. Look at it from a wholistic, or natural, or large systems perspective. More piled on unnatural substances may chase away a few symptoms for a while, but they do not clean up the inner piles of garbage that are the cause and breeding ground for the inevitable diseases in the first place. And it's still piling up and still piling up....

Our obvious problem is that the cells become so filthy that they become toxic and then they lose control, and then they eventually become stagnant and cancerous. Of course they all must do this in time since we are all sitting in a bowl of toxic soup. A toxic soup that makes us sick, makes us cranky, make us age, lose our temper, our sense of balance, and when we lose our ability to reason rationally our consciousness heads toward the reactive animal level.

I'm sure you've heard some of the stories about people on the freeways in Los Angeles, or New York, or Miami, jumping out of their cars in a road rage frenzy and shooting the guy next to them because, well, they *really* felt like it. Think about how the available oxygen is deficient in these cities and in their bodies, and how that unconsciously produces survival feelings of desperation.

Study Reveals Depression Treatment has Soared

On January 9th, 2002, Dr. Mark Olfson, a psychiatrist at Columbia University and the New York State Psychiatric Institute authored a study of two national surveys in the Journal of the American Medical Association. The studies showed that in just 10 years, from 1987-to-1997, the number of Americans treated for depression soared 371 percent from 1.7-to-6.3 million, and the proportion of those receiving antidepressants *doubled*. Do you really think that the number of our people with brain abnormalities suddenly and magically went up 371 percent?

The researchers attributed the sharp increases to.... [Wouldn't it be great if they came out and said low oxygen levels?] The emergence of aggressively marketed new drugs such as Prozac, the rise in managed care and an easing of the stigma attached to the disease. They found that the share of patients who used antidepressant medication climbed from 37 percent to nearly 75 percent. While an increase was not found surprising, Olfson said, "The *size of the increase* was larger than I think most people in the field expected," and studies since 1997 suggest the trend continues.

As the money gathers, you can bet that more and more of us will be medicated against our wills as more administrators read more and more 'studies' commissioned by medication manufacturers proving how it's normal that more people need to be medicated 'for their own good.'

Microbes and anaerobic fungi are silently living in people's brains. Tiny bugs living in the back of our throats can poison nearby spinal nerves and scramble nerve impulses traveling up and down and in and out of their brains? You see, under those circumstances, schizophrenic behavior, depression, and rage are entirely logical.

Chapter 3

Toxins

First Big Oxy-Truth:
The Ultimate Cause of All Disease is Toxins

Humans, animals, trees and plants have always been genetically equipped and raised within an oxygen-based ecosystem. We are oxygen-based life forms. That's why normal oxygen is so good for us and doesn't bother us. Oxygen only disturbs the things that evolved under low oxygen conditions. We thrive in oxygen. Unfortunately, may I remind you; man has polluted our ecosystem, cut down the rainforests, and ruined our oceans, the specific source of all the oxygen that keeps us alive and cleans us up. Because this pollution and destruction is going on, we are all oxygen-deficient at the cell and fluid level. Our bodies can no longer take out, or oxidize, our inner garbage in the way originally designed by Nature. Too much unnatural garbage is accumulating inside us daily, and disease thrives on it.

Go outside, and this time pop open your dumpster and take a whiff. Does that smell like fresh wholesome natural food? I doubt it. What about that little squirrel that was run over in the street a couple of days ago? If you go back, you'll see that flies are swarming around it and it's full of worms and maggots, and you don't want to get too close to the carcass because, 'Boy, does that stink!'

On a smaller scale, when the blood and tissue oxygen drops too low and stays too low inside us, we start looking like a good food source to uninvited microscopic parasites in the same way the flies liked the looks of that road kill. The same warm, wet, dark, low-oxygen garbage condition is advancing inside us, all due to the daily absorption of pollution and unnatural bad food surrounding us at every turn. What's going to live on this garbage? Bacteria, viruses, fungi, and other pathogens. They want to eat, reproduce, and live too. Under the right conditions, and according to their natural function, they're attracted to us and see us as food. They start eating our inner garbage, and us, and thereby complete their missions to clean up the dead items and decaying matter mess that we let accumulate inside our bodies. The only problem is that they eat our organs and tissues, and while they're doing that, their excrement is poisonous to us.

If you're full of garbage, you're low on oxygen; and the bacteria, viruses, funguses, pathogens, micro toxins, parasites, and disease will live on the sludge. The party they're having can easily be seen, live, in color, and in real time. I have witnessed many darkfield microscopic examinations of fresh living blood. You can actually see the holes through the blood cells

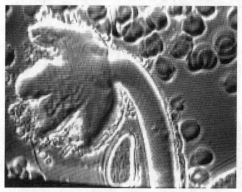 where parasites ate and bored their way through a cell. You can see all sorts of non-human bugs and organisms living and breeding in everybody's toxic soup blood.

Doctors are often taught that 'blood is sterile.' But how many of these strange things are living in you despite their teachings?

This *strange organism* —and more like it—was growing in a patient's bloodstream. Perhaps it was some sort of budding and burst parasite or growth. Definitely NOT a healthy blood cell! You can see the much smaller normal blood cells next to it. Darkfield microscopes reveal all sorts of microscopic toxin-spewing critters nesting and living in everyone's under-oxygenated blood. Examine your own blood under a darkfield microscope and see for yourself! Look at this book's back cover where anyone with eyes can compare the color difference between healthy and so-called 'normal' blood. We commonly see these disappear in properly oxygenated blood!

I have been telling everyone for many years that the ultimate cause of ALL disease is a LACK OF ENOUGH OXYGEN to clean our inner fluid environments. This lack of oxygen allows the two major causes of all disease to overcome us: Toxic build-ups, and the formation of various colonies of micro-organisms inside us, especially in the blood. The toxins poison us, and the organisms eat us while dispersing more toxins.

If our inner oxygen levels are not restored to proper levels and kept there long enough, the outcome is predictable, not pretty, and the results are all around us. Hospital admission rates keep the score.

The well known health gurus are all finally starting to come around and repeat what I have been teaching, because in the final analysis it's true— without plenty of fresh active oxygen, you can't clean out!

Baby's First Toxin

When I say toxin I am referring to any thing that does not belong in a healthy, clean body. There are two types of toxins. The first type starts out outside us, and comes from things such as man's wrong use of chemistry, and from natural earth sources such as the radiation in uranium. This type is the pollution around us and in our food, air, and water. The other type is comprised of the naturally occurring wastes left over from our interior life-generating processes.

Your toxic accumulations started early when you were in the womb and got them from Mom. Unless your normal cell formations were disrupted by injury or poisons somehow, at the time you were born your human body cells were the cleanest they will ever be. If your mother ate things that were tainted with pollutants, or if she smoked, drank alcohol, hit the crack pipe, or sprayed the roaches while pregnant, that was your body's first introduction to cell damage from toxins. Hopefully, your mother ate healthy, exercised, and never lived near the smokestacks or downstream from the chemical plant. And if some factory down the street didn't drop its effluent poisons into the river so they eventually made their way into the aquifer, or your well, ending up inside Mom's blood stream, you were probably born with nicely formed clean healthy cells. There is so much pollution we can't see. I don't think *anyone* is immune from this any longer. It's just a matter of degree as to how much YOU have.

Think about the minute damage that's been done to you since you were born. Every time you ate some food with a chemical in it, or a dye, a coloring, an unnatural preservative, or something not a food that's completely incompatible with your body, such as the hormones that are injected into our meats, what are these combined poisons doing to you? Since they aren't food, if you consume or absorb them you can't digest them. The body doesn't know what to do with them so it ends up stuffing them somewhere in the body, usually under a coat of slimy mucous.

Koch on Toxins

Around 1950, Dr. William F. Koch MD, Ph.D., a brilliant free radical chemist taught us that: "The toxicity is integrated into the cell walls themselves." The cell will actually integrate the surrounding toxic garbage within itself, because the soup is everywhere, can't be avoided, and nothing in the soup is immune. The permeable cell membranes will take in the dirt from the fluids surrounding them because they have no choice. They're trying to absorb oxygen, food, and fluids, and the garbage comes along for the ride.

Individual cells do the work in our organs. Your specific problems will follow patterns depending upon which cell a toxin has integrated itself into, and which organ the cell is a part of. The weakest cells will fall first, and in line according to your genetic miasms or some other factors. Under the piling up garbage, the cells will usually start to fail by either under or over producing something, as in poor digestion or auto-immune diseases for example, and eventually when the dirt stressors are too great; they become infected and stop working.

The Difference between Healthy and Diseased Cells

Let's think for a minute about a key difference between a diseased cell and a strong healthy cell. One of the things that a normal healthy cell makes is a self-protective antioxidant coating. The cells naturally protect themselves from oxygen potentially burning or oxidizing them. Diseased cells can't do this.

Antioxidants and their precursors are naturally occurring and found in our food. They don't 'stop' oxidation as their misshapen name implies. Instead, they regulate it. The healthy cells surround themselves with antioxidants as a protective coating.

Let's simplify it further, because it is important that you understand this process to prove how safe Oxygen Therapy is. Years ago, when I was a pup, there was a TV commercial about the invisible shield (Knock! Knock!) created by a tooth paste that coated your teeth. Just like the TV's invisible tooth decay preventive shield, the healthy cell's antioxidant shielding protects the normal healthy human cells from being oxidized, or burned up by oxygen.

Diseased cells have a big disadvantage. They cannot manufacture the antioxidant cell coating because their Life Force, and their chemicals, proteins, and electrical energies are being used to manufacture more viruses, more bacteria, more fungi, or whatever has invaded them. The diseased cells are used up feeding the pathogens and the mycoplasmas, therefore no protective coat surrounds them.

Only the Weak Cells Are Burnt Up and Washed Away

Any ancient cellular life form such as a bacteria, a virus, or a fungus, including any other human pathogens—or any body cell—in a weakened, diseased or dying state—have no antioxidant coating and no way to protect themselves from being burnt up by oxygen! Hence the title of this book, *Flood Your Body with Oxygen* (properly and safely) because; oxygen selectively attacks only *diseased* cells—due to their damaged or missing antioxidant shielding. No shield, no protection from being burnt up. Nature is beautiful in its design simplicity and efficiency.

Diseased Organs

Back to the organs. An organ is, mechanically speaking, a bunch of cells. So that means that at birth, if Mom's placenta did its filtering job, we should also have nice clean healthy organs. However, as time goes by we age and our toxins pile up and up. Our organs continually slough off dead, weak, diseased, and dying cells. Of course the bacteria, viruses, microbes, pathogens and all these little bugs and parasites come into the organs and start eating and living off these weak and dying cells. The diseased organ cells become food for the microbes to digest because they no longer have enough Life Force to produce protective mechanisms. The microbial scavengers are always looking for an opportunity to attack at the place of least resistance, so they invade the damaged cells first like a bunch of cannibals shopping for a meaty bargain.

Then the bugs eat this garbage and excrete more garbage of their own. Other bugs come along and start living on the garbage wastes from the first bugs. They in turn also start living on more damaged cells, and they, too, dump their excrement into your body fluids. Tinier bugs live on that, and they also dump their excrement in the mix. Pretty soon you're filthy; full of toxins—just like that dumpster—except that you're walking around looking better than the dumpster while heading for your eventual date with the doctor.

Proper Oxygen is Harmless to the Body

When there is a healthy antioxidant coating around a cell, any surplus oxygen in the vicinity is not attracted to that cell. The oxygen simply bounces right off it and continues on its way seeking a diseased cell or microbe to burn up. Our bodies love oxygen and are programmed to use it properly. Whenever our cells need some oxygen, the nuclei of the cells signal the cell membranes to open up their calcium ionic channels so that the cell can, in a sense, reach out, grab some oxygen and pull it in as needed for burning and cleaning. Here's the rest of the story which they don't teach in medical school:

Bipolar Theory of Living Processes by George Crile, an MD from Pennsylvania. Oxygen is electronegative (not bipolar), it's the most electronegative element known to man. When the cell becomes so electro negatively charged with the oxygen, in other words when the mitochondria and the Golgi bodies are sufficiently full of oxygen, the cell retains a charge that is basically electronegative. The cell then starts to attract and absorb nutrients and the rest of the electropositive elements like minerals. The cell flip flops its charge back and forth as the oxygen is used up. The cell loses its negative charge and becomes more positive which makes it attract and draw in or seek more negative oxygen, and the cycle repeats almost like a pump. Warburg spoke of this briefly. This has been a standard tenant in naturopathy which differentiates it from medical science.

The whole bodily oxygen usage process is regulated and safe. Are you getting the picture more clearly now about how simply and amazingly this all fits together?

Emotions and Cell Energy Production

You've got less oxygen around you than you're designed to run on, and, as we've learned, your fluids, cells, and cell walls are probably already not happy little campers. The mitochondria inside the cells are the electrical powerhouses of the body. You need a good supply of oxygen for the mitochondria to efficiently burn food and to generate energy. If you examine the mitochondria sitting inside the cell, they're also floating inside the dirty cellular fluid and swimming in the garbage. The mitochondrial areas are the sites where you burn your food to generate energy. They're also your direct link to consciousness, and studies show that depressed people are not generating enough energy in their nervous systems. The more balanced energy you have, the happier you will be. If accumulating waste products are gumming up your mitochondrial energy production, then less energy will mean less happiness, more sadness, and more depression. Reactions to infections in your spinal fluid or fungi in your cranium will also commonly be diagnosed as 'mental illness.'

The Winners Have the Oxy-Edge

So many people complain they don't have enough energy any more. It's all from a buildup of toxins and a lack of oxygen and water that you mistakenly call aging. It's also because Americans drink more caffeinated soft drinks than they do water while desperately seeking energy, but the caffeine dehydrates them and lowers the oxygen transport and that drains away their energy and only makes them need more. Great for repeat business, bad for you. All of the caffeine addicts are dehydrating themselves daily, and slowing down the oxygen transport.

Movie stars, construction workers, teachers, moms, mechanics, athletes, and especially business people are facing and managing stressful challenges every day that use up their energy and cause end-of-the-day exhaustion. You pound down the coffee and colas, and the bags under the eyes get bigger, and you feel you're always dragging around. You fall asleep as soon as you stop activity. Everyone I know who is correctly using oxygen supplementation and eating correctly no longer have these problems! They become full of *life* instead. In business, and in sports, at crunch time when you have to out-last and out-think, oxygen gives you the edge over all your competitors who haven't tried it yet. You compete with your clothes, and you compete with your grooming and car, now step it up and compete with your stamina and mental clarity. You would be wise not to let some oxygenated person get any edge over you.

How Does Nature Take Out the Garbage?

You're here, sitting in the garbage dump environment called society, and your cells are sitting inside that liquid garbage pile carbonized sack called your body. How are we going to live in such a place and yet still fix the problems? Luckily, Nature already figured this out long before we got here.

The process is ultimately very simple. The body gathers food, does its job, and excretes its wastes. You eat, drink, and breathe. If you live, you excrete, and your human vehicle is a machine involved in a process, a very orderly one. Consider what comes out of your body. If you were to analyze your waste—what you urinate out, what you defecate out, what you sweat out, what comes out when you spit, or blow your nose, what flakes off of you, and what comes out of your pores—you would find these compounds are usually composed of one of four basic chemicals. These four are also the same as the elements found in smog. They are Hydrogen, Carbon, Nitrogen, and Sulfur. There are other trace elements, but the Hydrogen, Carbon, Nitrogen and Sulfur are the four chemicals that make up all the waste products of life, so naturally they are the main components of everything that comes out of you.

Now the very interesting thing about these waste products is that you can't easily take them out of the body. So now you've got a real problem on your hands. By themselves, they can't come out! But Nature figured out a way to solve this chemical problem. She first combines them with oxygen and turns them into H_2O, CO_2, NO_2, and SO_2. In these forms they easily come out of you all the time.

Blow on your hand. The gas coming out of your mouth is warm, and it's wet. You're warm because your little mitochondria furnaces are burning your food. And your breath is wet because it has water mixed in it.

That wetness was your body sending hydrogen combined with oxygen (H_2O) out of your mouth. Did you feel the pressure of the gas? Well that's your body sending out carbon combined with oxygen (CO_2). You follow every breath you take by immediately exhaling two of the major waste compounds that your body must eliminate constantly. They can finally come out because they were combined with oxygen. It's the same for all the other wastes. The way you remove your waste products is to first combine them with oxygen. If you don't combine them with oxygen they stay in the body. If you don't constantly have enough oxygen available, how much waste will come out?—Not enough!

Plentiful Oxygen Removes Toxins from the Body

When garbage stays in the body uncollected, free radicals, toxins and microbes have a grand time damaging everything. When we start safely and properly flooding the body with active forms of oxygen, we slowly— over many days—fill the fluids and cells with it. Obviously we must gradually be surrounding our organs with oxygen as well, and at this point, we must have also surrounded the free radicals, toxins, and microbes with it.

Now what happens? O_1 is used by Nature out in the environment to dehydrogenate, pull apart, break down, and clean up the toxicity of unnatural substances. Surprise! Oxygen Therapies have done exactly the same thing in the human body over the past one 100 plus years.

Individual results may vary, but when doctors, naturopaths, and other healers use some delivery system to add O_1 to the body fluids, molecule by molecule the oxygen will bind with and pull out all of the toxicity. The cells—finally having enough oxygen—will use the opportunity to create enough *ENERGY* to push out the garbage, and push out all of the unnatural junk, the stuff that is not supposed to be in there.

> **Your body has always been trying hard to take out the toxicity, but you never had the oxygen needed to do it!**

The body and the oxygen don't care if the substance entered through your food, your water, your air, or whether it was absorbed from the clothes you wear on your back or even the things some people stuff into their veins. Because of the chemistry and active oxygen's negative charge, oxygen can't help but take anything in the body that doesn't belong there and oxidize it. The toxins and bacteria are all positively charged so they are drawn out to bind with, and be burnt, and be neutralized by the oxygen, the ultimate free radical scavenger. That means cleaner fluids, cells, muscles, organs, minds, digestive systems, immune systems, bodies, emotions, and consciousness!

If the warm, wet food-burning process is going on within your cells, but you're being forced by circumstance to burn the food and toxins in a low oxygen atmosphere, you're just like a campfire made of wet wood. You're not going to burn very well and so there's going to be a lot left over. You won't generate enough energy and you won't burn enough of the trash because you *can't*. Are you taking out all of the trash, or is the trash slowly taking you out?

Your cells need oxygen; they need it all the time. If you do nothing and allow the cells to continue becoming covered with garbage, you are simply letting needed oxygen be blocked from getting into the mitochondria. If this garbage piling up continues to such a degree that eventually 60 percent of the oxygen needed by the cell is not there, then that cell will be so damaged by the lack of oxygen that it's respiratory mechanism and enzymes cannot function.

In other words, if the garbage in the cells lowers the oxygen levels around the energy furnaces enough, the furnaces themselves are destroyed. Then our cells can no longer breathe oxygen to make energy.

But our bodies are programmed to survive. So when the cells can't breathe anymore, in order to stay alive they mutate and drop 31 steps in evolution down to the level of a plant type cell that just grows and grows and grows. To live in no or low oxygen conditions they lose of all their higher functions; they are no longer regulated members of the organized body community. They no longer make hormones, no longer digest your food, and no longer heal whatever part of the body disease has settled into. We need to breathe, and so do your cells, but because they can't get the oxygen necessary to breathe they mutate into a simple survival mode.

Nature has programmed this vehicle we walk around in to survive in adverse conditions. To do this during low oxygen conditions, it will keep your cells alive by switching them over to fermenting to get their energy. They stop trying to breathe the oxygen that's not there and start fermenting the sugar, the glucose that's surrounds them all the time.

Your body is full of glucose because your natural body sugar is where you get your quick escape and survive energy. The problem is that energy in the form of ATP made from oxygen is perfect, but ATP that's made from fermentation is inefficient. It's full of lactic acid and lactic acidosis builds up and damages more cells around it, causing the formation of more fermenting cells. It's all from a lack of oxygen. The fermenting cells with no higher function are cancer cells. There is no cancer cell that is not fermenting due to their damaged respiration mechanisms.

Speaking of cancer ultimately being a fermentative disease, while on tour I went to dinner with a friend whose brother is the head of one of the biggest cancer labs in California, the Fermentation Process Laboratories. Notice that name, head of Fermentation Process Laboratories. I explained the above cause of cancer to him as discovered by Dr. Otto Warburg, two-time Nobel Prize Winner, yet he ignored the information. He could not see what was right in front of him and did not want to see it, either.

You must clean out the garbage, the toxins, the critters, and the plant forms before you can get into good enough order to become consistently fit and healthy. ONLY ACTIVE OXYGEN can clean you out completely. To be sure you're not chasing false symptoms, always re-oxygenate first before trying anything else.

If toxins are constantly saturating your body, your immune system's cleaning functions become depressed, and therefore the toxins build up even more. You must oxygenate sufficiently and for long enough to clean out. Your body has not been able to get rid of the dead, feeble, diseased, weak and dying cells because it has been too busy trying to repair you, or keep you alive, or just plain keep you breathing. With all of your problems, where was it supposed to get enough oxygen to make enough energy to oxidize anything extra away?

After cleaning out, you will then have to rebuild your immune system because your immune system is probably shot from all this work of trying to repair and trying to defend itself in an environment of insufficient oxygen. The immune system must be rebuilt; that's the real art of medicine. Supplementing with nano-ionic full spectrum minerals is the key to doing this. You must clean up your food. Start using as much organic food grown on mineralized soils as you can find and afford. You could try supplementing with MinRaSol and OxyMune or Hydroxygen. Study my *Crown Jewels of Health* precepts. You must keep out any new toxic build up.

Your Energy Bodies

The other things that suppress your immune system—such as bad food, bad lifestyles, chronic bad relationships, hate, anger, or fear—you must make concerted efforts to remove these from your life. Perhaps take up a spiritual discipline, or at least live by a moral code. We all know how to live guilt-free to keep from depressing our own immune systems. I interviewed many criminals and asked them if they knew when they were doing harmful things and what the right thing to do would have been. They all admitted they did know, but felt impelled to make the wrong choice.

Studies are showing that whatever you hold to be your ideal in life, whatever that is, if you consciously live up to that ideal it enhances your immune system's ability to deal with any microbial or energy invaders. If you are guilty about not living up to your moral ideals, your spiritual ideals, your religious ideals, or your family ideals, if you don't live up to them, it suppresses your immune system. It's a known scientific and psychological fact. Along with washing out the waste metabolites and toxins, you mush wash out the mental and emotional toxins.

Cleaning Out

With normal active oxygen supplementation, cleansing may happen so slowly that you hardly notice anything. If you choose to go beyond supplementation and seriously flood the body with oxygen, at that point in time you're going to see all kinds of garbage appearing in your bloodstream, in your urine, and in your fecal matter. It's all your very own personal garbage. It will come out of your pores, your breath, your under arms, from your feet, everywhere. It's all leaving, it's going out. This is good news, but at times, from looking at the excreted waste, you'll wonder if that dead squirrel was living inside you.

Chapter 4

Anaerobes

Second Big Oxy-Truth:
The Majority of Diseases Are Anaerobic

It's so simple that it befuddles the great minds. Unlike healthy human cells that love oxygen, the vast majority of disease causing microbes absolutely cannot live in active forms of oxygen! This is not taught in medical schools.

Almost every virus, bacteria, fungi, mycoplasm, parasite, and other pathogen found in all diseases, including HIV, arthritis, heart disease, cancer, chronic fatigue, epstein-barr, candida, and every other disease you can name are the same as all other primitive lower life forms. They are facultative anaerobes. 'Anaerobic' simply means they cannot live in oxygen. The disease bugs simply can't live in active oxygen!

"But wait," a 30 year specialist in medical microbiology at a major Los Angeles hospital told me, "yeasts can ferment wine quite nicely but are still not killed by the oxygen in the air. They can grow either aerobically or anaerobically." Ah, yes, grasshopper, but air is only around 21 percent oxygen and that relatively smaller portion is not *active* oxygen.

During prohibition a family friend's father would make literal 'bathtub gin' in the backyard. Air didn't bother him. But, she said he used to curse whenever he was brewing and a lightning storm came up, because all the natural ozone in the air killed the yeasts, stopping the ferment and ruining the gin. Ozone releases *singlet active* oxygen, quite different from regular air.

The specialist also stated, "The vast majority of gut bacteria are strict anaerobes. For instance, many Lactobacilli are anaerobic. And many of the enteric (intestinal) pathogens are aerobic, such as Salmonella, Shigella, E. coli O157, and other enteropathogenic E coli, and these are not at all bothered by the oxygen in the air." Again, I'm not speaking of air but of active oxygen, quite a bit more reactive.

She continued, "Of course Campylobacter insists on reduced oxygen tension, and there are many opportunistic pathogens found in gut flora that are indeed anaerobes, such as Bacteroides, Clostridium, etc." When the oxygen drops the bad guys flourish. *The balance* must be maintained, and the poor necessary aerobes within us are often overpowered in modern society. If they behave, anyway—some anaerobes are our good friends. Toxicity and low oxygen destroy Nature's dynamic that we should preserve.

Within the definition of primitive anaerobism lays the answer to the majority of diseases that plague humanity. If the disease microbes can't live in harmless active oxygen, why not FLOOD YOUR BODY WITH OXYGEN? Properly and safely, of course, under the guidance of a skilled oxygen therapist. And notice that it takes a non-doctor without any universities, hospitals, drug companies or public relations organizations behind him to tell you this.

Disease causing bugs can't live in active forms of oxygen, and proper active oxygen is safe and harmless to humans! Put this all together and think about how profound and simple this answer for all disease is, and how it's been right under our noses all along! Now think about how much needless suffering is going on right now, and add to it the suffering that is about to happen. Almost all of it does not have to happen *if* we apply this knowledge.

The vast majority of all *disease-causing* microbes are anaerobic. Scientists agree that most antibiotics don't work any more. This is because most disease causing microbes are 'smart' enough to create chance mutations that survive being medicated and therefore escape being killed by antibiotics. Unlike this problem encountered when using antibiotics, it does not matter how many times the anaerobic microbes mutate away from the antibiotics. The fact remains that **ANAEROBES STILL CANNOT LIVE IN ACTIVE FORMS OF OXYGEN THAT ARE HARMLESS TO HUMANS!**

Microbes Specialize to Evade the Immune System

At one of the medical conferences I attended, they were showing us pictures of tiny parasitic mycoplasmas that exist by living on our blood cells. They have no cell walls and are therefore resistant to antibiotics. They're similar to the spiroplasms living on plant cells.

They nest on and eat your blood cells like fleas on a dog. Like the bacteria, the fungi, the protozoa, and all of the other pathogens they are all still simple anaerobic organisms. But how do such simple lower life forms evade our immune systems?

To evade detection, bacteria cluster together and then become highly specialized. In humans, we would call these cell cluster equivalents 'organs.' Amazingly, scientists have found that bacteria have enough collective or chemical consciousness to come together and form 'smart' colonies. These smart colonies of bacteria are called 'biofilms.'

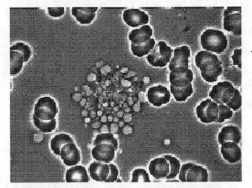

Darkfield microscope showing unhealthy blood cells near a pile of 'garbage' in patient blood.

"We tend to think of bacteria as primitive, single celled creatures," says Phil Stewart from Montana State University in Bozeman. "But in biofilms they differentiate, communicate, cooperate, and deploy collective defenses against antibiotics. Individual microorganisms in a biofilm act together like one multi-cellular organism." Human biofilm infections include dental cavities, gum disease, childhood ear infections, and some infections of the prostate gland and heart. Biofilms also underlie the devastating lung infections that occur in people with cystic fibrosis.

Biofilms account for two-thirds of the bacterial infections that physicians encounter. Many of these are caused by microbes that are common free-floating inhabitants of the body but become virulent as part of a biofilm community.

Free swimming bacterial cells alight on a surface and attach after aggregating on a surface. They produce copious amounts of a sugary, mucous coating. Within this slime, they can form complex communities with intricate architecture featuring columns, water channels, and mushroom-like towers. New genes are expressed to synthesize the slime.

After six hours, the bacteria start to communicate by making protein molecule signaling messages and releasing and exchanging them. Bacteria reproduce, and begin to form structures. Oxygen levels <u>decrease toward the center</u>. [They're not stupid, are they?]

These structural details may improve nutrient uptake and waste elimination, as blood vessels do in an animal's body. In the case of your mouth, teeming bacteria can in just a few hours erect the microscopic equivalent of a coral reef on your teeth. A prolific variety of environmental niches are formed, and the biofilm affords protection from antibiotics and toxins

As the bacteria make a simple layer, one cell deep, they begin to produce slime. It protects them from being washed away or drying out and also slows down antibiotics and other toxins that might seep in. Cells dissolve the slime and are released to float away and attach somewhere else where the cycle starts all over again.

Free-floating, or planktonic, bacterial lifestyles that are most familiar to laboratory scientists may be nothing more than a way for cells to disperse and colonize new habitats. Adapted from —Science News, July 14, 2001, Vol. 160, No.2, Pages 17-to-32, by Jessa Netting.

The scientists are starting to agree with what I've been advocating for years. As I told you, the bugs are inside us and competing with us for our space. Keep lowering the oxygen, feeding them garbage, and not disturbing them, and they'll organize and take over unto your eventual predictable demise.

Evolving Above the Level of Anaerobes

Normal healthy human cells love oxygen. Lots of oxygen. They aren't anaerobes, and, unlike cancer cells, don't ferment sugar to create energy. Our cells respire by breathing oxygen and using it to oxidize food for energy, making high quality energy storehouses of adenosine triphosphate, or ATP. Our friend *oxygen* serves a dual role because, as we've seen, it burns the food and then carries away the wastes left over from the burning. In Nature, the more oxygen available in an environment the higher the evolution possibilities available to the life form in that environment. This is important to understand.

Our bodies and the cells that make them up are highly evolved, and therefore oxygen loving. The organisms associated with disease are *primitive*, and do not like oxygen. Most microbes are ancient life forms, which evolved and prospered only in the absence of oxygen. In contrast, man is an oxygen-based higher life form created and evolved within an oxygen-rich atmosphere.

Ask yourself. What would happen to these anaerobic viruses and bacteria if they were to be completely surrounded, or flooded, with a very active, energetic and aggressive form of pure oxygen like O_1, or O_3, or even fresh O_2, for a long time? What if enough special forms of oxygen or ozone were to be slowly and harmlessly introduced into the body on a daily basis, and over the course of a few months? What if in order to be extra safe, we mostly by-passed the lungs, and yet eventually evenly saturated all the bodily fluids and cells with it? Wouldn't the disease causing microbes that can't live in oxygen cease to exist? Wouldn't they eventually all be burned up and washed out?

I once asked the head of Microbiology for the State of New York,

> *"What percentages of microorganisms are anaerobic?"*
>
> *"Most of them." He said. He was reluctant to give an exact figure.*
>
> *"Give me a figure, what's your best guess?"*
>
> *"I would say 60 percent."*
>
> *"OK. How about the rest of them?" I prodded.*
>
> *"Well, the other 40 percent, they're not strictly anaerobic."*
>
> *"Well, did the rest have an anaerobic portion of their life cycle?" I asked.*
>
> *"Oh, yes they do." He said*
>
> *"Then that means if you keep them surrounded with oxygen and wait long enough, you'll eventually catch them all?" I said.*
>
> *"Yes, you're right."*

This was verified and confirmed by the work of several other researchers I have investigated. They all reported that although not all microbes are strictly anaerobic, the ones that aren't strictly anaerobic become anaerobic at some point in-between their birth and death.

So that's why Oxygen Therapies work so well on so many diseases. We can safely say that at some point in time this destruction of anaerobes— when surrounded with active forms of oxygen—includes almost all of the disease causing bacteria, viruses, fungi, mycoplasmas, and any new smaller and smaller microbes that they can discover. They are ALL anaerobic, either strictly anaerobic or at some point anaerobic. Therefore, if you surround an anaerobic disease-causing organism with an active form of oxygen, how will it live?

And that is why it is so encouraging to hear the numerous accounts from people I interviewed who said their diseases simply went away without drugs. They properly flooded their bodies with oxygen and got rid of all the toxins, bacteria, viruses, and fungi.

Active Oxygen is the Key

I keep saying *active forms of oxygen* instead of just *oxygen* because there is an important distinction that must be adhered to in order for these therapies to work properly. All 30 or so major Oxygen Therapies, including ozone, work because they flood the body with Nature's oxygen atoms in singlet form. Although breathing oxygen from a tank (for example, firemen getting rid of hangovers) or using hyperbaric (high-pressure) oxygen to prevent quadriplegia after a traumatic accident both have excellent track records, they only use the stable form of oxygen, O_2. While it is known by some that the lung membrane splits the stable form of oxygen into singlet oxygens so that the body can easily use it, most diseases need more action and faster results than are afforded by normal oxygen in its stable form.

That's why I concentrate on singlet oxygen and its by-products. As I explained, these are very energetic oxidizers and they quickly 'burn up' waste products and pollution without harming normal healthy cells.

Singlet oxygen is the beneficial free radical. It oxidizes or eats up all of the harmful free radicals left over from the incomplete combustion in our cells.

Parasites

Just because we shower, brush our teeth, comb our hair, put on our suits, and make ourselves look shiny and nice on the outside, we assume that we are clean inside. Ask any colonic therapist for the truth. The older ones will tell you that parasites used to significantly infect about 60 percent of the population. Today they report that almost every single person is infected. One day you might feel something moving around inside you. Or a parasite that you didn't know you had might suddenly seem to appear out of nowhere.

You go to the bathroom and there's a little worm there and you go uh-oh, what's going on? You go rushing to get your colon cleaned, and a whole nest of different sizes and colors of worms fall out. The good news is that most parasites are just like the microbes; they're primitive life forms that can't live when surrounded with active forms of oxygen. I heard a tape where a top parasitologist, said "and if nothing else works, we just give 'em oxygen."

Free Radicals

Free radicals are any compounds with unpaired electrons looking to bond with something. The prime generators of free radicals in the human body are pollution and toxicity. They jam up the chemical processes and out of the pollution spins compounds with unbound electrons in its orbits. Doctor William F. Koch, MD, Ph.D., and Professor of Chemistry first espoused this fact at Wayne State University. Doctor Koch was also a clinical practitioner and free radical chemist par excellence whose teacher was Moses Gomberg, the discoverer of free radicals. The man obviously knew what he was talking about when he said, "The cause of free radicals in the human body is a *lack* of oxygen."

Why? Because he knew that if the human body was stuffed or 'flooded' with oxygen, then what's going to happen to any positively charged (they all are) polluting free radical the moment it is generated? Picture all those singlet oxygens cruising around inside you and looking for action, harmlessly bouncing off normal healthy cells and wanting to establish meaningful relationships wherein they can bond with something positive!

Singlet oxygen is negatively charged, just like healing clays, volcanic ash, humic shale, seaweed, and other historically effective healing agents. The pollution, wastes, and the free radicals being created from the garbage are positively charged. The two opposite charges instantly attract each other, and the oxygen 'scavenges' or scoops up pollution's harmful free radicals, neutralizes them, and whisks them away and then out of the body. We need more oxygen, the GOOD free radical.

If the body is kept full of its real—though seldom found in our society—normal quotient of oxygen, then any free radicals that might be created are instantly neutralized since they are always surrounded by hungry singlet oxygens. Most of our oxygen is spent removing ATP lactic acid acidosis.

Free radicals are caused by lactic acid and pollution. Harmful free radicals are not caused by oxygen as some Ph.D. might try to convince you as, meanwhile, he obtains a grant from the multivitamin companies putting out advertising trying to 'scare-sell' you into buying antioxidant compounds to 'protect' you from the horrible and completely mythical bogeyman 'ravages of oxygen.' The research-challenged repeat the lies.

"Singlet O1 is an especially reactive form of oxygen capable of rapidly oxidizing many molecules, including membrane lipids. Its formation in O2 generating systems has often been proposed, but *clear cut evidence for a damaging role of singlet O1 in such systems has NOT been obtained*." (My emphasis) B. Halliwell, Department of Biochemistry, University of London King's College, London, U.K.—*Oxygen Radicals: A Commonsense Look At Their Nature And Medical Importance, Medical Biology 62: 72, 1984.*

What About Antioxidants?

You're supposed to be eating naturally and getting your antioxidants from your food. You're supposed to be eating what used to be called Mom's cooking, but now we have to call it 'organic food' grown on mineralized soils.

Because you're not eating correctly, the vitamin companies want to cash in on that and sell you 'antioxidants.' The name is scary. But antioxidants do *not* 'stop oxygen' as the name implies. Labels use the word 'anti' and then combine 'anti' with the word 'oxidant' implying somehow that we need to be against oxidants or we need to 'stop' oxygen. This position is dead wrong, and the word 'antioxidant' is a complete misnomer. Dark green and yellow vegetables—containing vitamins like A, C, D, and E, and others—are antioxidant foods. The label 'antioxidant' is applied to substances that are, actually necessary components of your daily diet. These substances do not 'stop' oxygen, but instead are welcome regulators of the oxidation reactions and processes in your body. Nobody has time to explain all this in a brief advertisement.

Instead, the advertisers play upon your natural confusion created by the word 'anti' and followed by a word pertaining to oxygen. They falsely trade upon this confusion and say "stop oxygen damage, buy our pills," and tell you that oxygen is a horrible free radical producer, and that you must be afraid of oxygen and therefore of life itself. Therefore you must buy antioxidants to feel safe and protect yourself against Nature's terrible oxygen. They're lying to you by telling only half the story in order to sell you something. The other half of the story is that antioxidants and oxygen work great together, and that oxygen is in truth the good free radical that scavenges the other bad free radicals in the body, and that oxygen is the good free radical that also burns up and takes out the trash. Antioxidants regulate this process.

Dr. William F. Koch wrote in—*Neoplastic and Viral Parasitism, Their Basic Chemistry and Its Clinical Reversal,* "The cause of all harmful free radicals is a lack of singlet oxygen in the body to scavenge them."

Oxidation-reduction reactions, the process of atoms swapping electrons, are the basis of life itself. If we stopped the oxidation in our bodies (as implied by the word 'anti' in 'antioxidant') we would die. The sun's electron forces are delivered to us by oxygen and they travel across and through the minerals and oils. Nature's softened ionic minerals and our proper use of dietary oils enable our bodies to construct the electron pathways needed by life within our bodies.

Thinking only of profit while ignoring health issues, commercialized mass-market food that ends up on so many tables is often so low in minerals that many people really do need to supplement with ionic minerals and antioxidants to get the proper oxidation happening again, not to stop it. That is why many people not already supplementing their bodies with Nature's oxygen may feel a little better when taking antioxidants. Supplementing with antioxidants *increases* oxygen absorption. And any oxygen absorption boost is an improvement in an oxygen-starved person!

If someone is going to be taking a course of the Oxygen Therapies under the care of a competently trained health care professional, or even experimenting on one's own, he or she must, in my opinion, have an adequate supply of antioxidants in the daily diet. If these are not in the diet, they must, as German clinics proved, be supplemented in order for the wanted specific Oxygen Therapies to work on everyone. Some clinic patients were so deficient in these foods that their bodies were having trouble using extra oxygen at all. When they do receive supplements of the antioxidant enzymes, vitamins A, C, D, and E, catalase, glutathione peroxidase, and super oxide dismutase in their food, they finally respond to the Oxygen Therapy in question. In other words, their nutritional deficiencies needed to be corrected before the Oxygen Therapies would work well. That's when they started getting proper results.

The prime reason we come to the Oxygen Therapies, besides the existing pollution and oxygen loss, is because we aren't eating correctly in the first place. And that's mostly because of commercial interests putting profits ahead of people.

After food, some of the best all around antioxidant supplements around are spirulina, blends of superoxide dismutase grown from wheat sprouts, and the new high-antioxidant potential bottled oxygenated waters. As we go to press, I have become very impressed with the results available from Kevin Lockhart's 'E-3 Manna' a Liquid blue-green algae harvested from the center of Klamath Lake and shipped frozen overnight to insure maximum potency. This algae gets its mineral and vitamin content from the Cascade Mountains and is reputed to be nutritionally the best algae in the world. The same company also sells 500 mg. dry caps of 'Blue-Green Manna.'

Somatids, The Smallest Life Forms, Pleomorphism, & Oxygen

Whatever your viewpoint on creation may be, there exists the opinion that life evolved from out of a swamp somewhere. The original smallest Life Form, whatever that was, was floating in something. There was a change in the nutrients and a change in the pH around it, and evolution started.

Antoine Beauchamp, followed later by the confirming work of Canadian researcher Gaston Nassens, discovered that Somatids are tiny little particles on the border of energy and physicality floating in our blood. Under the right conditions and pH (low, acidic), they will start to bud and turn into spores. If you change the nutrients and lower the oxygen levels around them, the spores can bloom into tiny viral sized forms and all these different forms can evolve into mycoplasmas. Then the mycoplasmas may mutate into larger and larger bacteria, ascending up the evolutionary ladder of life.

The point here is that the seeds of life are quite resilient, microscopic, and everywhere. LIFE always assures its continual existence under any conditions by taking on tiny indestructible forms. Under the right conditions the smallest forms will evolve into higher and higher forms. This is not taught in medical school. This is not taught in science class.

Why this is important to us? It's because all this pleomorphism (microscopic bugs changing into many forms) only occurs when the oxygen levels drop and the garbage levels rise! So if you don't clean out and raise oxygen levels, tiny eggs of life spontaneously occurring everywhere will start growing and forming microbial colonies within you.

Please keep in mind that for all the problems I make you aware of, there are always solutions. Instead of just being the victim of your surroundings, you can become a cause point of your own reality! You can properly and safely totally purify your body. When used correctly, active Oxygen Therapies are blatantly nontoxic. The process has been around 100 years in medicine, and it's always been a part of life.

There are many people such as therapists, naturopaths, and doctors who now know how to use oxygen properly for supplementation and disease treatment. People everywhere successfully and safely use Oxygen Therapies and supplements on a daily basis. Appropriately trained doctors worldwide know how to prescribe active forms of it, and how to properly apply it according to time tested protocols.

Individual results may vary, but according to the historical and clinical records, active oxygen always enhances the immune and every other system, including the digestive, eliminative, energy generation, mental faculties, etc. It raises the consciousness. Above all else, oxygen removes free radicals, eliminates toxins, and it's deadly to the majority of the microbes and garbage within us.

Part Two

The
Solutions

Chapter 5

Clean Food, Clean Water, Clean Air

What is this chapter doing in a book about oxygen? Actually the food, water and air around you are very important to your body's ability to *use* oxygen. In our pure natural state life can supply us with everything we need, and in abundance.

It's only natural that our Creator should supply us with our medicine in abundance. It's all around us. We call it Nature. We are part of her. The farther away we get from Mother Nature, the sicker we get. Her solutions that give us the best results replenish our body vehicle with her ingredients that are missing from our lives.

It took a long time for us to get out of balance; so that's why we need to apply the proper corrective solutions continually over the long term.

Food

> **You eat constantly. But you have to eat right, or you can't take up the oxygen!**

As well as our concerns about cellular oxygen levels, and pollution, it could very well be that the *only other reason* we're having this conversation at all is directly related to all the wrong food you've been unknowingly shoveling into yourself for your entire life.

You must closely study what you put into yourself. Start with the book, *Eat Right For Your Type*, by Dr. Peter J. D'Adamo, and live by it. Chemistry and thousands of clinical tests and certified medical results prove that your immune system responds to your food as if it was a pile of invading germs. The chemical signatures of some of your food choices create allergic reactions specific only to you, according to your blood type; and each of the blood types O, A, B, and AB (and subtypes) requires a different type of diet, dictated by your ancestral heritage.

When seeking health and/or energy, first determine your blood type, and then, referencing the book, only eat mostly the foods marked 'highly beneficial' for your blood type, and none of the foods that you are allergic to. We're all chemically different. This is not theory; this is finally pure provable scientific fact. It's also the big reason why there's no such thing as one health diet that works best for everyone. You can also look at raw foods as well.

Now let's add more solid science to your new food choices based upon your blood type. Udo Erasmus is one of the world's experts on fats and oils in our diets. I advise you to closely study his work as well, and follow his advice. There are good reasons why I say this.

According to the work of West German Physicist Dr. Johanna Budwig, Ph.D., the red blood cells in the lungs give up carbon dioxide and take in oxygen. The blood cells are then transported to the cell site via the blood vessels where they release their oxygen into the plasma. This released oxygen is attracted to the cells by the resonance of the pi-electron, oxidation enhancing, fatty acids. Without the right fats and oils in our diets, oxygen cannot work its way into the cell.

Dr. Budwig published technical papers as far back as 1951 proving that electron-rich, fatty acids play the decisive role in respiratory enzymes, and are the basis of cell oxidation.

Accordingly, we should eat these essential polyunsaturated, fatty acids to enhance oxygenation. "Electron-rich fats resonate with the wavelength of the sun's light and control the entire scope of our body's vital life functions!" says Dr. Budwig. Conversely, if we eat anything hydrogenated, it defeats the purpose of oxygenation. Dr. Budwig gives the example of a hydrogenated food that can't transfer oxygen's electrons: margarine.

Margarine is in my opinion a non-food that only plugs us up. Margarine is created by superheating oil and bubbling hydrogen through it in order to solidify it. All nutrition is lost. Does the label of what you are about to eat say 'hydrogenated?' Is there an alternative, or can you skip it? Hydrogenated foods are one of modern man's pervasive silent killers. Even flies won't eat margarine, put some out and watch. Try Udo's oil, and consider why those eating raw natural fats and dairy do so well.

The Best Oil

Udo Erasmus, is the author of the modern classic *Fats and Oils*. He has traveled extensively and interviewed all the commercial oil producers. He teaches that the best single oil (as opposed to blended oil) for human consumption is freshly prepared flaxseed oil.

I first met Udo while we were both lecturing in Toronto. Since then, he has improved on and gone beyond his earlier recommendations. Combine his following precepts with the blood type food information above, and you'll be in the right eating ballpark for your body. Or, as my English wife would say, "You'll be well away!" I asked Udo to give us the high points of his collected writings. Here's his kind response, exclusively for you.

Udo's Contribution to Mr. Oxygen's Book

Health has 14 physical components. Oxygen is one of them. The others are greens, good fats, protein, minerals, vitamins, antioxidants, phytonutrients, fiber, digestive enzymes, probiotics, water, light, and fuel. To be healthy, you need each one of these.

All of these components of health come from food, water, air, and light. In fact, Nature has been using food, water, air, and light to construct the bodies of creatures for 4.5 billion years. The genetic program built into each of our 70 trillion cells knows how to build a healthy body if we give it the building blocks that come from food, water, air, and light. Given how simple it should be to build a healthy body, why are so many people sick? What are we doing wrong? And how do we fix it? Here's my list.

1. Eat *less* fuel.

Sweet and starchy foods are fuel foods. The problem with excessive fuel consumption is that if you don't need all the fuel you eat; the extra fuel must be turned into fat and stored in preparation for the next famine. That is the law in the body.

Most overweight conditions in North America—now totaling more than 55 percent of the population—come not from eating too much fat but from *excessive* consumption of sweets and starches. These include sugar, honey, syrups, bananas, fruit, breakfast cereals, bread, pastries, potatoes, yams, sweet potatoes, pasta, muffins, crackers, potato chips, French fries, corn, polenta, popcorn, grains (wheat, cous cous, oats, rye, barley, rice, tabouleh), beans. It also includes highly processed junk food in crinkly bags, and prepared foods such as ketchup, sauces, yogurt, and prepared meat products that are flavored with sugar or extended with starch.

2. Eat more good fats.

While most people think about eating less fat, health requires the consumption of more fats. They just have to be the right kinds of fat.

Every cell in the body requires good fats. We cannot live without them, cannot make them in our body, and therefore must get them from foods. These good fats, known as n-3 and n-6 essential fatty acids, are sensitive to destruction by light, oxygen, and heat. Industrial processes, applied to extend oil shelf life, damage or remove them.

N-3 is missing from most people's diet. N-3 and n-6 are insufficiently supplied by low fat diets, which—contrary to popular belief—are bad rather than good for health, leading to dry skin, low energy, and deterioration of every cell, tissue, gland, and organ.

The good fats increase energy and performance, improve oxygen metabolism, improve brain function, make skin nice, speed recovery from fatigue and injury, speed healing, and lower cardiovascular, cancer, diabetic, immune, bone mineral, obesity, inflammation, and hormone imbalance risk factors.

3. Eat more *greens.*

Whether you like them or hate them, green foods are the most nutritious food on the planet, and are therefore the most important food. Green foods manufacture (from scratch) most of the nutrients essential to us, including *vitamins, amino acids* (proteins), *fatty acids* (fats), *fiber, enzymes, antioxidants, phytonutrients* (herbal medicines), *fuel,* and *oxygen.* They also trap *sunlight* for us, hold *water* in the soil, and provide that to us, and suck on rocks to absorb *minerals* for our benefit. Finally, they also supply us with *probiotics* that cling to them as these green plants push up through the soil stretching toward the sun. Green plants alkalinize the body, preventing acidity that leads to disease. Green plants even make steaks and dairy products—cows and milk are made from grass.

4. Get more mucilage fiber.

This kind of fiber helps maintain bowel regularity, keeps probiotics healthy and populous, stabilizes blood glucose, carries cholesterol and bile acids from the body, escorts toxins from the liver to the toilet, and removes heavy metals from the body. It works against many degenerative diseases.

5. Avoid low and no fat diets.

Low fat diets can hurt and even kill us. A no fat diet will kill anyone who uses it long enough. The quality of fat eaten is more important than the quantity.

We do not need high fat or low fat diets. We need the Right Fat diet. The fats we eat should provide us with the good 'essential' kind without which we cannot live. The n-3 and n-6 fats should be taken in the right ratio. And these fats should be made with care to protect them from damage done by light, oxygen, and heat. Less than 15 percent of calories from fats get close to not enough. Up to 60 percent of our calories could come from good fats without causing fat-related degenerative conditions.

6. Eat more health oils and fewer shelf oils.

Most people don't know that the common oils they buy have been treated with drain cleaner (NaOH), window washing acid (H_3PO_4), bleaching clays that turn oil rancid and impart bad odor, and are then deodorized by being heated to frying temperature before they are bottled. During these processes, which are used to increase shelf life, some molecules are damaged and others are removed. These are not healthy oils.

We should switch to oils made with health in mind. These are found in brown glass bottles, in refrigerators, in health food stores. The 'brown glass' oils are pressed from organically grown seeds, and remain undamaged by being protected from destruction caused by light, oxygen, and heat. One, called the Perfected Oil Blend, is additionally protected from all light by being enclosed in a box. The blend contains nine ingredients that provide everything we need from fats and nothing we should avoid.

7. Eat less cooked and fried foods.

Cooking kills probiotics and destroys enzymes. Frying kills probiotics, destroys enzymes, and damages molecules, changing them from natural and healthy for us to unnatural and toxic to the body. When we eat cooked foods, the probiotics and digestive enzymes lost in cooking processes should be replaced. Digestive enzymes are best taken with the food. Probiotics are most effective if they are first dissolved in the mouth. They survive best if taken after meals, when stomach acid is lowest.

8. Eat less margarine, shortening, and convenience-junk food made with *hydrogenated* oils.

Trans-fatty acids in hydrogenated products double risk of heart attack, increase diabetes, kill 30,000 Americans every year and cause many other problems. They are consistently advertised by implying better cardiovascular health. This is blatant advertising misrepresentation without research support.

9. Consume optimum amounts of all 14 components of health.

We cannot substitute any one for another. We need them all. Optimum health requires optimum intake of all 14.

10. Pills cannot optimize health.

Health must be built from foods and nutrients eaten as close as possible to fresh, organic, and whole—the way Nature made them. Supplements can help, but only **in addition** to food and lifestyle improvements, not instead of them.

11. Be physically active.

If there's nothing to do, we need no body. We could just float about as disembodied spirits. The body is made for action. Even if it is fed properly, the body cannot stay healthy without activity.

12. Make time for rest and for play.

Most of us are stressed, because we do too much and think too much. We have become human doings and have forgotten how to be human beings. We should waste some time doing nothing and having fun. It's good for health not to be serious all the time.

13. Make time to enjoy life and be grateful.

Remarkable processes conspired to allow three buckets of water and a handful of minerals to have the very enjoyable and quite temporary human experience. Most people don't know what it *feels* like to be alive. Feeling adds quality to life. If each person on this planet were busy enjoying his or her life, would anyone have time to hate or hurt others?

14. Make gratitude your attitude.

We have a lot to be grateful for. Why gratitude? It feels good, and it is stress-free.

Udo's dietary precepts are based upon his book: *Fats That Heal, Fats That Kill.* Udo wants you to have a free audiotape about his products entitled: *Perfected Health Plan.* (See resources section.)

What's Wrong with Us That the Largest Section in the Drugstore is Always 'Digestive Aids?'

Programs, meal replacements, quick fixes, potions, pills, and other methods are not going to the real root of the problem. You're out of balance. Most of the digestive problems occur as we get older. We eat just plain wrong non-foods and fill with toxins, the oxygen levels drop, bugs move in, and systems go out of balance. We don't digest well because of the lack of the necessary raw materials for digestion, and that always leads to increasingly poorer digestion. As we fill up with toxins, we lose the ability to make enough stomach acid to digest our food. As a result, the undigested food and toxins turn into particles for the bugs to live off. The rotting toxins pile up in the fluids, and to protect the inner organs from them, the body pushes them further away out where as a last resort they get stored out in the skin (and in the fat cells under the skin).

Most people don't eat correctly, so they're always out of balance. Our bodies store wastes that it can't remove any other way in the skin. Because the toxins are so poisonous and there's usually so little water and oxygen to dissolve them, the body is too scared to let those stored toxins back into the inner organs and into circulation—no matter what you do to lose weight. And even if you manage to force the fat out of you, without changing your diet to being natural the weight eventually comes right back on anyway (sound familiar?) because you never removed the underlying causes of waste storage in the skin; no matter how many meal replacements you buy.

As we get older and continue eating unnatural food in the wrong proportions, we have more and more problems. Our under siege bodies eventually lose the ability to digest food due to toxic pile-up.

The primary defense against heartburn and weight gain (after slowly switching over to eating real unprocessed food) is the simple expedient of **chewing, chewing, chewing** each mouthful of food. Make a point of it. The hectic pace of modern life causes us to drift away mentally when we are eating, and to ignore our eating if we talk while we eat. Then we forget all about chewing. If you just chew each mouthful more, then you will definitely saturate the food bolus (chewed mouthful) with the necessary mouth produced enzymes and stomach enzyme precursors, and you will digest far better, feel full faster, extract maximum energy from each morsel of food, and need to eat less.

You can throw oxygen, minerals, and herbs, etc., at your problems and maybe get a reprieve, but unless you eat as Nature intended, consuming 75-to-85 percent natural organic fruits and vegetables daily, then in the long run your problems probably will all come around again and again.

Natural food at Donsbach Hospital.

May 28, 2002 (AP)—"ONE or two extra apples or oranges a day could *cut the risk of dying early from heart disease or cancer by 20 percent* scientists said today. Researchers found low blood levels of vitamin C, related to low consumption of fruit and vegetables are associated with higher death rates from heart disease, strokes and some cancers in men. Professor Kay-Tee Khaw, from Cambridge University reported: 'The findings indicate that modest increases in fruit and vegetable intake of just one or two servings a day may be associated with large benefits for health.' The results hold true regardless of age, blood pressure, or whether or not a person smokes. The findings are from the European Prospective Investigation into Cancer and Nutrition study."

Salt and Vegetable Sodium

There are two types of people. Those who are genetically low on their output of hydrochloric acid, who need enough natural vegetable sodium to push it out of their stomachs, and those who have enough sodium and make enough hydrochloric acid in their stomach, but do not have enough natural vegetable sodium to create an alkaline environment in the small intestine to complete the digestion. So no matter who you are, you need more *natural vegetable* sodium to digest food properly. It's not salt; we only get natural sodium from plants.

Doctors and TV ads don't recognize the chronic societal lack of natural vegetable sodium and potassium, so they throw acid blockers at everybody. One thing that our society needs to learn more about and to promote is that a lack of natural vegetable sodium and potassium are two of the most common underlying dietary problems plaguing us today. Since sodium is the great balancer of both types of people, we need to get enough natural vegetable sodium in our diets long-term. We also need more potassium foods to overcome the imbalance from all the unnatural salt found in our diet.

Lots of good dietary advice like this is available from master iridologist and nutritionist Ron Logan, M.H. ID (See resources section). He sells books, herbs, the Bernard Jensen natural vegetable sodium broth and other supplements, and he's a delegate to the World Health Organization's conference on nutrition.

My friend Jim Brown in Boca Raton, Florida, is a phlebotomist, phase contrast microscopist, ozone enthusiast, and avid researcher. Like Ron Logan, but through the different route of performing many darkfield blood analyses, Jim discovered that—contrary to mass thinking created by mass misinformation—most people's sodium-potassium balance is off because of what appears at first to be a *lack* of salt in their diet.

The logic he came up with here is that when people take natural salt they improve. But over the long term, and when used too often, even natural sea salt can throw off our sodium/potassium balance without our first having the proper potassium to sodium ratio in our diet. Extra salt instead of natural vegetable sodium can lead to edema. Sodium is normally on the outside of the cell and potassium is on the inside, and they must be in balance with each other to generate the electricity needed to produce the energy to power both the cell and the organ it serves. Any disruption of the sodium/potassium pump causes cellular stress, disease, and cancer. It is critical that we have enough water and oxygen and that the lymphatic system is operating flawlessly to prevent the disruption of the sodium/potassium pump and to maintain life.

Salt is in our food and on our tables, but our American table salt is mined from ancient seabeds and then aluminum and dextrose (sugar) is added to it. Aluminum and aluminum cookware has been linked to Alzheimer's. To get away from the additives, people have been switching to natural evaporated sea salt.

Everyone eats salt, we must have it to make bicarbonate and stomach acid to digest our food, but shipwreck survivors adrift on life rafts who drink seawater soon die *because of the salt*. Sea salt may be natural, but when evaporated from the seawater its sodium is still bound with chlorine (NaCl). Evaporated seawater source sea salt may be purer than processed table salt, but its sodium is still bound to chlorine. It's still not the separate vegetable sodium and chlorine matrix created by photosynthesis that we get naturally from Nature's bounty of *vegetables*.

On the too much or two little salt question we see statements like this, "Ethnic groups that don't use salt have low blood pressure." And yet those tackling America's chronic dehydration say the opposite, that we need more salt, because salt is necessary "…to retain the water, otherwise, the  water will pass through too quickly." From Dr. Batmanghelidj, author of *Your Body's Many Cries for Water*. The good doctor further advocates using more natural salt and says, "One woman I heard of drank plenty of water and still had high blood pressure, because she refused to take *any* salt. She was too frightened from all the misinformation she heard. Unfortunately, the health movement is fanatical against salt, which can make it impossible to cure dehydration. I saw one woman with asthma who was drinking water, but who did not get a lasting cure until she started taking some salt."

Agreeing with Dr. Batmanghelidj, there are other studies saying that some may need more salt. In *Unconventional Wisdom* by Emma Ross, via (AP) *Low-Salt Diet a Risk?* London—"A low-salt diet may not be so healthy after all. Defying a generation of health advice, a controversial new study concludes that the less salt people eat, the higher their risk of untimely death. The study led by Dr. Michael Alderman, chairman of epidemiology at Albert Einstein School of Medicine in New York and President of the American Society of Hypertension, suggests the government should consider suspending its recommendation that people restrict the amount of salt they eat. 'The lower the sodium, the worse off you are,' Alderman said. 'There's an association. Is it the cause? I don't know. Any way you slice it, that's not an argument for eating a low sodium diet.'"

When I interviewed Dr. Batmanghelidj he made the case for salt by reminding me that "We lose a quart of water by breathing, and at least a pint and a half in gentle perspiration, and more in urination. You basically need two-to-two and one half quarts of water daily, and this amount of water will take some salt out of the body. That's why you need salt in your diet. The body has a mechanism of retaining salt if it is not in the food." He also said, "Cows eat lots of vegetables, but still need salt."

Confused by all this? Do we get too much salt, or too little?

There is a reason I'm going on about salt and sodium in an oxygen book. The balancing your sodium question is important to our study of oxygen, for if we get edema (as in swollen legs) due to an imbalance in this sodium-potassium-salt area, the oxygen has a harder time getting into the cells because the lymph fluid carrying it is *stalled*. Drugs aren't the answer. My mother's arthritis drugs kept making her swell up and down just before her body got so worn out from the drug side effects that she passed away. Her lymph stalled and toxic overload ensued. One quarter of the body's available energy is used to maintain the sodium regulating pump mechanism of our cells when at rest. Obviously the body considers the question very important.

We do know that a potassium-sodium imbalance caused by missing potassium, sodium, and calcium is the cause of high blood pressure, and the main cause of stroke, and contributes to heart disease, memory decline, osteoporosis, asthma, ulcers, stomach cancer, kidney stones, cataracts and others.

> The MOST critical elements to life and a powerful immune system are first **water,** then **oxygen,** and then **sodium and potassium**. In serious cases, you have to remember that water and oxygen are primary nutrients, sodium and potassium are secondary nutrients, and diet is a third. For health, they all must be in balance.

Potassium is the Missing Key

Ron Logan sent me a report from Finland that may surprise us all. It appears everyone is looking at the salt question, but that's the wrong place to be looking. The Finnish health department people took away everyone's salt 20 years ago and replaced it with 'PanSalt,' or what we call 'Solgar Heart Salt.' This salt has 40 percent less sodium, as they have replaced 28 percent of the sodium with potassium, and 12 percent of the sodium with magnesium (we also don't get enough magnesium).

Let's compare this PanSalt to the natural sodium-potassium ratio found in Nature when left alone. Richard D. Moore, M.D., Ph.D., stated that *The New England Journal of Medicine* published a study about how our Paleolithic cavemen ancestors got around 11,000 mg of potassium and only 700 mg of sodium daily. Modern hunter-gatherer ratios are the same, 16 portions of potassium to one of sodium universally occur in a natural diet.

And what are our average supermarket and fast food diet ratios today? The natural 11,000 mg of daily potassium has shrunk to less than *one quarter* of that, to 2,500, and artificial sodium has increased more than 5 times from 700-to-4,000 mg daily. We went from around 16 times more potassium than sodium in the *real* world, to our modern plagues like heart disease and blood pressure reflecting everyone having the ratios reversed with almost twice as much sodium as potassium in our diets. Hmmm!

Humans have lived for thousands of years with loads of potassium and a little bit of salt, and suddenly within one 100 years we reversed this ratio. We're not getting away with it are we?

Almost every unprocessed fruit, vegetable, or legume, will have 20-to-100 times as much potassium as sodium. Top potassium foods are bananas, oranges, apples, rutabagas and cabbage. Potatoes are especially one of the richest sources of potassium. When the animals are allowed to graze, 99 percent of meat, fish, fowl, eggs, and dairy products have the natural ratio of three-to-five times more potassium than sodium.

The key is to eat whole, unprocessed foods. Dr. Moore states that our 'modern' processed foods have added salt (so you will buy more) and depleted potassium. An example is white flour having three-quarters of its natural potassium removed while huge amounts of salt are added. Are the naturopathic natural diet teachings making very good sense to you yet?

Any whole food that has *not been processed* is loaded with potassium. If people ate only whole unprocessed foods drank enough water and used only mineralized natural salt very modestly, there would be no potassium-sodium imbalance.

So, the Finnish had a good idea 20 years ago but their PanSalt sodium ratio is still too high, below a one to one ratio, and not the natural recommended four-to-sixteen parts potassium and one part sodium. But, it is a step in the right direction. Food processors, restaurants, diners, just about everyone in Finland uses this salt. The Finnish results speak for themselves. Strokes and heart attacks were reduced by 60 percent throughout the nation! Look at what Americans eat. Heart disease and strokes are two of the top three causes of our deaths.

The Finnish results were due to potassium being an element vital to hydrating the cells. There are two microscopic oceans of water in the body; one ocean is held inside the cells of the body, and the other ocean is held outside the cells. Good health depends on a most delicate balance between the volume of these two separated oceans, and this balance is maintained by the interplay between our sodium/potassium pump system and the sodium/calcium pump system and the sodium/hydrogen exchange system. The same equivalent amount of sodium outside the cell has to be matched by the amount of potassium inside the cell, and since there is more volume inside a cell than outside it, the cell has to be filled with potassium to hold the pressures equal. Without potassium, none of the cells will hold enough water. That's why we need lots of fruit and vegetable potassium in our diets. Where will you get the proper ratios you need to have good fluid pumping and a good oxygen uptake? Natural food.

Meanwhile, therapists are using salt supplements to quickly and temporarily try and solve imbalances, and their patients are usually careful to use the purest natural salt, like the gray clumpy mineralized evaporated sea salt from the health food stores. Salt alone is not a long-term supplement solution. One needs to drink plenty of water so the body can wash out excess salt.

Too much salt—even the natural mineralized kind—without enough water can make your limbs swell. Salt is chlorine bound to sodium—in the proper ratio it's better than no sodium—but we are physically set up to get most of our natural sodium from vegetables. Eat correct ratio natural foods and try Dr. Jensen's veggie broth long-term as a seasoning in order to boost your potassium levels. Researchers have calculated we need a minimum of 3-to-4 grams of salt a day along with plenty of water.

This reversed ratio of the natural proportions of 1 sodium-to-16 potassium in our daily diets greatly contributes to Americans not being able to digest our food properly. By the way, the 'experts' have also inverted the recommended natural calcium and magnesium ratios (we need more magnesium), right along with advising stomach acid should be suppressed. Acid production can be encouraged with betane HCL supplements.

It's a fact of chemistry that our cells can't work correctly if all these ratios are wrong. With poor digestion, undigested food enters the blood to become a rotting feast for the anaerobic bugs. Here's the good news. You do not need to understand any of this stuff. Just remember to eat 85 percent live (not canned) vegetables (salads and juices) and be alert for any beneficial long-term changes.

Considering Changing Your Diet?

If you wisely change your diet for the better, then realize it's going to take *time* for your digestive process to adapt to your decision. Many have gradually moved to dropping the majority of starches and eating lots of natural raw vegetables and fruit and then lost weight, gained energy, acquired serenity, and have had many physical problems disappear. What you eat and how you breathe is really the basis of it all. As I say, if people did these things right I'd be out of a job.

My good friend Carl is 74, he mostly eats vegetable juices and meats, no sugar, and only a small amount of simple carbohydrates like bread. He does wind sprints on a rebounder because "a half hour of just bouncing is too boring," oxygenates with deep breathing exercises while walking, and is often found out in the Bronx pouring concrete and doing construction work all day while also running many businesses in-between. He has no arthritis, no aches and pains, and says "I do the work of 10 men, and I love to go dancing." I also hardly ever hear him hold a grudge or complain. He is no special superman; he simply eats, breathes, and thinks correctly. So can you.

Many people feel dislike of others that are healthy and might view them as show off physical culturists—because it's easier to ridicule than make changes in their own personal lives. Suppressed feelings of envy have their roots in the error of falling for the lie of hopelessness and thinking that it would be too hard for you personally to ever change. Don't buy it. You can do one small new thing every day—and after a while you're somebody else in a whole new place.

A large percentage of your digestive balance (and therefore your health) is wholly dependant upon the type of beneficial intestinal flora—digestive bacteria habitually maintained by the body. The body *eventually* adapts to almost any changes we throw at it, like moving to the Artic Circle or the Sahara Desert, or eating the latest currently available food source. It also adapts to changing its habitual intake from a constant assault by processed food to the soothing balance of natural food, but your intestine must be shown your sincerity through repetitively eating the same things.

Deep body processes like digestion are important and therefore stable, so the intestine must be given time to be 'convinced' it really is confronting a permanent change in diet. Eventually it will change the ratios of the old bacteria and grow the proper new bacteria needed so it can digest the new food. This may take a long time, even months, so switch over slowly. Until then, expect body noises and less than full efficiency during the changeover. That's why some complain they tried changing their diet, but did not get an immediate reaction. My friend Diane just told me: "We switched to sprouts and wheatgrass overnight and ended up in bed for two days." This is the classic, instant, and rapid detox. Patience, grasshopper. You must understand that competing flora change slowly in the body.

Water

The next leg of Nature's primary triad is cool, pure, refreshing and cleansing water. Los Angeles, California had one of the highest throat cancer rates in the world; they traced it right back to their drinking water supply being full of trihalomethanes (THMs.) After conducting a five year study to find the best way to eliminate these THMs, they chose ozone water purification because it was the only method where they could prove they could actually get rid of these cancer causing THMs. What's coming out of *your* tap? Chlorine, fluorine, trihalomethanes, pesticides, and other poison isn't what we want in our bodies is it? Get a good filtering system and always oxygenate and ozonate your water before use.

Since it makes up two-thirds of us, water is obviously a *big* concern in our search for healing and energy. The water molecule has an angle of 105 degrees, which is in the golden mean proportion. Of course the more pure and natural your daily water is, the better off you are. There are three emerging aspects of water that are getting more attention: pH, structure, and subtle energies.

Our acid/alkaline pH balance encourages or suppresses microscopic disease bug and cancer growth. Our poor diets cause everyone to be too acid and having a chronic slight acidity in our blood and tissues sets us up for disease. The body's number one method of neutralizing acidic pH is through using oxygen to turn carbon monoxide into carbon dioxide. We must eat naturally, oxygenate, and drink lots of water to dilute our carbon dioxide waste so we can be slightly alkaline. I repeat, the body's *number one method* of neutralizing acid wastes is through using oxygen! Cancer cells and many other imbalances cannot survive in a hydrated oxygenated alkaline environment. It's that simple!

Looking deeper into the sub-atomic levels of water, remember we are mostly water. Now think about how the structure and programming imprinted upon your body's water crystals will determine your aggregate inner frequencies, and how the energies imprinted upon the water and minerals that your body is made of are re-radiating into you continually.

There is a lot of research work that is being undertaken by Dr. Mona Harrison, former Assistant Dean at Boston University School of Medicine, and others, on the properties of water. Dr. Harrison was very kind in her praise of my work in her foreword for this book. From her clinic in Yelm, Washington, she advises us to explore alkaline and acid waters. We can be either too alkaline or too acid, and water conditioning machines (many of which are being sold out of the Orient) are being used to adjust the properties of drinking water and eventually bring the pH (acid/alkaline ratio) into balance. It's a two way street. Being in balance will allow better utilization of your cellular oxygen, and of course active oxygen also helps balance pH by removing toxins and lactic acid.

To really get ahead in the game we must all look deeper into the new fields of 'structured' water, 'clustered' water, 'ionized' water and of water charged or imprinted with beneficial subtle energies. Cutting edge home oxygenating supplements are starting to implement these new water technologies in their manufacture.

Japanese scientists have photographed microscopic water existing in crystallized forms. Crystals store and radiate energy. The Japanese photographic documentation proves that pollution, and amazingly enough,

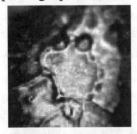

even negative thoughts can damage the structure, function and energy storage and radiation capabilities of water crystals.

Water's microscopic crystalline forms are similar to snowflakes, and no two are alike. Polluted city water crystals exposed to negative emotionally charged

environments photograph as brown, damaged, and lacking distinct structure.

Pure mountain, spring, glacial, and structured waters when photographed show beautiful natural crystalline patterns. Water crystals are thereby proven to contain the imprinted 'recordings' of their previous environments' energy. The sample water crystal pictured here was prayed over.

'Structuring' is to remove or erase the stored aberrant energies. Filtering, ozone, magnetism, distillation, spinning vortexes and other methods are used to accomplish this task to varying degrees. In the field of 'subtle energy' we know that ion-sized tiny and pure natural liquid minerals are easily and immediately absorbed into our cells, and that these ionized minerals can be nicely added to ultrapure structured water. Then, using advanced micro sized antennae and appropriately designed ultra short wave transmitters; beneficial frequencies can be imprinted or stored within the water crystals and minerals. OxyMune is an example of a product using this technology.

Do you think that the cleanliness, structure, and vibrational state of all these billions of miniature crystals floating as water in you, and comprising two-thirds of you, is important to address during your search for health?

Drunvalo Melchizedek lectures on water. He said that in the early 1990s at the University of Georgia they discovered that every cell in your body that is diseased—or that is harmed in one way or another—is surrounded by something called unstructured water. Whatever the disease is, it's always surrounded by this unstructured water. He says they also discovered that every healthy cell, no matter which cell it is, is always surrounded by structured water. I called the University of Georgia Library and was told there were no available papers on the subject. Available papers or not, the concept makes sense to me, especially after viewing microscopic pictures of different waters. Papers are useful, but they do not create our reality.

The difference between structured and unstructured water is in the number of electrons that are in the outer orbits of the water atoms. Unstructured water atoms are missing outer orbit electrons, and this 'incomplete water' condition is said to foster disease. Water atoms that have these outer electrons in place have completed shells, and are said to promote health. It makes sense, because the movement of electrons is the sun's energy delivery system. Water and oxygen deliver electrons. Without enough electron life-energy in our food and water we end up in poor health.

New forms of 'Super Ionized Water' are beyond having the complete numbers of electrons in their outer orbit electron rings. The new waters have *extra* electrons. The new Super Ionized Water is said to be able to clean up all pollution by donating electrons and breaking it down into its base elements, just like active oxygen and ozone, but these waters are in a liquid instead of gaseous form. I have concluded that therefore Super Ionized Water appears to be a new electron delivery system that acts just like the active oxygen I have promoted for so many years. We know active oxygen works by delivering the sun's energy via extra electrons, so one of oxygen's healing actions has to do with resupplying and adding the missing electrons to the outer orbits of water atoms clustered around disease in the body. By adding electrons either through active oxygen or super water supplementation the water inside and outside of diseased cells would thereby become naturally re-energized 'healthy' structured water having a full electron compliment.

Melchizedeck explains that newborn babies all have bodies full of structured water, and also that water flowing through rivers and lakes in natural unpolluted situations is structured. However, water that goes through a pipe, as does almost all of our water that goes into our bodies, is unstructured. It takes just a few feet of moving through a pipe under pressure and the water can't rotate the way that it wants to. Instead, it is forced to move in concentric rings. And these concentric rings rip off the outer electrons and form unstructured water.

He concludes that we are kind of making the same type of mistakes as the Romans did a long time ago when they ate off lead plates. We are drinking out of high-pressured water pipes that eventually contribute to disease. Do some research on Viktor Schaumberger. He wrote about this subject of vortexes being natural and pollution free and even designed natural piping for us to use.

Drinking pure structured water charged with beneficial subtle energies and oxygen is essential to putting Nature's tiny, original, and naturally patterned 'energy seed' water crystals into you. Drink it consistently and expect to eventually replace the damaged or polluted water crystals making up two-thirds of you with life affirming ones. Re-patterning your inner water and mineral programming energy radiation matrix back to an original newborn design of perfection will obviously have positive effects reaching far into health and consciousness. Many of my tie-it-all-together for you original theories will eventually be proven when sensitive enough measuring instruments become commonplace.

Dehydration

Did you know that tiredness, anxiety, agitation, short temper, cravings, and depression are perceptive feelings that your brain generates as signs of dehydration? Your brain needs hydroelectric energy from water, and these are some of your body's many cries for it. Other chronic drought management signs are allergies, hypertension, and diabetes. Your big crisis call for more water is pain. With the help of Dr. Batmanghelidj, I'll give you a quick overview explaining how body dehydration works in general to ruin your health.

Water carries oxygen into the body, and we can't mention water without also discussing what happens to you if you don't get enough of it— cellular drought, also known as dehydration. Eighty percent of Americans are *chronically dehydrated* to the point their thirst mechanism actually shuts off. This means none of us are appropriately thirsty anymore! Older people in particular are thirst challenged. Experiments have been conducted to demonstrate that older people cannot recognize their own dehydration. If you keep them without water for 18-to-24 hours, and if you put a jug of water next to them, they still wouldn't satisfy their bodies' water needs.

In addition, we drink more soft drinks than water. Too many people are hooked on the high endorphin rush of sugared (or even much worse, artificially sweetened) caffeinated drinks and sodas that squeeze water out of the kidneys and dehydrate us further. For many, a caffeinated and artificially sweetened beverage is all they drink, and I have known soda addicts. They show up in front of me with cancer. These beverages should be imbibed in moderation—and not a 'source' of water.

If you must drink a cup of a caffeine-containing beverage, you should drink an additional glass of pure water to compensate for it.

In my opinion, a lack of oxygen AND the water to transport it are the primary causes of all disease. If you don't have enough water in your lungs, you can't transport oxygen from breathing into the bloodstream. I'm an oxygen expert, and one of my hydration/water expert counterparts is Dr. Batmanghelidj who, as I told you earlier, wrote *Your Body's Many Cries For Water,* as well as his new *ABC of Allergies, Asthma and Lupus.* We met at the IBOM oxygen doctors meeting. He discovered that our chronically unnoticed lack of water is the other cause of many of our illnesses. It makes total sense, since water carries oxygen into and throughout the body. Without enough water, the oxygen can't get anywhere in the body.

"Ice sticks to your skin. Through sheets of Hydronium ions, H_3O^+, water actually holds your cells together. It forms a natural molecular structure and maintains the integrity of your cells. Whenever you become dehydrated, your cells die.

"Water is one of the main sources of energy for the brain and the entire body. Proteins and the enzymes of the body are less sticky and function more efficiently in solutions of *lower* viscosity (lots of water). To modern medicine, water is incidental as merely a solvent or a packing material and a means of transport. But this view is incorrect.

"Water is a source of energy. One of the ways water produces energy is through hydrolysis. Hydrolysis means the splitting of water into its two components: hydrogen and oxygen. Whenever this occurs, energy is released. The energy generated by water helps produce ATP, a compound that stores body energy. Just as there is solar energy, there is hydro-electric energy.

"Certain parts of the brain draw most of their energy from water. Neurotransmission is heavily dependent on energy from water. What occurs along the nerves is an exchange of charged minerals, called cations. Cation exchange gets its energy from water.

[Reference 'A mechanism of ATP-driven cation pumps' written by Phillippa M. Wiggens *Biophysics of Water*, John Wiley and Sons, 1982, pp 266-269.]

"Water affects nerve transmission in one other way, too. There seems to exist small waterways or micro-streams along the length of nerves that float the packaged materials along biological structures called 'microtubules.' These waterways transport the products manufactured in the brain to their destinations in the nerve endings." —Dr. Batmanghelidj

I interviewed Dr. Batmanghelidj and combined his answers with some other interviews he had done, (for example, with Sam Biser). The good doctor Batmanghelidj explains in detail how stomach pains, depression, high cholesterol, liver and kidney damage, fatigue, exhaustion, and more all stem from a lack of water. According to his research:

Allergies —Once in a state of dehydration, the body's cells release more histamine to conserve water, leading to the allergy symptoms. With a proper amount of water going through it daily, the body regulates the histamine and the allergic reactions are held back. [Anyone with an auto-immune disease like lupus take note:] There is also no neutralizing factor to neutralize antigens (toxins or enzymes) that stimulate the production of antibodies. The body has to wash out the pollen or whatever the offending substance is to get it away from the mucosal surfaces because there aren't any antibodies to neutralize offending (allergy-causing) substances.

Asthma —This is the body sealing the lungs off and constricting the bronchials with swelling producing histamines in order to conserve water loss.

Breast Cancer —Chronic dehydration is a stress on the body, which raises the level of the hormone prolactin, which has been known to incite breast cancer. Also, the dehydration would alter the balance of amino acids and allow more DNA errors during cell division, another factor promoting breast cancer. Dehydration suppresses the immune system—especially interleukin-2 and interferon production, the two elements needed to activate the killer cells against cancer cells.

Edema —Due to acidity and dehydration, the veins become leaky and the blood plasma proteins leak out and become trapped in the plasma, causing swelling. This is the body's way of preserving water when it doesn't get enough. When water is available to get inside the cells freely, it is filtered from the "salty ocean" outside of the cell and injected by osmosis into the cells that are being overworked despite their water shortage. This is the reason why in severe dehydration we develop an edema and retain water. The design of our bodies is such that the extent of the ocean of water outside the cells is expanded to have the extra water available for filtration and emergency injection into vital cells. The brain commands an increase in salt and water retention by the kidneys. This is how we get an edema when we don't drink enough water.

When we drink enough water to pass clear urine, we also pass out a lot of the salt that was held back. This is how we can get rid of edema fluid in the body; by drinking more water. Not diuretics, but more water! In people who have an extensive edema and show signs of their heart beginning to have irregular or very rapid beats with least effort, the increase in water intake should be gradual and spaced out, but not withheld from the body. Naturally, in the beginning salt intake should be limited for two or three days because the body is still in an overdrive mode to retain it. [Please consider the proper ratios explained earlier, and a possible potassium deficit masquerading as a 'lack of salt.'] Once the edema has cleared up, salt should not be withheld from the body.

Heart Attacks —Histamine in the lungs regulates prostaglandins, which are powerful in extremely tiny concentrations. Some of the prostaglandins are constrictors. If some of these spill over to the circulation that is going directly to the left side of the heart, the same prostaglandins that are regulating the lungs can also cause vasoconstriction in the heart. And that's how sudden heart arrest can begin.

High Blood Pressure —This is caused by a water deficiency inside the cells and an excess of water outside the cells. The result of eating too much processed food that doesn't have enough potassium, magnesium, and calcium. There aren't enough of these vital minerals to run the pumps and hold water in the cells—so the fluid collects and stalls outside the cell (along with whatever oxygen there is) while the brain tries to force it into the cell by raising the pressure. People who are given diuretics are being sent to their doom gradually. He says "Doctors are killing people."

I know this is true. My own mother was slowly killed this way. It took a few years for her to weaken and die, and the list of drugs she took got longer and longer as she got worse and worse. It played out exactly as Dr. Batmanghelidj says. High blood pressure at first from dehydration (and socialized improper sodium/potassium and calcium/magnesium ratios), followed by diuretics causing more dehydration, arthritis, swelling up, lots of pain, and finally culminating in her quadruple by-pass followed by her further suffering and eventual premature death.

Dr. Batmanghelidj: "When [due to a chronic lack of water] the body has to force water into tissues, it will do this by raising the pressure to try and force the water past the water blocking cholesterol. The cells cover themselves with cholesterol so they can retain their water during dehydration. This pressure that is required is called hypertension. Hypertension is caused by water deficiency. One of the stupidest things in medicine is to go and give the person dehydrating diuretics.

"We are killing people. People who are given diuretics are being sent to their doom gradually. The doctors think salt is the problem. The salt is NOT the problem. The body retains the salt in order to keep more water in the tissues. Your body requires salt [in the proper ratio] to retain the water. Otherwise, the water will pass through too quickly. So the medical thinking is 180 degrees wrong.

"By giving diuretics, you are forcing water out of a body that is already *starved* for water. Diuretics in my mind are so dangerous that I foresee that soon treatment of hypertension with diuretics will become a legal issue in the courts.

"The person who is being dehydrated is gradually losing oxygen and functions in certain parts of his body and is being made sicker. That is why hypertension patients don't come off of their drugs. First they are given diuretics, and then a beta blocker, and a calcium blocker, and eventually bypass surgery. Hospitals give patients soda instead of water.

"The body needs water all the while, and it is being forced into this situation through sheer ignorance. Drugs do not bring water into the body. You are drinking some water with your medication, and *the water is more important than the medication*. Even city water with chemicals added to it is better than the alternative."

High Cholesterol —This condition is an adaptation to a low water intake. Cholesterol, like the waxy substances on cactus, aloe, and other desert plants, is deposited in tissues because of dehydration. The cells put out cholesterol to close themselves off—so they do not lose any more water.

Immune Cell Deficiency —Without enough water the bone marrow cannot produce enough antibodies and immune cells.

Ulcer Pain —This is caused by dehydration. The pancreas produces a watery bicarbonate solution in order to neutralize the acid, because this acid cannot be allowed to go into the intestines. But when there is not enough water, the body creates a spasm in the sphincter muscle between the stomach and duodenum to block the acid from traveling to the intestines.

pH, Acid-Alkaline Balance —Dehydration causes acid pH. Most people with breathing problems are acidic; most people with cancer problems are acidic.

Most people with other problems are acidic. With enough water and oxygen the kidneys will concentrate and flush out the acidity.

How Much Water Should We Drink?

Calculate your weight in pounds, take half of that, and call it ounces. That's how many ounces are needed during the day. It works out to 6-to-12 glasses per day. [This is the target amount you work up to slowly as you give your body time to adjust.] Try drinking the most water when you first get up, and a half-hour before meals. Spread the rest out. You will need more water if you are overheated or drinking caffeine.

You should have been eating enough natural vegetables and unprocessed foods your whole life to get sufficient vegetable potassium and sodium. But most are so dehydrated they need a temporary quick solution enabling more water and oxygen uptake. If you don't get a quarter of a teaspoon of salt for every quart you drink, your body will not be able to retain the water—and you may still remain dehydrated. You also need iodine from sea kelp and dulse to balance the salt. Powdered and liquid versions of kelp are available at health food stores, and via mail-order. Here are the doctor's teachings in summation.

> Drink enough water for your weight. Increases in water intake must be slow and spread out until urine production begins to increase at the same rate that you drink water, especially if you have kidney problems.

> Spreading the water out over the day, the most water is to be taken upon rising, and one half hour before meals. The water we drink is immediately recycled and prepared and back to the stomach ready to digest the food in half an hour. We have no reservoir, we use what we get. The water you drank two hours ago is already on its way out.

> If you are addicted to anything, including coffee or soda, or food, or in pain, reach for your water first. Give it a few minutes, and see if you still need the thing you craved.

> For every four glasses of water you drink, you should take a quarter of a teaspoon of natural mineralized salt (it can be in the water). Especially during heavy exercise which sweats out your salt.

> To support the increased salt intake, consider natural iodine supplements in the form of kelp and dulse.

> Get enough potassium. Potassium is a vital food necessary for hydration of the body. Potassium is concentrated in potato, banana, fresh orange juice, vegetable broths, etc. We have no important body mechanism for potassium retention.

> If this still has no effect, the amino acids tryptophan and tyrosine may need support so you can 'use' the water. Dehydration harms the liver that manufactures amino acids and upsets their sequencing. Replace with a balanced variety of amino acids. The body doesn't like just one amino acid.

> Walk, walk, and walk. You need to walk to open up the capillaries to hydrate and oxygenate the tissues.

Caution: You should always consult with your health care professional before making any drastic changes. That applies to anything in this book. People who are on medication should not cut off their medication abruptly. The doctor advises people get the full details from his books and to begin increasingly drinking more water. Then they should consult with a health care professional and test to see if they can tail off their medication. People who have heart or kidney problems should always consult with their health care professional (one skilled in oxygenation and hydration) before increasing water intake.

Now what? So, you're dehydrated. Something doesn't work, and something hurts. Your urine is dark. But you're not even thirsty. And when you try to drink water you can't—because you fill up quickly—and maybe you get a little nauseous whenever you drink a great deal of water. You've got all the symptoms of classic toxic un-oxygenated dehydration. How do you get the water in you? The *way* you drink water is just as important as drinking it! I will repeat. Calculate your needs and spread your requirements and intake out over the day, once every hour, all day long works best.

Sam Biser interviewed Nathaniel Clevenger, a Dr. Batmanghelidj water drinker.

Nathaniel said, "The first couple of weeks you have to get used to the fact that you are going to be in the bathroom all of the time, and you are. But what happens is after a few weeks, your body expands and begins to hold it. But in a typical day, I have to excuse myself six or seven times, but it's still worth it. It's not too high a price to pay for these benefits. It makes total sense, doesn't it?

You were conceived in water. You jump in a pool when you want to cool down. Everything we do that comforts us has to do with water. It is a natural thing for your body to love, but we don't get enough of it.

You need to drink as much as you possibly can. It saves your life. It has for me."

Oxygenated Water

Water is one of the most vital materials known to man. Without water, man can only survive for four days. The quality of pure life giving water has dwindled over the centuries and the fresh, active form of water rarely exists.

> **The only way most of us will get enough oxygen is by drinking it**

Most European and many American cities purify their municipal drinking water by bubbling oxygen/ozone through it to kill all the bacteria and viruses, etc. (See *Inactivation Kinetics of Viruses and Bacteria by use of Ozone*—E. Katzenelson, et. al., American Water Works Society, 1974.) Most bottled water in the U.S. and Canada goes through the same ozone purification methods. Since your body is two-thirds water (we are internally permeated with fluids,) the same purification principals would directly apply to us.

While I was on one of my tours, I picked up a glass from the back of the room at one of the lecture halls, and I poured some water into it from one of the pitchers. Sure enough, the chlorine smell seemed to almost jump out of the glass, it was so odorous. I went to the kitchen and dumped it. One of my hosts, Darryl Wolfe, from the Wolfe Clinic in Canada, noticed that I did this; so we took ozone and bubbled it through the same water. Not only did the chlorine taste disappear; but the water actually tasted good.

Oxygenated drinking water is now a specialty beverage. It's sold in retail supermarkets, health food stores and even through multi-level marketing companies. The appeal of this water or beverage is that it holds its super oxygen charge. This is almost my final solution dream come true. As I've often written, bottled water *will* be the new "oxygen delivery system for the masses." This is really neat. A bottled spring water, or any other beverage, sports drink, or herbal tea, or whatever that will actually be charged with oxygen is a dream come true. The technology is in the bottle, so to speak. And it's remarkably effective. Water is charged with oxygen at a much higher than normal rate, and treated in way that it holds its oxygen charge. I think that the market of the future is going to be oxygenated beverages. That may seem unusual right now, but let's face it. The emergence of bottled water for sale was once ridiculed, and who would have imagined that it would have been the source of such a lucrative industry?

I believe we are now in the time before the 'storm' or the 'main events' depending upon how you view changes that are surely coming to our personal and global lives. Right from the start, my years of preaching/teaching oxygen work had a two-fold intent to:

1. Prepare as many of us as possible for the rapidly diminishing global fresh oxygen levels.

2. Prepare as many of us as possible for the high possibility of a worsening of the existing and always newly emerging diseases that might become plagues within a short time.

They may not be called 'plagues' on TV, but the rate of increase in diseases is skyrocketing, not diminishing. So what else would you call them?

Mark My Words

I have been saying this for 14 years. Owing to the increases in pollution and the rapid decline of available clean oxygen on our planet—and the resulting loss of oxygen from our food, our air, and our water, I firmly believe we're all going to be buying oxygen supplements in the markets. This market surge will be exponentially linked to how fast the common man and woman are 'sold' on the truths in this book. Their purchases will directly tie into their physical deterioration coupling with their increasing understanding that most diseases can't live in properly oxygenated environments, and that our immune systems run on oxygen.

Those who have enough awareness to supplement oxygen on a daily basis will also be the same ones who best survive any bio-attacks from criminals or mosquitoes. If you're a manufacturer, beat the rush. Get on board and start selling them now.

> We will understand the importance of oxygen when it's close to gone, and then only Herculean adjustments will ease the suffering.

Air

Swimming in a Sea of Air

Between goings from our homes to the car to our office buildings, most of us are living in a controlled atmosphere 22-to-23 hours per day. The oxygen content of the air in these sealed places quickly becomes lower than 21 percent. It has always amazed me that the number of people present at my lectures—living in a polluted big city and claiming that they eat correctly, exercise, and take all sorts of supplements still wonder why they are sick.

Many people run or jog or do breathing exercises, but what are they breathing while they do so? Air has to be pure, clean, fresh, and natural, charged with the sun's energy, filled with oxygen not bound to pollution, and abundantly so. If you do all the right things while living in a low oxygen cloud of toxins, don't wonder why the rate of toxic build-up in your cells is overcoming your best efforts. You can't clean up without having more oxygen and clean water going into you than pollution. Oxygen is the main element of body sanitation.

A man weighing 150 pounds has 90 pounds of oxygen in his body. A resting man consumes 575 liters in a 24-hour period. During activity consumption rises to 700 liters per 24-hour period.

Hermetically sealed buildings are an engineer's dream, but they may become your worst health nightmare. Unnatural water coolers serve up chlorinated beverages, and fluorescent office lights give me and lots of others headaches after a while. If a building has little fresh oxygen and no ozone in the air, microbes have a field day breeding in the air conditioning ducts. Cleaning solvents, out-gassing carpet adhesives, plastics, and unnatural fibers all get their fumes recirculated round and round. Spending eight or more hours a day inside a sick building syndrome area can eventually make someone allergic to many things.

The point is—you don't want to be breathing all this stuff. Country air may not be that much better, depending upon what your piece of God's green Earth is downwind from. One of the ways you can clean the air up and then re-energize it is by both filtering it and adding safe low levels of ozone to it. We'll talk about ozone air cleaners later.

Now that you have fully grasped these preliminary concepts leading to solutions, carefully examine the following chapters on 'active oxygen' and ozone delivery methods. If you doubt what I have been telling you at all, try holding your breath one more time!

Chapter 6

Why I Say,
"Flood" Your Body with Oxygen

Whenever I say this I always mean *safely* and under expert guidance. Before I discuss the various Oxygen Therapies solutions in detail, we'll look at a few related areas. First, just so you really understand and *feel* what the solutions are all about; I want you to conduct an experiment. Please just do it, because I want you to get a left and right brain connection here. Make your body understand everything I have been talking about by just doing the following without thinking about it.

Just stop breathing right now. Hold your breath. Don't take a deep breath, just stop breathing. As soon as you feel any discomfort at all start breathing again. (I'll wait...)

Did you just *feel* how important oxygen is to your body? I'm trying to show you precisely what I'm saying in many ways in this book. Oxygen is the prime thing in your life. I could have taken away your shoes, your watch, your money, your eyeglasses, your car, your vitamins, or your self-respect, and you still would have lived. You would have survived.

When you took away your oxygen, you very quickly reordered your daily priorities. You couldn't manage anymore, your body started to scream for breath—oxygen. This experiment quickly imprints upon you how important it is to oxygenate your body, not just now and again but every second of your life. When I have conducted the experiment with audiences, I tell them to hold up their hands and stop breathing and then lower their hands and breathe at the first sign of discomfort.

The hands go down fast. Maybe in 5 seconds, 7 seconds, 10 or 14 seconds. At that point, they didn't need the latest enzyme, vitamin, herb, powder, shark cartilage, or flower essence. They didn't need their homeopathies, aromatherapy, massages, injections, pills, or whatever they felt would benefit and improve their bodies and take away their diseases. Would they have screamed to have any of them back in 14 seconds? No.

Hopefully, you now understand that what I keep going on about is the very basis of life here. By comparison, everything else is an add-on. The therapies I just made fun of all have their place and value, and they are all completely compatible with Oxygen Therapies. If you use Oxygen

Therapies you can incorporate the other things and it will only boost their effectiveness.

Remember who you are

Everybody has run so far down the road chasing each of the latest health fads for so long that most have forgotten who they are, what they're made of, and where they came from. We are Infinite Potential, we are made of Nature, and Nature is our home. One of my jobs is to get everybody to remember this about themselves. Only then will the confusion over the best way to heal melt away.

Always Start with the Oxygen Question!

"Do I have enough oxygen in my body?" Oxygen Therapy is *the* place to start. I'm saying it's absolutely essential to ask this question *first*. The oxygen question must be addressed right from the start.

I have always seen that no matter what therapy you or the practitioners are using, if some form of an Oxygen Therapy were added to it, the results were better and it worked faster. Many have seen the wisdom of what others and I have been advocating for many years. They've added Oxygen Therapies to their regimen and reported to me over and over again how good they now feel. I'm teaching you this one subject specifically above all others. Oxygen Therapies are **THE SINGLE BASIC THERAPY.** All the other therapies are secondary by comparison. I'm very serious about what I am saying. All the books in your doctor's office, and all the books and pills in the health food store don't mean squat if you're short on oxygen at the cellular level; and the situation's only going to get worse, my friends.

FIG. 4.—THE INHALING FLASK, AND METHOD OF USING.

You need oxygen throughout your body. Your whole body constantly needs oxygen. Every cell needs constant, constant, constant oxygen. It's involved in all of our bodily functions. It's the first thing you took in when you were born, and sending it out will be the last thing you do as you expire. The majority of us take for grated the use of oxygen and how essential it is for sustaining not only life but the quality of life.

We begin and end our lives with oxygen; yet in-between we ignore its importance for continued good health. Why does every patient's room in a hospital come equipped to administer oxygen? United States federal law requires that every emergency room in a hospital be equipped to administer oxygen to its patients. I would say it's there for a very good reason.

By themselves, the therapies are simple yet very effective therapies. People have tried them over many months and years and have found them to be successful. There are only a few exceptions, and these occur when incorrect protocols are used.

Occasionally someone has no idea of what I'm talking about. They might hear some of the more fantastic sounding stories or claims and they are overwhelmed and outside their intellectual safe zones. In reaction they sometimes just throw up this wall of ignorance with the statement "That can't be true," or "No one has that many answers." or, "No one therapy is that good." But my friend, it really is that good. And not just because I said so. We know Oxygen Therapies are that good because they have already been effective throughout history. It's all quite simple and it has been proven time and time again. Oxygen Therapies address the basic cause of all disease. Individual results may vary, but if you adopt common sense in applying the therapies the causes of all disease won't affect you the way they used to—because you won't have so many of these causes attacking you in your life and invading your body.

What are Oxygen Therapies? They are a process of slowly and properly saturating or **FLOODING** the body with special 'active' forms of oxygen. Oxygen Therapy is not only one of the oldest therapies, but, as the evidence shows, it is one of the most successful.

The oldest known Oxygen Therapies come in various forms and the process of flooding the body with oxygen is a gradual one of slowly increasing the applications and dosages. Historical data show that approximately three weeks to maybe six months later, depending on which type of therapy you're correctly using, your body at some point gets so much oxygen in it that it can finally heal itself of many things.

In a nutshell, Oxygen Therapy is about cleaning out the water. If you can totally take that 100 plus pounds of filthy water inside you and clean it out, I can show you how that's the answer to so many of your health and energy problems. That's why people are getting fantastic results with the various Oxygen Therapies. They are providing the missing link and putting back the missing oxygen. It's like turning on a light switch. The body wakes up and moves into action.

Chapter 7

Getting Regular Oxygen Inside Us

So, How Do I Get the Oxygen, Do I Get a Tank or What?

People always ask me that. The whole reason for using Oxygen Therapies is to slowly and properly oxygenate the body until two significant things happen.

1. All the pollution, all the bacteria and microbes, and all the toxins that are in the body are burned up and washed out.

2. Then a well-oxygenated inner body environment is sustained within your fluids and cells. This keeps the body so clean and pristine that none of the bacteria, viruses, microbes, fungi or other pathogens can enter your body or exist in it for long. And toxins are quickly removed.

The first question I am always asked is, "What do I do, get a tank of oxygen to breathe, or do I have to be like pop star Michael Jackson and sleep in an oxygen tent, or what?"

Since that news release about Mr. Jackson and his oxygen tent came out years ago, everyone remembers it. Several of the famous Jackson family members definitely found out and now understand that active oxygen forms like ozone are the key, and that breathing from an oxygen tank is not what I'm talking about. Tanked oxygen only supplies stable oxygen, and that's good but not exactly what we are looking for. We need to use the more effective active forms of singlet oxygen. I know the family knows this because a few of them took me to dinner to ask me all about it.

Many other Hollywood stars and famous musicians and sports heroes have quietly gotten rid of many serious diseases, even including (gasp!) cancer and very serious viral infections (wink, wink) using Oxygen Therapies. You don't often see such things announced in the media, since heroes think its bad for their image to show frailty. I know what they have overcome by using Oxygen Therapies, because people I personally taught have been supplying them with their oxy–needs for years.

In the following chapters I'm going to show you the details of all known ways of oxygenating the body. Then I'll show you the evidence that it's safe to do so and that it's all been documented for a long time.

Breathing the Complete Breath

That even *sounds* healthy! The everyman's Oxygen Therapy is the one that is free, effective, and can be used by everyone; and when there are no compounds to carry, or machines to buy. I'm talking about breathing, the first stop on our tour through the world of oxygen. We are walking around in our own medicine and it's right under our noses. Breathing correctly is still important, so I am repeating part of this section from my first book.

The National Institute of Health (NIH) conducted a five-year study at Stanford University School of Medicine. They compared 498 long-distance runners to 365 average people and found that the deep breathing (oxygenating) runners developed the usual age-related disabilities at a slower rate, had better cardiovascular or heart conditions, and weighed less. They went to see a doctor only one-third as many times as the average person and had better attendance records at work.

When I explained this oxygen subject to a colleague who studies yoga she exclaimed, "So that's why all the yogis who practice alternate nostril breathing for two hours every morning don't get sick!"

Sports

If you're athletic, and into running, boxing, lifting weights, racing horses or dogs, or if you're a supermodel, you should be very interested in raising your bodies' oxygen level with oxygen supplements. Everyone I've met who stays properly oxygenated always starts looking younger and has lots more strength and energy reserves. Horse trainers have whole stables on oxygen or ozone supplements. The horses and dogs go out and race and then come back hardly winded. Many champion athletes secretly (so no one else gets the same edge) use oxygen supplements. Even armchair athletes' report having more energy.

We already know the body is constantly producing a substance called ATP from food. It's the energy currency of the body. We need to produce it constantly because we're using it up all the time. The body is amazing. It can produce ATP in either an oxygen-rich body environment or in an oxygen poor environment.

Usually we go along 'burning energy' from the ATP that is produced in the presence of oxygen. No problem... suddenly, we have to run very fast to grab a child out of the way of a speeding car! In that moment, or under similar important or athletic competitive situations, we can't breathe fast enough to immediately process enough oxygen to make lots of this ATP. Our bodies are equipped for survival, so they make ATP beforehand and store it for flight or fight situations. When we are not fully oxygenated, our bodies have to make the ATP under low oxygen conditions. The problem, as you know, is that this anaerobic formulation of ATP is different from the ATP made during oxygen rich times. Anaerobic ATP has a much lower energy potential and leaves lactic acid and harmful free radicals behind and causes a painful burning sensation in the muscles. We then have reduced efficiency, and we're sore. Lactic acid causes the muscles to quickly fatigue and depletes our strength and endurance rapidly. By always having a high tissue and fluid oxygen content, higher quality ATP energy is always produced that lasts longer and gives us less fatigue. An example is long distance runners and other athletes supplementing with oxygen or ozone that don't have to take every other weekend off to recover. Ask New York health writer Gary Null about this. He's a marathon runner, and I taught him about the use of active forms of oxygen. At first, even he didn't believe what I was saying. Then he tried it on himself.

The amount of oxygen a person can consume is called 'aerobic capacity.' By exercising regularly, a person can increase this capacity up to 30 percent. Other cofactors that influence lung volume are airways, hyper-responsiveness, atrophy, childhood respiratory infections, air pollution, posture, subluxation of the spine, exercise, deep and superficial fascia, nutrition, occupational hazards, abuse and trauma, attitude, age, height, weight and sex.

About Breathing

Take a deep breath of air. Notice, how did you just take that deep breath? Before that deep breath were you shallow breathing, using only the very top of your chest and lungs? The routine act of breathing is taken for granted, but few of us get it right.

By breathing shallowly in the upper part of our chest we're signaling our nervous system that we're in a constant 'flight or fight' mode asking our adrenals to pump adrenaline. Our adrenals become burnt out, we hardly digest our food anymore, there are few enzymes in our body, and we're all mentally freaked out a lot, and it's all related to the lack of oxygen in the body from bad air and poor breathing.

Breathing is one of the most neglected bodily functions. Most people don't breathe deeply. We are conditioned to breathe incorrectly.

Just by watching the television news every night and seeing how scary it is, we can easily focus on the negative. As a result, many of us acquire a tenseness that permeates our subconscious. We tighten up because we live under stresses and pressures of all kinds. Parents yell at us, we yell at them, our boss yells at us. Mates yell at each other. Children have become abused and abusive. Then we add worries about keeping the job, keeping the romance, paying the bills, obtaining clothing, shelter, and feeding the children, and worries about what is going to happen next in our lives. It all induces high anxiety levels, so the body tightens up. We end up habitually living in 'fight or flight' bodily reactions, and subtly panting up in the top of the chest. We thus deprive ourselves of this free and very necessary oxygen support.

There is scarce clean, fresh, available oxygen left in our polluted 21st century that's not unnaturally isotopic, and the little available oxygen that is there is probably used by only one third of our lung's capacity, depending upon our habitual breathing style. For health, you have to take full breaths. That's the most natural and cheapest Oxygen Therapy there is. I always say you're walking around in your own medicine while looking for it. All you have to do is notice it and apply it. You're ignoring the available benefits because your mind is diverted elsewhere and you're conditioned to breathe at the top of your lungs.

We put machines on people to measure the oxygen content in their tissues. We found out that the best Oxygen Therapy that made the indicator needle shoot straight up was *deep breathing.* All the other ways take a little while to go through your system, except for maybe the IV ozone and the IV H_2O_2.

There was a book written called *The Oxygen Breakthrough.* At the time, the author, Dr. Sheldon Hendler knew nothing about these other products we have been discussing. Dr. Hendler taught his patients breathing—deep breathing. He got most of them off the majority of their medication, in time, just teaching them how to breathe. In other words, naturally flooding their bodies' with oxygen.

The Science of Breath

One of the first breathing exercises I learned was from a little book called the *Science of Breath* that was first published in English in 1904 by the Yogi Publication Society and written by Yogi Ramacharaka. I have found it very educational. At one time I practiced his method regularly and was thereby able to develop an ability to swim underwater for a great distance. So, from personal experience, I can attest that this technique increases one's ability to oxygenate the body. Athletes practicing this technique can increase their breathing capacity.

Breathing and the Life Force

The best all-around breathing exercises I know of are found in the system known as *The Ancient Science of Breath*. We're not discussing hyperventilation, which is an excited state of taking in more oxygen than needed on a short term basis and with resistance.

My own condensed version of these ancient breathing practices and beliefs follows in paraphrase: Physical life is a series of breaths. Natural man breathes correctly. Civilized man has become programmed with improper methods and attitudes of sitting, standing, and walking, which have robbed him of his birthright of natural and correct breathing. This lack of complete breathing has always been a contributing factor in disease. Air contains more than oxygen, hydrogen, and nitrogen; it also carries what we can refer to as the *Life Force*.

The *Life Force* is linked closely to oxygen and has electric, magnetic, and gravitic properties. Correct application of this is said to treat disease, fear, worry, and the baser emotions.

Oxygen from the air comes in contact with the impure blood in our lungs. Here a form of combustion takes place where the blood takes up oxygen and releases carbonic acid gas generated from the waste products and poisonous matter gathered up by the blood flowing through all parts of the system. The blood is then purified and oxygenated, carried back to the heart, and again becomes rich, bright red, and loaded with life-giving qualities and properties.

Avoid Mouth Breathing

Nostril breathing is far superior to mouth breathing. Our nostrils are an important filtering and straining system, purifying the air that reaches our delicate organs. Breathing can be classified four ways.

1. High in the lungs (collarbone), requiring the most expenditure of energy and the least benefit; it is practiced unconsciously by most of us. You're probably doing it right now.

2. Mid-lung (ribs).

3. Low in the lungs (deep).

4. Complete breath, effortlessly pulling down the diaphragm, which is far better than all the others.

The Complete Breath Exercise

The complete breath contains all the best of the other three, with their shortcomings eliminated. It must be performed with total relaxation. Never strain against any set of muscles. You can learn this easiest while lying down, and after practicing it can be done when sitting, standing, and then all the time. Usually when we take a deep breath we fill the upper part of our lungs, then the middle, then try and force more air in deeper. Try this now.

The complete breath is performed in the *opposite* order—from the bottom up—and without force. Breathing only through the nose. Imagine just below your intestines is the bottom section of an inflated balloon that fills your upper body completely in a large circle. The stomach, lungs, and all internal organs rest on it. Attached to the center of this balloon bottom is the end of a string that extends through the center of the body and down between the legs and feet.

Imagine a friend slowly pulling this string down as far as it is comfortable, parallel to the body and at right angles to the soles of your feet. You imagine it gently but completely being pulled down from below while all your internal organs and lungs follow. Try it. The lungs immediately have to take in more air to fill the vacuum. Mentally pull the string way down, and while still *totally* relaxed, start to fill what feels like the very bottom of your stomach with air. Your stomach will bulge out naturally. When it's full, only then do you allow any air to start filling your upper stomach and then your lower ribs. The whole process is done with one long gentle slow breath. Continuing the same breath, now expand your whole chest the same way.

One after the other you must physically gently expand and inflate each area. Use your hands to feel the expansion at first: The diaphragm slowly pulls down, then the stomach pushes out and the ribs expand, and finally the upper chest expands and rises. When you think you're at the absolute limit and can't fit anymore air in, then momentarily gently hold this position and shift your awareness to the muscles between your ribs. Notice any areas that are resisting further expansion, and *totally* relax them. The intercostal muscles between adjacent ribs are the final keys. They will give up their tension and expand as you slowly draw in more air, and you'll probably expand another inch or two. I have always found there's that last little constrictive bit that I physically cannot expand unless I melt the resistance mentally through more relaxation. But once I do, the complete breath must release endorphins, because it always feels fantastic.

Count

You are sequentially expanding a 360-degree balloon that starts at the bottom of the stomach and ends at the neck. After the technique is practiced, experiment by using the relaxation coupled with imagination to expand further. Mentally count slowly while inhaling to see how long this expansion takes. Try to take as long as possible.

Once this is accomplished and while still remaining totally relaxed, hold all the air in your body for as long as possible and start a new count. Then, slowly let the air out while counting again. Exhale everything in the reverse sequence: first compress the tops of your lungs, then middle, and then the lower. All the while have your imagined friend pulling up in the reverse direction with the imaginary string. You will find as you collapse your lungs, followed by your abdomen, you are exhaling residual stale air you didn't even know that you had been carrying around. At the end, as gently as possible hold yourself with your lungs totally empty for as long as you can. Count this exhale and hold. Repeat the cyclic process for 20 minutes.

The four movements are eventually expanded into a great circle of *inhale, hold, exhale, and hold.* At first, practice these four steps as distinct movements. This is not the ideal, however; they should become a single fluid, wavelike motion that comes with practice. Try to extend your counts with each session. As an approximation, the inhale and exhale should be about equal in time, and the inhale/hold and exhale/hold less than that, and equal. Count 10-5-10-5, increasing with practice to 20-10-20-10 and beyond. If you have any trouble sustaining any of the four counts, lower the others to match it while slowly bringing it up to their level. Balance is the object. Perhaps you could get to the point where all four counts are equal. As I said, it takes some time and effort—but when accomplished, it feels great!

Yogic Techniques

The consciousness explorers expand upon this exercise to induce a state where they imagine themselves as actually part of giving and receiving love to and from everything. But for beginners it is enough to use this to increase our cellular oxygenation. The average person has a lung capacity of 250cc, but with a few weeks of daily practice it is possible to double this. The yogis state that when the capacity reaches 750cc the intuitional powers develop more fully.

Spirituality Linked to Oxygenation

Contemplative spiritual exercises, calm heartfelt non-intrusive prayer, relaxation techniques, and meditation all reduce the body's need for oxygen. They increase the Life Force, which is the main energy we get from breathing, so less oxygen is required to sustain us.

For the more esoteric of my readers, it has been postulated that humans are evolving from 9 base to 12 base Chakra and Ray systems, and from 2 strand to 12 strand energy matrix DNA, enabling us to more closely reintegrate our higher Spiritual selves into our physical vehicles. We'll have to wait and see about this, but in any case, the point is that there is always more to healing than just the physical.

Regardless of your belief systems, at some subatomic purely scientific level, beyond the frequency range of our limited normal vision, we all actually look like this. It's just what *is*, the hidden *rest* of us.

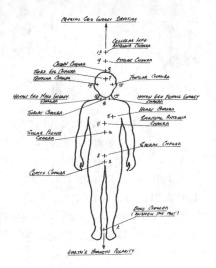

Evolving Duality Ascenders' new 12 Base Chakra system as seen by NZ's Clint Miller.

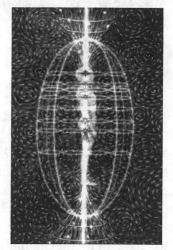

Universal Calibration Lattice ™
C. Peggy Phoenix Dubro,
EMF Balancing Technique® Originator.
www.EMFWorldwide.com

In those really tough health cases where nothing you have tried seems to work perfectly, go back and check your procedures again. Then consider that the source of all imperfection starts in our inner worlds first. Healing inner energies and other cleansing work may need to be done as well.

We are truly wondrous creations. People often report being healed after stored harmful emotions and energies are released and detoxified.
Unconditional Love = Quickest way into balance.

Traditional Bottled Oxygen Therapies

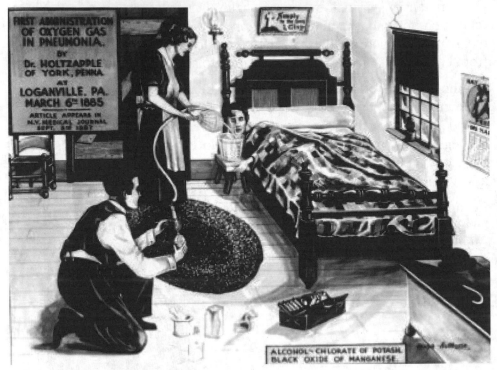

Traditional Oxygen Therapies beyond deep breathing are centered on breathing bottled oxygen. As mentioned previously, these are beneficial. However, compared to the Oxygen Therapies I refer to that use active forms of singlet oxygen for disease prevention and treatment, they are inherently held back by the stability of their atomic form. O_2 wants to remain O_2. The active therapies I promote use singlet oxygen, the active form O_1. These traditional methods of administering oxygen are mentioned so that you can understand the difference.

Bottled Oxygen

Green bottle USP 1072 oxygen is used in hospitals, emergency rooms, fire and police ambulances, and factories that hang it on the wall for emergency first aid. We do not advocate improper usage of oxygen. Military fighter pilot testing has proven that our tolerances for pure oxygen are far greater than commonly assumed.

Singers Britney Spears, and Posh Spice, actors Jeff Goldblum, Ben Stiller, and actresses Kirstie Alley, Uma Thurman, and Liz Hurley are rumored to use Opur oxygen canisters from Europe.

Oxygen Bars

All the rage now, especially in the polluted inner cities, are *oxygen bars*. New York, Los Angeles, Japan, West Palm Beach's 'Breathe,' London and Edinburgh's 'Academy,' 'Karin Herzog,' and 'Cobella Akqa' are all sprouting the latest form of urban gathering place to emerge upon the hip scene. Even Woody Harrelson from *Cheers* had one.

Most patrons belly up to the bar and pay per minute to breathe bottled oxygen through tubes. The oxygen is often flavored to increase marketability, and franchises are quickly being sold. Some bars use 'flavorings' that contain stimulants, so part of the pick-up you get may not be induced by oxygen.

Street patrons will often duck into these smokeless and smogless islands of tranquility, grateful for opportunities to grab hits of fresh oxygen while the air outside rages with pollution. The oxy-bars springing up in the nightclubs may have a novelty aspect, but they are usually the only place you can get smokeless fresh air during the party! These bars are often also selling oxygenated water, oxygen sprays, and other oxy-supplements and purifying their interiors through ozone air purification.

I recently walked into the fashionable local Aventura Mall, in North Miami and found this oxygen bar operated by 'O₂Cool.' They are selling their own state of the art oxygen bar equipment. Contact information for their products are available through Oxygen America.

One of the other stores in the Mall uses ozone air purification throughout, and when I step off the promenade and into this store it is like passing through the curtain created by a waterfall. Inside the store the air is cool, sweet, and clean, while just outside the door the air is warm, odorous, humid, and stuffy from so many people being in the mall. One of the new store salesmen was wondering why his nose was watering, and I had to explain detoxification to him.

In the metropolitan Japanese versions of oxygen bars you typically walk in from the street where you've been breathing smog. There's so little oxygen in the outside air that you might have been having trouble breathing, so you go into the bar, hand over your yen, and deep breathe a few liters of pure cool oxygen for a while. Blessed relief! Your head clears out, and you soon feel a little more energized.

Mini oxygen concentrator from O2 Technologies.

It's all the same principle at work, and it's commonly being used in football and other games where you see someone sitting on the bench breathing oxygen. They're breathing the oxygen because the body is stressed from lack of it and demands more. I've often pointed out that this O_2 form of oxygen, the same gaseous form used in hospitals, emergency rooms, ambulances, and at the oxygen bars may do a lot of temporary good. But in disease management cases, this regular old stable bottled oxygen isn't the most efficient Oxygen Therapy.

The oxygen bars are fun to use, and next time you're there breathing your fresh cool oxygen while listening to your favorite music on your personal headset, look around the store for oxygenated water and other oxy-supplements and ozone air purifiers to take home with you.

Air Travel

Ever ride in a commercial jet plane with its low oxygen recirculated air? Ever hate the air you're breathing on one? Here's why.

Travel Tip: On your next flight, politely head for the cockpit while on the ground, and ask the command pilot to turn the 'air pacs' (oxygen generators) all the way up. They usually disable them to save $60,000 per medium sized aircraft per year, your health be dammed. Don't worry, the stews are on your side. 737 Economy can get as low as eight cubic feet of air per minute, and upper 747 First class gets 60! There is a lot of controversy over blood clots forming in passengers while sitting in low oxygen cramped seats. The UK's Aviation Health Institute announced that more people died in their seats from ailments in 1998 than died in crashes. Make noise and change the system.

Hyperbarics

There is a whole specialty area in medicine devoted to hyperbarics. In this therapy people are temporarily sealed into tanks resembling diving bells, or the old 'iron lungs.' The air around them is pumped up to specific pressures, around 2.3-to-4.5 atmospheres. This forces oxygen deep into one's body. We readily absorb oxygen through the skin, and this method only uses stable air oxygen, not the active forms.

Ginger Neubauer, Courtesy Neubauer Clinic.

The doctors in the field have hundreds of studies backing up their work, and some major hospitals have hyperbaric chambers in their buildings. Although these chambers do not use active forms of oxygen, there are many more uses they can be put to.

One specific hyperbaric chamber use that needs more exploring boosts ozone delivery. Patients who have already been safely and properly freshly pre-stuffed with active forms of oxygen (like ozone) can then be placed in one of these chambers and pressurized. The rationale is that viruses can hide in the bone marrow and other deep cavities in the body, walled off from most immune activity. In some cases, the only way to get at these microbes is to use this hyperbaric pressure method to drive the freshly applied oxygen/ozone deep into the bones and cartilage.

Many ozone IV-using former AIDS patients owe their complete recovery to this combination method devised by retired MD, DC and ozone clinic owner Dr. James Boyce.

One of the world experts in the hyperbaric field, who has several chambers and a good deal of positive clinical data, is Dr. Richard Neubauer. His clinic is in Fort Lauderdale-by-the-Sea, in Florida.

"They Dive to Stay Alive." So said a typical article under health and cardiology in a medical bulletin on May 26, 1992. It showed a man in a hyperbaric dive chamber in a hospital in Western Australia. The hospital is putting their patients into decompression chambers, increasing the oxygen concentration to the heart, and reactivating damaged heart muscles. Gangrene, diabetic ulcers, strokes, and many other ailments are commonly positively managed through use of hyperbarics.

One of the best examples of the use of hyperbaric oxygen was the 1989 case of little Jessica McClure, the girl who gained national attention by falling into a Texas well. She was trapped for three days with her leg pinned back.

What the American people weren't told was that her injured leg that 'would have to be amputated' was saved because they immediately got her into a hyperbaric oxygen tank within two hours. The pressure pushed oxygen through the skin into the damaged and dead tissues,

Huge steel hyperbaric chamber for mass HBO treatments.

re-oxygenating them. That's the power of oxygen. No other drugs or stimulants were used, just oxygen. Not many heard the rest of the story. And no one ever bothered to inform the rest of the medical field that that's how they saved Jessica McClure's leg. Hyperbaric oxygen has been used for years in this country but basically few are familiar with it in the medical field.

One of the best uses for hyperbaric chambers is to immediately put any fall or car crash spinal trauma victim into one. Pressurized oxygen is forced past the torn tissues and into the idling spinal nerve neurons before edema can set in that would normally cut off the blood oxygen reaching and killing the nerves. Many people would have become quadriplegics (without the use of their arms and legs) if they hadn't immediately been given hyperbaric therapy right after their auto and ladder accidents. This therapy has been around a long time; thus many who are now crippled may not have had to go through that experience if the hospital administrators and doctors had stayed current, and read up on the existing hyperbaric medical literature.

New soft portable hyperbaric units roll up like a tent, as opposed to large heavy clinical units. Crash patients with spinal trauma could be kept in these portables on the way to the hospital.

The new portable chambers are easy to move, but not as roomy or durable as a full-sized clinical machine wherein patients lie down, or sit up, or congregate in pressurized oxygen atmospheres.

Oxygen and Exercise

From Nancy Adams'—WOW! O$_2$, or, Working Out With Oxygen. ©1993

"Dr. Manfred Von Ardenne wrote a wonderful book, Oxygen Multistep Therapy, based on his 25 years of research in Germany with oxygen and exercise. It has been known that the oxygen content of the blood decreases with age. When it goes below a certain point, death follows. Von Ardenne wanted to find a way to increase it. Exercise alone does not do that, and breathing oxygen alone did not do it permanently, either. He discovered that using exercise with breathing supplemental oxygen, along with certain nutrients (vitamin B-6, and magnesium and potassium aspartates) which, taken 30 minutes before exercise, increase oxygen utilization at the cellular level, did increase blood oxygen, sometimes permanently. His focus was on treating disease conditions, many of which responded favorably.

"Based on Von Ardenne's findings and from his focus on health and building it and keeping it, Dr. William Campbell Douglass, Jr., intrepid researcher himself, wrote a monograph, *Stop Aging* or *Slow the Process..* Get on your exercise bike or treadmill, and exercise for 20 minutes while breathing five liters or more per minute of oxygen in addition to the air you are breathing anyway. You can get either a canulla, that tubing with two prongs that go in the nostrils, or an oxygen mask. The simplest way is to get a concentrator. They can be rented fairly cheaply, though you will likely need a prescription. Be sure you get one that puts out five liters per minute or more for this application.

A doctor told me that they are using eight liters a minute, for 45 minutes a day with exercise, and getting results with MS in Switzerland. I haven't checked that out. I do know, from correspondence with Von Ardenne's son, also a doctor and now running the institute in Dresden, that a separate clinic is there that works with cancer patients." Check out the new super large 'Integra Hi-Flo$_2$ Pro' 10 liter per minute oxygen concentrator. It's perfect for this and any other type of oxygen exercise regimen application.

Integra 10.

Urban Oxy–Myths

> ➢ *Oxygen causes free radicals.*

<u>No</u>, not the way you're thinking of them. That's only a half-truth. There are harmful free radicals that come from things like toxins, etc., and then there are the all important beneficial 'good' free radicals.

Without free radical reactions we would die. The good ones from oxygen eat—or 'scavenge'—the bad ones. Active Oxygen Therapies work because they're good free radicals neutralizing harmful ones.

> ➢ *Can't you get too much oxygen?*

Water is good for you, right? Have you ever heard of drowning? You can drown by drinking water. An extensive trial and adjust process during the application of millions of dosages of various Oxygen Therapies stretching throughout history has already taught us what I always call the *'Safe window of effectiveness'* for each application.

> ➢ *I'm confused, if oxygen kills harmful microbes, why aren't they all dead already?*

Air is not *active* oxygen! There's a big difference between *active* oxygen (O_1) and the 21 percent or so of air that is regular stable air oxygen (O_2). 'Experts'—who do not realize that they *don't know* enough to recognize the difference— commonly get confused and say inaccurate things while assuming (or misleading) that Oxygen Therapies are only about 'air,' or 'bottled,' oxygen. Some experts consequently vaingloriously pontificate in error.

To add to the confusion, breathing bottled oxygen is also called 'Oxygen Therapy.' This book is mostly about active, or singlet oxygen delivery methods, a whole different world.

> ➢ *If the ancient air atmospheric oxygen content was really higher, the forests would have all burnt down, right?*

All those Cretaceous era plants and trees living under the ancient atmospheric mantle of extreme humidity existed within a wet greenhouse effect—ancient forests were simply too wet to burn.

> ➢ *Didn't I hear something about babies in incubators getting too much oxygen and having vision problems?*

The blindness has instead been traced to too bright artificial lighting shining into the eyes of premature babies. Too much *dry* oxygen dries out tissues; oxygen must be moistened to be absorbed well.

While we're near the thought of oxygen and babies let's peek at a critical review of the subject. Contrary to the myth that oxygen blinds babies it's actually the *lack* of oxygen that's causing lots of problems and permanent injuries in babies. Here's a summary of a recent paper.

Oxygen Starving Practices and Experiments
Distributed by: Prevent Blindness in Premature Babies

"ABSTRACT: Many premature babies with immature lungs die or suffer brain damage from <u>lack</u> of supplementary oxygen because medical schools have for decades taught to give them as little as possible of this life-saving gas. Too much oxygen is said to afflict preemies with retinopathy of prematurity (ROP), a sometimes blinding disease of their equally immature eyes. The resulting practice of breath-starving preemies at the edge of asphyxiation still kills and maims many of these each year to try and guard some among them from blindness. Moreover, there is <u>no evidence</u> that withholding oxygen works to prevent blindness.

"…the rate of **deaths from the cure** was about eight times higher than the rate of gross visual defects from the disease.

"…**16,000 extra deaths per year** in the United States from the oxygen withholding.

"…during its early and most uncompromising years, **the oxygen starvation doctrine was steadily slaying more than twice as many Americans per year than the Vietnam war ever did**.

> "No one has attempted to look further at this hidden part of the iceberg, the additional toll in physical and mental damage from the 'better-dead-than-blind' therapy." —Peter Aleff

More information available at:

PREVENT BLINDNESS IN PREMATURE BABIES

Margaret Watson, president
P.O. Box 44792
Madison, Wisconsin 53744-4792
Tel: 608-845-6500
Fax: 608-257-4143
prevent@execpc.com

Chapter 8

Oxygenated Liquids

When I speak to you of oxygenated water I am sometimes speaking about freshly ozonated water, but in this instance I am referring to the super-oxygenated bottled waters. All bottled water that passes over state lines must be ozonated for purity, but the ozone disappears in 20 minutes. Beyond that, stuffing oxygen in-between water molecules is the quickest and most effective way to bring oxygen to the masses. There are several bottled oxygenated water companies around, Oxy-Water, Oxy Up, HiOsilver, Rainfresh, Aqua Rush, and Avani, to name a few.

In 1992, I wrote: *"Good clean pure water is being stuffed with oxygen in a variety of ways, by many companies. The methods used to do this, and the resulting oxygenated liquids are very diverse, but as the technologies improve we will see the parts per million climb, and these waters will be sold everywhere, fulfilling my predictions that oxygenated products will be in every home."*

Why did I make this forecast? Again, from 1992, *"Due to the increases in pollution, and the rapid decline of available clean oxygen on our planet, and the resulting loss of oxygen from our food, our air, and our water, I firmly believe that in the near future the common man will be flocking to the marketplace to get enough oxygen for his needs. This will directly tie into the increasing understanding that diseases can't live in properly oxygenated bodies, and that our immune systems run on oxygen.*

> **Eventually, most of us will get the extra oxygen we need by drinking oxygen charged water!**

For the sake of the human race, it is incumbent upon producers to place these emerging oxygenating technologies into their production lines as quickly as possible, so that they become widely available.

Everyone will buy mass produced oxygenated water and other oxygenated beverages and oxygenated products in order to feel good again and doing so will confirm the well-founded belief that they will help to prevent disease. Oxygenated drinks are the next wave of popular beverages."

Hydration

The first time I drank a bottle of oxygenated water I felt the surprising oxy-rush and started cleaning the house and writing, because I did get a continuing burst of energy. With oxygenated water—as it is with any oxygen supplement or therapy—it's hit or miss as to who feels that 'oxygen rush.' After a few bottles, the big noticeable energy boost differences calm down and turn into slow sustained energy. These drinks will replace caffeine by supplying healthy natural long-term energy without artificial and questionable stimulants. Professional athletes, airline pilots, and long haul truckers are rushing to get these products so they can stay at the top of their game without burning out their adrenals with stimulants.

Remember, the body actually robs water from cells to process liquids such as coffee, tea, and soft drinks. We have to have lots of clean fresh water to carry our oxygen, and you see a disturbing picture developing that results in today's low cellular oxygenation levels leading to high mass disease levels.

Oxygenated Water, the Differences

People are putting oxygen into bottled water in various ways, ranging from bubbling ozone through it so they can meet legal purity standards—all the way up to many parts per million of oxygen being bonded and stabilized into the water itself.

Ozonated (ozone purified) bottled supermarket water is only 'ozonized,' and does not really qualify as 'oxygenated' or 'oxygenized' water. When the bottlers bubble ozone through their water tankers to purify the water on the way to the bottling plant, the ozone/oxygen quickly dissipates. That's great for purification, but after an hour there is no appreciable quantity of oxygen left in water that is only 'ozonized' before it ends up on the supermarket shelves. You have to be careful because legally, the supermarket waters can still declare and label themselves 'oxygenated' when all they are is purified through temporary ozonization.

How does the consumer sort all this out and be sure the water is not only purified but holds extra oxygen way above normal? Simply ask, "How many parts per million of oxygen is in the bottle *as it sits on the shelf*?" Some companies tout the level at bottling, which may be quite different than the oxygen content after warehousing and transport.

Tap Water	0-4 ppm oxygen
Bottled Water	2-7 ppm oxygen
Good Spring Water	4-7 ppm oxygen
Distilled Water	7-10 ppm oxygen
Oxygenated Water	12+ ppm oxygen
Good Oxygenated Water	30–50+ ppm oxygen

('ppm' is parts per million of oxygen in solution)

Oxygenated waters are energetic, and that's why we drink them; but that also means they will react with whatever touches them. Did you ever buy a gallon of water that came in one of those soft plastic milk jug type containers? Personally, I can't stand the plastic taste they leave in the water and I don't want to slowly fill my cells with plastic molecules. The plastic has leeched into the water, and any oxygen has slowly leaked out of the water through the gas-porous walls of the jug.

Antique Liquozone oxygen water bottles.

Note the container that your oxygenated water comes in. Glass bottles were used as water packaging around 1900 in the antique bottled oxy-waters like Liquozone, and Hydrozone. Glass is still the best packaging method available that does not affect the water's taste or quality. Running a close second to glass bottles are the 'P.E.T.' type hard plastic bottles. As stated, the soft polyethylene bottles are the worst.

> You don't need your water to be dumping more garbage into you.
> You need it clean so it can be taking the garbage out.

Average 'normal' tap water that has gone through the municipal chemical dumping so-called 'purification' process has from zero (usually) to 4 parts per million oxygen in it. Tap water is considered 'dead' water. It smells, tastes bad, is discolored, has bad vibrations, no zeta potential, no

Ozone shower with extra ozonated water hose outlet.

antioxidant properties, and no oxygen. And you're letting that liquid garbage make up two thirds of your physicality? Think it over, friends. You need drinking, cooking, and bathing water that is pure, of clean memory, and full of oxygen so you can continually flush out your toxins.

Besides the obvious 'quality of drinking water' question, do you know that you absorb your bath and shower water right through your skin? If it has chemicals in it, so do you! I filter and then ozonate my shower water.

Generally, a quality oxygenated water starts with pure spring or distilled water and has at least 30 (or beyond) parts per million of oxygen in it. Exaggerated high ppm claims are being made in the marketplace that play on the public's lack of oxy-water knowledge. Some companies use 'mathematical extrapolations' to make their high ppm claims. Everyone in the industry I spoke with said that's not the proper way to do it. And we also have to keep in mind that there is more to good water than simply the amount of oxygen in it. The source water quality, the processes used, and the final purity are very important. For example, you can add chemicals to get the oxygen level up, but the water is no longer pure.

Broadly speaking there are two ways to oxygenate bottled water. The traditional method uses refrigeration to cool the water so it can hold more oxygen, and then pressure is employed to stuff oxygen in-between the water molecules. High oxygen levels are also claimed for the use of chemical bonding, microbial action, enzymes, radio waves, magnetism, and vortexes, just to name a few. They all can be used to achieve our goal of higher parts per million oxygen kept in solution.

Ozonated water must be used immediately for oxygenation benefits, but what we look for in the milder case of oxygen escaping from bottled waters are the methods that let the oxygen stay in the water whether or not it is in a sealed (capped) container. The longer it stays in, the better the chance that most of the oxygen goes into whoever is drinking the water.

Good distilled water provides a great base to build a good oxygenated water and mineral supplement from. The whole point of drinking clean water is drinking *clean* water. Municipalities dump chlorine and fluorine into the water due to heavy lobbying efforts from chemical plants needing to sell chemicals. Ground and spring water can become polluted from pesticides and runoff. The only way to build the best oxygenated water is to start with water that is nothing but water. The best way to remove everything (meaning your total dissolved solids—'TDS') from your water is through distillation. I'll use one of the current oxy-waters called 'Oxy-Water' as our example.

Oxy-Water, the Only Steam Distilled Oxygenized Water

The oxygen content of Oxy-Water is five-to-seven times higher than that of regular bottled water and up to 10 times higher than that of some tap water. Reverse osmosis membranes and mere filters are less than perfect when it comes to removing *everything* from water. That's why the Oxy-Water people use more than just these techniques and further use steam distillation to create their base water. Not only does distillation remove everything but the water from the water, but it removes any atomic 'memories' as well. I often discuss water as possessing the proven quality of 'memory.' Filters and reverse osmosis membranes cannot erase the structural memories that the Japanese scientists are photographing in water crystals, but distillation can blank water back to its pure unrecorded and unadulterated state.

The Importance of Distilled Water

Triple distilled water is so clean that it is classified as 'ultrapure.' From the Oxy-Water people, quoting Dr. Theresa Dale, Ph.D., ND—"The distillation process removes every kind of bacteria, virus, parasite, pathogen, pesticide, herbicide, heavy metal, and inorganic material, and also prepares water to accept and retain the maximum amount of oxygen.

"When ingested, water attracts and dissolves toxins and inorganic mineral deposits stored in our joints, organs, arteries, and nerve tracts, pulling them into our blood to be carried to our excretory organs for discharge. Distilled water is an excellent natural solvent."

Note: You should know that the human cells *cannot* absorb inorganic materials (rocks) available in plain tap or spring water. These minerals have to be converted first to an organic state for human assimilation. This is achieved only after they pass through *plants, not water*.

> A recent study concluded each hemoglobin molecule in the blood requires 40 molecules of water to transport efficiently one molecule of oxygen to the cells.

It is my opinion that well-intentioned souls who tell people to stay away from distilled water because it 'leeches minerals out of the body', only tell half the story. But you do not have to choose between the options of distilled or non-distilled water. The answer is simple, drink oxygenized ultrapure distilled water and take a full spectrum of plant derived minerals in your diet or as supplements daily! That way you get the best of both.

Back to Dr. Dale —"Simply put, we need the purest water possible to carry what the body's cells need plenty of—oxygen. An inadequate amount of water can cause the hemoglobin molecule to actually *repel* oxygen. [Remember that 80 percent of all Americans are estimated to be clinically dehydrated!] Pure (distilled) water 'wets' the hemoglobin molecules, allowing them to more effectively pick up and transport oxygen throughout the body. The tiredness many people feel after eating is a result of oxygen being consumed during the digestive process, which diminishes the oxygen available to the rest of the body.

"It seems as though some questions have arisen regarding the ability of the human body to absorb oxygen in water. The oxygen in the Oxy-Water is absorbed instantly through the mucosa of the mouth just as homeopathic remedies and nutrients are. Actually, the moment the water touches the inside of the mouth (especially under the tongue), the transference of information present in any substance begins to absorb. I recommend anyone who wants better health to drink Oxy Water.

Sports

"Oxygen is critical for muscle function. Winners have lots of oxygen in them all the time. Proper oxygenation allows the body to produce and supply

aerobic ATP to the muscles, the *good* ATP, giving them strength

Oxy-Water sponsors Billy Boat Racing.

Gung Le
World Champion
Martial Arts
Instructor

and elasticity. A lack of oxygen (hypoxia) causes the body to produce the 'bad' oxygen-deficient form of ATP and lactic acid. This lactic acid reduces the efficiency of the muscles and can lead to cramps, pulls, strains, etc.

"Unfortunately the popular heavily advertised energy drinks contain ingredients that trigger the digestive juices, causing oxygen to be transferred to the digestive tract to enable digestion. This *reduces* the amount of oxygen available to the rest of the body, especially the brain and muscles.

"Other ingredients in energy drinks, such as sugars, cause a short-lived energy burst. The 'crash' which occurs after that initial surge can leave an athlete with less energy than if they had not had the drink in the first place. Couple that with oxygen loss due to digestive processing and it is understandable why most serious athletes choose only pure oxygenated water.

"Pure (distilled) water requires no digestion. The distilled water used in Oxy-Water is the absolute purest available and is completely compatible with the hydration demands of the body. The oxygen in Oxy-Water is quickly absorbed into the body from the instant it comes in contact with your mouth tissue, and continues throughout the digestive tract."

The Oxy Water Process

The Oxy-Water people start with filtration and ozonation, and then add distillation, and then refrigeration and high pressure oxygen saturation and other methods to build their water. These multiple methods are more costly and not used by most oxygenated water bottlers. However, the Oxy-Water people feel there is no substitute for quality. They then go even further and use a few more proprietary methods with their oxygenated water building process in order to keep the oxygen content high in their bottled water. Unlike the products from some other companies, all you get with Oxy-Water is water and oxygen.

Here is the summation of a study of Oxy-Water presented to The American College of Sports Medicine in June of 2001. Abstract #945, *Effects of Oxygenized Water on Percent Oxygen Saturation and Performance during Exercise*—A Jenkins, M. Moreland, T.B. Waddell, B. Fernhall, FACSM, The George Washington University, Washington, DC.

"Individuals who are highly trained may benefit from the use of oxygenized water to increase percent oxygen saturation during acute bouts of intense exercise and possibly prolong time to fatigue. Even small increases in oxygen saturation may be significant in highly trained individuals and elite performers."

Oxy-Water FAQ

From the manufacturer: *"Isn't there already oxygen in water?*—Yes, but…Oxy-Water has been able to develop a proprietary method of infusing oxygen into water without causing a molecular change. We can cause this physical change in the water without the addition of any chemicals.

"How can you prove that the molecular compound of H2O can hold a gas?—The presence of this oxygen has been supported by an independent laboratory that, prior to testing Oxy-Water, did not believe that oxygen could be trapped in water. The oxygen found in Oxy-Water is the same oxygen that you receive from medical personnel in the event of illness. We use NO chemicals to create an alleged oxygen increase.

"How can H2O be physically altered to hold more oxygen?—What some 'experts' have the tendency to over look (many times to their embarrassment) is that between the H2O molecules there are voids that can be filled with other microscopic substances. With the proper equipment and strictly controlled process, it is possible to remove the 'stuff' between the molecules and replace it with oxygen gas. Using a special technique, Oxy-Water is then able to trap that gas between the molecules so that it will be immediately bio-available to the body. This process has been called 'oxygenized' to differentiate it from 'oxygenated' or 'super oxygenated.'

"How difficult is it to put oxygen into the water?—It is NOT difficult to put oxygen into water. The difficulty is to *keep* the oxygen *in* the water so that is available to those who consume it. Oxy-Water's method of increasing oxygen in water has been found to be one of the most stable methods which can be used. This guarantees the consumer a beneficial oxygen supplement at consumption. Some companies claim high oxygen content at the point of bottling, but are unable to maintain it for any appreciable length of time. Oxy-Water has tested its product after being bottled for 16 months and found it to still contain over three times more oxygen than normal bottled water.

"Do you have independent analysis that confirms the presence of extra oxygen?—Yes, the Atomic Absorption Laboratory tested Oxy-Water and found the oxygen content at 34 ppm, over two weeks after the bottling date. They were so surprised that in a letter to us they said, "We didn't

think it was possible to increase the dissolved oxygen in water but, you proved us wrong." Since then we have improved our technology and are carrying 40-to-45 ppm of oxygen.

"How long will the oxygen stay in the bottle?—We have tested no production water that contains more the 20-to-25 ppm of oxygen after several weeks in the bottle, in spite of their claims. The excuse that their oxygen cannot be measured should be looked on with great skepticism."

Oxygenated Water, the Savior of Mankind

Just like the other Oxygen Therapies, bottled oxygenated water is proving what is possible when we re-oxygenate the body by any and every means possible. And what could be easier than drinking oxygenated water? It's real and available to you right now. Oxygenated waters, especially the better ones, are certain to dramatically improve the health and well being of millions of people over the next decade.

I already mentioned that I have a favorite dream that I have promoted for a long time. My dream is of the masses all drinking oxygenated water. I have been advocating for more than a decade that the answer to many of humanity's ills lies right here in this one idea! People don't even have to know about it to reap the benefits, they can just drink water. I'll bet that someday in the future we'll look back and consider the absolute beauty and simplicity of oxygenating all drinking water in order to oxygenate all the people as one of the most beneficial ideas of its time.

Oxygenated Water Coolers

Perfect for home, frat houses, the neighborhood gym, hospitals, igloos, or the office. O_2 Technologies has succeeded in integrating a miniature oxygen generator with the household water cooler mechanism, resulting in a super oxygenating water cooler system contained in single body. This patented system increases the dissolved oxygen level in the drinking water, which is normally 7 ppm, up to 45 ppm at the maximum—an increase of over 600 percent! As a comparison, oxygen-rich mountain streams can have up to 15 ppm.

The Millennium™ Oxygenating Water Cooler

(shown without bottle cover) Featured in the Hammacher Schlemmer catalog. From the catalogue: Bottling O_2 Water vs. Constant Saturation Coolers. "There are many brands of bottled oxygenated drinking water being sold today, creating a very fast growing market. However, dissolved oxygen in the water may dissipate out of the bottles while they sit on the shelves; some bottlers use porous bottles that cannot keep the oxygen molecules trapped for a very long time. The extra oxygen in some water, when placed in an open cup will dissipate into the air within 30 minutes."

With O_2 Technologies oxygenating cooler system, a very high level of oxygen is automatically maintained, as the oxygen generator inside the cooler keeps saturating the water with fresh oxygen. Simply use any bottled water or filtered tap water, and you can have an almost endless supply of oxygen-saturated water at a very low cost."

Oxygen Shot

New oxy-product ad: QUICK RELIEF IN THE PALM OF YOUR HAND! Oxygen Shot is a new patented product in a small spray can. Water and oxygen under pressure. On a hot day, after a workout, or if you just need a boost. Just spray a cool blast of Oxygen Shot oxygenated water spray on your face and body and you'll feel the difference, immediately. All natural, 25 percent purified sterile water and 75 percent oxygen gas. Seen in Major League dugouts during the ball game. Ahhhhh….

Ozonated Bottled Water

Surprise! The bottled oxygenated water business isn't new. Ozonated bottled water was enjoying a splendid distribution out of Chicago, New Orleans, and other cities during the first part of the twentieth century.

Antique Liquozone Bottles

Chapter 9

Hydrogen Peroxide

Have you ever been to New York City, downtown Toronto, or London, and ridden in the subway? That heavy, stale, unpleasant odor in the air that you breathed in was in fact particulate matter occasionally mixed with human waste droplets. Or what about sitting in a taxicab or on a bus or jet with somebody coughing and sneezing and huffing and puffing all over the place? Whatever comes out of them is being transferred to you, and you're breathing in all that unpleasant stuff.

With all your particulate and pollution and gas-inhaling going on, why is it that you're not going down with every disease in the world? It's because of your immune system. Your immune system works enough to get you through life and keep you around but it doesn't work so great that it prevents you from getting ill or suffering from a serious disease. This has a lot to do with whether or not you get enough oxygen.

Picture a soldier on the battlefield working around the clock each and every day, relentlessly without any rest. He would eventually collapse through exhaustion. With the constant bombardment of pollutants from every possible area of our lives, our immune systems are becoming like that soldier, they are so over-worked that it's almost impossible to keep up with the workload. But the odds can be increased. How? By flooding human bodies with oxygen, of course. At the turn of last century doctors used common hydrogen peroxide to do this because all it is, is H_2 water, plus O_1 oxygen, and our bodies naturally produce it by combining water with oxygen.

What is Hydrogen Peroxide?

Hydrogen peroxide is simply water with an active singlet oxygen attached to it. That makes it an oxidizer of microbes and toxins when the active singlet oxygen breaks off of it and it turns back into water. Peroxide is a clear, colorless liquid. Its smell is not that detectable, but it does give off a faint bleach odor. Almost velvety to the touch, it quickly becomes an irritant when used in its undiluted form.

How is Hydrogen Peroxide Made?

The chemical manufacturing method reacts barium peroxide with cold diluted sulfuric acid. Hydrogen peroxide is also manufactured by electrolytic and organic oxidation processes. The electrolytic production of a peroxydisulfate intermediate is followed by steam hydrolysis to H_2O_2, with regeneration of the original sulfuric acid or ammonium bisulfate raw materials. Bubbling ozone through water only creates trace amounts of it.

Hydrogen peroxide is a water like liquid manufactured as an aqueous solution of 35-to-90 percent H_2O_2 by weight, and essentially anhydrous H_2O_2 has become commercially available. H_2O_2 is not combustible. Water is a safe dilutant and coolant to use on it.

Hydrogen Peroxide in the Atmosphere

Hydrogen peroxide was recognized as far back as 1863 as a naturally occurring substance existing throughout our atmosphere. Due to the atmospheric saturation of widespread pollution, it now exists to a much lesser degree. Science attests that the hydrogen peroxide in the atmosphere combines with the sulfur dioxide pollution we put into the air to form sulfuric acid. This is the acid rain, a way of the atmosphere cleansing itself.

Small traces of hydrogen peroxide are also found in rain, and snow, and in the oceans. It is found in the waters of most 'healing springs,' such as in Lourdes, France. Natural water, fruits, and most of our food chain show traces of this remarkable substance.

Hydrogen Peroxide's Discovery

French chemist Louis-Jacques Thenard is credited as discovering hydrogen peroxide in the year 1818. The 'oxygenated water,' as he named it, opened up new avenues of its use. As hydrogen peroxide's curative properties were developed and enhanced it become more than a simple bleaching, disinfectant, and oxidizing agent.

Hydrogen Peroxide's Role in the Body

Talk about natural! Every cell in your body is trying to manufacture its own peroxide on a regular basis. It is part and parcel of your immune system. It's as much a part of Nature as you. If it wasn't supposed to be there, why would every cell in your body be trying to make it at a consistent rate in harmony with the whole body?

Scientists are usually trained in the drug viewpoint where they seek to poison the body that's sick in order to make it well. Huh? The people who think like this will tell you that hydrogen peroxide is a waste product of cellular metabolism; that it's just a leftover. However, Nature is a model of efficiency in its perfection. Nature would not have every cell in the body producing something if it wasn't needed or wanted, or if it wasn't supposed to be there. Why would your body be wasting your time creating the energy and the enzymes and all the other important processes to make a chemical that's waste? It doesn't make any sense. Nature makes efficient use of its resources. Here's an example.

Antibodies That Burn Water

Recently it was shown that antibodies, upon exposure to light, generate hydrogen peroxide through the conversion of photogenerated singlet oxygen. Wentworth and associates now show that this process is catalytic, and that the electron source for this reaction is most likely water—not photo-oxidizable residues such as tryptophan, metal ions, nor chloride ions. Water would be oxidized by singlet oxygen to create H_2O_3 as an initial intermediate. Crystallographic studies with xenon, a binding-site mimic, suggest that singlet oxygen may be bound near conserved Trp and Tyr residues.

Peroxide is necessary for our immune systems. Did you know that your body makes hydrogen peroxide to fight infection? It is a fact that hydrogen peroxide must be present for our immune system to function correctly. White blood cells are known as leukocytes. leukocytes are divided into three classes: granulocytes, lymphocytes and monocytes. Granulocytes are themselves divided into three classes: neutrophils, eosinophils and basophils. Neutrophils produce hydrogen peroxide as the first line of defense against toxins, parasites, bacteria, viruses and yeast.

Peroxide works well because it is always giving up O_1 when faced with four enzymes, which the body also manufactures. When our peroxide hits those four enzymes, through a complex mechanism it releases the O_1 and turns back into water. The hydrogen peroxide produced in the body is a defensive mechanism attacking the bacteria, viruses, funguses and pathogens. It's the soldier standing on duty ready to defend the body from unwanted invaders.

Now here's the problem. If our body is given the task of making hydrogen peroxide out of oxygen and water for survival, and we have less than half of the oxygen that we need, and no clean water, aren't we going to make less than half of the peroxide we need? There goes the immune system.

The peroxide is made within the white blood cells, and these white blood cells engulf bacteria. Bacteria are eaten by lymphocytes which first surround them. But how do they kill them? How do the white blood cells harmlessly eat the disease causing microbes?

Peroxisomes, the Body's Peroxide Factories

There's an area in the white blood cell known as a peroxisome. Have you heard the word 'perox' before? The peroxisomes are areas in our cells programmed to make hydrogen peroxide. The white blood cells engulf bacteria, viruses, or other unwanted invaders and squirt hydrogen peroxide on them. They simply burn up, or oxidize, the offending microorganism. That's your immune system in action. The very fact that all your cells make peroxide defines this process as your very first line of immune defense. But you have to have enough oxygen around to make it in the first place.

The peroxisomes make our peroxide continually, but normal, healthy cells are left unharmed. This is extremely important for you to grasp, for it is the reason we can put Oxygen Therapies into the body without hurting the normal cells. When our own H_2O_2 becomes water, and gives up the O_1, if that O_1 comes into contact with a normal cell it will bounce off and not be attracted to the antioxidant coating around the cell.

Bacteria, viruses, fungi, pathogens and ancient microorganisms, are the same today as if they just crawled right out of the primordial swamp. They're still the same forms of life they've been for centuries. They do not have any higher functioning. They do not and are not a higher multi-celled organism. They are not specialized. They have no antioxidant coating. So when oxygen from our immune system's natural hydrogen peroxide surrounds any one of these guys that do not have an antioxidant coating, it burns them up because they can't protect themselves. They are ill-equipped enemy soldiers without any armor.

Peroxide: Called Oxygen Water, Dioxogen, Alphozone and Hydrozone

For the past 100 years, many people have tried imbibing or orally ingesting dilute solutions of pure hydrogen peroxide. They diluted it in water and drank it. This old folk remedy that goes back to the 18th century has a remarkable ability to kill bacteria, and is unique in its simplicity. When hydrogen peroxide's effectiveness as a bactericide, virucide, and oxidizer became apparent, it was put to use medicinally to treat a whole host of ailments that afflicted the human body.

Peroxide's Past Role in Medicine

What is so special about this form of water plus oxygen? When it enters the human body, it breaks down into harmless oxygen and releases singlet oxygen! No drugs, no side effects; just oxygen and water. Our immune system is already constantly making it naturally, so in the correct percentages it is welcomed almost everywhere by the body. Because almost every cell except the lining of our blood vessels makes it, we come equipped nicely to handle it. Our bodies' naturally make their own peroxide and then use it to safely transport the oxygen intact to the parts of us that need it after transport. Once the hydrogen peroxide arrives, our bodies utilize four enzymes they manufacture to specifically break down the peroxide: catalase, glutathione peroxidase, glutathione reductase, and superoxide distmutase. The body constantly makes these four enzymes because it needs to separate the active singlet oxygen away from the water. The result is that our bodies turn out to be perfect oxygen therapists, because for centuries they have been using the resulting singlet oxygen to kill bad bacteria.

In 1888, consultant physician Dr. I. N. Love was working at the City Hospital in St. Louis. Doctor Love reported: "Oxygen-water cured cancer of the womb, scarlet fever and diphtheria in my clinic." What a statement! In 1888 a cure for cancer was announced in the American Medical Association's Journal. So if they could cure it in back then, why can't they cure it now? Could it be that no one remembers what was stated back then, so they went off in other directions? For most of today's intellectuals who obviously need to do more research, the sheer simplicity of it all just seems too good to be true, but more of that later.

Documented Findings

Love documented his findings in the March 3rd AMA Journal issue entitled —*Peroxide of Hydrogen as a Remedial Agent.*—He related success in treating many conditions that had afflicted his patients and mankind in general. Dr. Love commented: "From its very nature, this agent (oxygen water) should be a powerful antiseptic and a destroyer of microbes. Anything that accomplishes oxidation so rapidly, if it can be applied safely, must be an excellent application to purulent surfaces for its cleansing effects.

Asthma, diphtheria, hay fever, nasal catarrh, scarlet fever, tonsillitis and whooping cough were all successfully treated. The method used to administer it involved a dilute solution of the peroxide being syringed into the nostrils. This method would appear somewhat crude and almost barbaric to skilled surgeons of the 21st century, yet Dr. Love observed, "The beneficial effects of the application was apparent, all the distressing symptoms were much abated and within three or four days they had passed away."

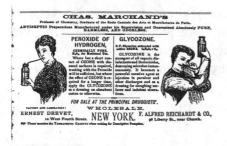

Although using it to treat conditions like diabetes and gonorrhea proved ineffective at the time, today's different active singlet oxygen delivery methods are routinely accomplishing this task. Degenerative diseases—including cancer of the womb—were being cured by a simple, inexpensive, yet extremely powerful preparation. Preparations of hydrogen peroxide used on uterine cancer caused Dr. Love to remark upon it as a "cleanser, deodorizer and stimulator of healing." What the good doctor referred to as *oxygen water* is H$_2$O (water) plus O$_1$ (active singlet oxygen). Love's oxygen-water was hydrogen peroxide, H$_2$O$_2$.

This aroused the curiosity of other doctors and they began to experiment with hydrogen peroxide's versatility. In the latter part of 1888, Dr. Georgia P. R. Cortelyou reported his clinical experiences with hydrogen peroxide. His use of a fine spray of hydrogen peroxide in treating throat and nose conditions such as cough, diphtheria, chronic laryngitis, rhinitis, sore throat, and tonsillitis were recorded as being successful for many.

The recognition of hydrogen peroxide was still scattered and in its developmental stages, yet its popularity was growing in the medical and industrial communities with enthusiasm. From 1880-to-1904, a Frenchman by the name of Charles Marchand published 18 books on the subject of hydrogen peroxide and the effectiveness of ozone.

In 1904 he published one of his most notable books. It was catalogued in the Library of Congress and bears the Surgeon General of the United States' stamp of approval. It is handsomely shown as *The Therapeutical Applications of Hydrozone and Glycozone*. Hydrozone was a solution of hydrogen peroxide. Glycozone was glycerin with ozone bubbled through it so that the glycerin would hold the ozone gas in solution. The effectiveness of these preparations became apparent, and their commercial uses were utilized widely. They were being sold in the United States in 1904 with great eagerness.

Walter Grotz made sure this book was reprinted for modern times. It is distributed by The Family Health News in Miami. This book, two thirds of which are reprints of the medical journals of the day, is still in existence from more than 100 years ago. It is strikingly effective in its detail. About two inches thick in size, it fills every single page with doctors in America stating all the disease conditions that they had completely turned around by using hydrogen peroxide on their patients. "Therapeutic Applications of Hydrozone and Glycozone," is a true old-world medical treasure.

The Word was Spreading

As word spread and interest peaked in peroxide's value, other countries became vigorously engaged in researching it. About 1924, Doctors Oliver and Murphy—British Army physicians—were running a hospital while serving in India. People were dying en masse during the big epidemic after World War I which, before its conclusion, mercilessly took the lives of 20 million people. Oliver and Murphy were working tirelessly on the sick. A plague of influenzal pneumonia was rampant, and the Indian Ghurkas soldering for Britain were getting violently ill and dying with an 80 percent death rate of those stricken. With alarming suddenness, the soldiers would develop a fever, fall over and die.

One of the soldiers was profoundly ill, and had been delirious for two days with a high fever, while hovering at the point of death with no recovery in sight. Dr. Oliver conceived the idea that maybe they should do an infusion containing dilute pure hydrogen peroxide because nothing else was working. The man the doctors had chosen had lain delirious for two days. He had to be tied in bed owing to the wild state that his delirium produced. Instead of watching this continue, Oliver's instinct led him to pronounce:

> "I wonder what would happen if I injected hydrogen peroxide into this fellow over here that is about to die?"

> "I don't know, but he does look like he is about to die, so it really doesn't matter, let's go find out." Said Murphy.

They set about injecting hydrogen peroxide into the man. They used medical grade 3 percent hydrogen peroxide (2 oz.) diluted with 8 oz. of saline solution. They infused it intravenously, very slowly over a 15 minute period. The stricken Ghurka immediately shook violently, sweated profusely, and fell into a deep sleep. In a few hours he awoke from his exhaustion and asked for food. The next day he didn't die. He simply went home. After his initial shock, the diluted hydrogen peroxide had traveled throughout his veins and into his body. While doing so, it had killed the offending viruses, fungi, bacteria, and mycoplasms, all the things that were causing his illness to worsen.

Dr. Oliver continued to treat critically ill influenza pneumonia patients by injecting the hydrogen peroxide directly into the veins. The common death rate for this infectious disease of more than 80 percent did not affect Dr. Oliver's patients. Their mortality rate was reportedly only 48 percent. Success rates all depend upon at what point you get them and what the antioxidant/nutritional picture is.

On February 21st of 1924, Drs. Oliver and Murphy wrote up their experiences battling influenza and pneumonia with peroxide. At the time, these diseases were one of the world's largest killers of people.

Interest in hydrogen peroxide's outstanding properties became more prevalent. During the 1920s peroxide's development progressed in leaps and bounds, and a remarkably brilliant man, William Frederick Koch, took interest in it. He was a most noted chemist and physician who started to look into peroxide as a treatment for cancer patients. Koch had studied under Professor Moses Gomberg. The knowledge he acquired from Gomberg was highly beneficial to an understanding of peroxide, since in 1910 Professor Gomberg had discovered and announced the nature of free radicals that we hear so much about today.

Dr. William F. Koch

Dr. Koch earned his doctorate in chemistry as a Free radical specialist, and later became a medical doctor. He was a Professor of Chemistry and Histology at the University of Michigan Medical School and Professor of Physiology at Wayne State University. No doubt attributed to his brilliance, was the fact that he came from a family of merit. His Uncle, Robert Koch was a Nobel laureate who discovered the tuberculosis germ. The head of Dow Chemical Laboratories, Dr. W. Dow, called Koch "The greatest living doctor, a modern Pasteur." Dr. Koch was further praised by Dr. A. R. Mitchell (once the chairman on the Board of Trustees) of the American Association for The Advancement of Science.

Dr. Koch went on to perfect his Koch Therapy, using a mixture he called a homeopathic dilution of Glyoxylide. Glyoxylide is believed to cause repeated cascading of the same oxygen found in hydrogen peroxide, except over the course of many months. Homeopathic dilutions of Glyoxylide were introduced into the body by

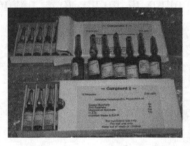

means of intramuscular injections, and following the proper dietary guidelines, over several months the Glyoxylide would scavenge all viruses and toxins from the body, de-hydrogenate them, and turn them all into 'anti-toxins.' The anti-toxins would then attack the remaining toxins and convert them into anti-toxins as well. The real Koch Therapy is rare, and with the present daily toxic load on humanity, it is harder to get the cyclic cleansing reaction to endure. The patient can't smell gasoline, perfume or get drunk, etc.

Koch explained this chemistry in several books. In clinics and field trials, his therapy was documented curing many human and animal diseases, including cancer and cattle herd diseases, and was well written up for about 30 years. Dr. Koch was murdered in Brazil after being persecuted and hounded out of our country.

The history of the medicinal use of peroxide from that period is somewhat scattered. In the early 1960s, however, major studies in the medical use of hydrogen peroxide were conducted at the Baylor University Medical Center in Texas. Research at the University showed conclusively that inter-arterial hydrogen peroxide dissolved plaque in large arteries.

Dr. Charles H. Farr

Stepping further into the 21st century, we turn our attention to yet another remarkable man, Dr. Charles H. Farr. He held doctoral degrees in both pharmacology and medicine, and was the founder of the International Academy of Bio-oxidative Medicine (IBOM). Amongst others, he held memberships in the American Medical association, the Oklahoma Academy of Science, Sigma XI, the Southwestern Section of the Society of Experimental Biology and Medicine, the New York Academy of Sciences, the American College for Advancement in Medicine, the International Academy of Preventative Medicine, the Academy of Metabiology, and the American Holistic Medical Association. Having authored more than 31 scientific and medical publications and having been awarded many honors, he stood out as a man of brilliance in the field of medicine.

Dr. Farr (now deceased) took a very different viewpoint in dealing with peroxide, advocating caution about taking it by mouth, preferring only intravenous methods. Having had the pleasure of many a conversation with Dr. Farr, I count him as another oxy-man of genius who should not be forgotten. We were once having lunch at an International Bio-oxidative Medicine Association meeting (IBOM) of international physicians in Texas. He leaned over and told me that much of the growing success of his IBOM organization was due to my always speaking of it during my lecturing. He thanked me and said, "We owe it all to you, Ed." It's heartwarming to be appreciated.

We all owe it all to him as a debt of gratitude and appreciation for the volumes of scientific papers left to us on the subject of peroxide. Physicians are advised to get the official IBOM proceedings and training, if they wish to pursue this area. IOMA has his writings, see *Resources*.

The IBOM Newsletter, Volume 1, No.2, April 1987 states: "Not only is medical peroxide important to the body's normal function, but its oxidizing ability can be used to oxidize weak old white blood cells, destroy immunocomplexes, kill bacteria, protozoa, and yeast, inhibit viruses, oxidize fatty deposits on the arterial walls, increase oxygen tension between cells, stimulate oxidative enzymes, return elasticity to the arterial walls, dilate coronary vessels, and regulate membrane transport.

Doctor panel fielding questions at IBOM annual meeting

Dr. Farr's Peroxide Experiment

In the early days of his research, Dr. Farr came across an article by Gymon. He discovered that the oxygen content of the body could be measured by using hydrogen peroxide. The peroxide was given in the artery and a probe was inserted to measure how much oxygen was being produced. Some very interesting things took place. Of the three people measured, three patterns emerged. There was no consistency to it. The arterial amount and tissue amount were inconsistent, yet sometimes they corresponded very well. The conclusion of the Gymon test was to not fool around with peroxide, just give people oxygen with a mask. Dr. Farr started thinking about this experiment in more depth. The Gymon experiments were brief and only lasted for around 35-to-40 minutes.

Dr. Farr stated, "This led me to other articles that explained that there is a substance in the body called Cytochrome-C. The Cytochrome-C mixes with hydrogen peroxide and forms a stable complex. That complex is there for about 40 minutes or longer before it starts breaking down. After about 30 minutes Cytochrome-C starts acting like catalase. What does catalase do? It breaks down hydrogen peroxide. So what happens is, after 35 or 40 minutes, you're getting some boiling action from catalase as well as Cytochrome-C breaking down the hydrogen peroxide. So, it's not such a simple thing. **You don't just take the hydrogen peroxide and split up the oxygen and water. It's a much more complex action**. The oxygen and peroxide get involved in all kinds of metabolism in the red cell alone. The old ideas were that peroxide was simple. It broke down very easily into water. If you put it into the body, it breaks down very quickly. In fact most studies show that it might break down in as easily as two seconds or less. When broken down, there is no hydrogen peroxide, you have nothing but oxygen and water."

[I personally often speak of peroxide in this slightly inaccurate simplistic way to make it easily understandable, since most people do not have the ability or time to understand complex biological reactions. The simple explanation usually given, that peroxide turns into water and oxygen, is basically true, but our inner chemistries surrounding that are incredibly complex.]

"With Cytochrome-C, we actually found that the injected peroxide lasted for 40 minutes. It came back to the heart and lungs. What happens to it there? It 'boiled' off. You have oxygen in your blood. Excess oxygen comes around to the lungs; it comes off. To get into the lungs, it comes in through the pulmonary artery into the alveoli, the little air sacs. Everyone assumes that the oxygen comes out there, that it doesn't go any farther. That's what everyone is thinking. We said, 'Something is wrong.'"

Peroxidation

First, there is the evidence that it lasts for 30 minutes; not just 12 or 2 seconds. Secondly, if that oxygen was coming off the lungs, there should be some way to measure that. What do you do when you measure oxygen consumption? You measure metabolic rate. By measuring the amount of oxygen breathed in and subtracting from this the amount breathed out, we can measure metabolic rate. You would think that if the hydrogen peroxide's oxygen was coming out into the lungs, it would be exhaled and we would have a negative amount in our equation; more coming out than going in. The opposite happened."

Here's What Dr. Farr Discovered

1. The metabolic rate was significantly increased.

2. A stronger concentration would make the metabolic rate go up quicker than the weaker ones, but they would both level out to the same rate.

3. Mitochondrial action was stimulated, even doubled.

4. Unlike oxygen's properties of contracting the vessels, dilation of the arteries of the body occurred.

5. The oxygen produced by the dilute H_2O_2 injections did leave the body through the expired air from the lungs, but as it bubbles up it is reabsorbed.

Dr. Farr continued "We're saying this is not oxygen by itself performing all these fantastic things. This is something else. This is a peroxidation. The burning up activity is doing many things in metabolism that we don't understand yet. We measured temperature because when you race an engine, it's going to produce heat. The same thing happens in the body. Your temperature goes up and stays up while you're being infused with hydrogen peroxide.

Summary of Experiments

"The first effect of hydrogen peroxide that we proved from these experiments," Dr. Farr noted, "is that oxygen (from peroxide injections) doesn't leave the body, it comes out in the lungs and goes back in. That effect has proven to be medically beneficial."

Dr. Farr Explains Hydrogen Peroxide Cleansing

"The second effect of hydrogen peroxide is that it provides singlet oxygen, which, in turn, transforms biological waste products and industrial toxins into inert substances by oxidizing them. This enables the kidneys and liver to handle them more easily. It doubles the rate of enzymatic metabolism in the mitochondria within each cell, thus *enabling the body to cleanse itself* of toxins and still have plenty of energy to handle daily functioning. This increase in metabolism probably accounts for some of the anti-bacterial, anti-fungal, and anti-viral effects of hydrogen peroxide.

The effect of singlet oxygen in the human body is two-fold.

1. It kills, or severely inhibits the growth of, anaerobic organisms, bacteria and viruses which use carbon dioxide for fuel and leave oxygen as a by-product. On contact with the anaerobic organism, this action is immediate. Anaerobic bacteria are pathogens, the organisms which cause disease.

2. The aerobic bacteria, which burn oxygen for fuel and leave carbon dioxide as a by-product as humans do, are found in the human intestine and come in the form of friendly bacteria which aid digestion. These organisms thrive in the presence of hydrogen peroxide."

Cleansing Reactions

As Dr. Farr emphasized, "Without cell death there can be no healing!"

At some point you are going to get a cleansing reaction. This is a process where the body eliminates old cells and toxic material. All you will know and feel is that you are getting a cold, or a rash, or loose stools, or a fever. The cleansing reaction means the oxygen has hit the spot where you start burning up, or oxidizing, the bacteria, the viruses, the funguses, and the pathogens.

The first temporary cleansing reaction that I experienced was three days of diarrhea. I kept taking the peroxide, waited for two more weeks; then I had a fever for two weeks. I waited another two or three weeks; then I started expectorating, large amounts of phlegm out of my lungs. As a child, I had bronchial pneumonia seven times which put me in the hospital. The debris from that was still sitting in my lungs, sitting in the bottom of the alveoli in my lung air sacs, waiting to be excreted. The body had never had enough oxygen before this to complete this task, so it just left it sitting there.

At the age of 38, I finally had active oxygen in my body, and the debris started breaking up. I could actually feel the toxic matter bubbling and being broken down. Oxygenating my body also caused a small tooth fragment that was left in the gum from an earlier extraction to surface and come out.

Another characteristic of a cleansing reaction is that your urine will have a very strong odor. Likewise, your fecal matter will have all sorts of unusual things being released in it. Given the correct tools, the body flushes out toxic matter. If you don't oxygenate slowly and therefore detox slowly, you'll have odors coming from your feet, your ears, your under arms, and at times you will smell like a skunk! It's the garbage leaving.

I've talked to some people who excitedly told me that they woke up in the morning, and found that their sheets were stained a green, blue, or yellow color. Well, if they worked in a factory for 30 years, and encountered all those chemicals, just imagine what they will have breathed in and absorbed through the skin all those years. The peroxide forced it all to resurface, and body wisdom pushed it straight out of the skin. If the body finally gets enough oxygen, the dry hardened mucopolysaccharide (mucous) coated garbage no longer has the capacity to attach itself to anything. It all gets pushed out through the nearest orifice, or directly through the skin.

Medical Applications of Hydrogen Peroxide

Dr. Farr on Peroxide Spot Injections

Hydrogen Peroxide

New Discovery, New Therapeutic Tool, New Mechanism of Action

Adapted from: *The Use of Dilute Hydrogen Peroxide to Inject Trigger Points, Sports Injuries and Inflamed Joints with a Postulated mechanism of Action*—by Charles H. Farr, MD, Ph.D February 19, 1993.

In our hands the local injection of 0.03 percent hydrogen peroxide in joints, soft tissues, nerves, tendons, ligaments, etc., has been successful in relieving pain and inflammation and has been without systemic or local side effects or other complications. Other physicians, who practice sports medicine, have reported to me rapid healing and return to normal function when injured tissue is treated with injections of hydrogen peroxide. Based on our increasing knowledge of the mechanism of healing induced by the cytokines stimulated by hydrogen peroxide, we can understand why rapid recovery should occur when treating these sports injuries.

Steroids, anesthetics, antiseptics, analgesics, sclerosing agents and anti-inflammatory substances are all used for local or infiltrating therapeutic injections. The common denominator for these agents is they are classified as drugs and are capable of being toxic to the patient. *None* of the agents used today for injection into torn, fractured, strained, sprained, bruised, inflamed, infected, damaged or otherwise messed up tissue are capable of inducing healing and repair. *None* are non-toxic or naturally occurring.

This is a brand new use of a very old product. In addition to trigger points in soft tissues, we have injected nerves, tendons, ligaments, muscles, bursa, tendon insertions and anything else we thought might benefit. We have also injected joint spaces and surrounding tissues in both osteo and rheumatoid arthritis. There have been no side effects or complications as a result of these injections.

Athletes Take Note!

Many areas which we injected had been previously treated by other physicians using local anesthetics and/or steroids with limited or no success. In most of these cases, when we used hydrogen peroxide, the pain or inflammation resolved within a week and often as quickly as 24-hours. We occasionally have someone in to whom we have to make a second injection but none, so far, more than twice. In those cases where the response was not as good as we expected, we feel it might have been our technique instead of the peroxide. If presented with the need to inject an area more than once we would have no hesitation because we have given up to 100 intravenous infusions to a single person with no adverse effect. We feel there is no practical limit to the number of local injections you could safely give to any one person provided you space them out with a few days between injections.

Most of the responses we have observed, for the relief of pain or muscle spasm, occurred within two-to-three days and were long-lasting. Sometimes relief occurred within 12-hours but occasionally it would take several days before relief was obtained. If the initial injection reduced the pain or spasticity, but did not completely relieve it, then we'd give a second or third injection. When treating inflamed joints and tissues the response is much slower. In these cases, measurable response is more often counted in weeks, instead of hours, for the inflammation to resolve.

CAUTION:—Hydrogen peroxide is a highly reactive molecule. Please **do not add anything to your peroxide solution**. If you do, one of two things may happen: (1) The peroxide may be consumed in a chemical reaction and no longer exist, or (2) The peroxide reacts with the substance you added to the mixture, creating an entirely new substance. You really do not know then what you are now using in your patient and it could be toxic.

We hope physicians who are skilled in injecting materials into joints and soft tissues and also those who practice sports medicine will explore the use of hydrogen peroxide. Since the action of hydrogen peroxide is basic to significant metabolic pathways necessary for tissue healing and repair, the discovery of its usefulness in injuries and disease is limited only by the imagination and curiosity of the investigator.—Charles H. Farr MD, Ph.D.

Infusions of Hydrogen Peroxide

The first question people have asked me concerning direct IV infusions of hydrogen peroxide is "How safe is it?" To which I reply, "There are hundreds of thousands of people who have been injected with peroxide and they don't seem to have any complaints." Doctors have been doing this for many years, but please ensure that you find a physician who is trained with the procedures if you decide to try the therapy.

How is IV hydrogen peroxide administered? Doctors use dilute 0.0375 solutions of medical grade hydrogen peroxide, which comes already bottled for them. Then 5ccs of this pharmaceutical grade 3 percent peroxide are put into 500ccs of 5 percent of water as a carrier solution, and

2 grams of magnesium chloride are added to prevent sclerosis of the vein. This mixture is then slowly infused into the circulatory system through a vein in the arm by drip. Doctors always use this very dilute, very weak solution of hydrogen peroxide, never any stronger! It is infused slowly over a period of 90 minutes. The patient sits comfortably in a big soft chair, with his arm resting on a pillow and only feels a sensation of warmth. Hydrogen peroxide has a remarkable clearing effect on the skin. After only a few intravenous treatments the skin takes on a translucent clarity usually seen only in children.

Hydrogen peroxide therapy is not an isolated therapy. It's one of the many beneficial Oxygen Therapies that people have taken. And it's very safe when used correctly. In the early days they used too high a concentration during IVs. There were horror stories that were recounted, but the practitioners finally figured out where they went wrong and made adjustments to the concentrations given.

How Many Treatments?

Your doctor will guide you on this according to your individual health requirements. On average, treatments are one-to-three times per week, occasionally five times per week for an acute illness. The number of treatments depends on the nature of the illness that you are dealing with. From 10-to-50 treatments will get the job done in most cases, and then you can revert to other Oxygen Therapies or the occasional intravenous infusion as a means of maintenance.

7,000 Hydrogen Peroxide Articles in the Available Literature

Seven-thousand medical articles provided the basis for the International Bio-oxidative Medicine Foundation to begin advocating and training physicians in the IV use of hydrogen peroxide, and also in using ozone and other Oxygen Therapies. There are five doctors in the U.K. who are currently using treatments by means of IV peroxide infusions. The success that Dr. Farr was having treating chronic fatigue syndrome was written up in the medical papers.

Look into the literature. It's all there, pre 1966 and up to the present date. IBOM has the index of the literature; more than seven-thousand articles have been published on hydrogen peroxide. The subjects are not all human. They're not all positive, but the available literature contains an amazing amount of beneficial knowledge if you take the time to read it.

Oral Ingestion

As previously explained, Dr. Ian Love was published in the AMA Journal in 1888, as successful with this form of treatment, curing 'cancer of the womb, scarlet fever and diphtheria.' How were his patients taking it besides spraying dilute solutions of it into their nostrils? They were drinking very dilute solutions of it. The oral method requires that drops of medical grade 35 percent hydrogen peroxide be added to a glass of water and ingested two-to-three times daily. It is as simple as that. The water used to dilute it should be the purest and freshest water that you can get if you wish to try this method.

For extra safety, some physicians caution against adding peroxide to water containing iron because the combination of peroxide and free iron produces a high number of free radicals which is likely to cause a stomach upset and, over a prolonged period of time, maybe even cancer. It is also recommended that iron supplements be taken only one hour after oral ingestion of the peroxide. Distilled water is recommended.

During one of my lectures one man told me that he had a very serious condition called silicosis. He decided to try the hydrogen peroxide and started with only three drops three times a day on a daily basis. Think about how small that is. With only that small dosage amount on a daily basis, over a period of months his body threw out his silicosis poisoning, which is presumed incurable. Ask any doctor, what will happen if you get silicosis.

"Well there's nothing we can do." They say.

"Will I die, doc?"

"Well, everybody else has."

And yet in a few weeks this man had coughed it all up. This man was pretty sick, and those who are very experienced in this protocol give the advice to not go faster than that, just to take a little bit everyday because if people are really sick you don't want to rush the detoxification process.

If you rush it, you're just going to have more complications and stir up more garbage and feel worse and compound the possible cleansing reactions like diarrhea, cramping, swelling, bloating or rashes that come from going too fast by giving too much at first.

> **Less active oxygen does what a lot can do, it just takes longer.**

I've had situations where somebody attends one of my lectures, listens to this information and then thinks, "Nah I don't need to do that, I'm going to start right in at 20 drops three times a day." They went home, and, glug, glug, glug, down it went, "I'm a real man." The next day they are complaining, "I feel so sick, I'm throwing up … I've got a fever, this diarrhea is killing me … I've been sitting on the toilet all day what can I do?" There is no macho when you're talking peroxide. Everybody should only go here in consultation with their competently trained physician skilled in the oxidative modalities. Use wisdom, listen to the experts. Although not as forceful, the newer slow oxygen release OxyMune and Hydroxygen supplements are far safer to begin a detoxification with.

How Long Does One Stay on An Oxygen Therapy?

People stay on their Oxygen Therapies as long as they've decided to at low levels, or safely increase the levels under an expert's care until whatever symptom or disease they wanted to address goes away. They are choosing to eliminate the problem rather than just suppress it.

Let's use an illustration of a glass of muddy water. If you let it settle, what happens? The mud sinks to one place, at the bottom of the glass. Our bodies are very much like that; the metaphorical mud, or toxins, will settle in an area and sit there in a comfort zone. But when you shake your figurative glass up, all the water in you is now combined with the mud; there is no clear water. If you are aggressively using an Oxygen Therapy it's like shaking up the mud and putting it into solution. You have to keep it flowing and wait for it to clear.

If you have ever had the opportunity to have your blood examined under a darkfield microscope you were probably surprised to see what was growing in you. Your blood, if it was like everyone else's, will be filthy. It will be brown, purple, or even black. A sick person's cells will be thick and clumped together. Parasites will be floating in the blood and eating holes into the cells. The resulting cells will be sluggish, irregular in shape and size, and the fluids they float in will be horrible looking. Once you start cleaning it out with oxygen, the blood will temporarily look *worse*. Many an uneducated therapist has incorrectly diagnosed oxygen cleansings as damage, and advises the patient to stop using oxygen. Big mistake! If they had continued past the flushing out phase the cells start looking fresh, alive and vibrant. At that point their inner environment is a good clean healthy one, so they can function in it with ease.

I've talked to people that made the mistake of going against sound advice and stopping the treatment abruptly. They related, "But I did the peroxide, (or) I did the ozone, but I've got all these new problems now." So I questioned them. "Well, tell me how you took it, and how did you stop it?" "Oh, I was at a high level, and just cut it out." they replied. The answer was simple; they didn't detoxify properly; they didn't stay with it long enough and they stirred up all their inner garbage, but when they just stopped, the stirred up solution of old toxic sludge all settled into different organs and gave rise to new symptoms. Logical, isn't it? There's no mystery to this oxygenation/detoxification idea in healing.

After the initial cleanse, it's wise to slowly reduce the dosage and then apply a maintenance program. People keep themselves clean and oxygenated this way.

For political and safety reasons, and because so many alternatives including soothing peroxide-aloe blends are now available, I no longer recommend oral dilute food grade peroxide usage. And I reserve the right to make an exception to my rule in the case of poverty stricken people facing imminent death who have no other alternatives. There is still a lot of unsafe misinformation out there surrounding peroxide's use. Because of that I must continue to discuss it.

Comparing Peroxide Strengths

For the chemists among us, Mike Davis worked out the different peroxide strengths available and how they compare as to total delivered peroxide expressed in grams. He posted this on the www.oxytherapies.com list years ago.

"Use food grade 35 percent peroxide or even purer. Injectable or pharmaceutical grade is the best to use, the 6 percent probably the next, and the 3 percent the best known here due to its availability. They are always diluted. [Supermarket 3 percent is not recommended for internal use due to impurities – stabilizers, heavy metals, phenols and quinines]. Most people can't handle more than .07-to-1 percent and will stay at that level or stop all together.

<u>3% H_2O_2 diluted to .5%</u>
.5 oz = 150 ml
Total HP = (150 ml x .005) x 1.45 g/ml = 1.09 g

<u>6% H_2O_2</u>
2 teaspoons = 10 ml
For 5 oz: 10 ml x 6% /160 ml = .375%
For 8 oz: 10 ml x 6% /250 ml = .24%
Total HP = (.06 x 10ml) x 1.45 g/ml = .87 g

<u>35% H_2O_2</u>
25 drops = 1.25 ml @ 20 drop/ml
For 5 oz: 1.25 ml x 35% /151.25 ml = .29%
For 8 oz: 1.25 ml x 35% /241.25 ml = .18%
Total HP = (1.25 ml x .35) x 1.45 g/ml = .63 g

Note: In the previously published dosage schedules, the more concentrated amounts were built up to over a period of a few weeks. However, a slower period of a couple of months may be wiser. This approach allows the body to produce more protective enzymes through the process of enzymatic induction so you will experience less collateral damage from the next round of oxidative stress (You benefit during the in-between times) and probably come out even in the end. Jumping right into the higher amounts is probably not wise."—Mike Davis

Peroxide in Poverty and Disease-Stricken Countries

The Clinical Director of Abha Light in AIDS Ground Zero, South Africa, is Didi Ananda Rucira. She wrote to me on April 19, 2001:

*"My patients live in very filthy environs (slums)-no sanitation, no water, and no electricity-with poverty-no food, no jobs. **The official statistics put one in every eight persons in Kenya infected with HIV, but in the slums this number is higher.** It is beyond comprehension what Africa will look like in five or ten year's time. To complicate things, it is usual for the hospitals, clinics, doctors and nurses to NOT tell the patients any diagnosis or anything at all. Usually the patients never know what they have or what they've been treated for.*

I see many HIV+ patients, but it's never 'about HIV'. It's always: 'I've got a cough, a headache and a bellyache. I've got diarrhea, I don't have any appetite and I'm losing weight. I feel weak.' HIV is never mentioned. TB is so prevalent here it's practically epidemic among the PLAs. PCP, a lung infection common to HIV patients, is prevalent too. The presentation to me is nearly the same as with TB-a painful, dry, hacking cough with bloody expectoration.

They are already 'under siege' and it's difficult for me to discern what a 'cleansing, healing process is' or what's a critical attack of some disease that afflicts these patients. My project runs on a shoestring, so I don't have more funds myself to buy your book. I need also more advice on how to best use H_2O_2. Could you:

    A. Assist with my needs in any way through further contacts or sending me some of the materials I need?

    B. Donate your book to my project?

Thanks, Didi A. Rucira, Director, Abha Light"

Abha Light is a non-political volunteer organization staffed by nuns who minister to the needy at the sites of disasters and emergencies worldwide. I sent her books, and set about to devise a peroxide dosage schedule for their specific client patient population. I advised her to also add a small amount of peroxide to all the drinking and washing water. I told her that individual results may vary, but experience clearly shows that her patients will get wonderful results if she simply follows correct directions, tailored for both using the locally available six percent, and the depleted state of her overly sick and poor patients.

In rich countries like America, I advocate only skin absorption of peroxide by bathing or soaking in diluted solutions of it. But let's get real, dirt poor starving people, humans, just like we are, with no resources, living on sugary cola drinks and white bread and dying from AIDS in some third world slum with no proper infrastructure, are  another story altogether. They need something cheap and available RIGHT NOW. Let's keep them alive first, and then examine if oral ingestion of diluted low strength pure peroxide has any irritation or other unwanted side effects that millions of previous applications have not produced.

Let me be clear once again. I no longer advocate oral use of peroxide for Americans with jobs, sanitation, and the best medical infrastructure in the world (even if it does need tuning). I did so back 14 years ago when there were few choices of home supplementation oxygenators available; but with so many new choices around us, it is time to move on up to modern levels of safer sophistication!

As I teach, products like Hydroxygen Plus and OxyMune or Homozon, or the sodium chlorite formulations like Dynamo are much smoother  elements to include in any oxygenation (or oxygenation pushed by reduction) program. Hydroxygen and OxyMune ALSO provide minerals, enzymes, and amino acids! For around 20 bucks you get a lot more than oxygen while slowly oxygenating; and they do not give the same disruptive strong cleansing reactions as high dosage peroxide. But for these suffering African slum dwellers, the 20 bucks you don't think twice about spending on beverages might be a few months pay. Rather than worry about perfection and ignore the problem, the following is what I advised Didi to do with her patients:

Local Food Grade 6% Peroxide for the Less Advantaged.

"Use your locally available 6 percent peroxide. Is the drinking water you will put the 6 percent peroxide in clean? If not, the peroxide will react with the water's pollutants and create smells in the water that lead to nausea. On top of all the other conditions that they have to endure, war and strife torn impoverished countries are probably facing parasites and water borne pathogens everywhere—as the municipal infrastructures crumble. I strongly advocate treating all questionable drinking water with peroxide.

"SLOWLY have them work up to 50-to-75 drops of 35 percent food grade peroxide per day, and have them stay there for three months, or until the problem goes away, then SLOWLY come back down. If you can't get food grade 35 percent easily or cheaply locally, then you will have to convert all we say over to your local 6 percent peroxide.

35% Food Grade Peroxide converted to safer 6%

1 drop of 35% = 6 drops of 6%
10 drops of 35% = 60 drops of 6%
50 drops of 35% = 300 drops of 6%
75 drops of 35% = 450 drops of 6%
100 drops of 35% = 600 drops of 6%

"Let's explore what I mean by 'SLOWLY.' Start your patients off with 2 drops of your local 6 percent in 8 oz. or so of clean water twice a day, between meals, for three days. If that tiny amount makes someone nauseous, it is probably more psychological than anything else, so you can always dilute the same amount of drops into more water until they can't taste or smell the bleachy odor. (Healthier people absorbing better and having better diets can always start higher up on the dosage schedule without any problems.)

2 drops of 6% HP in 8 oz. of clean water, 2 times a day, between meals
3 days later add one more drop = 3 drops 2 times a day
3 days later add one more drop = 4 drops 2 times a day
3 days later add one more drop = 5 drops 2 times a day
3 days later add one more drop = 6 drops 2 times a day
3 days later add one more drop = 7 drops 2 times a day
3 days later add one more drop = 8 drops 2 times a day
3 days later add one more drop = 9 drops 2 times a day
3 days later add one more drop = 10 drops 2 times a day
3 days later add one more drop = 11 drops 2 times a day
3 days later add one more drop = 12 drops 2 times a day
3 days later add one more drop = 13 drops 2 times a day
3 days later add one more drop = 14 drops 2 times a day
3 days later add one more drop = 15 drops 2 times a day
You are now at 30 drops of 6% per day = 5 drops of 35%

3 days later reduce the number of drops
But now give them: 11 drops 3 times a day
3 days later add one more drop = 12 drops 3 times a day
3 days later add one more drop = 13 drops 3 times a day
3 days later add one more drop = 14 drops 3 times a day
3 days later add one more drop = 15 drops 3 times a day
3 days later add one more drop = 16 drops 3 times a day
3 days later add one more drop = 17 drops 3 times a day
3 days later add one more drop = 18 drops 3 times a day
3 days later add one more drop = 19 drops 3 times a day
3 days later add one more drop, = 20 drops 3 times a day
You are now at 60 drops of 6% per day = 10 drops of 35%

This is the 'maintenance' amount one slowly drops back to and stays on once the cleanses, the rashes, fevers, wet sinuses, loose stools, etc. eventually stop.

3 days later reduce the number of drops
But apply them 4 times a day = 16 drops 4 times a day
3 days later add one more drop = 17 drops 4 times a day
3 days later add one more drop = 18 drops 4 times a day
3 days later add one more drop = 19 drops 4 times a day
3 days later add one more drop = 20 drops 4 times a day
3 days later add one more drop = 21 drops 4 times a day
3 days later add one more drop = 22 drops 4 times a day
3 days later add one more drop = 23 drops 4 times a day
3 days later add one more drop = 24 drops 4 times a day
3 days later add one more drop = 25 drops 4 times a day
3 days later add one more drop = 26 drops 4 times a day
3 days later add one more drop = 27 drops 4 times a day
3 days later add one more drop = 28 drops 4 times a day
3 days later add one more drop = 29 drops 4 times a day
3 days later add one more drop = 30 drops 4 times a day
You are now at 120 drops of 6% per day = 20 drops of 35%

This has taken 114 days, or 16.28 weeks, or 3.8 months to do. Depending on how the patients feel, or how brave they are, and how patient they are, they can either stay at this dosage or keep ascending the staircase and:

3 days later reduce the amount to 21 drops of 6%
But apply it 6 times a day: = 126 drops per day
3 days later add one more drop = 22 drops 6 times a day
3 days later add one more drop = 23 drops 6 times a day
3 days later add one more drop = 24 drops 6 times a day, etc.

"Then every three days continue increasing by 1 drop per dosage, until you reach 50 drops of 6 percent, six times a day, or 300 drops per day, which is the equivalent of 50 drops of 35 percent per day.

"That level is fine for a few months with an antioxidant rich dark green and yellow vegetable diet, but it is probably not advisable to have someone live at this dose continually. It is better to treat for six months, reduce and rest at maintenance levels, 60 drops of 6 percent spread over the day (put in water supplies) then treat again, as and if needed.

Considering your patient population, and the lack of antioxidants in the local food, you probably do not want anyone going above the level of 300 drops per day of 6 percent. By increasing the dosages, walking up the staircase with this extremely slow method, you will give the greatest blend of efficacy and patient comfort.

"As the patient hits a cleansing reaction, the equivalent of a plateau, or a 'landing' on the dosage staircase, they can sit on the landing, 'rest' while continuing at that one dosage, or cut back to a lower dosage for a while. When they feel good enough, they can return to their ascent.

Do Not Stop Abruptly!

"Stopping abruptly causes whole new problems as all the toxins and bugs stop flowing out and settle into (plug) new areas. Only stop by weaning them off, by reducing dosages very slowly. Just remember, as a rule of thumb, at any point on the stairs, at the very start of any uncomfortable cleansing reactions, cut the dosage in half, and stay there till the reactions stop, and then continue up again."

And what results did they start seeing in their volunteer clinic after I gave them my Mr. Oxygen's™ *Low Dose Foreign World Oxy-Solution*?

September 20, 2001

"Ed,

Did I thank you verrrrrrrrrrrrrry much for the recommended adjusted dosages for the weak. I've started giving out a lot more H_2O_2 and giving it in your suggested way. I won't have a reliable response for some time, as it will take 3+ months to get dosages 'up to speed'... I found a food-grade source here in Kenya, so that problem is solved. Till later, then, Didi."

January 14, 2002
(Four months later)

"The Abha Light health project in Nairobi Kenya uses mainly homeopathy as its therapy to treat HIV/AIDS patients. However, (peroxide) H_2O_2 is a

mainstay as part of the treatment. In the past eight months we have introduced H_2O_2 to over 50 HIV+ patients. The results have been mixed, but overall, have been very excellent. Some patients have literally been *dragged back from the brink of death* from the use of H_2O_2.

"At least 30 of the patients report a remission of many of their problems. In all cases, patients have expressed an increase in stamina and overall well-being. Of course, the homeopathic and naturopathic treatments have played their part in the healing process. But it is clear from our experience that H_2O_2 has had a *tremendous impact* on the health of our patients."

Sincerely, Didi Ananda Rucira Director
Abha Light visit: http://home.pacific.net.sg/~rucira/alf

Wait until they take it for a year! Why did she say, "The results have been mixed," but still praise the results? If you read the first part of the book you'll remember they all have to live on Coke and white bread—definitely not enough antioxidants or clean water in the diet! You can't be eating only overly processed white-floured bread and white-sugared carbonated drinks and expect your body to be able to fully uptake and utilize the oxygen and heal immediately. Oxygen will go as far as it can; but nutrition and water are key factors in its uptake.

> "Some patients have literally been <u>dragged back from the brink of death</u> from the use of H_2O_2."-Didi Ananda Rucira Director, Abha Light Clinic (Nairobi slums)

I was just sitting here working on this section, and it hit me in a moment of astonishment and amazement. Imagine something from my mind, typed into a keyboard and sent via email through telephone wires that actually manifested itself on the other side of the world as a tin walled clinic in a slum becoming a community based peroxide distribution point. And as the oxygen spreads out into the community people come back to life! It blows my mind!

Peroxide VS. Ozone

Others and I are of the opinion that history shows IV, RHP, and oral Ozone Therapies are far more effective and safer than peroxide. The reason stems from the chemistry of each. Peroxide contains hydrogen, which has the property of burning, and ozone is pure harmless cooling oxygen, so there is less risk with no hydrogen. Both methods of oxygenating are beneficial, but I would advise doctors to learn about and try ozone first. Peroxide bathing in dilute solutions is the safest way to go.

Bathing in Hydrogen Peroxide

Although I do not recommend oral use of peroxide, I have no problem with bathing in it or spraying diluted solutions of it on the body to be absorbed directly through the skin. Bathing in a tub full of water and hydrogen peroxide is a highly effective, less controversial method of getting extra oxygen into the body. This method has been around for a long time, and it avoids the possible irritation that I experienced from

taking too many high dosages orally during my initial peroxide experiments years ago. The heat of the bathtub water opens the pores in the skin and enables the capillary bed of blood vessels below the skin to absorb the peroxide.

For bathing in peroxide, the general rule of thumb is to pour one pint of 35 percent food grade hydrogen peroxide into your bath. I have personally pushed the envelope by experimentally adding two gallons

All of Hospital Santa Monica's colonic, pool, bathing, and drinking water is oxygenated from 55 gallon drums of peroxide. Courtesy Dr. Kurt Donsbach.

of 35 percent peroxide diluted in a bathtub full of water. I tingled all over, and there sure were a lot of bubbles coming off of me!

People usually soak in a warm tub full of water and some peroxide from between 20-to-45 minutes depending on tolerance and the condition. Peroxide and water bathing is often recommended as an adjunct to other therapies, or as part of a regime including ozone.

How often do people bathe in water diluted peroxide? Depending on the condition, the general rule is one-to-three times a week. In severe cases it is used daily. Others have reported to me that they use it every other day with excellent results.

Caution! Strong Peroxide Can Cause White-'Burns'

Direct application of dilute hydrogen peroxide to the skin by spray bottle is another method that is employed by many. It will however, temporarily 'white-burn' the skin if the concentrations are too strong. It will oxidize any pieces of dead skin because the dead cells of your calluses, on your palms, for example, will turn white and sting temporarily if you spill undiluted 35 percent peroxide on them. The antidote is to flush the area with lots of water and/or vinegar and just wait until the discomfort subsides. Some people get scared if this suddenly happens to them, but usually the stinging goes away soon without further harm.

But the interesting thing about hydrogen peroxide is, since it's made in your body, it's natural to your body. It has little effect upon normal healthy cells. If you take a drop of the strong 35 percent food grade peroxide and put it on the back of your hand, it usually doesn't burn. It will 'white-burn' only on the palm, but not on the back of the hand unless the hand is dirty and there is a lot of dead skin on it. The palm is full of dead skin, calluses, and bacteria. What does peroxide do when it sees bacteria or dead cells? It oxidizes or eats them, as we discussed earlier, and leaves normal healthy cells alone. So peroxide can safely be diluted and used on the skin. The temporary whitening or white-burning and stinging of skin when accidentally exposed to more concentrated H_2O_2 is due to trapped oxygen beneath the epidermis, caused as a result of H_2O_2 permeating through the skin and being decomposed by epidural enzymes.

The Effects of H_2O_2 Bathing

During the course of my research, I discovered that Dr. Boyce, who ran one of the HIV clinics that I visited, had sero-converted (turned HIV negative) 118 people. One of the methods he used along with other oxygenating methods was to take a gallon of 35 percent peroxide, put it into a bathtub full of water and have his patients soak in it for 45-to-60 minutes every day. If anyone knows about peroxide, a gallon of 35 percent is potent stuff. I mean, you are going to know about it. Your skin is going to tickle, and it's going to bubble all over your body.

One man that I interviewed said he took two pints of 35 percent peroxide, threw it in the bath, jumped in, and subsequently had a green grunge come out of his pores which floated to the surface of the water.

On one of my Australian tours, a man stood up during one of the lectures and proudly said. "Look at my hair," "It used to be gray, before that it was black, and before that it was gray." "Wait, what are you talking about?" I asked. "I bathe in dilute hydrogen peroxide, and my hair turned from gray back to black after a couple of months of bathing in peroxide." Then I stopped, and it turned gray again. So I got back in the tub and it turned black again."

Remarkable Inconvenient Results

I have seen so many remarkable results with hydrogen peroxide. It is surely worthy of your consideration as an oxygenating treatment. Many people have said to me, "Well you know it gets inconvenient taking the peroxide, all that soaking in the bathtub and feeling nauseous at times." So I tell them disease is also inconvenient, and so is death.

Dilute Peroxide – Enemas and Colonics

Suggestions for Safe Use of Hydrogen Peroxide (H_2O_2) in Home Enemas © 1993, Nancy Adams, revised 2001 (Used with permission.)

"In a program of cleansing, hydrogen peroxide can be a very useful tool. In the proper amounts, it has been observed to accelerate the rate of cleansing, and aids in the ecology of the colon by introducing oxygen into the environment of the colon, thereby helping to encourage the friendly flora, what I call the 'Good Guys'—acidophilus, bifidus, and others. They love oxygen, and discourage the anaerobic (putrefactive) bacteria, which cause so much trouble, and thrive in a low oxygen setting. Most disease-producing microbes are anaerobic.

"It is extremely important not to use too much! There is what Ed McCabe rightly calls a *Window of Safe Effectiveness* with this God-given substance. Less, and you get no results; more, and you could burn yourself. This paper sets forth my opinion based on 10 years' experience as a colon therapist as to what may be experienced as safe amounts. You may find it desirable to read basic cleansing information by others and me. Peroxide is not the whole story in cleansing! It is only a useful adjunct.

"I recommend starting VERY GRADUALLY on all matters pertaining to building health. You didn't get in this condition overnight, and it isn't going to go away that fast either. If you are very sick, you should be under the care of a competent health care practitioner. Please make sure your doctor is aware that you intend to take responsibility for your health and what measures you are taking and planning to take for health building. As you cleanse and share your results, if you run up against resistance and closed-mindedness, your concept of what constitutes 'competence' may change.

"All the amounts given here are using 3 percent hydrogen peroxide. To dilute from 35-to-3 percent, mix 1 part of 35 percent hydrogen peroxide with 11 parts of distilled water. This will actually give you 2.916 percent, not 3 percent, but it's close enough. Measurements are approximate, of course.

"Into a 2-quart enema bag, place the water FIRST. The following chart is a guideline. Amounts are assumed to be added to 2-quarts of water. Stir well. Distilled water should be used with H_2O_2.

Gradually Increasing Dosage Schedule

Square One:

1 teaspoon   3% per 2 qt water = ±   78 ppm
1 ½ teaspoon 3% per 2 qt water = ± 117 ppm
2 teaspoon   3% per 2 qt water = ± 156 ppm
2 ½ teaspoon 3% per 2 qt water = ± 195 ppm
3 teaspoon (1 Tablespoon)   3% per 2 qt water = ±   234 ppm
1 Tablespoon + ½ teaspoon   3% per 2 qt water = ± 273 ppm
1 Tablespoon + 1 teaspoon   3% per 2 qt water = ± 313 ppm
1 Tablespoon + 1½ teaspoon 3% per 2 qt water = ± 352 ppm

(Note: ± is read 'plus or minus', and ppm means parts per million)

"I suggest that you do not exceed this amount. *Parts per million* may seem tame—it isn't, believe me.

"Suggestions for use: Start every cleanse with plain water enemas for at least one or two days. Then start at *Square One*. Use distilled water if possible. Use that concentration for two-to-three days, and then increase to the next level no faster than every second or third day, working up slowly. I have found that doing this allows the body to adjust more easily. There is no need to go all the way up to 350 ppm, if you are comfortable with less than the maximum amount I have given. It's your body, and your program. These are just suggestions.

EXTEMELY IMPORTANT!

NEVER USE STRAIGHT 3% PEROXIDE FOR AN ENEMA!

"You may notice cramping. This may occur in the first few days of any cleanse! It seems to be a little stronger with the peroxide added. I find that my own experience and that of clients—as Ed McCabe has pointed out—waste must be bound with oxygen to leave the body. When you introduce more oxygen, waste products which may have been 'waiting in line' so to speak, due to oxygen debt within the body, seem to want to leave all at once. This is why I say start slowly.

"**A deep breath** —After spending a day in a place full of smoke, a peroxide enema seems to help clear the lungs. Why is that? It seems odd, doesn't it? Until you remember that both the colon and the lungs are, along with kidneys and skin, major organs of elimination and they are all working on the same project. As waste leaves the body and oxygen comes in, you may find your breathing feels lighter. When this happened to me, I said, oh, my gosh, I thought I was breathing before! It is a delightful feeling. One of my clients has asthma; she says that peroxide in her enema is the only thing that keeps her breathing freely, other than her 'pump.'

"**Feelings of joy and contentment** —You may notice this in the later stages of a cleanse—after about five-to-six days of daily enemas or colonics with peroxide. This is a direct result of more oxygen in the body! It is allowing the waste to bind with it and leave, and after that's in process, you have some as a reserve.

"**More mental clarity** —Comes with better, more efficient oxygenation of tissues and of the blood going to the brain (which, by the way, comprises 60 percent of cardiac output). Will last longer as your cleanse progresses. See other sections of this book on increasing oxygenation by other means as well.

"**Nausea** —If you experience a little nausea, it is likely due to an overloaded liver. The water going through the colon wall enters the portal circulation to the liver and stirs things up. This reaction could happen without peroxide too. Proceed at a comfortable pace for you. Lots of nausea may mean too much peroxide. Give it a rest for a couple of days, and then go back to *Square One* and start over. As you cleanse, things will probably get easier. If you have been eating a mostly cooked-food diet, you have more work to do to cleanse.

"**IMPORTANT NOTES**: Please understand that 35 percent hydrogen peroxide is a powerful oxidizer. If spilled on the skin, it can cause a burn. Use extreme care in handling 35 percent food grade reagent hydrogen peroxide. If you spill it on your skin, wash immediately with lots of water. Food grade peroxide contains no stabilizers and breaks down over time releasing oxygen gas, which builds up pressure and may burst the container. Keep it properly labeled and out of the reach of children in the refrigerator or freezer to prolong storage and NEVER IN GLASS BOTTLES!

"Why do I suggest distilled water? I was told by a doctor at the 1991 conference of International BioOxidative Medicine Foundation that H_2O_2 forms hydroxyl radicals with transitional metals which may be in tap water. These are bad guys in the body! Again, please keep in mind that this is a powerful technology. Use it wisely! You wouldn't pour a whole box of soap in with your laundry—you use the least you can to get the job done right, and you measure it. This is the same idea. There are thousands of published reports on the medical uses of this God-Given substance, yet it is not a drug, just a simple, effective, and, when used correctly, a safe natural method of introducing more oxygen to the body. I wish you well with it.

Rectal Drip with Peroxide

"Once I went and paid $80 for an IV drip, about 250 ml (that's just over one measuring cup) of a dilute solution of peroxide. I guess it was 350 ppm. I wanted to see if the effect was the same as using a drip to the colon, since both ways it ends up in the blood. I felt wonderful for about three days. About two weeks later I did a rectal drip, not having done any cleansing in between. Guess what? I felt wonderful for about three days. You can try this if you like. Again I caution you about strength. Remember Ed McCabe's reported *Window of Safe Effectiveness* for using any Oxygen Therapy culled from history. I offer a kit with a small bag and a drip tube so you can control the flow, along with a specially designed rectal tip that won't fall out."

Inner Radiance Colonics and Nancy Adams assume no responsibility for your results either positive or negative. This writing is offered as education and information to prevent using too much.

It is important to note that many people have made mistakes using hydrogen peroxide internally and externally. For example, every time someone comes out with the safe drinkable version of it, mixing it with aloe vera, the FDA pulls it off the market. As long as they deny people safe access to it, I have to go into details about what's generally acceptable and what's safe and not safe about using it.

Personal Hygiene and Peroxide

Spraying the body with a solution of hydrogen peroxide is a must for the serious peroxide user. The strength can vary according to need. Most people use 3 percent but a 17.5 percent is also popular. Get out of the bathtub and take a solution of 6, 12, or 17.5 percent, and spray it on your body and leave it to dry. The ready mixed solution can be stored in the icebox in a sprayer. If you want to go a different route and use peroxide to apply to your skin in more of a time release fashion, the folks at Aura cosmetics have it all worked out for you through professionals.

Advanced Oxygen Skin Care Products

"AURA Research invites you to feel and see the difference in your skin with your first application. Experience more clarity, smoothness, and luminosity. Exclusive oxygen emulsion formulas are designed to give your skin optimum assistance in maintaining natural beauty through time. *Oxygen is essential to our lives . . . and the life of our skin.*

"And now, Aura has scientifically harnessed the essence of life for your skin in pure, non-irritating, easy-to-use products to solve all of your skin care concerns.

"What makes Aura Research products unique? Oxygen, both inside and out, is basic for beautiful, healthy looking skin. Aura Research has developed an exclusive emulsion system that brings and binds oxygen to your skin. Your skin has maximum opportunity to use the revitalizing oxygen!

"Working with cosmetic chemists in an FDA registered laboratory, the hydrogen peroxide emulsion has been refined and reformulated. Various ingredients with known benefits to the skin were added to the emulsion.

"The hydrogen peroxide (H_2O_2) in the emulsion is applied to the skin where catalase reacts turning the peroxide into water and oxygen. The natural oils in the emulsion do not let the oxygen escape into the air and the only direction the oxygen can travel is into the spaces between the cells. This intercellular space is where the cells are nourished . . . and through Aura Research technology they now can receive maximum, life assisting oxygen.

"Why does young skin look so...*young?* Well, in part because it simply works better. As we age, the skin cells, especially in the face, retain less water, receive less oxygen, get less nourishment, and become less efficient. Hence, older skin looks *older*—lined, dry, and wrinkled blotchy, pallid." Aura skin care products are only available through skin care specialists and physicians. See resources.

How else can peroxide be safely used?

Deodorant

Peroxide is a useful solution as a deodorant. Spray a dilute solution of peroxide directly onto the armpits to decrease perspiration and maintain freshness.

Detoxifying Bath

Add 6 oz. of 35 percent peroxide to ½ tub of water, ½ cup of sea salt, ½ cup of baking soda, or Epsom salts and soak for a minimum of 30 minutes.

Douche

A 1.5 percent dilution of peroxide (3 percent cut in half) is recommended by doctors for a vaginal douche. Dilute with warm water and with a ratio of 20:1 in order to use it this way.

Facial

Use a 3 percent solution on a cotton ball as a facial freshener after washing or steaming the face.

Feet

Add ½-to-1 oz. of 35 percent peroxide to one gallon of water in a bucket and soak. For athlete's foot, soak feet nightly until condition abates. Those who suffer from foot odor have related how successful this simple routine has been to resolve the problem.

Mouthwash

Use 3 percent peroxide in a glass of water. Aloe Vera or chlorophyll may be added to flavor. Rinse mouth with solution a number of times.

Showers

Keep a spray bottle of 3 percent peroxide in the shower Spray your body after washing to replace the acid mantle on your skin that the soap removes.

Teeth

You can either use baking soda adding enough 3 percent peroxide to make a paste, or dip your toothbrush in undiluted 3 percent peroxide and brush. Hydrogen peroxide is good to sterilize your toothbrush. Saliva is a source for H_2O_2 decomposing enzymes, hence the foaming when brushing with H_2O_2 containing tooth paste. Oprah, the number one TV talk show queen has a magazine that runs ads for 'Rembrandt Plus' tooth paste with peroxide in it.

Contraindications

Never get peroxide in your eyes because it might give you a cataract. It doesn't mix well with the cells on the eye. With every other place on the body it doesn't bother you at all, if it's used correctly, and that means in dilution. Don't spray hydrogen peroxide on your hair because it will bleach it, but it can be sprayed on your body. It also bleaches towels and carpets, so use an old towel and avoid spraying it near clothing or furnishings.

There are also newer formulations claiming to be 'powdered 35 percent peroxide' sold in dry powder forms. For these products, when you want peroxide, you 'just add water.'

The best way to sum up our personal and medical peroxide section is to remember; the next time you are breathing recirculated common air while riding in a subway or plane, or while you are under a droplet deluge from some person sneezing in your direction, Nature has already provided a defense for these challenges. Nature's immune system produced peroxide is already inside you and always ready and waiting to step in and give you a helping hand oxidizing the invaders. Assuming, of course, that you have enough clean water and oxygen in you so that you can have a healthy immune system that manufactures enough peroxide internally.

Science is just starting to discover what I have been teaching. Not only does your immune system produce peroxide —which has been well documented, but your antibodies are producing ozone and ozone subspecies as well!

Reference: "Ozone is Produced by Antibodies During Bacterial Killing" The Scripps Research Institute, La Jolla, California November 14, 2002.

Chapter 10

Commercial and Domestic Peroxide Use

Hydrogen peroxide comes in various concentrations or strengths. For healing purposes, people only purchase and use food grade hydrogen peroxide. The peroxide sold in the drugstores in the U.S. is three percent peroxide, (the rest is mostly water) and in some countries you can get six percent. In the U.S. our three percent has additives—heavy metals, phenols, and stabilizers in it that are used to keep it from expanding and blowing up the bottle on the shelf because it's so active. That's why the supermarket and drugstore peroxide is never used internally, because they have to put inedible stabilizers in it. People in the U.S. wanted to replace

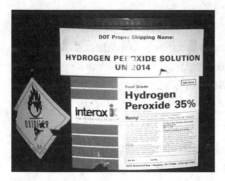

the missing oxygen in their bodies by drinking extremely dilute solutions of peroxide, and spray it on food, etc., yet they didn't have a pure source of it because of all the additives. The solution was to obtain food grade hydrogen peroxide. Food grade peroxide is readily available, and is safely sold through mail order in six percent solutions by companies like The Family Health News.

Grades of Hydrogen Peroxide

3% Drug/grocery grade. It also contains 'stabilizers' like phenol, acetanilid, sodium stanate, and tetrasodium phosphate.

6% Beautician grade. Stabilizers unknown.

30% Electronic/technical grade. Stabilizers unknown.

35% Food grade, also 50% food grade. Used in food packaging.

90% Rocket fuel.

99.6% Pure experimental grade.

> **To make a 3% food grade peroxide solution,**
> **mix 1oz. of 35% to 11ozs. of water.**
> **Only use distilled water if possible.**

The Many Uses of Peroxide

Hydrogen peroxide is widely used in industry and the many household items that the average consumer purchases will have at one stage been processed with it. Hydrogen peroxide applications include commercial bleaching, dye oxidation, manufacture of organic chemicals and peroxy chemicals, and power generation. Bleaching outlets consume more than one half of the H_2O_2 produced. These outlets include textile mill bleaching of practically all wool and cellulose fibers, as well as major quantities of synthetics, and paper and pulp mill bleaching of ground wood and chemical pulps.

Storing Peroxide

Because it's so reactive, peroxide will react with any container you put it in and release its stored oxygen. Use fresh peroxide, and cold store it clearly labeled so no one mistakes the clear liquid for water, and always wear gloves when handling strong peroxide. Storing it correctly is the best way to assure that one knows the strength is still in the bottle. Contact with metal or any other reactive substance or impurities and/or heating does increase the rate at which peroxide 'dismutates' or breaks back down into water by releasing its oxygen. The rate of breakdown depends upon the strength of the concentration and the type of metal it comes in contact with. Here's an approximate range of reactions and situations.

To keep peroxide pure and inert, using plastic containers, keep it in the refrigerator on a daily basis and for long term use keep it in the freezer. This will extend the life of the peroxide best. Normal, cool, clean, bottled 35 percent peroxide supposedly loses one half of a percentage of its strength each month. The concentration would drop from storing it in a hot environment or in a contaminated bottle.

I used to bathe in Johnny T's (the father of the Family Health News) oxy-spa, which was a hot tub that had solar collectors hooked up to it that was full of peroxide. When we turned on the circulator so that the peroxide water would go through the pumps and up through the metal in the solar collectors to be heated, the peroxide would disappear in a few days. This was twice as fast as if we had just left it in the tub.

The extreme example of metal breaking down peroxide would be the rocket fuel for the Nazi buzz bombs that rained upon England in World War II. The buzz bomb propulsion unit was a simple 90 percent plus peroxide solution sprayed upon a catalytic screen mesh of pure silver. The resulting metal and peroxide reaction was so violent, liberating the oxygen from the peroxide, that it propelled the rocket. There was no pollution since the by-product was only water and oxygen. Optimized, this method of propulsion could clean up our air.

Aseptic Packaging and Peroxide

Hydrogen peroxide is also used for the aseptic packaging system. It's not a food. It is used to protect food. Among other things, it's used for cheese, eggs, whey, and wine. Being used in the food packaging industry, they spray it inside long shelf life containers and in those little square boxes of drink that don't require refrigeration. At that point, they put it in the fruit juice or the milk and seal it up hermetically. The peroxide would then kill all the bacteria, the viruses and the funguses. It does it in the box; then you can put the box on the shelf because there is no way it can spoil. There are no microbes alive in that fluid. This grade is from the food packaging industry. There is a wine called blush wine which is more common in the United States. A bit like rosé wine, blush is made with hydrogen peroxide.

Hydrogen Peroxide Around the Home

The use of food grade hydrogen peroxide has many applications. If you can successfully use it on your body, why not also put it to good use around your home?

Bathroom

Keep a spray bottle of 3 percent undiluted hydrogen peroxide to wash surfaces, the toilet bowl and bath.

Dishwasher

Place 2 oz. of 3 percent peroxide in the dishwasher with your regular formula.

Humidifiers and Steamers

Mix one pint of 3 percent peroxide in one gallon of water

Kitchen

A bottle of undiluted 3 percent peroxide can be used to clean kitchen surfaces and appliances. It can be put to good use inside the refrigerator, to disinfect the area and thus help prevent contamination to food.

Laundry

Replace bleach by adding 8 oz. of 3 percent peroxide to your wash.

Hydrogen Peroxide and Food

Marinade

Place meat, fish or poultry in a casserole (avoid using aluminum pans.) Cover with 3 percent peroxide. Cover and place in refrigerator for 30 minutes. Rinse well and cook.

Sprouting Seeds

Add 1 oz. of 3 percent peroxide to one pint of water and soak the seeds overnight. Add the same amount of peroxide each time you rinse seeds.

Vegetables

Add ¼ cup of 3 percent peroxide to a full sink of cold water. Thick-skinned vegetables like cucumbers should be soaked for 30 minutes whereas light skinned vegetables should be soaked for 20 minutes. Drain, dry and refrigerate. This helps prolongs life and freshness. If time is of the essence, spray vegetables and fruits with a solution of 3 percent peroxide. Allow to stand for a few minutes, rinse well and dry.

Hydrogen Peroxide in the Garden

Insecticide Spray

Mix 8 oz. of white sugar with 4-to-8 oz. of 3 percent peroxide in one gallon of water.

Plants

Place 1 oz. of 3 percent peroxide in one quart of water, or add 16 drops of 35 percent to one quart of water. Feed plants or spray mist them and watch them flourish.

Hydrogen Peroxide and Animals

For small animals like dogs and cats, many slowly over time work up to using 1 oz. of 3 percent peroxide to one quart of water.

Dolphins don't get cancer. Starfish don't get cancer. Octopi don't get cancer. In fact, none of the creatures of the sea (except those living in polluted water) get cancer. Sharks can live in polluted water and still are cancer resistant. The most common denominator is that all these creatures swim in seawater, which is rich in H_2O_2 and minerals.

Animal Husbandry, Agriculture, and Hydrogen Peroxide

I've interviewed farmers and approached commercial growing operations all over the United States and wherever I went to where they oxygenated the water, the animals, the livestock, the horses, the pigs, the sheep, and the goats—they're all perfectly healthy. Their veterinary bills consequently dropped to virtually nothing.

I've talked to commercial growers who take their pigs, sheep, horses, and goats down to the station for slaughter. The agricultural inspectors look at the organs, exclaiming, "I haven't seen an organ that clean in 30 years." I video taped an interview with a farmer about a conversation he had while bringing his oxygenated pigs and cows to market.

> "I don't believe this!" The meat inspectors exclaimed.
> "What do you mean?" The farmer asked.
> "Well, I've never seen a bacterial plate count so low, with so few bacteria. And these organs, they're pink, they're like young animals. How old is this cow?"
> "Five or six years."
> "You're kidding." Said the inspectors, "What are you doing?"

They had given oxygen to their stock in all their drinking water, and the inspectors were completely incredulous. Why? Most livestock organs are not clean and healthy anymore, so when faced with a lone farm presenting clean animals and organs, the inspectors think there is some sort of 'trick' going on because they're so clean! Normalcy is questioned when the average sinks low enough.

Likewise, organic farmers using peroxide or ozone in their water have gone down to their local agricultural station and amazingly had their milk rejected. "Why are you rejecting my milk?" "Because that's impossible!" the inspector shouted. "No one has a bacterial plate count that low." Once again we see, as in other areas, what we now routinely consider average is actually sub-normal.

Bacteria in milk (from low oxygen cows), is now considered every day 'normal' in the marketplace. If you use a modicum of intelligence and oxygenate your livestock you can get them to a normal, healthy state again and you can cut your vet bills way down and get much more yield, but until you train your local inspector you might be questioned at first. Obviously, by harmlessly oxygenating the cows and the milk—along with the processing and bottling plant wash water—you also don't need to destroy the wholesome natural enzymes we need back in our diets by super heating dairy products with pasteurization.

Manure Odors Removed by Peroxide

I used to live outside of Louisville, Kentucky ('Lou a vulll' to the initiated) and sometime I would pass by the stockyards downtown. The stench was overpowering, and I was just driving past it. But the *National Hog Farmer* magazine on December 15th, 2000 in the article *Pit Additives Reduce Hydrogen* reported that University of Minnesota researchers were testing the odor removal ability of several chemicals. They found that peroxide agitated into the manure brought the H2S rotten egg smell of hydrogen sulfide gas from off the scale readings of over 250 ppm down to less than 1 ppm in only 30 minutes! It was also the most cost effective solution coming in at a cost of $0.074/cubic meter of manure.

Crops and Invisible Pesticides

How about the crops? Our crops and lawns and their pesticide and insecticide runoffs, and therefore our water supplies, are all sprayed and laced with tiny deadly mutagenic pesticides. There is a much larger problem going on here than anyone suspects. Our species is getting weaker with each generation, and the occurrence and seriousness of undiagnosed sterility and birth defects are rapidly increasing throughout our society.

Here's why this epidemic of undetected pesticide toxin proliferation is going on right under the noses of our best doctors. Despite his other problems, during my interviews with Lucas Boeve I found his philosophy on environmental toxins simple and accurate. His findings were a combination of disciplines gleaned from what he learned while watching water being purified by ozone with the teachings by the German doctors successfully curing diseases with ozone:

"*The proliferation of the chemical industry's pesticides, herbicides, and other harmful chemicals is rampant. The molecular weight of these pesticides is so low that they can pass right through our skin pores, and right through the pores of our stomachs or intestines and get right into our blood stream and DNA. When these low weight molecular organics or chemicals, get into our blood stream **they can't get out again**. The liver and kidneys are such that items of such a low molecular weight **can't be filtered out**. So how do we get rid of these toxins, how do we get them out of our blood stream? If you have allergies, chances are that they are* *caused by one of these low weight molecular organics. **They won't show up on a blood test.***"

Even if you feel healthy, this stuff is silently collecting in your tissues. Unless you *oxygenate* and *clean out* now you will pay for it later. The bottled drinking water industry routinely oxidizes these low weight organic pesticides away with ozone or peroxide. The chemistry and scientific principles are exactly the same, so why should the watery fluids in our bodies be any different?

Played Out Soils

Let's look at the soils that grow our food. All our food is grown on used over and over again demineralized soils. Fertilizers are used in typical American 1950s plastic society thinking to artificially make the plants grow fast–but without any substance. Our soils are overworked by the farmers within the super-mega-agricultural business to such a degree that they no longer have much of a nutritive value left to go into creating the foods that are delivered to your table.

Industry has sprouted a glut of factories dumping their toxic waste into the water. We now have filthy smoke coming out of the factories combining with and severely compromising the oxygen in the air. We have apples and other produce being sprayed with pesticides creating all kinds of toxic problems in us. Our health picture starts to look a little bleak with the cumulative effects during the passage of time. Is it any wonder that we're not really healthy and natural anymore? Fresh minerals, and clean, sweet smelling oxygen are supposed to go into our plants, but they are being bound up and used up long before we can get them.

Commercial Growers and Plants

One huge commercial grower operation that I videotaped, The Jolly Farmer in East Lempster, New Hampshire is way ahead of the curve. They give 65 ppm peroxide to their seedlings during their entire existence.

It is supposed to be highly unusual, but all their tiny seedlings arrive on the other side of the country after being in the back of trucks for days without perishing! Their competitors and the county agents all assume there is no way to prevent these losses and suffer the economic drain. These same peroxide watering methods also got rid of all the greenhouse diseases that are, according to same county agricultural agents, officially 'incurable.' All the other growers suffer the bottom line damaging diseases, but The Jolly Farmer doesn't. Why? In Nature, the plants are supposed to be drinking peroxide in the rain and ground water. The Jolly Farmer organization is simply supplying the missing natural ingredients. The plants get robust.

And there is another benefit, sort of the icing on the cake. Since they inject the dilute peroxide into all of their water, their barnyard animals are fed with the same water and healthily growing right along with the plants. Just like we humans, animals and plants are oxygen-based life forms that have evolved in an oxygen-based ecosystem, and they're trying to exist on one-half of the oxygen they were 'designed' to run on. If you give them the missing oxygen, they suddenly spring into life and grow and heal like Nature intended.

From Dr. Boyce On Growing Oxygenated Tomatoes:

"We planted 3,000 individual plants in the bags. To protect the plants from insects we used 8 oz. of 35 percent food grade hydrogen peroxide to 16 gallons of ozonated water. Clean water was drawn from a deep artesian well and then ozonated. The bees loved it, and the bugs and caterpillars hated it. It caused the production of alcohol and the caterpillars would eat it off the leaf and die

"We sprayed the plants using 8 oz. of Sonic Bloom to 32 gallons of ozonated water. Dan Carlson also made an electronic bird chirper that turned on by an electric eye sensor when the sun came up and remained on for three hours. Then it reset itself for a repeat performance the next morning. The electronic bird caused the stomata, microscopic mouths beneath the leaves, to open wider (allowing for greater air foliation) and increased photosynthesis. Thus you get a better tomato. The tomatoes that we produced had more bricks. Bricks are a measurement of the carbohydrate level. They contained more meat and less seeds than the store bought tomatoes. The hydrogen peroxide not only acted as a natural insecticide but gave longer shelf life to the tomato. The tomato skins were firmer but not tough. The tomato had lots of meat with a small amount of juice. The hydrogen peroxide helped produce a uniform tomato and the stems and vines were strong. The plants grew to a height of 12-to-16 feet and produced astounding 30-to-40 tomatoes per plant. We then took the suckers off these plants and grew their offspring. Thus proving ontogeny produced phylogeny. The only genetic manipulation that was done was by cross pollination from the activity of the bees.

"We did not ozonate the feeding water. We only used ozone as an insecticide and with Sonic Bloom. All of our production went to the French Market in New Orleans. It was contracted for sale by one produce distributor who would take it to the local restaurants."

Chapter 11

Oxygenating Liquid Supplements

Beyond simple natural peroxide are commercial blends of various oxygenating compounds and liquids. Here is a review of the most notable in the field. They all have qualities and efficiencies that make each one suitable for different tasks. We'll start with the liquids.

Boracil

Boracil is a hard to find rare blend of zinc and borates. It was sold around Chicago as a dry crystal concentrate that could be reconstituted (Just add clean water) to make an oxygenated pink liquid, and it could be used orally and topically for just about everything peroxide was used for.

My sample of Boracil came as a pink liquid in a one-gallon jug. Boracil was being promoted by an elderly sightless doctor in the Chicago area whom I remember well because he used to walk into my Chicago land lectures and lean on his cane while shouting, "Where's McCabe?" Unfortunately, I lost my sample of it, and Dr. Kenneth Rowell has passed away. He told me that he knew Dr. Koch back in the days when Dr. Koch was a professor of chemistry at Wayne State University.

Oxy Dan and Oxy Max

These two products are manufactured and sold exclusively by The Family Health News in Miami.

1. **OXY DAN** is a stabilized form of oxygen solution based upon the excellent Electrozone/Chlorozone technology I have spoken of that is pH balanced, and chlorine and sodium free.

2. **OXY MAX** is the concentrated form of Oxy Dan with the addition of 24ppm of msp (milled silver protein, an organic—not ionic—form of colloidal silver.) I've often said that colloidal silver's ability to draw oxygen is a perfect adjunct for any Oxy-Therapy, so this is a great combination. People are reporting that Oxy Max is knocking out colds and flu's in as little as 24-hours. It is an improvement over peroxide. *The Family Health News* is the exclusive distributor for both of these products. They also sell Colosan, a magnesium peroxide based bowel cleanser.

Stabilized Electrolytes of Oxygen, Stabilized Oxygen Sodium Chlorite, Chlorine Dioxide Solutions

Stabilized Electrolytes of Oxygen, or SEO solutions, mimic peroxide in that both have one millivolt of charge; but peroxide is negatively charged, and SEO solutions are positively charged. Also known as Stabilized Chlorine Dioxide, these purifying agents were developed back in the 1930s. In use on humans for 50 years without any problems, they are also known as 'Stabilized Sodium Chlorite,' or by the misnomer, 'Stabilized

Oxygen.' These formulas are all proven bactericides, virucides, and fungicides. Not only do they purify water but professional world class athletes use them regularly, and double blind studies have concluded they greatly assist human and animal organisms to fight disease.

Common brand names: DynamO$_2$, Genesis 1000, Aquagen, Halox, Aerobic 07, Aerox, Aerobic Oxygen, Oxyfresh, ASO, and NaClo. New manufacturers and private branders of original formulations are springing up daily, so watch for lots of new names that have appeared.

They are essentially a formulation mixing a solution of mildly buffered sodium chlorite (4-to-6 percent volume) with deionized water (94-to-96 percent volume). Steve Kraus's Aquagen is reportedly 15 percent sodium chlorite, and is being marketed by the famous designer Vidal Sassoon's wife Beverly and son Elan. Interesting oxy-history note: The Sasoons were taught all about Oxy-Therapy by my old friend Richard Murray who used to have dinner with them.

He brought them my first book and taught them all about oxygen years ago. These products are weakly buffered high alkaline pH, but unlike highly buffered drain cleaners or other strong alkaline solutions, they immediately loose their alkalinity upon contact with any substance that is of a lower pH. Bacteria, viruses, the acid mantle of human skin, and the hydrochloric acid in our stomachs all react with them to immediately render their alkalinity harmless to humans.

They are generally colorless and odorless, but a drop rubbed between the fingertips will drop the pH and release the characteristic chlorine dioxide odor. Many people mistake this odor for the smell of bleach, but it isn't. If the wrong analytical tests are performed on these solutions they will give a false positive for the presence of household bleach. Oxidative chemistry is an exacting science, and only specialists in the field are knowledgeable enough to select the correct test procedures to analyze these substances, $NaClO_2$ and $NaClO_3$.

Sold in concentrate form and always diluted before use, they are called 'Stabilized Oxygen' by most people, although this is technically a misnomer, since it implies that somehow oxygen has been turned liquid, stable, and bottled up, which it isn't. These formulations work by releasing some molecular oxygen, chlorite ions, and trace amounts of sodium and chlorine dioxide into the blood plasma. All this happens when the ingested diluted stabilized oxygen hits the stomach acid, becomes highly unstable and breaks down immediately. Once the ClO negative charged species are in the body, they create natural enzyme/chlorite combinations that supply the immune and energy production enzymes with some basic chemicals they need to function. Modern living has robbed us of an adequate supply of these chemicals.

The newest version of these compounds is DynamO2. Its inventor Dr. Abe Chaplan states that its formulation is better than the 50 year old technology used by the other brands. DynamO2 is *double buffered,* and that "Most are buffered only to survive the acid environment of the stomach." This new formulation of DynamO2 is double-buffered with zweitter ions to also release oxygen into the alkaline environment of the small and large intestine. Each buffering system allows the incorporation of significant amounts of oxygen bearing electrolytes; therefore the manufacturer states that they can "almost double the amount of oxygen bearing electrolytes they can put into the system and be assured it will reach the colon and produce oxygen in its environment."

Why the manufacturer claims Dynamo2 is different: "Regular stabilized electrolytes of oxygen products are still not optimum because they release all of their oxygen gas in the stomach. It took the 'top guns' of the biochemical world at the Milhauser Labs at New York University to figure out how to get oxygen release where the anaerobic invaders actually live —in the alkaline duodenum and small intestine, and in the colon (which may be acid or alkaline depending on the diet).

"Getting a single solution of dissolved mineral salts to release oxygen gas continuously starting from the stomach through to the colon wasn't easy. It was like getting a glass to hold both very hot and very cold water at once (rather than just turning the mixture into tepid water). But the biochemists were able to solve the problem, and more! Since each of the two buffering systems carries its own quantum of minerals, the *double buffered* solution has nearly twice amount of oxygen dissolved into the same amount of liquid!

"Dynamo2 is the only *double buffered* stabilized electrolytes of oxygen product on the market today. It starts out with more oxygen than other stabilized oxygen products can carry, and is the only one that releases oxygen from the stomach onward for maximum benefits to the entire body."

SEO Marketing

New Zealand 'Chemists' (Pharmacists) and grocery stores all sell lots of SEO.

The SEO products are not sold as drugs, but as general adaptogens, dietary supplements, or water purifiers. These colorless liquids are usually sold as small pocket size bottles but occasionally they are purchased in case lots or in bulk containers. They have been successfully used by thousands of international consumers and businesses over the past 50 years. Currently they are sold in the U.S. by mail order and in health food stores. They are commonly seen in all of the leading health magazines and trade shows.

In New Zealand, the Genesis SEO product is sold in almost every chemist's shoppe (pharmacy) and health food store in the nation, as well

World's <u>first</u> oxy-TV commercial appearing on New Zealand TV.

as in the 'natural' section of large department store mass retailing chains. Bruce Smeaton, the originating ex-owner of the Genesis 1000 brand, along with his ex-business partner John, holds the singular distinction of having the first Oxygen Therapy in history to make its way into the mass consciousness via television and radio advertising. The TV slogan used was notable: "Oxygen, the Spark of Life!"

SEOs are marketed internationally for use on humans and animals as alternatives to antibiotics and drug therapy, first-aid antiseptic, germicidal face wash and cleanser against various skin disorders, throat gargle, douche, mouth wash for dental problems, vegetable and fruit wash, water treatment, disinfectant and sterilizer, and all purpose germicidal cleanser. One of the main benefits touted is the ability of SEOs to "increase the dissolved oxygen content in the body fluids."—per quotes from common sales literature. SEOs are sold as meat, fish, and vegetable washes since decay is caused by bacteria, and pesticide residue lingers on vegetables. The SEOs can oxidize these problems away. In the same manner they are used as food preservatives.

Colonic therapists are routinely advised by researchers to add 30 drops of their favorite SEO to a quart of distilled water for use as a colon rinse and douche.

SEO Safety

When I went to New Zealand to lecture tour, I studied a whole nation that only had one Oxy-Therapy, the SEO called Genesis 1000. All I saw there

were a lot of happy people returned to health from using it for years. Just as with the other Oxygen Therapies, barns full of Chooks (New Zealand for chickens) were no longer getting common barnyard diseases and ended up weighing more when they got to market. Race horses were being cured of supposedly 'incurable' equine encephalitis, stroke victims were returned to health, old dying dogs were brought back to playful puppyhood life, and Olympic Athletes were all improving their personal best times using Genesis 1000 successfully without harm.

NZ Trainer Jack Stewart cured champion Vanatua's 'incurable' encephalitis with Genesis 1000.

A debate over the long-term safety of using these products in high dosages has emerged from the naturopathic community. The first tenant of naturopathy is 'First, do thy patient no harm.' Because the SEOs are essentially reducers instead of oxidizers, some contend that the SEOs should be used only temporarily, or as water purifiers, since much better oxygenating alternatives without the 50 year old chlorine technology already exist. Essentially the message is that critics prefer to primarily energize the oxidizing side of our body chemistry, not the reducing side.

It is also claimed that long-term chlorine absorption can 'electroplate' any fluid contaminating THMs (trihalomethanes) to the walls of our blood vessels at our 98 degree F. body temperatures, and thereby contribute to hardening of the arteries and other vascular problems. The manufacturers of the SEOs state that their products have no—and emit no—free chlorine. I have personally seen no harm from their use as long as the label directions were followed. Thousands of people in New Zealand were on them for years and it was all good.

Quick Carry Oxygen Generating Home Multi-Supplements

Unlike other Oxygen Therapies, OxyMune and Hydroxygen are similar in that they both slowly generate oxygen out of your own internal pool of water. AND, unlike other Oxygen Therapies they ALSO give you ingredients needed for repair, such as ionic minerals, enzymes, and amino acids. Quite a lot in a small easy-to-carry inexpensive package! In this class of mild oxygenating system balancers I now consider only these two brands worthy of mention. I rate the proper use of these supplements at the top end of your home supplement choice list. These formulations are designed for everybody, for those who don't have any problems and want to stay that way, and also for those at the other end of the scale who are extremely ill and need what they offer as part of a larger program that includes serious Ozone Therapy. I like them so much I have helped both companies with their marketing so that as many as possible can hear about them.

Hydroxygen Plus

Why It Is Essential to Oxygenate and Restore Your Missing Oxygen, Minerals, Amino Acids, and Enzymes.

Hydroxygen supplies humans and animals with basic building blocks necessary to life, good health, and repair. Its action is seriously enhanced by its micro ionic delivery system readily delivering the metabolic building blocks to every cell. Because of these benefits, I have never seen so many people respond so quickly and as positively as they do to this wonderful supplement.

Many people are full of unnecessary pain and look much older before their time, all of which reflects their inner imbalances. In my opinion many people, and even animals, are diseased unnecessarily. From the interviews that I have conducted, I would say that Hydroxygen Plus is a powerful free radical scavenger that helps your cells 'breathe' better while rapidly detoxifying them at the deepest levels.

Most people spend all day within their unnatural environments. Concrete, plastic, steel and unnatural materials surround us, and most of us are breathing unnatural polluted air. And for the sake of convenience, we too often eat Life Force and mineral depleted 'dead' processed food. How can we *fix* the health problems we are creating by living this way?

All living things need fresh, energetic re-supplies of the natural elements that their bodies are made of. We must put back into our bodies the natural elements that God and Nature expected us to be re-supplying ourselves with. This is a wonderful, simple, natural product that delivers ALL of the major necessary building blocks for healthy life, and it's as simple as putting a few drops of a pleasant tasting liquid in your drinks.

I have interviewed and witnessed testimonials from many, many people and doctors who are using the Hydroxygen Plus natural oxygen, hydrogen, sulfur, mineral, enzyme, amino acid and vitamin supplement, and their living healthy truth is very impressive!

The Secrets of Hydroxygen Plus

Look at your body. What are your hands, or your arms, or your feet made of? Minerals, pure, natural minerals that amino acids and enzymes have shaped into 'you.' If you want to repair, you need these elements to do it with, but unfortunately, until now, most supplements have never been absorbed completely.

Hydroxygen Plus not only contains all of these elements but precisely delivers these building block foods to the cells in the exact form and size our bodies require for immediate absorption. That's why Hydroxygen Plus works fast. Hydroxygen Plus delivers energized ions into our cells after linking them to micro-sized mineral carriers. Our bodies can immediately put these tiny, powerful natural food elements to work as needed.

We are all two-thirds water that should be constantly oxygenated with fresh active oxygen. Hydroxygen Plus delivers oxygen into our bodies by actually using our internal body water as the generating source of the oxygen, causing oxygen to be time-released over and over from inside us for up to at least 12-hours! This is a more natural oxygen delivery system, an improvement on other Oxygen Therapies!

Other brands of non-oxygenating products may claim they are delivering electrons or Life Forces to the body, but only Hydroxygen Plus [Note, and now OxyMune as well.] contains the energizing electrons, ions, AND also the necessary oxygen, hydrogen, sulfur, amino acids, enzymes and minerals that make it all come together all in just ONE bottle, and affordably!

It's easy for you to receive all of these benefits by putting a few pleasant tasting drops of Hydroxygen Plus in your everyday liquids on a consistent basis. You'll be amazed at the rapid cleansing and rebuilding processes your body can accomplish when it is consistently given the proper ingredients. The body completely reconstitutes itself every 11 months. A continual three times a day addition of Hydroxygen Plus to your diet may result in an entirely new, more energized and oxygenated body! It sounds simple, but Hydroxygen is strong. The continual long-term application of this proprietary blend of structured natural secrets is good stuff.

I wrote the following Hydroxygen and cancer article for Well Being Journal 2000's Cancer Special Magazine.

Hydroxygen Plus and Cancer

Dallas TV 11 had to repeat the Francis Guido story because there was so much demand for it by viewers. Francis loves to garden. One day she no longer felt good. Pancreatic cancer had struck. Blood work and CAT scans every four months confirmed its continuing growth for a year and a half. Finally her doctor told her there was "nothing more they could do," and sent her home to die. She had no energy, couldn't get out of bed, or brush her teeth. Food would stick in her throat. She prayed over her dilemma, and the name of a local naturopath popped into her mind. By this time her throat was so closed off she had stopped drinking water. Dr. Revis helped her to drink an eight ounce glass of water with drops of the supplement 'Hydroxygen Plus' in it. Over a month, she worked up to 72 drops per day. She had slowly started eating and drinking again. When she returned to her medical doctors they scanned all of her organs and bones, from head to toe, and couldn't find any trace of cancer! A year and one half later she looks so much younger, alive and vital that no one recognizes her at the supermarket. She reports: "I'm not dying! The Hydroxygen works, but you have to *want* to live, and continue taking it."

Like Francis, your natural physical building blocks need constant re-supply. Regaining control of your life and health starts with deciding to re-supply the missing basic elements; oxygen, minerals, hydrogen, amino acids, and enzymes, and not by chasing health fads. For example: Your body and cells are made of minerals. But our food is grown on farmed-out mineral depleted soils! If your cells mutate into cancer cells your body tries to repair or replace them, but how can your body repair mutated cancerous cells without each of the necessary missing minerals?

We can actually measure the natural increase of someone's oxygen saturation 30 minutes after Hydroxygen drops are taken. The 'Plus' in 'Hydroxygen Plus' indicates the delivery of oxygen and hydrogen, PLUS minerals, electrons, amino acids, and enzymes, without charging extra like other companies. I think it's a good buy. Having all of this in one product is revolutionary.

The manufacturer's website also has a user's testimonial section, check the experiences out at http://www.globalhealthtrax.org/19753 Let's take the case of Joe Ritter, age 66, from Fallbrook, California. Throat cancer and surgery combined with 40 radiation treatments left him burnt and scarred inside. He heard about and started taking the Hydroxygen supplement, and when I interviewed him he was working out at the gym four hours a day, and considering himself way ahead of others who had the same surgery and radiation a year ago. He has the blood pressure of a 20 year old, the lowest his has ever been.

For a variety of reasons not everyone gets the same results with any supplement; but it's hard to pass by another of my interviews, this one with William Lee. He's 69, he loves fishing, hunting, yard work, and working with flowers in South Dakota. He had four heart attacks, and after a stressful period his cancer which he had six years ago flared up in his colon, stomach, blood, and prostate. His naturopath had guided him, working him up to taking high dosages of the Hydroxygen several times a day for the past seven months. He had the usual classic cleansing reactions as the body finally obtained the building block raw materials needed to purge itself of unwanted diseased tissue. He reported lumps of foul smelling substances that looked like 'rotten hamburger' flushing out of his colon; his urine was strong and dark, and he had occasional headaches. As he said, "The Hydroxygen is doing its job!" His doctor reports he is free from the cancer, his heart is improving tremendously, and all previous heart pains have disappeared. William concludes: "It's a fantastic product."

No one makes any promises about supplements, but since it's an all-natural product I don't see any harm in supplementing whatever you are already doing with it. Please consult with your practitioner who can, if needed, talk it over with the manufacturer's chemists. I urge you to take advantage of all this modern knowledge and experience coming into focus.

Hydroxygen - Clinical AIDS Test
Johannesburg, South Africa

What better place is there to stringently test a product than in Africa? Dr. Priscilla Rowen is a Johannesburg Africa healer running a naturopathic practice specializing in Bio-oxidative medicine, i.e. ozone, and IV hydrogen peroxide. My oxy-friend Roland Fritz in Johannesburg found her for me when I said I was looking to put a manufacturer's product to the hard core test in Africa, AIDS ground zero. I wrote her, and here are her replies.

"I would be interested in doing a trial on your suggested Hydroxygen Plus. I am a member of the International Bio-oxidative Medicine Association started by Dr. Charles Farr, and I was in the USA in 1999 to attend their congress. Cost is a problem among my patients, and I am always looking for an affordable oral treatment.

"I have suddenly been flooded with persons who I believe would make good subjects, reliable etc." [I love synchronicity!]

I explained to the President Everett Hale, and the Vice President Lorin Dyrr, at Global Health Trax that I found a clinic serving low income AIDS sufferers in Africa. Most of their patients could barely afford to see a doctor, and it was a perfect place to do some good deeds and collect some front line data on Hydroxygen's performance 'under fire' using label dosages. Global Health Trax was kind enough to donate a few boxes of their product to our clinic project, and here are the first results:

"While the world is reeling with the World Trade Center disaster of 11/9/2001, 30 million people dying in Africa have paled to insignificance. Our biggest problem in South Africa is the diet of the urban black; it is impossible to get results when they have a diet of white bread, white maize meal and coca cola! Fortunately some folk are out there showing them how to grow their own veggies, and I believe they will be the ones to survive.

*"My AIDS patients are doing **exceptionally well, they are all working, and very positive that they will remain well**. My biggest problem in compiling data for you is that they cannot afford the testing. However, I can tell you about one lady from Harare, Dumi Hanyani. Her PCR (Test*

for the AIDS virus) **has dropped from 87,000 last Nov to 25,000 in July this year (eight months.)**

"Please, could I have some more Hydroxygen? I only have six bottles left; I currently have 10 patients to whom I give two bottles per month. Please keep in touch and thank you for your support on behalf of myself and my patients, Regards, Priscilla [Dr. Priscilla Rowan]"

In January of 2002 she wrote again: *"This report is dealing with patients who have received Hydroxygen and returned for follow up during the period Jan 2001 and Jan 2002. Some patients have been monitored during the whole period, while others have only joined the program for four or five months.*

"Patients not included in this report are those who consulted with me, and did not return for follow-up, and consequently I have no idea of their

progress although they were given Hydroxygen to start treatment and they totaled eight patients." [This is ALWAYS the problem in trying to collect meaningful data without having millions of dollars. Individual results will always vary, but here is what happened.]

Patient 1—Mrs. Patricia Kaliki

"Age 35 years, known AIDS patient, Ill since 1996. Her second child died at age five months of pneumonia related to AIDS in 1996. I first saw the patient in 1998, and treated her with Colloidal Silver and an oxygen supplement. She also had ozone intravenously on two occasions. [Two ozone shots only clean up a little bit. They just start the process.] She had erratic therapy, and financial problems were for the most part the reason she did not return frequently. She returned in August 2001 with her husband who was moribund with AIDS and severe weight loss (wasting syndrome), diarrhea, and gross dehydration. I attempted to treat him but he died a week later. At this stage Mrs. Kaliki and her eight year old daughter were started on Hydroxygen. To date [five months later] they are both well and doing exceptionally well on the therapy. Due to financial constraints, no formal testing has been performed on either patient.

"Mrs Kaliki had a large open wound over the sacrum, which is slowly healing. P.S. I will send photos of the patients when they can visit my rooms; they are currently visiting relatives in Zimbabwe."

Patient 2—M/s Prudence Tsotesi

*"She came to see me in August 2001 after being refused insurance for an educational policy she wanted to take out for her three year old son. She started on Hydroxygen and an herbal preparation, which she took regularly. In Dec, 2001 [four months later] I performed blood tests on her at Lancet Laboratories. Her CD4 was 1170, and her CD8 941, **and her PCR (AIDS virus) count was <u>undetectable</u>.***

"We have an interesting legal problem at the moment because the insurance company refuses to release her original blood results. The question arises was she positive initially. Certainly when I first saw her she was tired, loosing weight, and complained of breathlessness. By December she was well, and had just received a promotion at work as a result of good performance."

Patient 3—Ephriam Mhalakasa

*"Age 36 years. He had been sick since 1991. When I first saw him in June of 2000, he was a **full-blown AIDS patient** with a **PCR of 84,000**. I started him on electromagnetic resonance therapy, Colloidal Silver and nutrients. He responded reasonably well. But in January, 2001 he began with Hydroxygen. He works as a gardener of one of my patients and she collects his medication monthly without fail. The result has been that he has gone from being bedridden to where he now works five and a half days per week in people's gardens doing manual labor eight hours per day.*

"He has had no infections in the last year; the only blood tests we have on him are as follows: 1/8/00. Total T-cell cd3 count 1381. CD4 204. PCR 39,000." [His viral load has therefore gone down by more than half since starting the Hydroxygen.]

Patient 4—Mr. Albert Gil

*"Age 30 years. This patient was referred to me by a colleague in Cape Town, known **AIDS with full blown symptoms**. He was treated aggressively with multiple nutrients, cleansing, and intravenous hydrogen peroxide. Initial bloods were as follows: CD4 89. PCR 50,000. He started Hydroxygen in March, 2001 and his progress has been good. **He is clinically well and back at his job** but I am confused by his follow up bloods; they vary with a PCR of 'undetectable' to 140,000 viral load. I have spent many telephone calls to the lab. Questioning the validity of their testing to no avail. I find it difficult to persuade a patient who is*

clinically well to spend R1,400 (Rands) ($123) doing tests that are so inconsistent." [Has politics replaced truth at the lab?]

Patient 5—M/s Dumi Hanyani

*"Age 42 years. She came to me in November 2000 and her original test showed that the severe shingles and bout of malaria she had recently suffered was exacerbated by her AIDS status. **The original viral load was 87,800 and has now dropped to 25,400** in August 2001.* [Only 29 percent of what it was.]

"The patient has transformed from a sickly person unable to cope with rearing two children to a dynamic person who was about to take on the world, and would like to tell the world that AIDS is treatable. She is due to return from Zimbabwe in the next few weeks and I am sure her follow-up bloods will be near perfect."

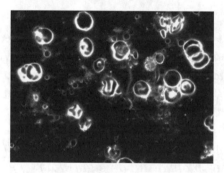

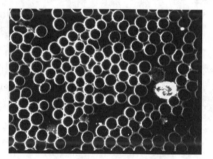

Darkfield blood sample from severely ill African AIDS patient showing cloudy plasma and degenerative red blood cells.

Same patient, but healthy blood after six months of steady treatment with Hydroxygen Plus in Africa.

Please note: These products are designed primarily as system balancers through cleansing and supplementation of missing nutrients only. I am not claiming or intending to claim these products or any of the other products in this book are going to kill viruses or cure any serious diseases like AIDS.

The fact that people got great results with them is wonderful, but it remains my professional opinion that any serious viral illness like AIDS always needs to be attended by an ozone-trained physician providing at least direct IV ozone or better. These products are always used *along with* ozone in serious cases. Same with special diets. You have to ask yourself, "How do I harmlessly *kill* the viruses — with vegetables?" I don't think so.

OxyMune - the New Kid on the Block

OxyMune is sold wholesale and retail out of New York and Hydroxygen is sold through multi-level network marketing from California. The OxyMune formula has a secret ingredient no other oxygen product has! Here are some excerpts of my writings on OxyMune:

OxyMune is a synergistic blend of powerful nutritional supplements, providing the body with oxygen, hydrogen, major and trace minerals, enzymes, and amino acids. OxyMune liquid concentrate is offered at a breakthrough in value and in price, and is available in an easy-to-carry, pocket-sized dispenser.

Illness and Energy Loss —The cause of most illnesses and the lack of youthful energy can ultimately be traced to a lack of oxygen, minerals and cleanliness. Unless we continually take in plentiful amounts of oxygen and minerals on a daily basis, none of our normal body maintenance, repair, or immune processes will function correctly. Without sufficient minerals, taking vitamins becomes futile; and without enough oxygen we cannot eliminate the body's waste. But by daily supplementation with OxyMune liquid drops served in pure, cool liquid, depleted oxygen and minerals will be replaced in the body. Unlike single-objective supplements, OxyMune provides a broad spectrum of basic, missing, vital ingredients to help the body easily and naturally recharge its energy reservoirs, digest food properly, rebuild its immune systems, repair tissues, and overcome disease.

Oxygen —From the first breath taken at birth, having enough oxygen is the most vital factor in our lives. The environment's ability to produce oxygen is now severely compromised by the rapid destruction of the oxygen-producing rainforests and the constant polluting of the air and water. Today, air and water commonly contain oxygen-consuming pollutants that are robbing the good health of every living thing. We clearly do not get enough oxygen into our bodies.

Using OxyMune three times daily promotes a catalytic action over a 6-12-hour time period that helps the body to steadily generate oxygen out of its own natural internal water. Each portion of OxyMune that is taken will increasingly help to cleanse the body so that the body can store more oxygen in its cells. OxyMune is vital for good health, especially in today's low-oxygen, overly polluted environment.

Minerals —Naturally occurring soil minerals that should be in our daily food are critically low, or missing entirely. Our food no longer contains adequate minerals. Food grown in soil that was played out long ago, and food rushed unripe to the grocery shelf may look inviting but greatly lacks nutritional value. Our health and bodily functions directly depend upon getting these absolutely essential minerals daily. OxyMune supplies a full range of missing ionic and trace minerals in super-absorbable form. Your vitamins will not work without oxygen and minerals.

Enzymes —In today's high-energy society, people eat unnatural and excessively altered food. Even healthy and nutritious food is frequently eaten in a rush. Digestive problems over a period of time lead directly to illness. We need plenty of supplemental enzymes to completely digest our food so that undigested food particles do not enter the bloodstream. When undigested food particles enter the bloodstream, they become a food supply for infectious microbes. OxyMune provides our bodies with a blend of essential enzymes needed for thorough digestion.

Amino Acids —The body repairs itself using amino acids to build new protein structures. OxyMune delivers a daily boost of essential amino acids that assist the body with its necessary self-replacement and repair activities.

OxyMune is a well-rounded and powerful blend of important nutrients commonly lacking in today's diet. Unlike the oversized molecules found in pills that are not absorbed well, OxyMune's tiny nano-ionic liquid delivery system insures full absorption in the body.

OxyMune is a serious, powerful and natural long-term supplement that supplies the body with essential ingredients commonly missing from our daily food; allowing the body to detoxify, rebuild its immune systems, and restore its youthful energy.

A stable synergistic proprietary blend of nano-ionic clean sea source major and trace minerals, hydrogen isotopes, amino acids, enzymes, fruit acids, and enhanced ultra-purified water. For both internal and external use, OxyMune comes in a one ounce unique, easy-to-carry flat pocket-sized bottle dispenser. *"OxyMune has been demonstrated to measurably increase oxygen content in the blood via transcutaneous pulse oximetry testing in those with relatively low blood oxygen saturation and has been found to improve fatigue and endurance, at least subjectively. OxyMune has many potentially exciting health and performance applications."*—Martin Dayton, MD

MSM and DMSO

Again we address more missing elements that should be in out diet. This time it's natural sulfur. We're 6% —and nobody has enough. I kept hearing about how in addition to its other properties MSM is a non-toxic sulfur-based oxygenator (liquid, granules, or creams) that users were reporting as helpful. According to successful users the whole key with MSM is to take it *consistently* (6-8 months) at *high enough dosages* with vitamin C and possibly taken along with some DMSO. The most success we've seen is at 500 mg of MSM for every 30 pounds of body weight twice a day.

I was first intrigued by MSM when I heard on a tape that it was found concentrated in ocean plankton, one of our two planetary oxygen sources. I reasoned that if it was that close to the oxygen and chlorophyll creation action while floating in pure ocean minerals, then surely it might have some special properties. If so, it might make an excellent adjunct to any oxygen supplementation regimen. It turned out my intuition was right.

Most of us don't get enough sulfur from naturally occurring DMSO and MSM because the fertilizer used on our mega-crop agriculture is no longer natural manure. ***The sulfur cycle has been broken***. Chemical fertilizers are not providing sufficient sulfur for the plants we eat. Researchers studying our body oxygen transport mechanisms are even declaring this lack of natural sulfur/DMSO/MSM in our diets as a *major* contributing factor to disease. The total lack of study in these areas is surprising and unsettling. All our illnesses hinge on oxygen availability and food quality. Now name a food conglomerate that will fund a study to rat itself out.

Without enough dietary sulfur our cell walls become inflexible, they harden and can't 'breathe.' Increasing loss of cell wall permeability means low oxygen transport *into* and low waste removal *out* of the cells! You put the oxygen in the body but less happens because you can't absorb enough of it through the hardening cell walls. Dietary deficiencies of sulfur, potassium, minerals, and other elements explain the rare souls that try ozone but get less than the full effect.

From Keith Ranch, Planet of Health: "Methylsulfonymethane (MSM) is naturally occurring biological active sulfur and serves many functions in the body. It is one of the *third* most abundant elements in Nature and within all living organisms. Its chemical formula is $(CH_3)2SO_2$ and is formed when microscopic plants called Plankton from the ocean release sulfur compounds called (Dimethylsulfonium) salts. In the ocean these salts are converted into dimethylsulfide or (DMS), and then rise into the upper atmosphere to meet ozone and very high-energy ultraviolet light. This is where the magic of Mother Nature takes place.

"When DMS collides with ozone and high energy ultraviolet light, the DMS is converted into MSM and DMSO (dimethyl sulfoxide) and returns to earth in the rain and enters into all living plants. This is one of the reasons why we all should be eating raw uncooked fruits and vegetables, because cooking and heating destroys MSM beneficial properties in fruits, vegetables and all plants.

"MSM is so important in maintaining good health and flexibility in the tendons, mussels, ligaments, skin and nerves and serves so many other vital functions in the body. Just to name a few:

- Oxygenates the blood so that red blood cells can deliver more oxygen and nutrients to the cells.
- Improves the lung and all cell permeability to increase lung capacity and increase oxygen intake to the body.
- Helps neutralize allergenic reaction to pollen, different types of foods, animals and fat allergies.
- Increases bile-function and has a cleansing effect in the digestive tract.
- Protects against the harmful effects of toxins, radiation, and pollution.
- Disinfects the blood, resists bacteria and protects the protoplasm of the cells.
- Helps to normalize, allergic reactions, stress, drug hypersensitivity, inflammation of mucous membranes, and inflammatory disorders including all forms of arthritis, muscle cramps, and infectious parasites of all kinds.
- Is excellent in maintaining flexibility in arteries, improving oxygen flow and keeping the arteries clean and clear of arterial plaque from forming.

"MSM is truly one of the most important supplements and naturally occurring nutrients in the ozone soaked heavens. Nature has magically provided MSM to all of us living beings on earth to keep us happy and healthy." Thanks, Keith.

By the way, and for your information, 'Lignisul' is the only manufacturer made all-natural MSM product from pine trees. It's sold under different brands, look for the big 'boomerang wing.' To help verify the user results reported about all the supplements I mention, check out the testimonial lists in the 'Evidence' section in the back of this book

Chapter 12

Ozone the King

We are always being told that smog is ozone and ozone is smog. Surprise! The media and their experts are wrong again. I've seen many articles that state ozone is the 'main component' of smog. That's pretty strange journalism as far as I'm concerned. It's also highly inaccurate. Ozone is measured in parts per million in the atmosphere. How can something so tiny that it has to be measured in parts per million be called the 'main component' of any dirty atmosphere? Especially when the air is black with soot and laced with tons of thick hydrocarbon based particles?

Isn't Ozone Smog?

You see, if you say ozone is the main problem you can divert public outrage away from the captains of industry. They then won't have to take responsibility for the green, orange, gray, pink, blue, or whatever color mix the air is today with all the tons of hydrocarbon particulate matter in it. You can just blame it on all that pesky ozone.

Negatively charged ozone is naturally attracted to positively charged air pollution. Ozone is always found near smog because Nature is desperately trying to use ozone to clean up the smog that's left over from human greed.

I just picked up a copy of a supposedly hip slick magazine known for marrying hippie ideals with corporate decision strategies—*LOHAS*. Somebody with money to burn took out a four page glossy centerfold ad for a European device to get rid of room ozone. It loudly proclaimed: "Bad ozone... harmful indoor ozone... highly toxic and damaging to your respiratory system... toxic oxidant... serious birth defects and accelerated aging." I'd be frustrated if I wasn't so disgusted. Again painting this wonderfully useful gas as toxic in order to scare-sell you something. Again the lies that ozone is created from auto exhaust and industrial emission pollution instead of only pure oxygen, and that's why your eyes and lungs hurt in dirty hydrocarbon smog clouds. Oh, it's not the tons of unburned gasoline and dirty carbon particulate matter that hurts when it gets inside you—it's the ozone! The saddest part is all the young trend followers who will only see such lies and quickly join the ranks of the anti-ozone brigade thinking they are doing good environmental deeds. They are their own worst enemies.

Funny how most of us who have been cleaning out for a while can inhale deeply while only inches away from the output of an ozone air purifier without any distress. That's the real world bottom line that disproves the erroneous science mistakenly being bandied about.

Environmental Cleansing Oxidation Technologies

If you were to fill a room with cigarette smoke and put properly sized ozone generators in the corner of the room, within five minutes the smoke would be gone. That's Nature in action, dehydrogenating pollution with her ozone. If you break down all the toxicity and all the harmful unnatural things through oxidation technologies you'll have a clean environment. I already told you of an oxidation technology that can break down any old or new toxic hydrocarbon spill site; gasoline, coal tar, dry cleaning solvents, whatever. It's cheap and available and no one uses it. But billions of your dollars are spent every year on 'testing and measuring and consulting' over toxic sites. They all feed at the environmental money trough while posing as concerned, and 20 years later very little has changed. We have the capability right now to clean up the air, the water, and the soil. Let's do it.

Ozone is not a problem as some would have us believe; at proper levels it has proven to be a wonderful air purifier and the safest medical therapy ever devised. Proper ozone is blatantly non-toxic.

Ozone is a bluish water-soluble gas. Its name is derived from the Greek word 'ozein' which means smell. This odor is very distinctive, and it possesses unique characteristics. The wonderful thing about ozone, it's a marvelous anti-bacterial, germicidal and fungicidal agent; it leaves no toxic residues or by-products and is non-carcinogenic. All that comes in one small package!

Ozone is classified as an oxidant or a substance that converts organic materials into their base compounds. To simplify the understanding of the term 'oxidant,' and the process of oxidation, it can be described as something which makes other substances consume oxygen. Fire is an example of this process, because it consumes oxygen. The amazing thing about ozone is that it's a very fast-acting oxidant. After oxidizing substances, it simply reverts back to clean harmless oxygen.

Ozone is Unstable

Ozone is an energized form of oxygen with extra electrons. Ozone only lasts for about 20 minutes, maybe longer in Nature if you get a long chain form of it like O_{17}, or O_{21}, beyond O_3, which is found in the upper atmospheres. You'll find that maybe it sticks around for a couple of days and finally breaks down and turns into oxygen again. O_2 can always be turned into O_3, O_4, O_5, O_6, O_7, or O_8, and beyond. Some of the weather planes have found it existing in forms up to O_{21}. These long chain molecules include oxygen, oxygen, oxygen, oxygen all strung together. They give up one O_1 and reduce down. They give up another one and reduce down, give up another one and reduce down. It's a cascading effect, the longer the chain of ozone the more of a cascade release of O_1 is delivered into the body. That's the underlying basis and structural composition of Ozone Therapy.

Non-Toxicity of Ozone

Well, if ozone's not smog, is it toxic? Is ozone really toxic, as some have suggested? A chemist, Richard Kunz, wrote in 1997, under the title, *Fighting Odors with Ozone*, "Ozone gas, O_3, is a very strong oxidizing agent —similar to chlorine gas. It is one of the harsher irritants in smog and causes burning eyes and nose and throat mucosa. One should never, never intentionally generate ozone in a living space."

This statement could be made accurate if you qualified it. If the above was true then a stroll outside would be an irritant to millions of people. What he should have said was, "High levels of ozone in occupied spaces can be an irritant to toxin-filled people, human antibodies naturally produce it."

Are concerns over ozone toxicity levels warranted? The following correspondence is from another chemist, Dr. George Freibott who works in up to 50,000 times the allowable standards of ozone in a confined area and has been for more than 23 years.

"If the *toxicity* levels are too high in one's body, then breathing or using ozone may appear to be an irritant or cause irritating effects to the *overloaded toxin levels* of that same body. This is not the ozone's fault that the body toxin levels are too high so as to cause healing reactions and effects (sometimes called a healing crisis.) The ozone causes the same action in lower concentration levels as in higher concentration levels, that is, to do nothing more than *oxidize waste and assist in the forming of compounds ready for elimination.* The level of intake should be moderated by the practitioner/therapist/patient so as to not cause too drastic a therapeutic/patient reaction. The chemical/natural compound called ozone is neither toxic nor harmful if used correctly.

The Armour report so often cited as the authoritative report on the damaging effects of the toxic gas ozone, *neither shows it to be harmful or toxic, when read in context.*"—ANA President Dr. George Freibott

What Dr. Freibott refers to in the Armour report is where they put tumor laden mice in high ozone environments and after a few days killed half of them and examined their lung tissues. Naturally they were in the middle of detoxifying and in the worst possible shape. But the *rest* of the report goes on to describe how the *remaining* mice were left in the ozone to see what would happen to them, and <u>all their tumors disappeared</u>. "They became strangely ozone-resistant."

The oldest oxidation organization in the world was founded as the Institute for Oxygen Therapy in Berlin in 1898. They commissioned Dr. F.M. Eugene Blass to open the Eastern American Association for Oxygen Therapy, which continues today with Dr. Freibott at its head under the name of the International Association for Oxygen Therapy, Priest River, ID, USA

The body naturally produces peroxide, and active singlet oxygen, and ozone! The body metabolizes the amino acid tryptophan to produce superoxide of anion (some consider this equal to ozone). This is a process that makes oxygen available to stagnant areas. Tryptophan has a high ability to absorb oxygen. Tryptophan traps a lot of oxygen inside itself for later release. Our antibodies also produce ozone.

Ozone Is Not Smog
Ozone Is Good and Natural

Ozone is one of the most beneficial substances on this planet, and the BAD science you hear quoted on the news every night is causing you to subconsciously be afraid of Nature, and therefore, of a part of life itself. They tell you that somehow hydrogen plus nitrogen or sulfur equals ozone. $H + N + S = O_3$? Not on this planet it doesn't!

In the region of the ozone layer, our rising oxygen is bombarded by the sun's photochemical energy in the form of ultraviolet (UV rays). The UV energy bombardment changes the oxygen from O_2 —two atoms of stable oxygen, into O_3 —three atoms of unstable active oxygen. We call this pure form of oxygen 'ozone.' The using up of the UV rays to create ozone is how the ozone layer shields us from their harmful effects. This is all part of the natural process of life on this living biosphere called earth. The chemical formula for this is $3O_2 > UV > 2O_3$. Ozone has always been with us in Nature, and the fact that ozone gives off that single oxygen atom is a significant factor in life, in medicine, and in toxic waste cleanup technology.

It has been proven extensively that O_3 will kill bacteria, viruses, fungi and molds by attaching to them and oxidizing and eliminating them, oxidizing means to burn without giving off light or heat. These bacteria and other microbes are lower life form organisms and are mostly anaerobic. That means they can't live around activated oxygen/ozone. Doctors using the proper concentrations and correct medical protocols have achieved substantial positive clinical results with ozone.

Far from being a poison, ozone, when used properly, has been shown repeatedly to kill pathogens —yet remain harmless to normal cells. This is because disease causing pathogens do not have any strong enzyme coatings to protect them —as do all the higher life forms like us, for example; pure ozone is available to purify all our country's stored blood supplies. There is no reason why people have had to come home from the hospital with AIDS or hepatitis from blood transfusions. European doctors and respected New York University researchers all state that ozone has been used to eliminate HIV in human and animal blood tests, and without any side effects. Why don't we see this on TV? Why isn't it being used?

Breathing ozonated air or drinking ozonated water (at the safe legal concentrations that are already conservatively laid out by the government) are two of the ways of getting activated oxygen into your body. Being blatantly non-toxic, these methods of killing viruses and bacteria in humans have been in use in European medicine for more than 50 years.

Most European and several major American cities have been purifying water, sewage and toxic dump sites with ozone, some for more than 70 years. When the British first built the London Underground mass transit subways, they had a big problem. From the book, *Rails Through The Clay, a History of London's Railway, by Desmond F. Croome & Alan Jackson*: "But it was not all milk and honey for the Central London. For all its brash modernity and achievement, it rattled and it *smelt*." They tried huge fans to get rid of the smell, but they didn't help. "...but complaints (from 'delicate people, ladies and others') continued until an elaborate pressure system, with 6,000 cubic feet per minute fans injecting filtered and 'ozonised' air was fitted in 1911. Ozone, incidentally, was also provided

for the enjoyment of other tube passengers. In 1913-to-1914 plants were installed at Edgware Road, Euston, Goodge Street and Charing Cross and later at other stations." I'm still waiting for New York City Subway authorities to wise up. Any sort of underground installation can clean up its funk with ozone. There are air ozone purifying machines placed every 50 feet along the damp escape tunnels beneath the Presidential White House that are purifying the air ending mold.

Modern lunchbox reproduction of London Tube ad.

Ozone based systems can even break down PCBs and all other industrial chemical wastes both organic and inorganic. This is possible because ozone-based systems are able to create enough of these singlet oxygen atoms to oxidize anything unnatural found in our air, water, sewage, and sediment. Ozone can do this and yet is so safe that it is used on humans and animals in the water purifier at Marine World and in the Olympic Swimming Pools. The White House has ozonated pools. Think of it, our presidents swim in ozone. Beijing Aquarium in China is the world's largest fish tank, and they ozonate their water. Shark tanks can only use ozone and seawater. If anything else—such as chlorine—is used, it will kill the sharks. Furthermore, published scientific papers show ozone will kill anthrax in air and water.

So why do they call ozone 'smog?' Bad science and bad reporting. A political misrepresentation! By publicly calling Nature's oxygen 'poison,' and diverting your attention away from the real polluters, no one has to clean up the environment. Did you know that your automobile emits its own weight in pollutants into the air every year? Television tries to position itself as 'concerned' and wastes your time arguing over what type of shopping bag you should lobby for at the local supermarket. Meanwhile the factory next door continues its deadly course of spewing tons of poisonous pollutants into your breathing air and drinking water.

While you are constantly watching news pieces designed to make you feel guilty about normal every day living, the media won't give any significant air time to cover the far more dangerous local industrial polluters who are too cheap to put scrubbers on their smokestacks. Why do the media ignore the real problems? Because the corporations might be 'offended.'

Go into a city, look up, and taste the dirty air you're breathing. Try and tell me that the brown/gray/yellow color that you see is ozone. It can't be. Ozone is clear! All of this particulate soup, the real danger, can be cleaned up with colorless, clear ozone.

Ozone-based systems are able to purify 99 percent of every liquid, gas or toxic substance coming out of any industrial operation. The engineers even tell me that we can include radiation in the list since the radiation is carried by something. Why aren't the ozone systems being used everywhere?

What got us into this pollution mess was the old idea that the earth and water and air magically combined into one giant 'sponge,' where we could just 'toss it out' and it would all disappear. Well our sponge is now full. Remote Arctic seals are full of PCB's, and polar bears are being born with both sex organs from pollution caused mutations! Their blood samples show the highest pollution levels in the world (it collects in blubber). The ocean polluters and rainforest clear-cutters have significantly removed Nature's cleaners—our oxygen, and the natural atmospheric sterilizing ozone the sun creates from it.

The newscasters and scientists try to blame your respiratory problems on our friend ozone and call it names like 'smog.' You can almost hear them thinking: "Well boss, here's 5,000 pounds of toxic hydrocarbons and nitric compounds coming from our factory, and those pesky environmentalists are starting to notice it and demand things and make noise.... Let's see.... There's less than .12 parts per million (or only 12 millionths of a pound) of ozone in the air.... I've got it, Mr. Burns! We'll blame everything on the tiny little air ozone molecules—so the sheeple won't notice our toxic soup, the real cause of their dead trees and of their lung, eye and throat irritation!" "Excellent Smithers, that's brilliant! Let's do lunch at the club. By the way, how's your daughters rash?"

Polluters pay big public relations firms big money to 'manage' public concerns and make these concerns go away. The PR firms maintain files on everyone on both sides of any issue. They fund their friends and drum up the media in favor of their side and to suppress and ridicule the enemies of their corporate position. They quietly pay scientists to produce reports that publicly rivet our attention on—and blame—ozone (Nature.) Then they secretly fund their own 'friendly name' sounding captive environmental groups to direct public opinion away. Through this chicanery, the huge polluters are never forced to take responsibility for their continuing use of cheap outmoded dirty engine and factory designs. That way they never have to pay to incorporate any of the already invented clean energy sources. What they call 'ozone smog' in the press is their own toxic soup.

What they don't tell you is that, as noted, Nature's ozone is trying to clean up the toxic soup, and makes up only a very tiny portion of the smog they report. The more chemicals dumped into the air, the harder Nature will try to clean it up, so the more negatively charged ozone that will be attracted to the positively charged chemical pollution. Therefore the reported ozone levels will always be *higher*. They also don't admit that ozone is strictly, always, only O_2 plus O_1—pure oxygen, and never anything else.

The mixed blessing ray of hope here is that the media professionals and the federal, state, and corporate decision makers and their families are *themselves* coming down with all manner of chronic illnesses and new mutant diseases. The traditional vacation hideaways of the rich are already spoiled. It is no longer an 'us' versus 'them' class struggle. We are all stewards in this boat called Earth. This forces change in business as usual.

As to their claims that ozone is a poison, I can refer detractors to internally clean people who work in very high concentrations of pure ozone all day long without ill effect. In fact, they commonly report a healthy invigoration from exposure to ozone. Where the ozone-as-irritant scare stories come from is the following typical scenario. When a smoker, or junk food, or drug addict —a person whose body cells are loaded with toxins— finally gets near enough to an activated oxygen (ozone) source, the ozone will start to oxidize the toxins integrated within its cells, in an effort to finally remove them. As a result, the pathways out of the body become temporarily filled with cellular debris. They swell up, become irritated, and filled with fluid. Often this is uncomfortable, but only for a few days, while, and until, the oxidized toxins leave the body. The health professionals skilled in medical ozone usage call this a typical harmless cleansing reaction. Those selling competing products call the cleanse 'damage.'

Most air ozone 'studies' are halted at the point when the detoxification causes discomfort and irritation, and not after the full cleansing has concluded. Therefore 'damage' is erroneously reported in the scientific literature. By contrast, any properly conducted experiments are allowed to continue past this cleansing point—and then report how the body replaces the weak, old, diseased, dying, and feeble cells with new and very healthy oxidative stress-resistant ones.

At times an isolated and questionable report will surface in the scientific literature, telling of animals exposed to ozone that developed lung scarring. These studies were usually done with super high ozone concentrations way beyond the typical medical protocols, and relate to impure ozone made from high amperage electricity and air, which is 80 percent nitrogen, not pure oxygen. Nitrogen plus heat plus moisture plus ozone equals nitric acid. Acid will definitely cause lung scarring. Again, this is not the way of Nature's pure natural ozone.

I've actually cornered a few scientists and reporters and asked if they knew that they were not being scientifically accurate when in the press they equate ozone with the toxic soup of smog. They admitted (in private) that they knew they weren't, but keep up the charade "because everyone else does."

What about the holes in the ozone layer? Consider the gluttonous 'clear cutting' of the oxygen producing rain forests and the disappearance of our own oxygen-producing national forests. Where is our oxygen going to come from so Nature can continually turn it into ozone? Then add the constant selfish polluting of the oceans and the greedy discharge of industrial pollutants, nuclear radiation and electrical energy into the atmosphere. These electrical, electronic, and radioactive discharges further scramble the elements in Nature's air. At home, chlorine gas comes out of your water faucet and rises up into the sky. More and more, our oxygen is either missing or bound up in toxins. What we're experiencing is an increasing shortage of atmospheric oxygen that's available to be turned into ozone in the first place! That's why there is an ozone hole at both poles, and the rest of the ozone layer is starting to look like Swiss cheese. Greed is the problem, not ozone.

The ozone layer is constantly changing; almost a living boundary, paper thin, and missing at night. When the oxygen is all bound up with toxins, then there will be no ozone layer. Without available oxygen, the sun's unfiltered ultraviolet light passes right on through to the ground without being absorbed in creating ozone. As a result, we are seeing increased cataracts, skin cancer, blindness, and the burning of vegetation. Our bodies and our food supply—therefore our very existence—is in danger, unless you personally do something about turning back the rampant greed and its effects that are destroying us.

What can you do to help preserve and re-supply the missing oxygen in your life? Stop those sub-humans who think, "We're all going to die anyway, I'm going to get all I can now, so cutting down all the trees doesn't matter." Convince the factory managers to get off their duffs and install existing devices like ozone-based smokestack scrubbers, factory discharge point ozone-based-purifiers, and to fund existing ignored clean energy sources, and toxic soil and water cleaning technologies. Quit looking at the toxic dirt and water and letting people suffer from it. The new technology exists to clean it up on site and for far less money right now, and then your future liability is over forever. Clean it all up so we can all live together on this planet, and enjoy the very best of health!

Oh, by the way, we can clean up most pollution coming from our vehicles by installing ozone generators in them. The car's electrical system can easily power small ozone generators that inject ozone into the engine air intakes and purified air into the passenger compartment. This would create complete combustion, higher efficiency, more power, and better gas and

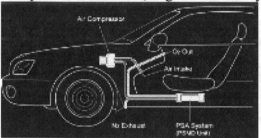

diesel mileage. I told you I would give you solutions, and this is a big one. I'll show you the photo of one such device later on in the book.

Most air pollution is unborn hydrocarbons. Feeding the right amount of metered active oxygen into the car engine air intake eliminates them. Use too strong an ozone concentration, and you'll burn up the engine, so if you start experimenting in this area, use less expensive engines to figure out the correct window of effectiveness with maximum burn and minimum melt for your cubic inches. A power boost without the bother of nitrous. All existing cars can be retrofitted once this is perfected. As an example of the motive power of an oxidizer, all WWII German Buzz Bombs were powered by 90 percent peroxide.

Chapter 13

The Commercial Use of Ozone

Want to help the planet in a BIG way? Do you want to stop a huge source

of groundwater pollution? Do you want to end 'sick building' syndrome? Get behind this chapter's information and you will.

Ozone saw its big entrance into the world of commerce in the 1800s when it was first used in a slaughterhouse in Nice, France to control odors. Ozone keeps the families of fish vendors happier when the vendor comes home from work. Endowed with the ability to disinfect and produce an odor free environment,

Ozone purifying fish department air for grateful employees at Key Foods in NYC.

ozone's popularity has increased in commercial industry but at a too-slow rate owing to well funded competitive pressures.

Air Ozonation Politics

How do you personally continue to treat your body? Since half the cause of all disease is toxicity, then wouldn't you want to start addressing your personal home environment? Smog from cities all over the world is being blown past your door and in your windows every night. Let's clean the air. Let's remove all the offensive things from the air that are bothering us. That's exactly what industrial ozone air machines are designed to do for multiple families and office workers and executives in large buildings.

Ozone can be put to use to convert airborne pollutants such as ammonia, sulfides, and other organic chemicals into non-odorous byproducts. It has the capacity to make chemicals and microbes inert. Put ozone-purified air in a building and it sterilizes the air. It kills bacteria, the fungi, the molds, and the pathogens, even on contact surfaces. It also deodorizes the home. No more garbage, mold or pet odors. It's a very fresh sweet smell, like the seashore or after a lightening storm. That's Nature's ozone.

Ozone has been used successfully for air purification in school systems (illness related absenteeism dropped), hospitals, fish markets, millions of homes, in gymnasiums, nursing homes, and to fix smoke-damaged theaters, hotel rooms and rental cars. In my own experience I have repeatedly seen it used it in many homes for years without harm. I am breathing it right now.

I also personally use it to clean rental cars and hotel rooms previously sprayed with insecticide or used by smokers. I saw it in use in a NY morgue and didn't smell a thing. The same was true of a fish market. With smoke damaged buildings nothing would get rid of the smell, but ozone did the job perfectly. I interviewed people in Bethesda, Maryland, who found the U.S. Navy literature showing they used it in their submarines. Manufacturers have sent me a numbers of lab studies proving that it disinfects. Funny how just little ol' me can find all this stuff but the regulatory agencies can't.

These common uses of ozone for air purification have been going on in the real world for more than 70 years. In truth, the only unsubstantiated and negative statements about it are coming from the agencies and the lung associations who have no practical experience but simply review the research literature paid for and put out by competing device manufacturers with plenty of money to publish 'studies.' Reading that, how could the officials come to anything other than untrue conclusions about ozone?

As I have said, the situation is probably all fueled by well-meaning and smart college kids being brainwashed in school that ozone is 'bad,' graduating, and bringing their beliefs into civil service without real world experience and true investigation. Added to this the regulatory employees don't dare go after the politically protected big money real polluters, and since they have to justify their jobs and look like they are doing something, they try to go after those without big political clout. That's you and me, the little ozone users.

Unfortunately people get sucked in by this stuff. The disparagers of air ozone have woefully inadequate 'real world' knowledge if they think that ozone is not a wonderful air disinfectant when used at safe low levels over time. If anyone believes that government employees are not influenced commercially by competitive interests to suppress many truths, they better have a sip of reality tea. In my expert opinion, some previous consumer magazine ozone testing was stacked against ozone. Lies and omissions in the report and tests were common. As it happened, I personally went through this with two manufacturers. Follow the money.

Typical Misinformation

The following article is typical misinformation, probably written by college kids who went straight to working for the government and immediately started reading terribly one-sided literature without having any 'real world' experience. How can reviewing 'the literature,' which is all that the government employees seem to do, be truthful if only one side (the big money competitors to ozone) of an issue get to publish anything and also control the so-called 'testing?'

What's the Real Story on Ozone?
[Always ask yourself, judged by whom, how, and why?]

OZONE GENERATORS. The California Department of Health Services advises the public not to use so-called air purifiers' that are specifically designed to generate ozone indoors. These ozone generators are sometimes marketed as emitting 'trivalent' oxygen, 'activated' oxygen, 'allotropic' oxygen, 'saturated' oxygen, 'super oxygen,' or 'mountain-fresh air.' These devices may also be combined with a negative ion generator. However, these devices are actually emitting ozone.

[These are the same Californian intellectual giants who have actually sent their agents to ticket homeowners using ozone pool water sterilization, claiming they were 'generating smog.' Meanwhile, of course, the local factory smokestacks are mysteriously 'exempt.']

*"Ozone is a harmful air pollutant that is the **main** ingredient of ground-level smog."*

[Hearing this dishonesty is always frustrating. How can something so tiny that it is measured in parts per billion be the 'main' anything, when the hood of my car is covered in non ozone black soot every morning? And if it is 'harmful,' why does Nature produce it 24/7?]

"Breathing ozone <u>can be</u> harmful, especially for children, the elderly, and people with asthma, emphysema, bronchitis, or other respiratory diseases. Ozone also irritates the eyes, nose, and throat. Long-term exposure to ozone <u>may</u> permanently reduce a person's breathing ability."

[Notice the underlined qualifiers 'can be' and 'may.' They can't go too far with their nonsense. What they call 'harm,' we all know as temporary detoxification of the cells covered with black soot that line the throat, sinus, and lungs. Any irritation disappears after ozone has time to oxidize the black soot. Until it is all expectorated or dissolved, yes you will feel it dissolving, but no permanent harm is done since the body quickly repairs itself once you clean it out. And while they're busy 'protecting' the elderly from ozone, perhaps they would care to explain the old lady who was stuck dragging an oxygen bottle around with her to breathe, who ozonated her home air and no longer needs the bottled oxygen?]

"Ozone at **safe levels** *does not clean the air. Independent studies by the U.S. Environmental Protection Agency, the Consumers Union, and others have shown that these devices do not* _effectively_ *destroy microbes, remove odor sources, or reduce indoor pollutants* _enough_ *to provide any health benefits."*

[Notice again the tricky usage of the underlined words. Artificially contrived 'safe levels' set way too low and designed to never let us see how good ozone is? Notice carefully that they actually admit ozone does destroy microbes, remove odor sources, and reduce indoor pollutants enough to provide health benefits. But they can't believe their own data because they would be fired if they did, so they 'sort of' lie enough by using 'effectively' and 'enough' to make you think the opposite of what happened in the tests by subtly minimizing the data. The so-called 'independent studies' are always performed using insufficient ozone and never applied long enough. Contact time is key. But, as I said, to be fair to these dangerous kids with the power to disrupt your life, if you didn't know what to look for you might read a stack of 'independent' one-sided tests handed to them by their bosses. They have been conditioned to think they are doing the public a service by 'protecting' all of us from ozone!]

"Ozone is used effectively in water to destroy microbes, but ozone in air must reach extremely hazardous levels (50-to-100 times the outdoor air quality standards) to effectively kill microbes."

Funny, I breathe those '50-to-100 times' levels all the time without harm, and so do hundreds of thousands of other people purifying their homes with air ozone. It's only seemingly 'hazardous' for a few days or weeks of adjustment if you're a 'fast food' eater, smoker, and possibly using bronchial chemical inhalers in a big city and sorely in need of an inner cleaning. If such a toxic person gets too much ozone too fast, he or she will definitely cough and cough as the body tries to clean. And everyone in the business always tells such people to go slow and keep it turned down until they do clean out.

During the manufacturing of the Homozon oxygenating intestinal cleanser, Dr. Freibott breathes ozone concentrations 52,000 times above those artificially set as 'safe' levels every time he manufactures it. These so-called 'totally unsafe' levels do not faze him at all because the reality is that they are perfectly safe as long as you're accustomed to it.

It is only the pollutants in our tissues that react with ozone as they are destroyed. The tissues are affected by the heat and foreign matter and gasses created by the reaction with the pollution. The body quickly cleans it all out, and re-grows the tissues as it always does anyway. If you did not know this, you would put a lab rat in a cage and force him to relentlessly breathe high levels of ozone, and then kill him and biopsy his lungs at the height of his detoxification while yelling, "Look at all the damage!" This describes their typical not so 'independent' test methods.

Use Your Brain

Let them remain unknowing if they wish. As long as we have free choice and free will you can ignore their misinformed opinions and move on. If ozone safely works for you, use it, but I suggest you use the REAL world to judge it, not intellectualizations.

There are products that have been given the CSA highest certification rating of 'Approved for use in occupied spaces' that generates ozone. They win this approval because they have a sensor that regulates the amount of ozone generated to be no more than .05 ppm, (the same level that exists naturally in outside fresh air). This level is not harmful but very desirable.

All of the major commercial ozone production units, such as those sold by Oxygen America, or in the Family Health News, or by Aqua Sun Ozone, Quantum, or Alpine, or Clinic Air, for example, (and there are many others) are good solid units backed up by their reputation. They all use a cold plasma electron field in which air is blown through this field to produce ozone. Nitrates are not a problem in any of the commercial units.

These ozone air purifiers naturally don't do much if the windows are open and the wind is blowing, or if the air is being changed constantly. However, in such a situation, assuming the wind dies down at night, you could have your properly adjusted (set low) ozone air purifier positioned above your bed. Since ozone is heavier than air, it would fall upon you. Therefore you would be breathing safe, beneficial levels of it at night as you sleep. Remember that the ozone generator has to be set at the proper level for *your* body—where it does not produce any irritation. If it's set too high, you'll wake up with a sore throat and you don't want that. I've interviewed many people who get used to—and clean out enough to—breathe ozone and love it, and they have reported how beneficial it has been.

Ignore the ignorant who tell you that breathing controlled levels of ozone is toxic, poison, or corrosive. Used correctly, ozone air purifiers—in my experienced opinion formed from interviewing many hundreds of actual users—appear to be highly therapeutic. Again, we're waiting for the clinical studies to definitely prove this. But it must be used correctly. So please educate yourself.

Air Ozone Manufacturer Reports

Individual results will always vary, but I received this report, sounding like so many others just like it, from Anthony Jamieson in South Africa.

"I have been manufacturing ozone generators and have had some unbelievable successes with people who have been suffering from asthma. My unit produces 50mg of ozone/hr, adjustable. One woman who is now in her early '40s suffered from the age of 16 with asthma. Within two weeks of using Puretech's ozone generator she had lost the ashen color of her skin. It returned to a healthy pink color and she was able to finally sleep comfortably at night. She can now walk around at night without discomfort or feeling fatigue. Ann is just one of many success stories. I initially started manufacturing ozone generators for the food industry. I just wish that I had discovered ozone's amazing benefits sooner."

Ozone Commercial Water Purification

Ozone Manufacturer DEL Industries - Partial Customer List

Life Support and Aquaculture:
- San Diego Zoo—Polar Bear Plunge, Hippo Beach
- Atlantis Paradise Island, Bahamas
- Mandalay Bay Casino, Las Vegas
- Underwater World—Mall of America, Melbourne, Beijing
- Alaska Sea Life Center
- Alaska Department of Fish and Game
- Canadian and U.S. Department of Fish and Game
- Biscayne Aquaculture

Water Parks and Resorts:
- Universal Studios Hollywood—Jurassic Park the Ride
- Universal Studios Florida—Jurassic Park the Ride, Suess Landing
- Atlantis Paradise Island, Bahamas
- Knott's Berry Farm—Wind Jammer
- Chun Ahn San Rok Waterpark

Commercial Aquatics Facilities:
- Four Seasons Aquatic Center
- Santa Monica Family YMCA
- University of San Francisco
- 92nd Street YMCA, New York
- Ahn San Sports Complex
- Schaeffer Rehabilitation Center
- Prince George Aquatic Center

Industrial:
- McKesson Water Products Company—Crystal Bottled Water
- Pacific Water Company
- Children's Hospital
- St. Clair Hospital
- Spokane Metal Products Co

Aquariums

According to the Sander Ozone Company in Germany, "Ozone is a very reactive form of oxygen that can destroy an enormous variety of liquid waste materials and toxins. In the aquarium, it offers a simple, highly effective method of maintaining a clean and stable environment for the algae and animals. Ozone is a safe and remarkably effective agent capable of killing a wide variety of microorganisms. Viruses, bacteria, spores, some chemical impurities, etc., are all attacked and destroyed by ozone. Additionally, toxic materials treated with ozone are nearly always converted into less toxic compounds, enhancing their absorption by bacteria, algae, and/or activated carbon.

"As an example, in the marine aquarium, ozone will steadily convert ammonia to nitrite and rapidly (on contact) convert nitrite to nitrate. Ozone promotes the formation of stable foaming compounds from otherwise non-foaming components, noticeably increasing the efficiency of protein skimmers in marine aquariums. When ozone is used with a protein skimmer, complex waste materials not removed as foam are further broken down to simpler component parts and passed off to the atmosphere or broken down into materials readily consumed by the bacteria and algae in the aquarium. We recommend that protein skimmers always be used with ozone.

"After 24-to-48-hours of protein skimming with ozone, the water in the aquarium will seem to 'disappear' as the small particles and colored materials are removed from the water. The clarity of the water is quite simply unequaled by any other system. Ozonizers can be used in both marine and freshwater aquariums."

Are ozonated tanks successful? One man reported to me that his fish grew so big in an ozonated aquarium, that it outgrew the tank.

Swimming Pools

The benefits of ozone in swimming pool water are widely known in Australia, Europe and Asian countries. Ozonated pools induced sparkling clear water. The typical heavy chlorine smell, red eyes, rashes and skin problems associated with chlorine do not exist when ozone is used for pools. Now what is your body trying to tell you if you get in a pool with chlorine and it reacts with rashes? You're obviously in a toxic environment. The skin is one of the body's largest organs and its absorption capacity is well documented. Why not absorb something of benefit to the body such as ozone, a non-toxic form of oxygen, instead of toxic chemicals? This is plain common sense.

What about Peroxide and Ozone in Pools?

The best situation for a pool is peroxide and ozone together. Ozone helps peroxide replacement, it keeps the peroxide replacement cost down, and the peroxide kills some organisms that ozone might miss in pools since ozone rapidly jumps out of the water. Peroxide test strips will tell you what concentration of peroxide you have in your pool. What we're looking for is to maintain enough concentration to keep us healthy as we soak in it, but not so much that it bleaches out your suit. Try 85-to-100 parts per million of peroxide in your pool or spa or bathtub. This is the concentration that professional pool maintenance people I have interviewed shoot for. If you don't have any peroxide test strips, try keeping the peroxide at the level where you can just about taste it. Depending upon the microbial and algae flora in your area, you may also want to look in to the new, emerging non-toxic adjunctive technologies of copper coils and copper-based or salt-water solutions.

Olympics

When the Europeans came to the U.S., their Olympic team would not swim in our California Olympic pool. They were used to swimming in ozonated pools. They said, "We're not going to swim in that pool. It's got chlorine in it, and that is an unfair competitive disadvantage since we are not used to absorbing and smelling chemicals in our ozonzted pools."

They made the Olympic Committee rip out the chlorine system and put in an ozone water purifier. I was talking to some of the coaches. Interesting what you find out when you go and interview people. The swimmers and coaches were taking the water back to their hotels and homes because it was cleaner than the water they got out of their taps in Los. Angeles. Now that's saying something.

As soon as the Olympics were over, they ripped out the ozonation system and put the chlorine back; because you don't want to offend the chlorine lobby, which is the chemical lobby which passes the laws making all the municipalities put chlorine into the water. ozone is used all over the world.

Firm Sees Ozone as Food Disinfectant

News release January 15, 2002. Chlorine is banned as a disinfectant for organic food in Europe and the U.K.—CNN's Diana Muriel

"So will ozone disinfection be the next big thing in the food industry? Bioquell seems to think all the right ingredients are there to make this technology a success story in 2002.

"For most of the past century, weapons of war have been used to clean our food. After World War I, scientists found a new use for stockpiled chemicals like chlorine—both as a water purifier and as a powerful agent to kill bacteria on food. Before then, ozone—a naturally occurring chemical compound—had been used to clean water. Now a small biotech company in the South of England is pitching the idea of once again using ozone in the food industry.

"Nick Adams, CEO of Bioquell, says it's what the public wants. 'People don't actually want to eat food that has been decontaminated using chlorine,' he says. 'Ozone, which breaks down to oxygen once it's actually done its job, is environmentally friendly, doesn't taint the food, and you can't smell ozone once its done its job.'

"Ozone can be used in gas form to kill bugs in a room, but it's more commonly applied as a liquid. Washing lettuce in ozone can kill bugs in a matter of minutes. They are dissolved into the ozone, which can then be recycled. But ozone cannot be stored—it must be generated on site—and that's expensive. An ozone generator can cost between $30,000 and $140,000. [Not exactly true.]

"In the long run, though, it could work out cheaper, says Adams. 'The up-front cost of installing ozone equipment will, after three years, give you an economic payback as compared with chlorine,' he says. 'With chlorine you have ongoing costs, and there are some quite significant environmental and safety costs which don't exist with ozone. In Europe and the U.K., chlorine has been banned as a disinfectant for organic food —a rapidly growing sector of the food industry. And last July 2001, the U.S. Food and Drug Administration approved ozone to disinfect food.'"

More Commercial Ozone Uses

Waste Water from the Paper Industry

Ozone reacts with the molecules responsible for coloration, breaking down those functional groups that have high electron densities. These include the humic acids, tannins, and lignins commonly found in vegetable matter.

The raw materials of the paper industry include water and many types of lignocellulose structures. As a result, the wastewater produced by this industry tends to be heavily contaminated and highly colored, especially when the industrial process makes use of artificial coloring agents. Clearly, this wastewater must be purified, and here conventional treatments often prove to be of limited efficacy.

Waste Water from the Dyeing Industry

The wastewater produced by the dyeing industry contains many artificial coloring agents. We can classify these using a number of different criteria:

- Application method: dipping, etc.
- Physical properties: soluble, suspended, etc.
- Chemical properties: acidic dyes, basic dyes, sulfide dyes, azo dyes, metallic complex dyes, etc.

Generally speaking, biological treatments are only capable of bringing about slight improvements in wastewater color. This is because the molecules responsible for coloration are not biodegradable. Aeration or oxygenation can, however, improve effluent color. One way decolorization might operate is through destruction of the double nitrogen-nitrogen bonds. This is confirmed by the appearance of N0 and N02 radicals following ozone treatment.

Treatment of Toxic Waste

The detoxification of industrial wastewater is an issue that has taken on marked importance in recent times.

The chemical industry, in particular, discharges many substances of high immediate toxicity; for this type of waste, mere dilution is not a sufficient anti-pollution measure. In mixtures containing toxic compounds, ozone will not always react specifically with the compound that we wish to eliminate. For this reason it is advisable, wherever possible, to apply conventional purification techniques first and follow up with ozonation as a second or third stage.

We should also bear in mind that concentration levels are a crucial factor when considering toxicity. Thus, ozone is increasingly used as a final treatment stage for eliminating all traces of harmful substances. The advantage here is that ozone treatment does not require the use of additives, which may prove more toxic than the compounds we wish to remove.

Elimination of Cyanides from Wastewater

Cyanides are a prime example of highly toxic water; according to the literature, these substances are toxic to fish at concentrations as low as 25 ppb.

They are powerful enzyme inhibitors and continue to find widespread use in many industries: froth flotation for mineral extraction; cyaniding of precious metal ores; gas purification in coking plant; synthesis of organic compounds (plastic, textiles); manufacture of pharmaceutical products (barbiturates); etc.

However, the largest quantities of cyanides are discharged from the degreasing tanks, electroplating vats and rinsing baths found in surface treatment plants. The cyanide concentrations in wastewater from these sources can vary from 20-to-100 mg/l. Chemically speaking, cyanide pollutants can be categorized in three main groups.

- Hydrocyanic acid

- Simple alkaline cyanides (sodium cyanide, potassium cyanide)

- Soluble complex cyanides (potassium copper cyanide, sodium iron cyanide, potassium iron cyanide, etc.)

Elimination of Heavy Metals from Wastewater

Industrial effluent often contains heavy metals, the toxicity of which is widely known. Like the cyanides, heavy metals may be present in the water in both free form and in complex compounds.

Broadly speaking, ozone oxidizes these heavy metals to form metallic oxides or hydroxides, which precipitate off and can easily be removed from the water. Ozone is used for eliminating heavy metals from the effluent produced by many types of industry, a typical case being that of mineral extraction plant.

Elimination of Phenols from Wastewater

Many types of industrial plant discharge phenolated wastewater; coking plant, oil refineries, petrochemical plant, mines, chemical and pharmaceutical process lines, foodstuff canning plant, paint stripping shops in the aeronautical industry, etc.

A number of purification methods exist for eliminating these toxic compounds from wastewater. On such method is oxidation by ozone.

Several highly detailed studies have been conducted into the way ozone reacts with phenols. Basically, ozone breaks down phenols to form oxalic acid and oxygen.

Studies into the kinetics of ozonation reactions show that many factors can affect the efficiency of the purification process. Predominant among these factors are the temperature and above all, the pH. Thus, by increasing the pH from 8-to-11 we double the rate at which phenols are broken down. At the higher pH value, ozone attacks phenolated compounds in preference to the other oxidizable matter, and this substantially reduces reaction times.

While it is theoretically possible to express the kinetics of the phenol breakdown process in purely mathematical terms, in practice we must also make allowance for complex relationships concerning the presence of other organic compounds in the effluent. Usually, higher amounts of ozone are required to break down phenols in the presence of other organic compounds than is required with pure phenol solutions.

However, this is not always the case; and we can even observe the opposite. One explanation for this is that the kinetic rate of the reaction is proportional to the concentration of the chemical substances involved; thus, with a high concentration of phenols the reaction is faster and less ozone is required. Another explanation may lie with the physical make-up of the phenol contaminants, since certain phenolated compounds are more readily broken down than the phenols themselves.

We usually consider that five parts of ozone are required to break down one part of phenol in wastewater. Generally speaking, then, ozone provides a valuable service in detoxifying wastewater prior to discharge in the environment.

Deodorization and the Treatment of Gaseous Effluent

Ozone reacts with many organic substances responsible for causing unpleasant odors. We could mention olefins, short-chain fatty acids, ketones, nitrites, ester, amines and sulphur-bearing compounds such as mercaptans. Owing to the very low concentrations involved, it is not usually possible to carry out economical treatment using conventional methods such as thermal or catalytic oxidation, absorption, washing or biofiltration. Ozone treatment is the only way to ensure rapid economical deodorization.

There are two types of ozone treatment for gaseous effluents:

Dry Treatment —The polluted air is mixed with ozone by means of an injection nozzle. The contact time, about a few seconds, is usually sufficient to effectively oxidize the gaseous pollutants.

Wet Treatment —This method, which involves simultaneous absorption and ozonation, is more efficient that the dry method and capable of treating higher pollutant concentrations. The polluted gas is purified by a counter-current flow of ozonated water. Certain condensable or soluble substances are removed by the water, while insoluble compounds react with the ozone. Usually, the water is recycled and resonated. In difficult cases, several purification units can be mounted in series. It is also possible to accelerate the oxidation process using catalyzers or combined ozone/ultraviolet techniques.

The ozone/ultraviolet combination is particularly powerful and capable of treating most of the substances dissolved in the rinse water. These are broken down into carbon dioxide and water. To illustrate the efficiency of this process, we should simply mention that it is capable of reducing the phenol concentration in air by a factor of 1,000.

Applications —Ozone treatment methods for gaseous effluents have many potential applications. In the construction industry, for example, ozone techniques are ideal for purifying the air conveyed by the air conditioning ducts. Besides the obvious aesthetic improvement, ozone sterilization for air conditioning systems also reduce bacteria-related health risks 'sick building syndrome.' Among the many other possible industrial applications, we could mention the elimination of smell from wastewater treatment and pumping stations, foodstuff processing plants, foundries, restaurants, etc.

Prison Laundries use Ozone

From *Ozone for Use in Laundry Washing,* by Nathan Schiff, Ph.D.

"When used in laundry wash water, ozone allows for shorter wash cycles resulting in significant energy and water/sewer surcharge savings. Enhanced soil removals, powerful disinfecting and reduced garment wear-and-tear are some of the added benefits. For this reason ozone laundry systems are gaining rapid international acceptance over conventional methods which require substantial amounts of hot water.

"Because it is so reactive, ozone readily attaches itself to fatty and other soils that bind dirt to clothing, destroying them rapidly. As one of the strongest known oxidizing agents, ozone is capable of breaking down virtually any organic soil into innocuous compounds such as carbon dioxide and water. Being a gas in solution, ozone penetrates and opens individual garment fibers, allowing faster cleaning and bleaching of garments with the use of fewer chemicals. The overall effect results in considerable reductions in the washing and drying cycle times, and whiter, cleaner and softer garments.

"Why are industry and government excited with this technology? Because of its powerful oxidation properties ozone dissolves soil on contact and does so at ambient water temperatures, instead of conventional 140°-to-160°F, thus drastically reducing heating costs. Some estimates indicate that *an 80 percent fuel consumption saving can be realized.*

"Ozone's powerful disinfecting properties allows for dramatic reductions in water and sewer surcharges. Since two instead of three rinses are usually required, quicker cycle times occur, which increases the productivity per machine, per hour. Labor savings, estimated at 20 percent can also be realized through shortened wash and dry times made possible by reduced chemical use and handling.

"The EPA and FDA have acknowledged that ozone is capable of reducing pathogenic bacterial levels on garments by 99.9992 percent. Ozone is also recognized for its powerful pollutant destruction capabilities in laundry wastewater and therefore in the purification of wastewater.

"Due to the reduced cycle time coupled with the elimination of chlorine and hot water, the life of the garment is often increased. Unlike chlorine with a long garment residual time—which may be detrimental to garments —all traces of ozone are instantly eliminated.

"Currently, ozone washing machines provide the best return on investment when used in large institutions such as prisons, hospitals or large hotels with at least 150 rooms.

"Running fewer rewash loads, whiteness has increased. Using a lot less chemicals and hot water, odors are gone and our customers are happy."
      —Shirley Howell, Laundry Production Manager of Diamond Linen

"The concept of washing clothes with the aid of ozone gas dissolved in ambient temperature water was first introduced to a highly skeptical laundry industry in 1991. Ozone Laundry Systems is a patented laundry wastewater treatment, wash process and wash apparatus.

"The spent wash water is collected, filtered and reused, *thereby eliminating waste water disposal problems*, resulting in considerable water and energy savings. Currently, we offer both open and closed loop systems.

"The Ozone Laundry Systems are attached to the front end of regular washing machines. They cannot replace the washing machines but assist in the chemical and mechanical processes to achieve better results"

Pollution from Commercial Cooling Towers Can Be Averted

You really want to clean up our environment? Specifically, do you want to stop feeling guilty about letting your faucet drip and instead focus on what's really going on? Look up at the top of any big building and notice the big pieces of rectangular shaped equipment perched on top of all of them. Many have big fans showing, and steam coming out of them. This equipment is called the 'air handling' equipment, or the 'cooling towers.'

You are looking straight at one of the most massive water pollution machines ever installed on the planet. Each one of these dumps tons and tons of polluted water into the environment annually. Now multiply this repeated pollution operation times every big building on the planet! The mind cannot comprehend the amount of water pollution we are discussing, where billions of gallons of polluted water are dumped annually, and nobody says a word. And guess what? Once again, it has already been used successfully, and you don't know about it, but it can ALL easily and simply be cleaned up and even prevented by using relatively inexpensive ozone technology!

We'll use a local building in Florida as our example to illustrate the mechanisms and quantities involved. Take a standard 800-ton cooling tower perched on a 22 story building. Standard cleaning maintenance on this one unit involves flushing the cooling tower water with an acid flush once or twice a year.

If you did not add chemicals to this system's water, you would have one big algae mess in the base. Normal operation consumes approximately $450 dollars per month in chemicals, and that adds up to $5,400 a year in chemicals, and by the way, the chemicals are slowly eating everything since the acid chemicals are corrosive.

The Tolling of Huge Environmental Damage

1. One of the first damaging effects to the environment noticed is that corrosive and polluted water is subject to 'blow-off' where water spray escapes the system from the fan propeller, and ends up on the building and in the environment. This totals about 150 gallons per day or 54,750 gallons per year.

2. Then there's the monthly cooling tower 'runoff.' Every month this discharges 1,500 gallons per month of polluted water into the aquifer.

3. Then, due to the sludge, mold, and corrosive pollution, there are the required two annual complete flushes per year, each discharging 15, 000 gallons into the aquifers.

Let's extend this out to all installations. In our example, southeast metro Florida probably has in excess of two 200 tall buildings; a quick guess might be that some 1,500 commercial buildings are like this.

Now let's average this to 18,000 gal per building discharging annually, times a conservative average of 1,000 buildings. This adds up to business as usual collectively dumping 18 million gallons of polluted corrosive water into the aquifer per year! And that's just southeastern Florida.

Our example does not count the roof drains that empty into drain fields (French drains) going straight into the aquifer without any purification treatment.

Let's compare what would be the instant beneficial result if we replaced the traditional pollutive chemical anti-fouling systems with Nature's ozone.

Our standard 800-ton cooling tower perched on that 22 story building would only need and use two pounds per day of ozone production.

Ozone cooling water treatment would employ one industrial ozone generator in the basement by the pumps, and two, one pound per day ozone generators up on roof at the cooling tower site.

Using an ozone generator with a 5-micron sand filter, by the time the filter becomes heavy with dirt and pollutants, it will actually bring the filtration down to only 1-micron particulate size.

Air Purifying Systems Ozone Cooling Tower Installation Hollywood, Florida.

The system automatically performs a one minute backwash flush—using 50 gallons of pure ozonated water—three times a day. What was once toxic corrosive water is now as pure as drinking water. By using the ozone generators with oxygen concentrators, or the new perfected generators out of South Africa, they are half the usual size but bring 10 times the production.

Ozone's Advantage over Chemicals

- **There is zero mold and algae growth in the coolers and pumps.**
- **There is zero growth in the pipes.**
- **Shows less than ½ of 1 percent on cooling tower spray in the edge area.**
- **Chemicals are harsh on pipes; they eat away the pipes −ozone doesn't.**
- **Chemicals eat away the galvanized metal cooling tower −ozone doesn't.**
- **Chemicals eat away the pumps −ozone doesn't.**
- **Ozone does not corrode any metal.**
- **The ozone system quickly pays for itself.**
- **Chemicals cost and pollute endlessly.**
- **There is no longer any reason to flush the system since it remains clean.**
- **We save all that water, no longer create 18 million gallons of polluted water in this one place, and we seamlessly maintain our standards of living while adding no pollution to the planet!**

Now imagine every commercial building on the planet using ozone in the air handlers instead. How much interior pollution did we just remove? Each system converted to ozone will save the building owner $6,000 per year. Many buildings have more than one system. How much did we just save the owners?

You have to ask yourself, should they be telling me to conserve water and not wash my car, or should the cleanup and enforcement energies be aimed elsewhere at the real polluters and wasters? Now how do you feel about letting your car go dirty and not watering your lawn?

These aren't pipe dreams. Complete systems are already in place, and working perfectly. I interviewed Carlos Jimenez, President of Air Purifying Systems in Hollywood, Florida several times, and visited his ozone installation sites to come up with this information. This is already reality, and simply needs implementation. It's your planet. What are *you* going to do about it?

Inside the Commercial Buildings

There are three top reasons to use ozone air purification systems in your condominium, hotel or business:

1. The ducts are kept clean, and the drains are kept clean. That's where 99 percent of your problems are.

2. Ozone in the airstream will go throughout your home, through the heating, ventilation and air conditioning (HVAC) ducts and be evenly distributed throughout the building or house with no one area having a high concentration.

3. The ozone will get into anything porous such as volatile organic carbons (VOCs) in carpeting, and its adhesives, and any outgassing formaldehyde, from furniture and Formica and their adhesives, and easily decompose their odors.

The FDA suggests using less than .05 ppm of ozone, and this large building application, because it is continual, requires only .001ppm.

People with allergies or other lung problems report definitely breathing better and easier.

Contractor Jobsite Use of Ozone.

Homemade contractor-use high-output unregulated air ozone generator made from mercury arc lamp, with box fan.

A contractor wrote to me explaining how they use a (converted lamp) super high-powered ozone generator to industrially clean their workspaces.

"We have lots of smokers for one thing. Some non-smoker workers are forced to share those areas. Our industrial ozone units are particularly effective against tobacco smoke as you are certainly aware. Often we use ozone on 'green' (chemically treated by pressure) wood of various types, as the fresh odor of the chemicals in these new installations gets overpowering. We often leave the generators on overnight in locked areas. We leave one on (using an on/off timer) all night in these unoccupied areas until morning, and when we come in and air out the room it has created a significant difference in air quality.

"The overall combined body odors of 5-to-30 sweating workers can build up in the smaller unventilated places we work in. We leave a unit on when the room is empty during breaks and lunch hours and it destroys the body odors immediately.

"Paints, the musty smell of fresh drywall installations, and the cat urine smell of new wall insulation materials all respond to ozone. We are extremely careful however to always address these things in UNOCCUPIED areas."

The federal government has published guidelines for the use of ozone in occupied areas. They are very conservative estimates, and assume toxic people are the subject breathers.

US DEPARTMENT OF LABOR
Occupational Safety and Health Administration
EXCERPTS FROM REGULATIONS
Occupational Health Guidelines for Ozone

INTRODUCTION
This guideline is intended as a source of information for employees, employers, physicians, industrial hygienists, and other occupational health professionals who may have a need for such information. It does not attempt to present all data; rather, it presents pertinent information and data in summary form.

SUBSTANCE IDENTIFICATION
- Formula: O_3
- Synonyms: None
- Appearance and odor: Colorless gas with a sharp, characteristic odor; **it can be smelled at concentrations below the permissible exposure level.**
- The odor is readily detectable at low concentrations (0.01 ppm to 0.05 ppm)

PERMISSIBLE EXPOSURE LIMIT (PEL) The current OSHA standard for ozone is
0.1 part of ozone per million parts of air (ppm) averaged over an eight hour work shift. This may also be expressed as 0.2 milligram of ozone per cubic meter of air (mg/m3).

EXPOSURE LIMITS
Several governmental agencies and organizations have established limits for ozone measured in parts per million (ppm). The following ppm values apply:
.12 Environmental Protection Agency for city, out-of-doors, air quality
.10 American Conference of Governmental Industrial Hygienists. Limit for exposure for eight hours a day with no side effects
.05 Food and Drug Administration restrict selling an ozone generator labeled as a medical device
.03 to .06 Value frequently measured in cities
.010 to .015 Odor threshold for most people
.005 to .01 Value measured in fresh country air

The Maximum Allowable Concentration (MAC) of ozone for human beings, as established by the American Council of Governmental Industrial Hygienists, is 0.1 parts per million by volume (ppm/v) for *continuous exposure* under occupational conditions. *This is a safe level;* exposure to ozone concentrations at or below this level can be tolerated. Concentrations above this level are intolerable only when exposure continues for certain durations.

HEALTH HAZARD INFORMATION
- **Routes of exposure**—Ozone affects the body by being inhaled or by irritating the eyes, nose, and throat.

Effects of exposure—When a person is exposed to very low concentrations of ozone for even a brief period of time, the person may notice a sharp, irritating odor. As the concentration of ozone increases, the ability to smell it may decrease. Irritation of the eyes, dryness of the nose and throat, and cough may be experienced.

The Marine World of Commercial Ozone
The Role of Ozone in Marine Environmental Protection

Ken D. Hughes, President, Delta Marine International, Inc.
and Hugh D. Williams, CDR , USCG (Ret.)

Ken Hughes is co-founder and President of Delta Marine International, Inc. of Fort Lauderdale, Florida. He got together with Hugh D. Williams, CDR, USCG (Ret), and wrote a paper on marine ozone installations titled: *The Role of Ozone in Marine Environmental Protection.* I am reprinting certain parts of it to honor their work trying to protect our environment from ourselves. They wrote it because they both were keenly environmentally aware and concerned over the impact of invasive aquatic species being flushed into the world's oceans from the ballast tanks of all large vessels.

Their real world very experienced conclusion is that "Ozone has an important but as yet largely unrealized and unfulfilled role to play in reducing environmental damage to marine ecosystems, as well as, improving the onboard environment and living conditions for all shipboard personnel."

Their full paper is available, and it describes the serious overwhelming problem of prolific secret ocean discharge of sewage and ballast water. I live near the Florida ports, and I can personally attest that stupid, greedy, lazy owners, ship captains and their employees are illegally dumping offshore. It is positively disgusting what is washing up on our beaches.

The good news is the solution exists, and can you guess what it is? Yep, it's ozone! From their Hughes & Williams paper: "Presently ozone is being used for the purification of potable water within the holding tanks on board two U.S. Navy vessels, (R/V KNORR and R/V ATLANTIS) and two National Science Foundation vessels, (R/V OCEANUS and R/V ENDEAVOR), as well as several luxury yachts. Woods Hole Oceanographic Institution operates the three former vessels, and the University of Rhode Island Graduate School of Oceanography the latter.

R/V ENDEAVOR,
typical ozone user.

"Ozone is also being used on board over two hundred and fifty yachts from 31 ft. to over 300 ft., including power and sail, for indoor air quality and odor control, as well as odor control in the headspace of black and gray water holding tanks.

"NASA uses ozone to remove toxins from launch pad cooling water after a rocket launch, and has replaced chlorine with ozone in its HVAC cooling towers at the Cape Kennedy Headquarters Building and Vehicle Assembly Building. American cities, such as Los Angeles, CA, and Costa Mesa, CA, and Newport News, VA, and over four thousand others worldwide use ozone either in their municipal water treatment, and/or sewage treatment plants.

"European nations, especially France and Germany, have mandated ozonation for swimming pools to eliminate chlorine that is not nearly as effective in such applications. Ozone is also used in Indoor Air Quality (IAQ) and odor control applications. The State of Florida recommends the use of ozone for such applications in warm, humid environments. Ozone is especially effective in overcoming what is known as Sick Building Syndrome.

"The Alaska Tanker Corporation invested a total of one million dollars to equip one of its crude carriers, (SS KENAI, that plies the Alaska – Japan oil trade route), with a totally ozone based ballast water treatment system. The methodology in that approach to ballast water treatment is a multitude of diffusers placed at different depths in the ballast tanks. A central ozone system is installed in a shipping container, from which a mass of stainless steel tubing is routed to the tanks to be treated.

And finally, "Lyntech Corporation of College Station, Texas is using an electrochemical process to generate ozone for ballast water treatment."

I interviewed the aforementioned Kenneth D. Hughes, President of Delta Marine International, Inc. in Fort Lauderdale, Florida about his marine ozone installation business. Remember as you read the following it is based upon Ken's past ten years of work where he outfitted his ozone equipment on most of the highest profile yachts and research vessels. His installations number over 2,000, he has distributors in Singapore and the U.K., and his equipment is often specified by name in yacht designs. He has compiled the following ozone fact chart he gives to customers, and he was kind enough to let me reprint it here.

I. OZONE PHYSICAL AND CHEMICAL CHARACTERISTICS

A. Made from three atoms of OXYGEN, Chemical symbol, O3.

B. Colorless in gaseous form, appears blue in "Thick Layers" {Blue Sky}.

C. Dark blue or almost black in liquid form.

D. Solid crystalline form is violet to blue in color.

E. Melting point ~ 80 K.

F. Boiling point ~ 161 K.

G. Has a very unique 'Electric' odor that can be detected as low as 0.012 PPM.

H. Highly unstable and not possible to store, all users generate it on-site, as needed.

II. FORMATION

A. Natural

    1. Lightning.

    2. Solar radiation. {Ultraviolet wavelengths A, B, & C}

    3. Photochemical Smog in large cities.

B. Man-made

    1. Corona discharge. {Usually in air or oxygen atmospheres}

    2. High voltage gas discharge tubes. {Germicidal UV lamps}

III. SOLUBILITY

A. Soluble in water.

    1. Affected by temperature, pressure and contaminant levels.

IV. OPTICAL PROPERTIES

A. Ozone has a peak absorption in the UV Spectrum at 255.3 NM, which puts its most effective band in the UV 'C' spectrum. Without this particular property, no life could exist on the earth's surface.

V. EFFECTS ON HUMAN BEINGS

A. In low concentrations, {Less than 0.05 Parts Per Million}, OZONE has a sweet pleasant odor. It is responsible for the familiar odor of a thunder storm due to the OZONE created by lightning. {High voltage electrical discharge}

B. Natural ground-level concentrations are typically 0.03 PPM.

C. O.S.H.A. 24 hour exposure limit, 0.05 PPM, {Parts Per Million}

D. O.S.H.A. 8 hour exposure limit, 0.1 PPM

E. O.S.H.A. Short-term exposure limit, 0.3 PPM

F. Effective as a disinfectant and sterilizing agent at levels of 35% to 50% of that specified by O.S.H.A. for 24 hour exposure levels.

G. Prolonged exposure to high levels, {^ 100 PPM}, produces headache and perhaps nausea.

H. Emergency first aid procedure is to remove victim from areas of high concentration, whereupon the symptoms will disappear.

VI. BENEFITS OF OZONE IN WATER TREATMENT

A. OZONE oxidizes and decomposes organic and inorganic contaminants at a higher rate that other reagents. Normally one or more orders of magnitude faster than Chlorine, the most commonly used reagent.

B.   OZONE has faster sterilization and disinfection rate over 3000 times that of Chlorine in water and it is far safer. That is why OZONE has been used in the Paris water system since 1903 and is currently used by Los Angeles, (1984) and San Diego, (1986).Los Angeles has the world's largest ozone generating system. (LA Aqueduct)

C.   Sterilization and disinfection rates are independent of NH3 and are not affected by pH as is Chlorine. Microorganisms that are normally resistant to Chlorine, requiring hours of contact time, are killed in seconds by ozone.

D.   There are no bacteria or viruses which are resistant to OZONE, as it acts as an OXIDANT of PROTOPLASM. That is why trace residual OZONE is accepted as a standard of reference for full disinfection of water. {Ref. Bottled Water}

E.   Other than microorganisms, OZONE decomposes organic and inorganic contaminants in water is into harmless compounds that can be easily separated or transformed by settling, filtering, etc.

F.   OZONE reacts favorably with material and compounds with which Chlorine doesn't react or reacts in unfavorable manner so as to leave problem by-prods.

    1.   Substances imparting color, smell or taste.

    2.   Iron and Manganese.

    3.   Cyanides, Phenols, etc.

G.   OZONE is the most 'Environmentally Friendly' oxidant and even, at least partially, decomposes Chlorinated compounds. {PCBs, Chlor. solvents, etc.}

VII. BASIC APPLICATIONS

A.   OZONE is highly unstable and will easily give up one atom of oxygen to combine, with, {Oxidize}, almost anything.

    1.   Naturally occurring OZONE and the OZONE formed from auto exhaust causes the premature aging and cracking of tires and wiper blades.

B.   OZONE is the second most powerful oxidant after FL, {Fluorine}, and the by-products are harmless as opposed to some chlorine by-products as:

    1.   Trihalomethanes.

    2.   Chloramines, {Chloroform, etc.}.

C.   OZONE is used extensively for processes requiring oxidation and disinfection

    1.   Drinking water treatment; purification, odor, taste, color, container sterilization.

    2.   Waste water treatment; disinfection, sterilization, BOD & COD reduction.

    3.   Cooling towers; scale removal, microbe control, corrosion control.

    4.   A/C Air Handlers; Odor control, mold/mildew elimination/control.

    5.   Aquaculture, fish and shrimp farming; ammonia removal, bacteria control.

    6.   Pools, spas, water parks, water displays; replacement of Chlorine.

    7.   Bottling plants; container sterilization.

    8.   Canning plants; container sterilization.

    9.   Breweries; removal of taste/odor and microbe control in process water.

10. Metal plating and finishing processes for waste water reclamation.
11. Bleach and detergent reduction in wash processes. {35% to 75% Typical}
12. Transportation and/or storage of fresh fruit and produce.
13. Fire damage restoration; smoke odor and mold/mildew removal.
14. Casinos and bars; elimination of tobacco smoke and stale wine & beer odor.
15. Rental car and truck agencies; removal of tobacco, pet, and body odor.
16. Hotels; elimination of tobacco smoke and stale room odors.

VIII. COMMON PROCESSES GENERATING OZONE AS A BY-PRODUCT

A. Sunlight.
B. Lightning.
C. Photocopiers.
D. Laser-Jet printers, scanners and copiers.
E. Surface treatment of plastic film.
F. Automobile exhaust + sunlight.
G. Fluorescent lighting.
H. Electric arc welding.
I. Electric-arc, 'PLASMA-JET' metal cutting.
J. Any high voltage process that generates a 'Corona discharge'.
K. 'Tanning' beds

IX. HOME APPLICATIONS

A. Odor control/removal.
   1. Pets.
   2. Sick Room and/or body odors.
   3. Tobacco smoke
   4. Food preparation and cooking, (Onions, garlic, fish).
B. Mold and/or Mildew in carpets and moist areas.
C. Removal of Formaldehyde, {Carpets, furniture, draperies, glued wood, etc.}.
D. Pools, "Jacuzzis" and Spas, elimination of dangerous and toxic chlorine.
E. A/C duct work sanitation and maintenance (kills mold and/or mildew in ductwork).
F. Extraordinarily effective in combating allergens.

X. MARINE APPLICATIONS

A. Fresh water purification.
B. Ballast water sterilization.
C. Mold and mildew that cause a boat's 'Musky' odor.
D. Tobacco smoke.
E. Black and gray water holding tank and vent odor control.
F. Bait and live well odor elimination.

G. Control of bilge water odor.

H. Extend fresh fruit and vegetable storage life in refrigerators and coolers.

I. Prevent odor and taste cross-contamination of foods in refrigerators and coolers.

J. A/C air handlers and duct work mold and mildew and indoor air quality control.

K. Diesel fuel odor elimination.

L. Food preparation and cooking odors.

M. Elimination of pet odors.

N. Destruction of substances that cause allergic reactions.

XI. OZONE GENERATION METHODS

A. Electric {Corona} Discharge.
   1. Basic Equation.
      a. a. $3 O_2 <-------> 2 O_3 <-------> 3 O_2$
   2. Theoretical Power Required.
      a. 285 K JOULES to generate 2 MOLES {96 Gram} of O3
      b. Equal to 0.9 Wh per Gram.
   3. Actual Power Requirements.
      a. 15 to 20 Wh per gram O3
      b. 7 to 10 kWh per pound of O3

B. UV Radiation.
   1. UV Radiation of Oxygen or other gases containing oxygen will produce OZONE
   2. Process is photochemical dissociation of oxygen.
   3. Typical concentrations achievable via UV radiation are below 1 gram OZONE per cubic meter of process gas. {Approx. equal to 500 PPM}.

C. OZONE generation is limited by the inherent power demands from any given piece of equipment; therefore, design limits impose the maximum amount of OZONE that can be produced from any single, specified source.

D. There has never been one single fatality that resulted from the generation and application of OZONE, since it was first isolated in 1840 in Germany.

XII. UNITS OF MEASUREMENT FOR OZONE

A. Mass of OZONE / Volume of gas, expressed as "g O3 / m3" AIR or OXYGEN.

B. Mass of OZONE / Mass of gas, expressed as "% Weight "{Lge. masses only}.

C. Volume of OZONE / Volume of gas, expressed in "PPM", {Parts Per Million}.

D. Volume of OZONE / Mass of gas {Not Used}.

E. Mass of OZONE / Volume of water, "g / m3 ", {Or PPM}.

XIII. CONCENTRATION CONVERSION TABLE

A. Physical Properties, Standard conditions P = 1013.25 MB, T = 273.3 K
   1. Density of OZONE , 2.14 Kg / m3
   2. Density of oxygen, 1.43 Kg / m3
   3. Density of air, 1.29 Kg / m3
   4. Density of water, 1000 Kg / m3

B. **USEFUL CONVERSION FACTORS.**

1. 1000 liters = 1 m3 = 264 US gallons
2. 1 g / m3 = 1 mg / L

C. **OZONE CONCENTRATION IN WATER.**
   1. 1 g O3 / m3 H2O = 1 PPM O3 in water {By weight}

D. **OZONE CONCENTRATION IN AIR BY VOLUME.**
   1. 1 g O3 / m3 AIR = 467 PPM O3 in air {By volume}
   2. 1 PPM O3 in AIR by volume = 2.14 mg O3 / m3 AIR

E. **OZONE CONCENTRATION IN AIR BY WEIGHT.**
   1. 100 g O3 / m3 AIR = 7.8% O3 in AIR {By weight}
   2. 1% O3 in AIR by weight = 12.8 g O3 / m3 AIR

F. **OZONE CONCENTRATION IN OXYGEN BY WEIGHT.**
   1. 100 g O3 / m3 O2 = 6.99% O3 in OXYGEN {By weight}
   2. 1% O3 in OXYGEN {By weight} = 14.3 g O3 / m3 O2

© Kenneth D. Hughes, President, Delta Marine International, Inc.,

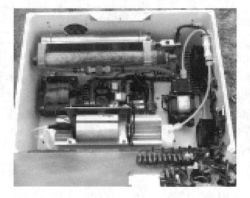

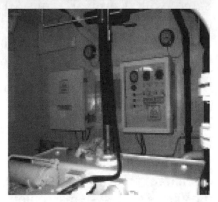

Top left: **Inside a Delta Marine ozone unit.**

Top right: **Delta ozone unit tucked in-between two huge 'Go Fast' boat racing engines.**

Right: **Ozone generators above the holding tanks purifying all black and grey water discharges back to clear water.**

The use of these ozone units in all boats will end the terrible environmental destruction coming from the daily ship ballast and holding tank flushing of non-local species of marine life and infectious disease spores into our local waterways. Without any local competitors the newly introduced offending species grow wildly, overcoming local marine life and plugging intakes. Have you heard of how the Zebra mussels have seriously clogged the Great Lakes? It was all inexpensive and avoidable if they had only required ozone holding and ballast tank sterilization.

The politics we need to overcome in the industrial pollution areas are daunting, but do-able. For example, Disney World advertises how environmentally conscious they are, but they wash 1.3 million pounds of laundry and it's all done with soapy wash water that has to go somewhere.

Why not eliminate the soap? Ozone could be used instead of soap just like many prisons are doing, but Disney plant engineers so far refuse to lower their costs and further protect the environment by investigating and implementing ozone systems that inexpensively and cleanly do the laundry and make any tainted discharges vanish. They have been offered ozone solutions many times, and yet they have not responded at all.

You say, "But why? It defies all logic that they would not want to save money and protect the environment to the best of their ability. A quick look at the huge number of big expensive houses surrounding Disney that are owned by their suppliers and their buyers in the Kissimee, FL area tell the story. The money is coming from somewhere. Just as in all the other areas where we have solutions that are not being adopted, everyone is getting and staying rich buying and selling business as usual while the environment belonging to all of us is damaged. Can you say the phrase 'greedy conflict of interest?'

Same with Carnival Cruise Lines and others like them. You may remember how Carnival recently got in trouble when passengers ratted them out for throwing garbage bags overboard in the middle of the night. Of course, they soon promised to stop due to the public pressure.

Many of the cruise line's management officers have been offered, at no cost, read 'completely free,' installation of an ozone purification demonstration unit in *any* of their huge ship ballast and holding tanks so they could see for themselves and on their very own ship how easy and inexpensive it would be to stop all the environmentally destructive discharges presently coming out of their liners. They were even told that if the free ozone systems couldn't fix their odor problems in 45 days and also completely clean up and sterilize their toxic and infectious wastes currently being dumped into the sea, then the supplier guaranteed that he would come back and pull the ozone generators out and return everything back to normal. What was the Carnival management response? Nothing, no response at all.

The fat cats in any company won't make any cleanup moves without public and media pressure forcing them to. That means *you*. Now it's *your* turn. What are *you* going to do about this?

Everybody thinks 'saving the environment' always means depriving ourselves and giving something up. Hopefully you now see how much simpler than that a lot of this is. We can clean up billions of gallons of water just by mandating the ozonating of cooling towers, industrial discharge points, and all ship discharges. And all *without giving up anything*! They will all *SAVE* money! And the rest of us will all live better and longer. Inertia and greed are the enemies. Maybe we could even evolve to the point that we get so educated and smart that we'll start adding just enough ozone to our car, truck, bus and plane air intakes to get rid of at least half of the millions of tons of unburnt gasoline and hydrocarbon air pollution that's slowly killing us as well.

Tiny 12 volt solid-state ozone module.

It would cost pennies per car to simply wire an ozone output controller to the gas pedal and put one of these puppies into the air intake of every car at the assembly line. Why aren't the 'concerned' environmental groups and agencies demanding that these be installed? Probably due to not knowing about them, probably due to hearing nothing but lies about ozone from the media, and probably lots of trough feeding in the 'business as usual' barnyard.

Chapter 14

Ozone in Daily Life

Industrially, and in the home, ozone air purifiers have been in use for decades. There have been no problems associated with their use, as long as they are used in average sized rooms, and the levels just below where someone feels discomfort. Enlightened hospital operating rooms commonly use ozone air purifiers to keep

everything sterile—the doctors and nurses aren't falling over dead with scarred lungs are they?

Ozone cleaning the air in a NY metropolitan morgue.

Ozone air sterilizer/purifiers/deodorizers are commonly used: by hotel chains to remove odors, by used car dealers to give old cars that 'new car smell,' by morgues to get ride of formaldehyde odors, by schools when they refinish a floor, this way they don't have to close the school because of the dangerous refinishing chemical odors, in bars, comedy clubs, and restaurants—so the majority of the population who are non-smokers can visit them again and go home without stinking like an ashtray.

These ozone products are also used in fitness and exercise clubs and gymnasiums where patrons don't smell body odor, but only fresh air instead while reporting increased endurance and strength. They are also used by grain storage building owners who report an end to mold and rot. Freezer operators, cargo ships and freight companies use them to keep their food fresh far longer than with just cold temperatures.

Owners of animal excretion-soiled stables, barns, veterinary kennels, and professional

Patrons enjoy New York's Helen Hayes Theatre. Ozone rapidly cleaned away heavy smoke damage a few hours before opening night.

dog and horse racing paddocks love them. If the animals could talk, they would probably echo this sentiment and describe the barn air as fresh as a day in the country.

Entrepreneurs even buy—at a discount—sick cattle that are worn out from antibiotics and drugs; remove the drugs and then ozonate their air and water and then sell the cattle as healthy, disease-free animals a year later at a profit. Plus, the consumer eats chemical-free meat. Do you have any smoke damaged goods? Fire damaged furniture? Put it in a room with an ozone air purifier running full tilt, and in a few days the useless items are restored. The applications are endless, wherever stale, polluted and toxic odors are encountered.

Factory and closed-up-tight office workers could ask management to install low level ozone air purifiers. Management would benefit at the bottom line, since happy oxygenated workers are more efficient workers, cheerier to customers, and history shows they don't need as many sick days. In 1922, ozone air lamps were

Alpine refrigerator ozone generator.

placed in the St. Louis schools and absenteeism dropped substantially. In 1920 English military hospitals used it on open wounds. Commercial clothes dryers came with UV ozone lamps in them. The federal government required their use in all government restrooms. If your home or work air stinks, think of ozone solutions.

If some sat only inches from an ozone generator and breathed deeply for a long time, they might have cell lysis (destruction) problems. But no one is advocating that, and product warning labels could handle the liabilities. No one deeply inhales oven cleaners, paint thinners or other common toxic household chemicals, and they are available everywhere without restriction. Why should

IOP Technologies O_3zonator.

ozone be any different? Jumping on the hyped media bandwagon, some government agencies want to regulate the amount of ozone emitted from an ozone generator. That's not the point, and way off the mark. The output shouldn't be regulated, because we never know the size of the room it will be used in, or the toxic cellular level of the room occupants.

'Ozone is a poisonous gas' is a great, quick 'one liner' for the media, but it is far from reality. This kind of instant unresearched journalism has created anti-ozone hysteria to such a degree that the 'Earth Day' environmental organizations even produce emblematic signs with well-meaning but uninformed anti-ozone slogans. In a twisted way, due to this stuff people are subconsciously made to fear the very act of breathing, so that every breath taken on a hot summer day in the city is tainted with a fear of life itself.

At the home level, many thousands of people are now exploring the many medical Oxygen Therapies and pollution control devices I wrote about in my first book, *Oxygen Therapies*. One of the simplest methods of using ozone at home is by installing a home ozone air or water purifier. They do a fine job for what they were designed for: general air and water purification. There are several brands on the market, and I use these devices at home with pleasant results all the time.

Many readers of my *Oxygen Therapies* book have even called and written to me about their own personal experiences. After installing air ozonators, they reported their 'house mold' went away,' that 'the 'odors stopped,' their 'emphysema became less,' or their 'lupus got better,' and one fellow actually told me 'the tartar fell off his teeth!' Sounds fantastic, but hearing

these stories first hand has *repeatedly* been my actual experience. Of course no one is making illegal medical claims for these devices, but the anecdotal evidence in this area continues to amaze us as it piles up steadily.

If your vehicle has dirt in its oil, has half its air supply cut off, and has never had an air or gas or oil filter changed, it will die after sputtering along for a while.

Our bodies are physical vehicles for our true selves, Soul. Your liver and kidney and lymph system are your own personal vehicle filters. Most die too soon with the body filters caked full of accumulated dirt as well. Compare the polluted situation today to the Bible dating some people in the Old Testament as being more than 900 years old, and how did dinosaurs get to be five stories tall? Back then there was no pollution — and lots of oxygen. You can guess why so many people are sick, and why so often in our supposedly 'modern' society.

I am convinced that what we're presently experiencing in our society is the rise of the age of toxins, diseases, and plagues all corresponding to the fall of our planetary and body oxygen levels. Fueled by greed and self-imposed ignorance, the phenomenon is sad indeed, and unless abated will drastically change or even eventually eliminate life on this orb.

Ozone Air Purifiers and Breathing Ozonated Air

Small modern office/home ozone air purifiers.

General Electric triple ozone lightbulb 'Art Deco' design older air sanitizer.

Having hopefully convinced you once and for all by using enough real world examples that ozone is not dreaded poisonous smog, but instead Nature's gift to humanity, we can now take our liberated minds and expand them even further to embrace the concept of actually putting ozone into the air. When we re-supply and re-introduce ozonated air with safe ozone levels into modern artificial closed spaces we commonly see sick buildings, mold, odors, and airborne allergies become yesterday's news. This fact has been recognized ever since ozone was first generated and named in the 1800s, and ozone air purifiers have been evolving ever since.

Although individuals have written to me with some fascinating healing stories that occurred from just continually using ozonated water or ozonated air, ozone air purifiers are not sold as medical devices. As I have always contended, just as when you breathe outside air during a lightning storm, breathing safe low concentrations of ozone from air purifiers necessarily gets some active oxygen into your lungs. And that oxygen ends up in the blood. Science dictates ozone air purifiers actually are a slower, smaller version of the wholesale flooding of your entire bloodstream through medical applications in a European clinic; and it therefore must have the same dual action of detoxification and microbe kill, but at a very reduced level.

The only questions remaining are, what size machine is right for your application, and how much can your own lungs be comfortable with breathing today?

Unlike in the clinics air ozonators don't use bottled medical grade oxygen as the source gas; they simply use room air. Room air is about 80 percent nitrogen, and when some of the remaining 20 percent (or less!) oxygen content of room air is turned into ozone by the ozone generator, the inefficient heat of some generator designs may make some of the nitrogen combine with some of the oxygen to form trace amounts of nitrous oxides. If the nitrous oxides hit excessive moisture in the air, they might turn into trace amounts of nitric acid.

If someone wanted to sell you a competitive type of non-ozonating air purification unit, they could rig a test to show how 'bad' ozone is. All they have to do is pay some Ph.D. to make this dirty type of hot spark ozone and force animals to breathe high amounts of it. Then they could keep out the competition by claiming that the ozone hurt the animal, the government would blindly follow right along, and the public would be none the wiser. Notice how the negative 'tests' of ozone don't describe which generator design, which concentration, or the input air quality or humidity levels present. And they're always ended before the lab mice can finish detoxing and adapt.

Natural ozone, at natural levels produced by emulating Nature, has never been toxic to lung tissue. There was some intellectual concern that lung tissue has a limited antioxidant capacity (Kelly et al. 1995), but that only tells us what we already know from the millions of units in service over the past one hundred years; that the lungs are very sensitive to ozone and we need to protect them from high dosages. Fortunately, almost every home ozonator sold today generates proper levels of clean ozone. UV bulb units are weaker than corona glow electrostatic plate units and the new high frequency pulsed units, but even weak ozone can be good ozone. It all has effect; the difference is how much and how fast.

> ## Do not expect to remove serious diseases with only air or water ozone units!

 People often ask me at what level should they set their ozone air purifier. Over decades of practical applications, many thousands of people have used ozone air purifiers properly. They set the output level of their machines just below the point where they can smell the ozone, and under these conditions no one has experienced any problems. Some slowly set it and higher after their bodies get accustomed to it—after more and more detoxing occurs.

You must keep in mind that air ozone will try to oxidize and de-hydrogenate any foreign particles back into their base elements. That's exactly why we use it. If those particles are sitting in piles in your lungs, then ozone will still do its job there, just as it attacks primitive mold spores or cigarette smoke floating around the room.

MR.OXYGEN'S™ CONSUMER SAFETY GUIDELINES
FOR HOME AIR OZONE GENERATORS.

1. **Is it a cold process corona glow ozone generator—the kind that doesn't create lots of nitric acid out of air nitrogen and moisture?**

2. **What is the ozone output and concentration compared to the size of the room it is to be used in?**

3. **Does the generator have instructive labeling saying: "Operate only at a level where no discomfort is experienced?"**

4. **Is the generator a quality design, using safe non-ozone reactive components?**

5. **Will the company give a warranty? All mechanical devices wear out. What then?**

Can Breathing Ozone Hurt me?

Toxic people, and if you live on this toxic soup planet I am including you, have been reported to start detoxifying in the presence of ozone. If the level of ozone entering the body through the lungs is too high, the dead bacteria, viruses, and oxidized toxic poisons leaving the body will plug up the channels of elimination making the chest feel tight, or perhaps making the nasal passages and mucous membranes swell up slightly. The worst thing that happens is that the un-detoxed will find it gets harder to breathe because they still have particulate matter in their lungs. When they leave the ozonated area, or turn the machine down or off, the symptoms soon go away. The real trick to the correct use of an ozone air purifier is to start out at the lowest setting and *slowly* over the course of a few weeks turn the generator up higher as comfort permits as you get used to it. When any un-detoxed friends come over, turn it down until they leave and then turn it back up to where it was after they depart.

Can breathing ozone hurt me? You are breathing it right now. It is part of Nature. We all agree that water is good, but get enough and we start to discuss drowning. It's the same for ozone and any ozone or other active oxygen delivery system you choose. *Extreme* levels of ozone might make you *feel* you are suffocating, because the lungs swell from the heat of the exothermic destruction of lung pollution and foreign particulates by ozone, but you normally aren't damaged. The body wisely responds by not transferring the oxidized foreign water and gasses created on the outside of the lung cell surfaces back inside itself. To protect the internal organs it seals itself off.

Conscious people turn the machine down or off *long before* that kind of level is ever reached. Normal people who accidentally breathe extreme levels of ozone in industrial accidents recover in a short time without permanent damage in all but the most severe cases. The *window of effectiveness* with all these therapies is all-important and must always be respected.

Which Air Ozonator Should I Buy?

My favorites seem to change so fast I dare not put them down in print. But that doesn't matter since most of the modern manufacturers listened to what I said in the first book, and consequently almost all designs are now perfectly acceptable. Work with your dealer to correctly choose the size of unit best suited to your particular room size and application.

Manufacturers have been selling thousands and thousands of these ozone generating products for more than a hundred years without anyone ever complaining about harmful effects. I have heard lately of some isolated idiotic California government agency people trying to cite pool owners using ozone pool purification—for 'generating ozone pollution!' These miscreants are either plants of the chemical industries pushing chlorine trying to get customers back, or woefully uneducated 'know it alls.'

In my first book, I only spoke of the rare occurrence of trace nitric compounds from air ozone generation so people would 'get it' that ozone tests could falsely be made to look bad by quoting some study showing animals damaged by extremely high concentrations of ozone produced from 'dirty' generators. My comments had far more impact over the years than I was expecting, and now the manufacturers all tout their lack of nitrous oxides. Sometimes I am amazed at what happens when I write.

Atmospheric Ozone Concentrations

0.001 ppm is too low to be measured accurately, even with elaborate electronic equipment; yet hypersensitive humans can actually detect this by smell. (The natural outdoor atmosphere typically has concentrations from 0.001 to 0.125 ppm. The level varies with altitude, atmospheric conditions, and locale.)

0.003 ppm to 0.010 ppm in clean air, the threshold of odor perception readily detected by most people. These concentrations can be measured with fair accuracy. They are typical in residences and offices equipped with a properly operating electronic air cleanser, when outdoor ozone is low. (Higher indoor concentrations may result, if there is infiltrating outdoor ozone.)

0.020 ppm is the threshold for odor detection by laboratory equipment (to the 90 percent confidence level.)

0.040 ppm (measured as a sustained concentration in the test room) is the CSA maximum limit for devices that are intended for household use.

0.050 ppm (0.0000715 ug/ml^3 or 0.0000003 volume percentage) is the maximum allowable ozone concentration recommended by ASHRAE in air conditioned and ventilated spaces. I disagree with this politically influenced limit, since I know people who regularly work in thousands of times the recommended daily limits, and have safely done so for many years. But, to be fair, they *are* very clean people!

35,000 ppm medical or industrial units only make 50 mcg/ml3 (35,000 ppm or 2.3 volume percentage.) It will literally dissolve on contact the

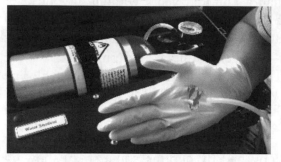

latex material of a thin surgical glove, yet _not_ hurt the hand in the glove. Latex is composed of dead cell organic rubber, and ozone attacks dead cells and therefore eats up the rubber instantly. The hand cells are alive so they are untouched. In this exact same way medical ozone instantly vaporizes microbes and dead, weak, diseased, or dying cells yet always leaves healthy cells untouched. All the words in this book, but when you see this demonstrated, you finally understand.

Air ozone strength concentration measurements always indicate higher concentrations the closer to the generator you take your sample. Look at sizing your ozone air purifying generator or series of purifiers to maintain a nice comfortable level of ozonated fresh air in the middle of your room, your boat, your house, or in any building or vehicle air you are trying to purify. For whole house, office, and condo systems, simply put ozone in front of your air intakes or inside your existing ductwork. You can have a contractor wire it in so it comes on only when the fan is running.

Do 'Ions' Matter in Ozone Air Purifiers?

Ozone is mega negative ions. Ion generators by themselves are a mere drop in the bucket when compared to the waterfall of a real ozone generator. They both emit negatively charged air particles, but the quantity is apples to oranges. No comparison. Using ozone purifiers results in removal of particulate contaminants and creates a cascade of negative ions and higher forms of oxygen. You may also experience the more relaxing atmosphere that comes from the smoothing out of erratic electromagnetic fields (EMF). Some manufacturers include both ion and ozone production in their machines, but they are not that necessary. Extras may have use, but just ozone is fine all by itself.

Environmental Illness, Allergies, and Sick Buildings

As I have been explaining for 14 years, allergies are caused by what I will symbolically call a pile of toxins in your body that get bigger and bigger each day as the irritating toxins of life here slowly build up and up and up within you. We're all different, and at some point the pile gets too big for you and one small exposure makes you go 'over the edge,' spiraling down into a cascading histamine or similar allergic reaction. Lupus and other auto-immune diseases are related examples of this. Each occurrence makes it easier and easier for this reaction to be set off again and soon it gets so bad that *any* irritant will easily set it off. At this point you have to start withdrawing from normal life.

I was allergic to many things as a child, owing to my (unknown at the time) poor diet of white bread and other bleached flour products, milk, and cookies (white sugar), few vegetables, etc. Ol' Doc Porter gave me all sorts of tests to find out what the causes were. It turned out that all grasses (pollen), feather pillows (chemical treatment), wool (chemical treatment), and various other things would set off a histamine reaction. In high school, we moved to the farm, and just one summer haying (over exposure) sent this body through the roof. Any pollen was trouble. I tried all the meds, to no avail. Of course not, medication does not remove the pile!

My eyes were continually sore, weepy, and bright red. Everywhere I went people would make jokes about hearing me coming from the sneezing. A trooper wanted to arrest me one night as we passed an accident scene. He took one look at me and asked, "Son, have you ever been in trouble before?" From my looks and red eyes he assumed I was on something. Luckily my friend Wayne's dad was the local Sheriff, and interjected, "Oh, that's just Ed, he always looks like that."

The reactions subsided once I moved away to college, but then I noticed beer and bad food would make it all start to come back. It wasn't until I was 38 when I oxygenated myself and ate right enough to 'remove the pile' that life returned to normal for me—and I became free of the constant allergies. We are under such a daily toxic load of pollution and irritants attacking us that we all need a good clean out.

Environmental scientists and knowledgeable ear, nose and throat doctors now call this becoming allergic to everything the 'spreading phenomenon.' They do not recognize my 'pile-of-toxins' metaphor because they are not trained in detoxification, only symptom management. They have a test called Provocation Neutralization Allergy Testing that can zero in on what sets off the reactions.

I have seen daily IV ozone detoxification therapy over the course of a few months remove so much stored toxicity that 'the pile' disappeared inside people; and nothing would easily set the person off anymore. Before doing ozone gas, one person I met was trapped at home due to allergic reactions, and after six months of ozone treatment she now lives a normal life. She no longer has to live in a protective bubble!

Building with slight mold problem! What's in your walls and ducts just out of view? Ozone permeates all.

Of particular concern today in our high tech world are the growing numbers of environmentally sick schools in which children can't learn and teachers can't teach. This is mostly due to airborne outgassing of vapors from carpets, their glues, and plastic building materials such as beams giving off formaldehyde. Sick office buildings with mold growing in closed ventilation systems are also damaging to health.

As an example of how this chemical damage can happen, my friend Billy had to sleep one night in a new building where urethane was drying. He woke up, and his throat was raw. Ever since then he has had a persistent cough because his saliva glands won't turn off and the fluid runs into his throat. But this reaction is mild compared to how bad it gets for some people.

Children and their teachers are trapped in new schools that literally drive them nuts! Through continual exposures being passed down and sensitivities being built up in each succeeding generation, a certain portion of our population is becoming especially sensitive to these mostly airborne toxins. This is why we are seeing new disease or allergy symptoms like the 'Bubble Boy' has which were unheard of before.

All sorts of people are now suffering terrible physical reactions when exposed to the 'Big Five' offenders:

1. **Dust**
2. **Molds**
3. **Certain Foods** —Colorings and preservatives in them can create Attention Deficit Disorder-like symptoms.
4. **Chemicals** (The most debilitating) —Ceiling beams out-gassing formaldehyde, carpets and their glue, paint, adhesives, etc.
5. **Pollen**

Once a child is exposed, he or she misbehaves, gets sick with a headache or bellyache, and/or his or her pulse changes. The child looks different, writes illegibly, and breathes differently. Children undergoing these conditions get red earlobes, bags under their eyes, red cheeks, restless legs, and runny noses. Dr. Doris J. Rapp is an expert on this and has a great video showing how exposures are followed by aberrant behavior phenomenon—from demonic rages, to listlessness—and can be repeated on demand. Reactions also range from hyperactivity to sluggishness, constantly falling asleep, or an inability to remember. According to the Allergy Research Foundation in Buffalo, NY, upon exposure one child's IQ dropped from 127 down to 57.

These chemicals are devastating the lives of families every time the child enters the building—or eats something in the artificial school lunches. Mom doesn't know any better and thinks she's a 'bad Mom' and/or that she has a 'bad kid.'

If the building air or the food problems being the triggers for the symptoms in the children ever come to light, the sufferers finally approach the administrators. But instead of cleaning up the building (which is wrongly assumed to be an expensive task) the parents are shuffled off to the uneducated school psychologists where they are told there is nothing wrong physically and that they need 'counseling' and anti-depressants.

The environmental-awareness-challenged administrators need not look at their budgets in fear or flip out when faced with this problem.

You don't have to tear down your new school, or bring in trailers, or install special rooms. Usually some properly installed filters and ozone air purifiers can easily eliminate most of these problems. Ozone can easily and inexpensively rid the air-handling ductwork of mold and bacteria! The side benefits might even be the same as when the St. Louis schools put ozonated air in their schools 80 years ago. Absenteeism from sickness dropped almost in half.

From Ozone in the St. Louis Schools —Edwin S. Hallett, St. Louis, Mo. Journal of Am. Soc. of Heat. Vent. Engineers, Jan 1920, and Oct, 1922— Fifteen months experience with more than 150 electrozone machines, in schools in 10 different towns:

"The use of ozone in English military hospitals (Rideal Ozone, 1920, p. 157) in treating open wounds reveals how baseless is the fear of harm from irritation from too high concentrations in ventilation…

"If a decision had been made in the first week of operation, rejection would have been certain. In some locations the machines could not be operated during working hours for the first week. Later, the machines could be operated with no perceptible odor, at full capacity, where, during the first few days, a fraction of the total capacity gave a most disagreeable odor... Peculiar and disagreeable are the mixed odors for the first few days of ozone application." [This is when all odors and their sources' surfaces are being decomposed, and are not quite yet inert. Simple to fix, just run the ozonators at night until they clear the building.]

So, we just solved another batch of life's serious problems; anybody keeping track of how many solutions we're up to yet? Who among my readers will have the follow-through necessary to bring these solutions all the way to implementation and stop the unnecessary suffering?

Who Is Against Ozone Air Sanitizing and Why?

Credible proceedings being published from institutions like the University of Florida. Their report dated July 14, 1993, from the "Symposium on Indoor Air Pollution," recommended, "Consider ozone generators as a method of reducing indoor air pollution. Studies have shown that modest ozone generation to levels of 0.05 PPM can reduce certain microbial agents, odors and mildew" and combat "Sick Building Syndrome." Despite such endorsements, air ozonation is still encountering the odd resistance, why?

According to Ken Hughes from Delta Marine in Florida, "Predictably perhaps, the major resistance to such applications of ozone has come from the A/C industry itself and companies that do mechanical ductwork. An ozone generating system would *cost as little as five percent of most duct replacement*, and more importantly, be only a *one-time installation* charge. Therein lies the rub, as there are virtually no residual or follow-up sales after ozone generating equipment is installed to take care of air conditioning mold and mildew and odor problems."

Devices mentioned in this book are on the Oxygen America website,
www.oxygenamerica.com

Ozonated Water in the Home

Clearwater Tech countertop ozone water purifier. Note the removable carbon filter. Carbon destroys ozone.

Aqua Sun Ozone water purifier pumps ozone through a tube into a submersible diffuser.

Thousands of municipalities such as Zurich, Switzerland; Florence, Italy; Brussels, Belgium; Marseille, France; Singapore, Moscow; Los Angeles, and other major cities all over the world purify the local water and/or sewage with ozone for good reason. It works.

Cities cleaning their water with ozone spend less because there is no expensive chlorine or fluoride to buy. The added benefit is that nobody has to drink it either. Their water is pure, but little of their ozone ends up at the end of the pipe in your home since it decomposes along the way. If you want more ozone for your family, you can do the same as the cities do right at home. You can make and use higher concentrations of safe ozonated water wherever and whenever you drink, bathe, clean meat or clean vegetables, or cook with water.

Bathing in O_3-charged water is a wonderful experience, as the skin is more a bridge than a barrier. The O_3 is absorbed through the skin, and on its way in it oxygenates and purifies, providing an excellent relaxing, yet energizing, soak. But be aware, because ozone is always jumping out of the surface of the water, yet it's heavier than air, so it

Municipal drinking water ozone purification system.

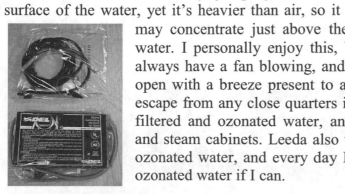

may concentrate just above the surface of the bath water. I personally enjoy this, but if it bothers you, always have a fan blowing, and/or a door or window open with a breeze present to allow excess ozone to escape from any close quarters if needed. I shower in filtered and ozonated water, and use ozone bagging and steam cabinets. Leeda also washes our food with ozonated water, and every day I drink oxygenated or ozonated water if I can.

To put the O_3 into water, a filter stone is used. The filter stone is similar to the bubbler in the bottom of an aquarium, except this filter stone has a special glue so that the ozone doesn't make the filter crumble to dust.

The ozone resistant filter stone is dropped into the water so that the ozone bubbles come up from the lowest point. The ozone is left to micro bubble through the water for five minutes for a glass of water, 15 minutes per liter, and one hour per gallon. At that point in time the water would be supercharged with ozone. You have to drink it all immediately to get the full effect of absorbing all the ozone in the mouth and stomach.

Half of the ozone in your glass will turn back into oxygen at about the one half-hour point. Gently swirl your containers before use, as the ozone rises to the top. Drink immediately. Although no one makes any sort of claims for these purifiers—not because we haven't seen a lot of good done—it's just that it is presently illegal to make a claim without spending millions of dollars on tests first.

By using high concentration ozone water purification methods you can make 10 megohm water. Ten megohm water is equivalent to triple distilled water. It's very hard to get water to this level of purity, but very possible by using ozone along with filtration, and it's much cheaper than distillation.

Really, the way to go is to install a whole house or whole building (or boat) ozone air system as well as plumb in an ozone water purification system. At the other end of the scale, manufacturers also make portable solar powered 12 volt (car battery) powered ozone generators for use in purifying drinking water on the spot in disaster areas. (Please see also the *commercial* section on ozone used in commercial building cooling towers.)

Yanco 12 volt ozone.

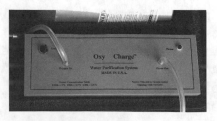

Why Do Certain Water Purifiers Cost So Much?

You have to be careful. The common home countertop ozone water purifiers always use air to make the ozone, and are shaped and sized in designs somewhat like small VCRs, or toasters and cost less than $500. You may be confused if you then run into an ozone water purifier that comes in a briefcase design, only uses pure bottled oxygen as its feed gas, and costs around $2,000 to $3,000—depending upon the bells and whistles built in. Yet both are labeled 'Water Purifier.'

What's the difference? They both purify water. One produces a much higher quality of ozone. What you have probably found is a high grade American or Canadian or European ozone/oxygen gas generator constructed of only the highest quality ozone-resistant and non-reactive components, and incorporating a closed bottled oxygen gas feed system. If legitimately of high quality, such a machine will generate pure oxygen/ozone gas that is equivalent in composition of output to the best medical grade ozone gas produced by the finest European medical ozone generators.

The practice of Ozone Therapy as used in medicine in Germany, Russia, Spain, Cuba, South Africa, and Italy commonly employs this highest purity gas made from pure bottled medical oxygen for injection and insufflation work. A high quality pure oxygen/ozone gas generator's output can also be bubbled through water to purify it. The FDA does not yet approve medical ozone generators for sale. Water purifiers are not used in medicine so they don't need approval.

Some people only want the highest quality drinking water and are willing to pay for it.

Solar Ozone Water Purifiers in Disaster Areas

Many ozone generators can be operated during and after disasters with only solar power, and of course they can purify any questionable water when normal clean municipal water services are not available.

Imagine, no more waiting for medicine to be flown in to a disaster site. Solar power or automobile and truck electrical systems can similarly be used to power ozone generators —instantly providing on demand treatment gas good for just about any situation or disease at the most remote or disaster struck locations on and off the planet. And you can purify your drinking water by removing all microbes at the same time! Every SEAL team should have one.

How did I come by all this knowledge? Thousands of conversations, interviews, and lots of research. Where did all this ozone stuff start? Who came up with the first uses? Stay tuned.

Chapter 15

The History of Ozone

How was ozone discovered? The tireless oxygen/ozone pioneers watched a new phenomena unfold before their very eyes. They left a heroic legacy in the world of industry and medicine and they richly deserve full recognition with their place secured in history. They're the pioneers of Oxygen Therapies.

Antique electrostatic ozone generating spark machine.

The Pioneers in Ozone History

Carl Wilhelm Scheele (1742-1786)—Scheele found that air is composed of two fluids, only one of which supports combustion. He was the first to obtain pure *'fire air'*—oxygen, (1771-1773) although he did not recognize it as an element. Scheele did not receive the credit for this because he hadn't published his work in a timely manner. Scheele's book *Chemical Treatise on Air and Fire* was not published until 1774, by which time European scientists were aware of Priestley's discovery of the same gas (dephlogisticated air) in 1774.

Joseph Priestley (1733-1804)—Priestley is generally credited with the discovery of oxygen in 1774, producing it by focusing sunlight to heat mercuric oxide.

Antoine Lavoisier (1743-1794)—Repeating Priestley's experiments Lavoisier sought credit for discovering oxygen. He proved water was composed of hydrogen and oxygen. Some say he founded the basis of modern chemistry. He lit the streets of Paris, and was beheaded in the French Revolution.

Martinus Van Marum (1750-1837)—Moving on into the discovery of ozone, we find this Dutch scientific dignitary as first reporting ozone's distinctive *odor* in the year 1785.

John Benjamin Dancer (1812-1887)—British born Dancer's ozone work is not generally recognized, but in fact he did contribute to the field of ozone. Using a Farady voltameter with large platinum electrodes he prepared gases (hydrogen and oxygen) from water by electro-decomposition. By slight modification of the conditions, he was able to prepare a colorless gas with a strong odor, which caused coughing. The strange gas was not named.

Thomas Andrews (1813-1885)—This Irish chemist and physicist demonstrated ozone's oxidizing and disinfecting properties. He was the first to establish the composition of ozone, proving it to be a form of oxygen. Before Andrews began to study ozone, it was postulated that the gas was either a compound of oxygen or that it was an oxide of hydrogen that contained a larger proportion of oxygen than water. Andrews proved conclusively that ozone is an allotrope of oxygen; and that from whatever source, it is one and the same body, having identical properties and the same constitution and is not a compound body but oxygen in an altered or allotropic condition. Ozone is triatomic, with molecules represented by the same formula, O_3.

Christian Frederick Schonbein (1799-1868)—Schonbein was of German birth. He began to investigate ozone in 1839, but 1840 is generally given as the year that he personally discovered ozone due to his making ozone more widely known. Working in a poorly ventilated laboratory, Schonbein noticed a pungent odor around some electrical equipment. He isolated the gas and subsequently named it 'ozone' because of its pungent odor. Schonbein was a prolific writer and published 343 scientific publications in 837 editions, including an 1832 book on the production of ozone entitled, *The Generation of Ozone by Chemical Means*. At that time it was referred to as *ozonized* oxygen.

Werner Von Siemens (1816-1892)—This brilliant chemist, inventor and entrepreneur of vision invented an ozone production tube in 1857. The Siemens Company is now huge and diversified.

Jacques Louis Soret (1827-1890)—Born in Geneva, this Swiss physicist and chemist, began his professional career as an assistant professor of chemistry. One of his most important works and discoveries was the confirmation of the formulation of ozone, that it is indeed triatomic oxygen.

Nikola Tesla (1856-1943)—Famous avid electrical inventor/genius patents his first ozone generator on September 22nd, 1896. In 1900, he formed the Tesla Ozone Company, and was using ozone medically.

Unknown to most people, Tesla also invented electric motors and radio! (He sued Marconi over ownership of radio's discovery and won.) He also invented the electromagnetic motors that have become the basis of all alternating current generators, a form of early television, electrical lighting and power distribution systems, standing waves, and the still unused wireless transmission of free electricity.

Many people make a case that their favorite nationality person was 'the discoverer' of Ozone and Medical Ozone Therapy. As you can see above, English, Irish, French, Dutch, Swiss, German, American and probably many more nationalities contributed to its inception and application. It truly was a worldwide effort.

Ozone's Introduction to Medicine

Oxygen Therapies have been used for more than 100 years, and have been used commonly for more than 80 years. We know this because a wealth of references and the old books that have been around can still be sought out. As a therapy, during the early 1900s ozone began to flourish here in America, but like many other successful natural remedies of the time it was discredited here by its chemical competitors, so ozone flourished in Germany where it was accorded wider acceptance.

When ozone gas was used for the first time in 1856 as a means of disinfecting operating rooms, it made its mark in medical history. Four years later, in 1860, ozonated water was put into use with the building of a municipal ozone water purification plant in the principality of Monaco, France. We now meet an interesting phase in history where the medical application of ozone begins to surface.

The Medical Pioneers in Ozone History

Dr. Day—In **1878**, Australia, Dr. Day wrote the first edition of papers on *Ozone Treatment and Scarlatina and Smallpox.*

Charles J. Kenworthy MD—In **1885**, the Florida Medical Association published *Ozone* by Charles J. Kenworthy M.R.S.V a doctor from Jacksonville, Florida who used ozone in his medical practice. Think about it, by 1885, this doctor was actually using ozone as medicine in America!

Dr. Charles O. Linder—In **1902**, Dr. Linder was featured in *Centennial Magazine* for injecting ozone and using state of the art ozone equipment in his Spokane, Washington clinic.

[Linder and Kenworthy's and Tesla's use of ozone proves ozone was in regular usage in the U.S. before 1885, and predating the 1906 Pure Food and Drug Act, its subsequent revisions, and the FDA as well. Therefore ozone's medical usage should be grandfathered in the U.S.]

Institute for Oxygen Therapy Healing—In **1898** the institute was started in Berlin by Thauerkauf & Luth, and the originator and founder of naturopathy Dr. Benedict Lust was practicing in New York.

Erwin Payr (1876-1946)—Payr learned of ozone, sitting as a patient in Fisch's dental chair. He immediately saw great possibilities for ozone in medicine. Payr and French physician Paul Aubourg were the first medical doctors to use ozone in rectal insufflation to treat mucous, colitis, and fistulae. He presented his 290-page publication, *On Treatment with Ozone in Surgery* at the 59th Congress of the German Surgical Society, in Berlin, 1935. In 1945, Payr's interest in ozone led him to inject ozone intravenously.

E. A. Fisch (1899-1966)—This dental physician and surgeon is credited with using ozone as a disinfectant in his dental work and called it 'Ozone Therapy.' Fisch had a wide range of experience with ozone and this was recorded in a large number of publications including Italian, French and German.

Joachim Haensler (1908-1981)—Working together with Hans Wolff, he took up medical work with ozone and designed a generator named the 'Ozonosan.' The large Haensler Ozone Company got its start here.

Hans Wolff (1924-1980)—After being a WWII American prisoner of war, Wolff started his ozone based medical practice in 1953 and devoted the rest of his life to it. In 1972, with Joachim Haensler, he formed the German Medical Ozone Society [Since 1993 renamed the Medical Society for Ozone Application in Prevention and Therapy]. He wrote the book *Medical Ozone* [Das medizinische Ozon] in 1979.

Ozone Begins to Take Off

In 1900, Tesla was using and teaching others to use ozone medically in the U.S. He was the first to use high voltage, high frequency, low amperage AC to make ozone. He was granted many ozone patents.

In Germany, Werner Von Siemens constructed a waterworks using ozone in Wiesbaden in 1901. This was followed one year later by another waterworks in the Westphalian city of Paderborn. In due course, Germany was to become one of the most prolific countries in the research and application of ozone.

In the 1900s, J.H. Clarke's London *Dictionary of Practical Materia Medica* described the successful use of 'Oxygenium' (ozone charged water) in treating anemia, cough, cancer, diabetes, influenza, morphine poisoning, canker sores, strychnine poisoning, and whooping-cough.

Recognizing its medicinal value, the Berlin physician Albert Wolff, first used ozone in 1915 to treat skin diseases. Ozone became available for the German army during World War I, and was used extensively to treat infected battle wounds and anaerobic infections.

Erwin Payr now involved greatly with ozone publishes *Ozone Treatment in Surgery* and presents it to the German Surgical Society in 1935. At the same time Paul Aubourg established the use of ozone enemas or rectal insufflation in the Paris Hospitals.

Not to be outdone, America, was now selling all sorts of big cabinet type medical ozone generators to practitioners all over the U.S. It was rapidly becoming big business, the 'talk of the town' and slowly becoming the 'in' thing to have.

By the time of the Second World War, Germany was filled with ozone institutes. Everybody was using it. However, as I explained earlier, the allied bombers then came along and blew up all the German ozone institutes, leaving only one building left standing, the IG Farben drug works. What a coincidence!

The IG Farben drug works company was the largest chemical manufacturing enterprise in the world on the eve of WWII wielding extraordinary political and economical power. Its representatives sit on the controlling boards of every pharmaceutical house of the world at this moment in time.

Due to the persecution by the drug trusts (pharmaceutical companies) at home, and the European destruction that the wars caused, ozone therefore disappeared from the public scene around the 1940s. The Germans still continued to use it, but the world in general no longer had such extensive access to this therapy.

Although the spread of German ozone institutes and the practice of Ozone Therapy was interrupted, medical ozone began to re-surface in other ways. During the Second World War the FBI would go to Brazil, seize Nazis and bring them to Ellis Island. As I already mentioned, Dr. Robert Mayer was a physician at the Ellis Island, New York prisoner of war camp. On becoming sick, one of the prisoners, a German engineer and chemist, went to Dr. Mayer and said, "Doc, I'm sick. Give me some ozone." Mayer declared, "Ozone, what's ozone?" That was the start of this German chemist teaching Dr. Mayer all about its medicinal value.

Because it worked better than anything else, Dr. Mayer continued to use ozone He specialized as a pediatrician using ozone for over 50 years on more than 14,000 children in his career before coming out of retirement to treat AIDS patients. That's when I met him. Before his retirement, he was working at Miami's Jackson Memorial Hospital and giving ozone to all the children as he made his daily rounds. After many years of this, the powers that be finally came by, and said, "Dr. Mayer please stop giving ozone to the children." He asked, "Why? It doesn't hurt them." The reply came, "I know, but nobody else is doing it, so you have to stop." Despite Dr. Mayer's success and by now expertise with ozone, he was thereafter prevented from giving ozone to sick children and curing them. Many Germans have retired to the U.S. and brought their knowledge of ozone with them, but Dr. Mayer was one of the few historical U.S. ozone doctors.

Ozone's safety when properly administered is proven. Remember the 1980 study by the German Medical Society for Ozone Therapy showed that out of all the millions of dosages given, the side effect rate was so low that only 32 of 5.5 million people had slight problems? The cleansing side effects of a few days of Ozone Therapy are sometimes; runny nose, a fever, diarrhea, some swelling, or some nausea, but always something that goes away in a couple of days. There's no permanent harm from the correct application of the proper protocols given at the right dosages.

Ozone was once limited until the advent of plastics in the 1950's because ozone gas attacks dead organic rubber hoses. It will eat up rubber surgical gloves and rubber bands, and that limited its usage in the old days. Today, the use of ozone resistant plastic components such as silicone or norprene in delivery hoses makes ozone easier to handle.

During the unfamiliarity days of ozone in the U.S., in 1958 Haensler introduced the first successful commercial ozone generator that produced specific therapeutic level ozone concentrations while using plastics in the design. In *The Use of Ozone in Medicine,* Renate Viebahn-Haensler stated "Haensler and Wolff paved the way for Ozone Therapy as we know it today." She further wrote "Constantly basing his research on the considerable number of publications by Payr and Aubourg, it was H. Wolff who subsequently introduced extracorporeal blood treatment [autohemotherapy] into medical practice; Werkmeister developed local treatment methods in the form of 'sub atmospheric ozone gas application', and Rokitansky—as a surgeon—presented the first comprehensive studies on the topical and systemic treatment of diabetic gangrene. Knoch then introduced rectal ozone insufflation into proctology, once more confirming its value in a controlled proctitis study."

Ozone Fever

Ozone's popularity and influence was not limited to medical use: Ozone also became of great commercial and industrial interest. Apart from the ozone doctors and clinics all over America during the early part of 1900, we find that old turn of the century historical records, pictures, advertisements, postcards, theatres, machines and bottles boasted the name ozone. In 1902, even a steamship proudly bore the name *Ozone*

Antique ozone button.

and old 1911 poster advertisements for the Central London (tube) Railway stated: "Every trip invigorates you. Father Neptune blows 80 million cubic feet of ozone through the tube daily!"

Ozone was further put to use as a food preservative for cold storage of meats. It was put into use to prevent yeast and mold destroying the fruit and meat.

1912 Ozol Ad

OXIDIZING THE BLOOD THROUGH THE AGENCY OF ELECTRIFICATION IN THE TREATMENT OF DISEASE ELECTRO-THERAPEUTIC GUIDE, 1912 HOMER CLARK BENNETT, M.D.

"PURE BLOOD MEANS GOOD HEALTH OZOL GAS MAKES PURE BLOOD

"A product of electrification, known as Ozol, by virtue of its merits, is now creating considerable interest in the medical world. The apparatus used in creating Ozol consists of an electrical ozonizer, operated by an alternating mode of high potential power—about 20 thousand volts—but the volume of mode used is very small, being less than two amperes.

"When the machine is in operation, a decided purple glow is to be seen in the ozonizers, and as the air is driven through the electrical discharge, there is liberation of ozone. A pressure blower is connected directly to a motor, and after the liberation of nascent ozone it passes through the crescent shaped tube, partially filled with a mixture of Pinus oils. Here the ozone and oils forms a new chemical compound, which has been given the name of 'OZOL.'

"The identity of the ozone is lost in forming this new compound, both as to its irritating properties and odor and also chemical reaction. In producing Ozol, none of the antiseptic or blood-building powers of the ozone are lost. These powers are increased to a marked degree through the agency of the terpenes.

"Ozol contains no ozone in its free state. While free ozone is a powerful oxidizing agent, those who have experimented with it find that it is too irritating to the respiratory organs to be practical, making it impossible to use enough of it to oxidize the blood to any great degree.

"The new compound gas is very agreeable to the sense of smell and the ozone is so disguised as to be unrecognizable in the product, Ozol. The pungency is so modified as to enable persons to inhale a 10 times greater quantity of the new compound than it would be possible to take of the free ozone. Consequently Ozol has 10 times greater oxidizing powers than ozone. [Via breathing, because you can inhale more.]

"I have witnessed the most phenomenal and unbelievable results from the inhalations of Ozol in the treatment of syphillis. Menstrual disorders of a functional nature yield to its influence promptly.

"The conditions that I'd especially mention as being most amenable to this treatment, are such as neurasthenia, melancholia, insomnia, anemia, asthma, hay-fever, bronchitis, early stages of pulmonary consumption, dyspepsia, constipation, headaches, inactive liver or kidneys, menstrual disorders, and syphillis in any stage, and I would say that it is a most valuable adjunct to surgical, electrical and other procedures, for the relief or cure of organic diseases.

"This method is most rational. It appeals to both the physician and the patient, and the results produced are almost invariably satisfactory, especially if the disease has not reached a condition of organic disintegration.

"A LOW TEMPERATURE IS NOT A NORMAL CONDITION IT'S A DANGER SIGNAL"

"So prevalent is subnormal temperature among people 'run-down,' that nine out of ten of them will show a subnormal temperature by actual thermometer test. The clinical thermometer is the best means of determining the existence of under oxidation and should be used preferably mornings. The temperature of one who is under-oxidized will be found to run from a fraction of a degree to three or four degrees below normal.

"ANTISEPTIC AND OXIDIZING POWERS OF OZOL"

"Scientists long ago recognized the great oxidizing and antiseptic powers of ozone, but owing to its irritating effects upon the respiratory organs, when used as an inhalant, little progress has been made with it in its free state in the treatment of disease.

"We have succeeded in perfecting a machine whereby we convert the air into ozone and then convert the ozone into a new peroxide compound ($C_{10}H_{18}O_3$) for inhalation purposes which we have named 'OZOL.'

"An all-important chemical change takes place and the new compound gas, Ozol ($C_{10}H_{18}O_3$), thus formed passes through flexible tubes and is inhaled by patients by means of moderately close fitting face masks. The chemical union of Ozol with the waste products of the body is produced thorough combustion, the body temperature is raised, waste products are properly eliminated, and a healthful condition of circulation is re-established.

"In the new peroxide compound, Ozol, none of the oxidizing, antiseptic or blood-building powers of the ozone are lost. In fact, these powers are increased to a marked degree. The new gas thus formed contains no ozone in its *free* state. The identity of the ozone is lost as to its pungency, odor and chemical reaction. A combination of oils of the Pinus group has been

selected (Eucalyptus, Pine, Thyme, etc.), which not only constitute an agreeable and effective inhalant, but are also recognized as possessing marked therapeutic value in the treatment of diseased mucous surfaces."

The ideas expressed in these snippets from the old history books tell us our forefathers knew all about ozone. It was not merely an isolated substance of little worth. Ozone was a quantum leap forward. It generated great interest in anyone who heard about it, and spread to many segments of society and to many countries. Bottom line. Ozone worked, and everybody wanted to jump onto the bandwagon. Now it's your turn to jump on. We'll start with examining all the ways that this wonderful gas can be safely put into the human body.

Glycozone

Chapter 16

Ozone Therapies in the Home

I have on file medical records of people who were treated for AIDS. They were HIV positive with the AIDS disease, and now they're HIV negative and they no longer have the disease or the secondary diseases. All I've done as a journalist within the past 14 years is interview thousands of people using ozone. I've been all over the world lecturing on this. I've visited scores of clinics and interviewed hundreds of doctors who have treated innumerable diseases, including AIDS: all of these doctors clearly told me that their patients got well by simply putting 'active' forms of oxygen into their bodies. When I visited those places, I videotaped, recorded, and wrote down as many of those interviews as I could.

During those interviews, hundreds of people who used an Oxygen Therapy told me that they no longer have cancer or some other degenerative condition. They somehow found out about ozone, got the treatment from a competently trained professional or treated themselves with the equipment and took oxygenating supplements. What did all of this tell me? Individual results will vary, yet these results were jumping out from the pages! In my well-researched and very experienced personal opinion, ozone can rid the human body of most disease by curing multiple maladies in a remarkable way.

According to the manufacturers like Haensler in Germany who divulge how many machines they sell per year and to whom, there are more than five thousand physicians in Europe who use Ozone Medical Therapy on a regular basis and who have been doing so for more than 50 years! How could ozone possibly be toxic? Ozone Therapy has been around 50 plus years and used on hundreds of thousands of people and millions of dosages have been given, so there's very little new information we need to learn about it. To comply with legalities only, some companies have applied for human testing in the U.S., but in the U.S. this is hardly encouraged or allowed. The tests are always rigged, or stopped through media, political, or economic 'problems' that magically spring up. We will get into some of the politics later, but for now I believe I have made my case. Ozone has been safely and effectively used daily all over the world on millions of humans. And strangely, as of the date of this publication, it's not officially tested on humans in the United States. I wonder why? Who's *really* running the show? And how much death and suffering can be laid at their doorstep from keeping ozone hidden from view?

Medical Applications of Ozone—the Official Position

In 1976, the Food and Drug Administration (FDA) strangely, against all logic and the massive amount of history and scientific evidence available with only a cursory examination, declared ozone (via publication in the Federal register) a 'toxic gas with no medical use.' They have taken it upon themselves to totally and publicly misrepresent reality. Does the evidence warrant such an opinion? With so much suffering going on, why would they mysteriously stand rigidly opposed to beneficial truth?

For the sake of simplicity we'll have a little review. Ozone is a natural substance, like air and sunshine. Medical and naturopathic ozone is a mixture of pure ozone gas and pure oxygen (0.05 percent to a maximum of 5 percent O_3.) Why was such panic terminology being used to describe ozone in a government publication?

Ozone is a very powerful bactericide, virucide and fungicide. In layman's terms, that means ozone has an extraordinary power to destroy and inactivate a wide variety of viruses, bacteria, yeast fungus and protozoa. It also oxidizes pesticides, chemical manufacturing wastes, poisonous compounds, and putrefactive matter and odors in air and water.

Lack of knowledge, stubborn resistance, arrogant know-it-all-ism, and a lazy, apathetic attitude toward the known facts has contributed toward the prevention of ozone's use. Evidence of its safety and benefits exist by the truckload, yet the official position is to hide under complacency and politics. The government keeps waiving the restrictions on stacking the agency advisory panels with vested interests, and it always ends up that more than half of the experts hired to advise the government on the safety and effectiveness of something have financial relationships with the pharmaceutical companies.

The drug industry is the most profitable industry in America, and secretly knows Oxygen Therapy is effective and sees it as suffering be damned competitive to them. There are more paid registered drug company lobbyists than there are Senators and Congressmen combined.

Naturopaths and medical doctors skilled enough to correctly give ozone and other Oxygen Therapies are found to be in scarce supply here in America. True old school naturopaths are the original oxygen therapists. Burnadine University (Burney) in Los Angeles California, has kept the original naturopathic Oxygen Therapy teachings alive in a joint effort with Eshoo University under the auspices of The American Schools of Naturopathy. Dr. George Freibott, is CEO and Vice President of

Burnadine. Some states like Nevada and Florida have been tricked into actually trying to outlaw naturopathy. There are, however, changes afoot in the legislative area. In many States consumers fed up with only being allowed to choose expensive therapies that don't work got together and forced passage of several Alternative Medicine Practice Acts. We'll examine them further on in the book.

Another bright spot appeared as another hole in the wall. June 26, 2001, 21CFR184.1563. The Food and Drug Administration (FDA) amended the food additive regulations to provide for the safe use of ozone in gaseous and aqueous phases as an antimicrobial agent on food, including meat and poultry! They may need some spontaneous prodding to extend their blessings to humans.

Oxygen supplements, therapies and related activities are historically inexpensive and proven effective when used as directed by competently trained healthcare professionals. We must use education, the media, and, with your help, aggressive American research as proof to silence whoever postures against oxygen, your very right to life itself.

Is Ozone Safe? - 5.5 Million Ozone Treatments Say Yes

As I will repeatedly refer to, the German Medical Society has reported in print that 384,775 patients were given 5,579,238 applications of ozone; adding that the side effect rate observed was only .000005 per application!

Now compare that to how many people die of drug-related side effects in this country annually. Statistics show that every year we lose two and a half times the number of people killed in Vietnam to the side effects of drug therapy.

By comparison, the side effects of ozone are minor inconveniences. When you start to oxidize your inner garbage, the body will move it out. If you ask it all to move out on the same day or within the same few days, you are going to be extremely uncomfortable. It will be just like jamming all the cars onto the freeway at rush hour. You can't get on the ramp, or down off the ramp, because everything's jammed up. That's what happens in your body if you ask it to cleanse too fast by using too high dosages.

That's why following known protocols and having a competent health care professional guide you through the process is very important. During cleansing you might start to mimic all your previous diseases. Imagine peeling an onion layer by layer as you go in toward the core.

That's what happens at the site of every cell in the body when finally enough oxygen reaches that cell. Layer upon layer upon layer of toxicity peels away—unearthing the leftover remnants of diseases, pollutants, drugs you took as a child, dead viruses and bacteria. As they are unearthed you may re-experience their symptoms as they leave. They're inert, but their chemical signatures might give you the same irritating symptoms as you suffered the first time around. When finished, naturally your health takes on a whole new dimension.

If You Choose to Self-Treat at Home

Please work with an experienced and competently trained ozone therapist or other healthcare professional skilled in the therapeutic use of active oxygen forms such as ozone if you make the decision to self-treat. Everybody, patient and therapist alike, please keep in mind that it is far safer to treat slowly with moderately increasing dosages spread over the long term. The patient is much more comfortable this way, as this gives the patients body time to flush and rebuild, flush and rebuild.

There are many options available for those desiring treatment and supplementation at home. The following pages will explain them to you in detail so you know what they are, how good they are, and what to expect from them. Please also review the prolific medical cites and testimonials in the back of this book.

Ozonated Pills, Powders, Potions & Lotions

Homozon

 Homozon stands in a class of its own. You cannot fail to be impressed with this product. Dr. F.M. Eugene Blass, the inventor published many works around 1929 concerning his magnesium, calcium, and sodium-based products that release ozone and hydrogen peroxide in the body. And today many use his name to sell knock-offs of his product.

Homozon was probably created through the collaboration of Nicola Tesla and Dr. Blass back in the '20s, '30s, and '40s. Incidentally, at one time they both lived in the same NYC hotel, so they no doubt would have spent many hours of conversation discussing their pursuits and experiments. Tesla was a genius and more than likely was the one who helped Blass optimize this wonderful preparation. For many years Blass made and sold Homozon and gave it to people. He had case history after case history, all showing the fantastic results that Homozon produced. At the age of 86, a black limousine pulled up outside his lab and two men got out and clubbed him to death in the street, according to his lab assistant. (A competitive ritual killing, oxy-Dr. Koch murdered same month and year.)

Homozon is based on a powdered form of pharmaceutical (not industrial or commercial grade) magnesium, mined from a specific site known for its desirable chemical properties. Homozon is completely legal for consumption and sale in the U.S. and has been in wide usage since 1898. Homozon consists of magnesium with ozone bonded to it by a special proprietary process. Magnesium with ozone is $2(MgO_3)_2$, Magnesium ozonide. All other copycat products use only magnesium, or magnesium oxide, or dioxide. The knockoffs of Homozon don't use the special pharmaceutical grade magnesium, and they only bond oxygen, not ozone, to the magnesium. Bonding ozone to a powder (a gas to a solid) is a trade secret process, that's why no one else has the same high quality product. Only Dr. Freibott and International Oxydation Laboratories can make it. That's why Homozon is superior.

There is no way for you or the many distributors of the knockoffs to tell the difference because so much literature and labeling is deceptive, and all the brands will give you the runs because magnesium is a laxative. Many will feel better from just colon waste liquification, but the huge importance to your health difference between Homozon and the imitators is that the Homozon gives the loose stools AND delivers super high oxygen levels. Almost all the Homozon knockoffs use Blass' name in their ads and claim to have high levels of oxygen, and some will claim to have ozone (O_3) but it's usually not true. Many sincere and well-meaning people have been parroting incorrect information for years about their products and assume loose stools are the proof of oxygenation. They will get mad that I said this. Only Homozon delivers the most ozone, and the ozone is what delivers the resulting high compounded oxygen levels.

The secret Homozon manufacturing process ends up producing a fluffy white powdered magnesium. When you open the lid of the blue cardboard can, you'll see the powder kind of float in the air. It's light, fluffy and velvety. It has so much oxygen in the powder that the agitation makes a little oxygen come out inside the can and hold the particles in the air. You can actually see it. It's unlike any other powder you've seen. Each can of Homozon has liters of oxygen bonded to the magnesium powder inside it. Ah, but you say, the can is only about six inches high, so how can it have liters of oxygen in it? Through chemistry. The manufacturing process bonds and compresses the ozone into the powder.

In Dr. Blass' 1929 literature, we find the following quotes:

1. OXIDATION IS THE SOURCE OF LIFE AND HEALTH
2. IMPAIRED OXIDATION MEANS DISEASE
3. CESSATION OF OXIDATION IS DEATH

"DISEASE is: Impaired health due to impure vital fluids. A clogged up and, in its normal functions, hampered organism giving food and lodging quarters for the various parasites, growths, etc., to an unhealthy body!

"As long as the digestive tract is not cleaned, there is little or no chance for the active oxygen to reach—passing through the capillaries and blood vessels—the other, especially the extreme, parts or organs of the body.

"Frequent stools (from use of the products) will cease as soon as the digestive tract is free of waste matter, then nascent oxygen (even of large and frequent doses of the preparations) will be available for the purification of the blood and lymph vessels and organs; a heavier urination will take place and is necessary to carry off the waste products of the oxidation process. As long as only Magozone or Homozon is used, watery stools will be present until the digestive tract is freed of waste matter. These stools are the logical result of oxidation (oxidation means changing solid waste matter into water and gas); it has nothing to do with diarrhea and does not weaken the body." (The Homozon oxidizes waste so well that it liquefies.)

If improperly digested food matter or unnatural chemicals stay in the colon too long, the body tries to protect itself from these substances by producing insulating secretions (mucus), and you've got a bunch of usually clear and seemingly harmless mucus coating the colon. If it stays there you've created a favorable condition for anaerobic putrefying bacteria. Unlike clean new baby poo, your stools start to stink and you know some food and toxins are simply rotting inside you, instead of being absorbed back into your blood. You're stopping absorption and increasing putrefaction, as in a stagnant swamp. Your bodily fluids, the lymph and the blood, end up always dirty, as though your body was a car that never had its air, fuel, or oil filters changed. A dirty machine doesn't function for very long and breaks down at the most inconvenient of times.

The oxygen and magnesium will eventually thoroughly clean the bowel. But as I said earlier, loose bowels are not in any way an indicator of oxygen delivery to the tissues and blood. Homozon is still the only product I know of that liberates many liters of oxygen per can. Having made that point, it is still true that any colon cleansing at all is going to increase some oxygen availability to the tissues of plugged up people, by making 'room' removing some cellular blockages. A lot of people and friends of mine have a lot of money invested in this market, so I'm taking a chance of ruffling a lot of feathers telling you the truth, but the people deserve to only have the best when paying good money seeking their health. Many who are convinced their oxygenating colon cleanser is as good as the original don't know what they're missing.

Manufacturer Protocol

Following directions on the label, mix the desired quantity of Homozon with water and drink. Then chase it with the juice of half a lemon. Lemon juice is used because most people have insufficient stomach acid so they are unable to completely break the ozone off from the powder to release all of the ozone. Lemon juice and stomach acid combine to break the magnesium ozone bond. The magnesium cleans out the colon, and the ozone oxygenates and ozonates the blood.

Homozon can also be sprinkled on food. Mixed with water, it is a light, chalky-tasting white liquid that releases a large amount of oxygen immediately. This may be one of the most efficient oxygenators yet. The first teaspoon I tried had so much oxygen in it that it made me giggle, and my head swoon. Yes, I guess it was love at first bite!

Homozon Cleansing Reactions

It's easiest to start with a heaping teaspoon before bed, and when you get up, keep a toilet brush handy! Plan on liquid stool releases 25 minutes after a meal, and learn to momentarily hold the liquid in during the day or else you'll keep soiling your clothes. When you start taking Homozon, I guarantee that you will smell the most interesting and rotten smells coming out of your toilet that you've ever encountered. You will remember that German beer you had years ago that tasted unlike any other beer you ever knocked back, and all of a sudden you smell it while you're sitting on the throne. Yes, some of that waste was still stored inside you. And you'll find little seeds and nuts coming out, charming pieces of things, along with strips of black matter that look like rubber tire pieces. Then you'll know your colon is having a good time! There's a mucoid coating inside the colon that gets too sticky if you maintain an acidic pH. It's all stored in pockets of old waste putrefying and auto-toxicating you every day until you get enough oxygen/ozone and water (or colonics) to clean the colon wall and throw the garbage out.

Most people don't have all these experiences, because most people never take enough of it. It's only the people who really have a disease or think, "I want to be sure I don't get a disease, and I want to clean my body out," who consistently apply the protocol several times daily for five or six weeks, or two-to-three months, and see how good the benefits are. They really clean out, and then they come back in a year or so to restart the procedure. This is the way to feel the true benefits of it. I once took it several times a day for months. I remember being amazed. One day I was so clean and had so much energy I felt 18 again, and my solar plexus was actually *radiant*. I had forgotten what that felt like, we all do in time. It's only the piled up waste that makes us feel and be old.

Ozonated Oils

Supermodels and supermodel wannabes take note! There are whole skin care lines based upon oxygenating your skin and keeping it young! Looking old comes from a build-up of toxins and a loss of oxygen in the connective tissues under the skin. In almost every case I have seen, everyone who does Oxygen Therapies looks younger and younger.

Ozonated Olive Oil

Therapeutic ozonated olive oil is produced by a number of companies. It is wonderful for a variety of ailments, delivering active oxygen right to the spot needed. Used externally, ozonated oil made from high-grade bottled oxygen/ozone is another medium to treat disease and infection. It is regarded by many as the most effective medium for retaining the gas, as olive oil is easily ozonated. Oxygen species are absorbed through the pores of the skin which enters the circulatory system and enriches the blood. It has been found especially useful in treating a variety of skin disorders. It has also been used in the post surgical treatment of wounds. Cuban physicians have used it to treat duodenal ulcers, gastritis, giardiasis and peptic ulcers. All the Ozone Therapies have advantages, but with ozonated oil the ozone keeps considerably longer than in water.

Among other things, users and practitioners have reported that ozonated olive oil has been used to successfully treat the following conditions:

Acne, athlete's foot, bacterial infections, bed sores, blackheads, bruises, candidiasis, carbuncles, chapped lips, cuts and wounds, dandruff, dermatitis, eczema, diaper rash, earache, fungal infections, impetigo, insect bites, fistulae, gingivitis, herpes simplex, hemorrhoids, leg ulcers, shingles, spots, stomach conditions, sunburn, whiteheads, wrinkles, yeast infections.

Doctor's report that the ozonated oil speeds the healing process, improves blood flow, alleviates pain, reduces inflammation; it has anti-bacterial, anti-fungal properties, heals cell tissues, improves circulation, stimulates blood flow and relieves pain among other things.

The oil can be taken on frequent occasions on a daily basis to heal the condition. It can be applied to the affected area and a half a teaspoon can be ingested to detoxify the liver and hasten the healing of the affected condition. The oil can be directly applied to the liver region, which is located underneath the rib cage to the right hand side of the body. A good ozonated olive oil when rubbed into the skin feels very soothing to affected areas. Apart from its therapeutic value, it is also energizing and refreshing.

Ingestion of Ozonated Oil

When properly stored, ozonated oil retains the oxygen/ozone gas for many months. The oil should be cold when ozonated but not to the point of congealing. This is then refrigerated. When used internally, ozonated oil has beneficial effects. A teaspoon to a tablespoon may be consumed one to four times a day. Capsules may be filled with the oil. Consult with your physician. The effects of ozonated olive oil have been extensively used in clinics and hospitals and documented in Cuba.

Ozone Equipment in the Home

Ozone Gas 'Body Suits'

You too can look like a doughboy in the comfort of your own home! After you try breathing safe levels of ozone, how about getting into a bag of it at a much higher concentration so that the ozone can create an "array of lipid oxidation products and reactive oxygen species, such as un-ionized hydrogen peroxide, [that] can pass through the transcutaneous barrier and, enhanced by the vasodilation that has occurred enter the circulation." (*Bocci et al. Eur J Appl Physiol*, 1999, 80: 549-554).

In other words, these O_3-created active oxygen species soak directly through your skin into your capillary network and are carried all over the body by the blood. Ozone bags are a bit like a spacesuit. You feed the high concentration (27-80 mcg/ml^3) ozone into the bag, and it puffs up. Instead of being immobile, you can walk around, talk on the phone, or recline as you receive your beauty and health treatment. Hot showers and leaving the skin wet are recommended beforehand to open the pores of the skin to allow easier absorption. Water carries oxygen into the body. After this form of treatment, people report that they feel relaxed and invigorated.

Limb Bagging

Instead of the whole body, how about just an arm or leg? Here ozone is primarily used for treating leg ulcers, gangrene, burns, slow healing infections, that sort of thing. Old open wounds often heal right up; the pictures in the literature are quite dramatic. In an emergency you can use a plastic trash bag wrapped airtight around the area and filled with high concentration ozone. Many clinics use this method and the ozone body suit as well.

Rectal Insufflation of O$_3$

I, and many, many, others commonly do this at home, but we'll put it under clinical use and describe it later.

Ozonated Steam Saunas

On top of breathing ozone, soaking in ozone, and drinking water full of ozone, why not try sitting in a sauna full of ozone and steam? There are two types of ozone steam sauna units available for the home; fold-up portables, and heavy-duty cabinets.

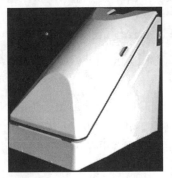

One of the cabinet models comes complete with its own oxygen concentrator and ozone generator! It's the world's *first* personal self-contained Oxygen/Ozone Sauna cabinet. All others have the ozone generator and oxygen concentrator (if there is one) outboard.

A truly revolutionary solution for beauty and health care. Steam inside the cabinet creates hyperthermia. As the pores are opened it raises the body temperature, allowing the highly

concentrated active oxygen species to enter the body transdermally. The oxygen gets into the blood and is carried into the cells. Oxidized

Steam generator on left. Center: Nylon sit-down ozone sauna. Ozone generator on the right.

matter is then released from the body through the open pores. The oxygen is utilized by the body to detoxify dermal cells and remove wastes with enhanced blood circulation, while ozone kills germs and bacteria on your skin. A steam sauna is a very relaxing and stimulating combination of water and ozone. When employing steam ozonation, do not exceed 39-to-40 degrees centigrade. Above 42 degrees centigrade the heat will destroy the ozone.

Differing Viewpoints

An industry colleague and I are adversarial friends. He doesn't think I should write this book or even promote Oxygen Therapies. He believes the scientists should be the only ones working in the field and that I should get out of it. He has a point, but only up to a point. I have a different viewpoint as you can tell. I feel that unless the people know about this subject and what methods are already proven and available, they can't demand it and it will stay hidden under the 'research' rock it's been under for the past 100 years in America. Until I did the 1,500+ lectures, no one cared. There was no demand. There was no big industry. Now there is.

My colleague describes his company as follows: "Our Company designs and produces high output ozone instruments for the therapeutic and steam applications. Our former clients refer almost all our clients to us. We are supplying custom designed and produced ozone systems and glassware to labs all over Canada and the USA. Our lab clients are also in Mexico, Brazil, Chile, Finland, Turkey, Malaysia, Singapore, and Australia. These labs are working on a vast range of the biological projects, ranging from skin graft treatments all the way to viral research. People working there generate an incredible amount of data, getting the ball rolling."

He requested that there are a few things, "which would be nice to put into a proper perspective in your book:

"1). People should be extra careful in respect to purchasing their equipment. They have to know where they will go with their instrument for service/repair/upgrade/support, once the original producer is out of business or out of reach.

[This sort of thing where start up companies come and go along with their service and support is common to any budding industry like our new personal ozone use industry.]

"2). As to ozonated water in the bathtub, prepared by using the diffusers.... I cannot even count how many people called me, still coughing, wondering what the heck is going on and why the ozone does not stay in the water. Most people told me the idea for this form of a treatment came from your book.

[As I noted in my 1988 *Oxygen Therapies* book*:* "OZONE BATHS...Ozone bubbled through warm water irrigates the skin. It disinfects and treats eczema and skin ulcers." Back in 1988, ozone was still new, and I was remiss in not providing more information like it jumps out of water. Nowadays most detail work is done by the salespeople for the ozone companies, and they are the ones who fail to give the proper warnings with the equipment they provide. No one is permanently hurt from breathing a little too much ozone, but the too-rapid detox can make you cough, as he said, so be careful!]

"3). Do not use air fed ozone generators for ozonation of olive oil.

"4). Do not use air fed ozone generators for insufflations." [I agree completely.]

In Search of Ozone

A man called Leonard wrote to me saying, "My wife and I are looking for an ozone generator just like we saw in East Texas." Apparently, they met a man there who had overcome stage four advanced lung cancer using one of these medical grade ozone generators. Unfortunately, these are usually not registered medical devices. Only the manufacturers approve their own devices, a conflict of interest, and we have been lucky that almost every person so far who got into ozone device manufacture was motivated by the sincere desire to help, and not to greedily take. These devices are strictly ultra high quality bottled oxygen fed water purifiers. If somebody takes it upon himself or herself to hook one up somehow and use it in some other way than water purification, it's their responsibility.

Undercover FDA Agent

For example, I know a New York man, Matthew Morton, who was approached by an undercover FDA agent who had a hidden tape recorder on him. The man sold the agent a water purifier and—because Matthew made medical claims surrounding the sale—was arrested for selling an unregistered medical device, and for mislabeling his product by virtue of the claims made. He went to prison.

So, be very careful when you're out there. You have to understand the politics of the situation in America today. Although there are more than 9,000 physicians in the world who have used Medical Ozone Therapy over a period of 50 years, in the U.S., the FDA has unjustly declared ozone a 'toxic gas with no medical applications.' And on the day they did this, these 9,000 physicians were merrily, safely, and effectively injecting ozone into people's bodies just as they had collectively been doing for those past 50 years. I think there just may be a disparity there between what is presently the law in this country and what is the actual reality of truth in the world.

Ozone Resistant Materials

Giovanni wrote to me at the Family Health News—"I request any information about what materials are resistant to ozone gas. If possible, the materials used in ozone generators."

Stainless steel is used in Europe in the construction of medical ozone generator tubes. In the United States, many of the generators use a double glass wall dielectric: in other words, over here many of the generating tubes are made only out of glass, because the users and manufacturers prefer to have the ozone come in contact only with the special ozone resistant glass, and never touching metal at all.

However, in fairness to the Europeans that make their machines with one half of the generator tube being glass and the other half being a stainless steel electrode, they've used that method for 50 years. No one has ever had a problem with it. So take your pick, but you're going to have to do a lot of research. Ozone generators for use in medicine specifically must only be made with ozone resistant materials. All the nuts, screws, and clamps, everything and anything that ozone touches must be made specifically of the resistant materials.

Ozone Services published the following list on the web:

> Materials with fair resistance to ozone which can be used for applications dealing with concentrations as high as 40 mcg: Stainless steel, Neoprene, EPDM, and Karlez.

> Materials with good resistance to ozone in concentrations up to 90 mcg: silicone, LDPE, Viton®, and Kynar (also known as PVDF).

> Materials with theoretically unlimited resistance to any ozone concentration: Teflon, and glass.

Please check with ALL manufacturers to be sure the ozone only touches the above components on its way through any machine or accessory.

The Violet Ray

The Violet Ray is also known as Tesla's Violet Ray and Edgar Cayce's Violet Ray. Nikola Tesla held the first patent on it, and a number of other 18th century researchers such as Frederick Strong, Thomas Curtis, Noble Eberhardt, and Chris Branston were doing a great deal of electro-medical research and experimenting with high frequency currents during the 18th and 19th centuries. Many patients reported successes through the use of their devices, but the electro-therapies were never popularized in modern times. Along with other electro-therapies such as Royal Rife's healing frequencies, or the Multi-Wave Oscillator, and all other non-drug therapies, these therapies were not properly supported and fell into obscurity.

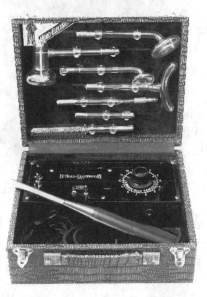

Author's antique French 'Holo-Electron' Violet Ray device with 8 different electrodes.

The Violet or the Neon Ray is a high voltage and low current source of high frequency (10,000+ Hertz). The Ray provides a pleasant tingling oxygenation (in the form of ozone O_3) in the blood. The Ray appears to also work by BALANCING and ENERGIZING the nervous system and moving the Chi.

Physicians in the 1920s used the Violet and Neon Ray devices successfully for almost every ailment. They commonly used 62 different forms of glass electrodes.

The generators were often used for cervical, throat, prostate cancer, breast cancer, and other diseases until the pharmaceutical companies eliminated respect for all other therapies. The doctors were forced to stop using the Neon Ray and the Violet Ray, and had to sell pharmaceutical drugs only. If they did not comply, they were labeled as quacks.

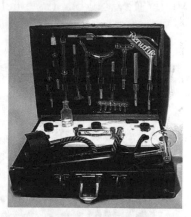

The Neon and Violet Ray devices are sold today only for cosmetic purposes, to reduce wrinkles and promote hair growth by bringing circulation and energy to the scalp and face. They produce mild ozone, and ozonation has also been proven to kill parasites, bacteria and viruses.

Like other Oxygen Therapies and supplements, the Violet Rays are reported to be useful in reducing pain. When you have pain, there is a corresponding build up of hydrogen in the area. We know the Violet Ray charges blood capillary beds with active oxygen, which then acts to replace the excessive hydrogen build-up. Where there is pain there is also a corresponding build up of positive ions. The Ray is said to replace these positive ions with beneficial negative ions, which is very healthy for the body. The Violet/Neon Ray may also cause electroporation of the cells, meaning the cells become more permeable (The cells 'open up' a little more than normal), which makes the toxins flush out easily so the nutrients and oxygen can rush in. This of course, can help detox the cells and improve the lymph flow.

Where there's pain, swelling, inflammation and disease, there is also a problem of low electrical charge in the same area. This charge can be measured in millivolts. The Ray devices are proving that in most cases

they can eventually raise the cell charges back to normal. A normal healthy cell has a charge of approximately 100 millivolts, an aged cell is 50 millivolts, and a cancer cell is 15 millivolts. The Violet Ray is thought to be a literal 'cell battery' recharger.

If this imparting of a recharge to our cells is one of the ways the Ray and other related electro-devices work, the reported healings would make sense, since at the most basic level we are all electricity. We are ultimately flowing atoms and electrons, and without electricity there is no life. I would love to see modern studies measuring this recharge phenomenon.

Perhaps proving the recharge phenomenon exists, high voltage aura photography already shows that the Ray moves our Chi, Prana, or Life Force, or at least our cellular electrical emanations. (See our Evidence section for photos). The Ray is said to circulate the stagnant blood. Many physicians believe stagnated [and therefore unoxygenated] blood is the cause of pain, swelling and disease.

The Violet Ray devices get their name from the bright violet plasma glow emanating from the excited gases inside the elecrodes. Inert argon and neon gases are trapped inside different shapes of sealed hollow glass, and when excited the argon glows purple, and neon glows orange. The gasses are excited into luminescence with high frequency currents produced by Tesla Coils. When held near the skin, or inserted into a body cavity, the electrodes transfer masses of electrons into the body. It's kind of like mild indoor mini-lightning coming off the electrodes.

The electron transfer at multiple frequencies is perceived by one's nervous system as a tingling, or a mild shock, especially if the electrode is not touching the body. To minimize this you can put the electrode glass in contact with the skin first, and then power up the device.

A similar ozone-generating device known as the 'Calozone' was also used in medicine. The main difference was that instead of having only a single Violet Ray electrode, the Calozone applicator was a bank of alternating orange and violet electrodes set into a movable hand-held rack configuration. The application tube stack was placed upon the body or moved across it.

At the site of all the electrons flowing into the body a low level of ozone is generated from the excited air oxygen around the skin and the electrode. The ozone 'sticks' to the skin. You can immediately smell the fresh, sweet ozone odor produced as the ozone and active oxygen sub-species are absorbed directly into the skin and picked up by the blood capillary networks. The active oxygen is then transported by the body fluids to whatever ails you. The smell of fresh ozone lingers on your skin.

Long-time expert and independent researcher Jeff Behary from The Electrotherapy Museum is also helping to keep interest in these forgotten devices alive. http://www.electrotherapymuseum.com

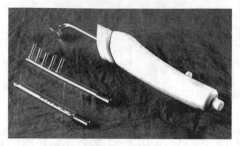

Modern version 'Neon Ray'
Supplied by Rev. Mary Seid.

Rev. Mary Seid is a certified Vibrational Therapist. She (see *resources*) supplied the Violet Ray testimonials in our evidence section at the end of the book, and some of the quotes and photos in this section. Here's what Mary has to say about the results of her use of the Violet and Neon Ray.

"Edgar Cayce recommended the Violet Ray to over a thousand patients. Of the 33 different devices that I have personally tested, I believe the Violet Ray is the most powerful and important healing device I own."

Old technical manuals and medical books proclaimed: "The Violet Ray has been known to relieve pain in any part of the body by bringing circulation to the area. The electrical currents penetrate every cell and tissue of the body to stimulate and strengthen the vital organs, build up the natural forces, steady the nerves, and develop the general health and conditions of the body. High frequency electrical currents carry oxygen to the blood, thus purifying the vital body fluid. The Ray equalizes the circulation in congested parts and restores bruised and inflamed tissues to normal. The applicator rubbed over fat will bring circulation and warmth to the area to metabolize fatty tissue. Wonderful for removing wrinkles.

"The treatment can be concentrated on any organ, muscle or nerve, however delicate or tender. By this treatment the impurities are washed from the affected organs by a rich, warm flow of blood which nourishes and stimulates the tissues and which is generated by the penetrating current of the Neon Ray, reaching the cells which supply life to the tissue, and causing them to vibrate. The stagnant pools of blood are thus set in motion and the poisons contained therein are eliminated by

the purifying processes of Nature. The blood thus being freed of poisons is able to take up oxygen from the lungs and nutrition from the digestive organs and return tissue building material to every part of the body. In this way new tissue is formed to replace that destroyed by disease and carried away in the blood stream. This process of stimulation is called a 'cellular massage'. "Because of the fact that the vibration or contractile effect is expanded upon the individual cells making up the tissue, rather than on individual muscles. This is an important reason why the high frequency current produces such marked effect on nutrition and general health. Cellular massage is much more penetrating and beneficial than muscular massage as produced by mechanical vibration."

Some modern experts in *other* specialties who have never seriously tried these devices have labeled such claims 'quackery,' and place the old electrotherapy machines in quackery museums as examples. I think they're throwing the baby out with the bath water. Some of the old electro-therapy devices such as shock coils hooked to batteries were obviously a lot more puffing and wishful thinking than productive. We know what ozone does as a sterilizer and detoxifier, and all high frequency machines produce it. There were a huge variety and range of devices invented and produced back then that remain as unmined gold. I hope more of these antique concepts will get a good (fair) modern sorting out with extensive testing.

As for the few who are proclaiming that all electro-therapy is not beneficial, I still have to ask, how can they truthfully claim enough expertise to judge something without ever having seriously used it? Diathermy and X-ray machines were once grouped with these devices, and they are now used everywhere.

We are composed of spinning atoms of electricity, magnetism, color, music, and sound. Doesn't it make sense that it just might be possible to balance our bodies with the same forms of energy once we can perfect their use?

Beck Box and Oxygen Therapies.

A minute ago I mentioned electroporation. The 'Beck Box' also creates this condition, and is another electronic device I highly recommend—especially when combined with an Oxygen Therapy cleansing.

Years ago I tried it on my own body, and two weeks after using it I had a fever come on out of nowhere without having any cold symptoms. What I experienced was that the body will try to burn up any dead microbes suddenly floating in its fluids when their quantity is large enough.

This small battery operated device has contact electrodes that are strapped to the skin on the wrist over the brachial artery. The box pulses low voltage electricity into the blood passing beneath the electrodes. Research in major hospitals showed such properly designed devices were capable of damaging the receptor sites of microbes in the blood badly enough to render them 'sterile' and unable to reproduce. That is, before all the research magically disappeared. In my case, it took about two weeks for the sterile bugs die of old age without reproducing, and then my body decided it had to burn the dead husks up with a fever.

I have interviewed others that have reported great benefit from the use of the Beck Box. One fellow was a microscopist, and he used to talk of his treated blood on a glass microscope slide being 'immortal,' after discovering it was remaining alive days after all the other blood samples in the trash had died. That's how clean and untainted his blood was.

Mr. Robert Beck, an engineer, developed this technology. He gave away designs for his box, and used to mistakenly mention in his lectures that "Ozone doesn't work," because a friend of his used it incorrectly. I showed him the error, and he then advocated that Oxygen Therapies always be used along with his device, especially lots of fresh ozonated drinking water. The body can only detoxify when it has enough oxygen to do the job after some other method is used to kill microbes. They were having too-rapid detoxification problems.

This device only treats the blood. Only active oxygen gets all the way into all the nooks and crannies beyond the blood.

Chapter 17

Ozone Therapies in the Clinics

Why a clinical section in a consumer book? Not because I advocate anyone do anything outside of being under the care of a competent professional, but in order for you to be a savvy international Oxygen Therapy consumer. When you go see the doctor here or elsewhere, you have to know if you are being treated correctly. The answer to that question lies between you and your ozone therapist, but this section is about empowering you with general medical and therapy guidelines so that you can use them to ask good questions and make intelligent choices based on conventional historical wisdom. Especially since these therapies are ignored officially in America, so how else are you going to know what's generally safe or effective if I don't tell you? I only do this because the system doesn't.

How Ozone Works in the Body

As I have said so many times, the basis of the Ozone Therapy is very straightforward. Ozone will oxidize or burn up any virus, bacteria, fungus, pathogen, toxicity or pollution in the body. When the body is given oxygen, it starts to clean itself out. It's that simple. In clinical use, ozone can blast holes through the membranes of bacteria, viruses, yeast, and abnormal cell tissues before it kills them, and yet leaves healthy cells alone, as long as it is used in above 27 and less than 80 micrograms per cubic milliliter concentrations.

Inactivation of Bacteria, Viruses, Fungi, Yeast, and Protozoa

In modern times, what we're really talking about is that ozone is a known virucide, bactericide, and fungicide as well as a killer of parasites. When you take bacteria or a virus, a fungus, or a pathogen, and surround them with ozone, the chances are that the ozone—in almost 100 percent of the cases—will eliminate the virus, the bacteria, and the fungi. This is the way it works. That's what the literature shows if someone would read it. The ozone is known to disrupt the integrity of the bacterial cell envelope through oxidation of the phospholipids and lipoproteins. With a virus, the ozone damages the viral capsid and disrupts the reproductive cycle by disrupting the virus to cell contact with peroxidation. The weak enzyme coatings on weak cells that make them vulnerable to invasion by viruses also make them susceptible to oxidation and elimination from the body by ozone, and the body quickly replaces them with fresh, healthy cells with strong enzyme coatings.

Enhancement of Circulation

In circulatory disease, a clumping of red blood cells hinders blood flow through the small capillaries and decreases oxygen absorption owing to the clumping that reduces the blood cell surface area. Ozone reduces or eliminates clumping; thus red cell flexibility is restored, along with oxygen carrying ability. Oxygenation of the tissues increases as the arterial partial pressure increases and viscosity decreases. Ozone also oxidizes the plaque in arteries, allowing the removal of the breakdown of products and unclogging the blood vessels.

Generally Accepted Effects of Traditional Ozone Therapy

Chelates heavy metals; works well in conjunction with EDTA
Cleans arteries and veins and improves circulation
Improves brain function and memory
Inactivates viruses, bacteria, yeast, fungus and protozoa
Normalizes hormone and enzyme production
Oxidizes toxins, allowing their excretion
Prevents and eliminates auto-immune diseases
Prevents and treats communicable diseases
Prevents and reverses degenerative diseases
Prevents stroke damage
Purifies the blood and lymph
Reduces cardiac arrhythmia
Reduces inflammation
Reduces pain; calms the nerves
Stimulates oxygen metabolism
Stimulates the immune system
Stops bleeding

13 Major Effects of Ozone on Humans

As Reported by Frank Shallenberger, M.D.,H.M.D.

Considered one of the leading authorities on medical ozone autohemotherapy, Dr. Shallenberger in Carson City, Nevada has conducted important work to support the hypothesis that ozone can have long-term positive effects on AIDS, including sending me a letter praising Polyatomic Apheresis Ozone Therapy. He also conducts workshops on the proper application of medical ozone. He reports he successfully treats patients with medical ozone. The 13 physiological effects of ozone he had seen are listed below and accompanied by a brief explanation.

1. Ozone stimulates the production of white blood cells. These cells protect the body from viruses, bacteria, fungi and cancer. Deprived of oxygen, these cells malfunction. They fail to eliminate invaders and even turn against normal, healthy cells (allergic reactions). Ozone significantly raises the oxygen levels in the blood for long periods after ozone administration; as a result, allergies have a tendency to become desensitized.

2. Interferon levels are significantly increased. Interferons are globular proteins. Interferons orchestrate every aspect of the immune system. Some interferons are produced by cells infected by viruses. These interferons warn adjacent, healthy cells of the likelihood of infection; in turn, they are rendered non-permissive host cells. In other words, they inhibit viral replication. Other interferons are produced in the muscles, connective tissue and by white blood cells. Levels of gamma interferon can be elevated 400-to-900 percent by ozone. This interferon is involved in the control of phagocytic cells that engulf and kill pathogens and abnormal cells. Interferons are FDA approved for the treatment of Chronic Hepatitis B and C, genital warts (caused by Papillomavirus, Hairy-cell Leukemia, Kaposi's Sarcoma, Relapsing-Remitting Multiple Sclerosis and Chronic Granulomatous Disease. Interferons are currently in clinical trials for throat warts (caused by Papillomavirus), HIV infection, Chronic Myelogenous Leukemia, Leukemia, Non-Hodgkin's Lymphoma, colon tumors, kidney tumors, bladder cancer, malignant melanoma, Basal Cell Carcinoma and Leishmaniasis. While levels induced by ozone remain safe, interferon levels that are FDA approved (and in clinical trials) are extremely toxic.

3. Ozone stimulates the production of Tumor Necrosis Factor. TNF is produced by the body when a tumor is growing. The greater the mass of the tumor the more tumor necrosis factor is produced (up to a point). When a tumor has turned metastatic, cancer cells are breaking off and being carried away by the blood and lymph. This allows the tumor to take up residence elsewhere in the body; or, in other words, to divide its forces. These lone cancer cells have little chance of growing due to the TNF produced to inhibit the original tumor. When the tumor is removed surgically, TNF levels drop dramatically and new tumors emerge from seemingly healthy tissue.

4. Ozone stimulates the secretion of IL-2. Interluekin-2 is one of the cornerstones of the immune system. It is secreted by T-helpers. In a process known as autostimulation, the IL-2 then binds to a receptor on the

T-helper and causes it to produce more IL-2. Its main duty is to induce lymphocytes to differentiate and proliferate, yielding more T-helpers, T-suppressors, cytotoxic T's, T-delayed's and T-memory cells.

5. Ozone kills most bacteria at low concentrations. The metabolism of most bacteria is on average one-seventeenth as efficient as our own. Because of this, most cannot afford to produce disposable anti-oxidant enzymes such as catalase. Very few types of bacteria can live in an environment composed of more than 2 percent (27 mcg/ml^3) ozone.

6. Ozone is effective against all types of fungi. This includes systemic candida albicans, athlete's foot, molds, mildews, yeasts, and even mushrooms.

7. Ozone fights viruses in a variety of ways. As discussed above, ozone also goes after the viral particles directly. The part of the virus most sensitive to oxidation is the 'reproductive structure.' This is how the virions enter the cell. With this structure inactivated, the virus is essentially 'dead.' Cells already infected have a natural weakness to ozone. Due to the metabolic burden of the infection, the cells can no longer produce the enzymes necessary to deal with the ozone and repair the cell.

8. Ozone is antineoplastic. This means that ozone inhibits the growth of new tissue because rapidly dividing cells shift their priorities away from producing the enzymes needed to protect themselves from the ozone. Cancer cells are rapidly dividing cells and are inhibited by ozone.

9. Ozone oxidizes arterial plaque. It breaks down the plaque involved in both Arteriosclerosis and Arthrosclerosis. This means ozone has a tendency to clear blockages of large and even smaller vessels. This allows for better tissue oxygenation in deficient organs.

10. Ozone increases the flexibility and elasticity of red blood cells. When one views a red blood cell under a microscope, it looks like a disc. In the capillaries, where they pick-up (lungs) and release (tissue) oxygen, these discs stretch out into the shape of an oval or umbrella. This aids their passage through the tiny vessels and makes the exchange of gas more efficient. The increase in flexibility of the RBCs allows oxygen levels to stay elevated for days, even weeks after treatment with ozone.

11. Ozone accelerates the Citric Acid Cycle. Also known as the Krebs Cycle or TCA Cycle, this is a very important step in the glycolysis of

carbohydrate for energy. This takes place in the mitochondria of the cell. Most of the energy stored in glucose (sugar) is converted in this pathway.

12. Ozone makes the antioxidant enzyme system more efficient.

13. Ozone degrades petrochemicals. These chemicals have a potential to place a great burden on the immune system. They also worsen and even cause allergies and are detrimental to your long-term health.

Ozone Concentration

Medical ozone is produced in varying concentrations. The quantity of ozone in comparison with the quantity of oxygen in the gas stream is called percent concentration. It is measured in micrograms (ug) of ozone per milliliter (or cc) of the mixture. A liter of oxygen weighs 1.4 grams.

Percentage of Ozone in Oxygen Gas
Converted to Micrograms Per Cubic Milliliter

0.5% ozone is = 7 ug/cc

1.0% ozone is = 14ug/cc

1.5% ozone is = 21ug/cc

2.0% ozone is = 28ug/cc

2.5% ozone is = 35ug/cc

3.0% ozone is = 42ug/cc

3.5% ozone is = 49ug/cc

4.0% ozone is = 56ug/cc

4.5% ozone is = 63ug/cc

5.0% ozone is = 70ug/cc

Five percent or 70 ug/cc is generally considered by most to be the upper limit of concentration for internal use of medical ozone. Some say there is no theoretical limit, as it only will destroy unhealthy cells. They say any cells showing destruction were unfit. For all practical purposes and to insure patient comfort, the standard mostly used is around 27 45 ug/cc, or mcg/ml^3.

How Doctors Make Injection Grade Ozone

I have included this little science section because I am continually asked, "How is medical grade ozone made?" This may not be of interest to the average Jill or Joe, but to the newbie ozone enthusiast it is very beneficial to get it all straight right up front.

To make clinical ozone, pure bottled oxygen (only) passes through a high frequency electronic corona field, and a little bit of the oxygen gets converted to ozone. When we speak of a concentration of '27 mcg,' we are referring to a 2 percent ozone, and 98 percent oxygen gas mixture, or 27 micrograms per cubic milliliter. Air is only 20 percent or so oxygen, and is mostly nitrogen and other inert gasses. That is why ozone made from air is suitable for water and air purification, but is NEVER used inside the body.

All medical ozone generators use only pure bottled oxygen. Also, the average corona discharge ozone generator will make different concentrations of ozone when powered at different frequencies, voltages, and input oxygen flow rates.

Here is a sample output diagram from a 6" long corona discharge tube with fixed voltage and frequency. The only variable is the oxygen flow rate. The oxygen flowing through the generator on its way to becoming ozone can be fast or slow. The slower the flow, the higher the strength of ozone created, up to a point. The length of time the oxygen is within the high frequency electronic corona glow inside the generator is called the 'contact time.'

The ozone and oxygen only touch interior components and tubing and syringes made of materials they will **NOT** react with. That's why most hobbyists can make ozone in their garages with an old neon sign transformer, or a UV tube, but it's not infusion grade ozone. Air has nitrogen in it, and susceptible hoses and components exposed to ozone get oxidized real quickly. Both will contaminate the produced ozone and fail. This is why the manufacturers get the big bucks for building only the best high quality ozone generators.

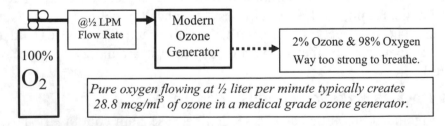

Pure oxygen flowing at ½ liter per minute typically creates 28.8 mcg/ml³ of ozone in a medical grade ozone generator.

Typical output at various flow rates for a 6" tube corona discharge ozone generator
At 1/16th of a liter of oxygen per minute flow rate, 5.0 % = 72.7 mcg/ml³, ozone is generated
At 1/8th of a liter of oxygen per minute flow rate, 3.5 % = 50.6 mcg/ml³, ozone is generated
At 1/4   of a liter of oxygen per minute flow rate, 2.5 % = 36.0 mcg/ml³, ozone is generated
At ½   of a liter of oxygen per minute flow rate, 2.0 % = 28.8 mcg/ml³, ozone is generated
(mcg/ml³ is the same as µg/ml)

Ozone generators split apart stable O_2 (O_1+O_1) into lots of singlet oxygen atoms: O_1 O_1 O_1 O_1.. O_1 O_1 O_1 O_1, etc.

Ozone, known as O_3, is created when three or more single oxygen atoms link together: $O_1+O_1+O_1 = O_3$.

But more than three oxygen atoms linked together is also called ozone:

$O_1+O_1+O_1+O_1 = O_4$ ozone, or $O_1+O_1+O_1+O_1+O_1 = O_5$ ozone and so on. The term 'ozone' covers many forms of active oxygen molecules, including O_3, O_4, O_5, O_6, O_7, O_8, O_9, O_{10}, O_{11}, O_{12}, O_{13}, O_{14}, O_{15}, O_{16}, O_{17}, O_{18}, O_{19} and more. The government weather planes have measured it up to and beyond 21 O_1 atoms of oxygen naturally linked together in the upper atmosphere. This is in the upper regions of the atmosphere where the sun is using up its UV rays creating our precious ozone during the striking of our planet's oxygen envelope. Structurally, O_{21} looks like this:

$O_1+O_1+O_1+O_1+O_1+O_1+O_1+O_1+O_1+O_1+O_1+O_1+O_1+O_1+O_1+O_1+O_1+O_1+O_1+O_1+O_1$.

Ozone is unstable, and that's good, because it works for us by giving up singlet oxygen. It's not the ozone, but <u>the singlet oxygen created by its decomposition</u> back into stable O_2 that does all the disinfection and sterilization work.

$O_3=O_1+O_1+O_1 \longrightarrow O_2 + O_1$.

$O_4=O_1+O_1+O_1+O_1 \longrightarrow O_3+O_1 \longrightarrow O_2 + O_1$ and so on, as more and more singlet oxygen is released. The cascading stops when all of the remaining singlet active oxygen atoms are released, or have combined with and dehydrogenated something like toxins.

This same principal applies to our immune system's peroxide giving up a single oxygen molecule and turning back into plain water. $H_2O_2=H_2O + O_1$.   In fact, all of the various Oxygen Therapies work not because they are ozone, or peroxide, or any other substance, but because they are all <u>ultimately singlet oxygen delivery systems.</u>

When we breathe as the O_2 in the air passes through our lung membranes, it splits the O_2 bond apart and delivers 2 singlet O_1 oxygen atoms to our body fluids and cells. This is exactly why Oxygen Therapies are so safe and effective, just like implosion and gravity based technologies, they are emulating the ultimate master of efficiency, Nature itself.

Stored Blood Purified by Ozone

This has been known for years and ignored. We can immediately and inexpensively purify the entire world's blood supplies. Never again should anyone get *any* disease from a blood transfusion. Who says so? NATO!

Each year thousands of people contact AIDS, Hepatitis, and other diseases through the transmittal of infected blood. Although alternatives to blood transfusions exist (just ask any Jehovah's Witness, they all carry cards on them naming the blood alternatives,) hospitals continue to transfuse unozonated blood, and they know disease is still spread through transfusions despite their testing. Ozone bubbled through stored blood could eliminate all this expense and risk. If ozone was always simply bubbled through the blood stored in the blood banks, nobody would get AIDS or hepatitis by means of a blood transfusion!

The End of Contracting Diseases from Transfusions

In September 1992, the Canadian government stated that blood can be completely sterilized with ozone. The Canadian Department of Defense knows that if you bubble ozone through blood you can kill everything in there that would cause a disease. Hepatitis, HIV, any viruses, any bacteria, any funguses, and any pathogens in the blood would be destroyed.

1993 April Journal Of The Canadian Medical Association—Medical Science News; 148(7) pg. 1155) *"Are Worry Free Transfusions Just A Whiff of Ozone Away?"* by Albert C Baggs, BSc.—"Scientists in the U.S. and Canada are investigating the use of ozone to destroy the HIV virus, the hepatitis and herpes viruses, and other infectious agents in the blood used for transfusion. The studies were endorsed by medical circles of the North Atlantic Treaty Organization (NATO) because of a concern that viral pandemics have compromised the ability of world blood banks to meet urgent and heavy military demands."...In a brief to the NATO Blood Committee the Surgeon General of the Canadian Armed Forces reported upon Canadian findings that "...a 3 minute ozonation of serum spiked with one million HIV-1 particles per milliliter would achieve virtually 100 percent viral inactivation. It was also found that the procedure would destroy several other lipid-encapsulated viruses, including simian immunodeficiency virus and various strains of interest to veterinarians." The journal report described the work of Mueller Medical and Medizone International.

They simply looked at all the literature and said, "This is amazing, this is incredible. No one is doing this, let's find out more." Why don't American healthcare people think like this anymore?

1983 World Ozone Conference Published Results

On May 24th and 25th of 1983, in Washington, DC, doctors from all over the world met. They then announced and published their *Proceedings of the 6th World Ozone Conference.*

What did they announce? "Ozone removes viruses and bacteria from blood, human blood [circulating] and stored blood." But no one storing blood listened. So let's finish disclosing what they published as the results commonly achieved. Ozone has been reported successfully used on AIDS, herpes, hepatitis, mononucleosis, cirrhosis of the liver, gangrene, cardiovascular disease, arteriosclerosis, high cholesterol, cancerous tumors, lymphomas, leukemia. Ozone is highly effective on rheumatoid and other arthritis, allergies of all types; ozone is effective against multiple sclerosis, ameliorates Alzheimer's disease, senility and Parkinson's, and is effective on proctitis, colitis, prostate, candadisis, trichomoniasis, and cystitis. Externally, ozone is effective in treating acne, burns, leg ulcers, open sores and wounds, eczema and fungus. Remember these are doctors from all over the world who have actually been successfully using Ozone Therapy daily, some up for up to 50 years, in their clinics!

And, let's not forget that in 1987 *The Use of Ozone in Medicine* was published. It lists more than 48 diseases commonly treated with ozone, including: abscesses, acne, AIDS, allergies, anal fissures, cerebral sclerosis, circulatory disturbances, cirrhosis of the liver, menopause, constipation, cystitis, bed sores, dermatology, fistulae, funguses, verniculosis, gangrene, hepatitis, herpes, high cholesterol, colitis, neurology, dental medicine, tumors, cancer, orthopedic, osteomyelitis, Parkinson's, rheumatism, Raynaud's disease, scars, inflammation of the vertebras, stomatitis, joint dystrophy, surgery, phlebitis, open sores, urology, vascular surgery, and wound healing. It works in gastroenterology, gerontology, proctology, gynecology, and radiology and has anti-viral effects. By the way, did I miss your disease?

Clinical Ozonated Water

A lot of people that I have interviewed and heard of have gained a great deal of relief by drinking ozonated water and/or applying it externally to their body. Let's explore this some more. Famous German ozone experts and authors Rilling and Viebahn-Haensler listed the following uses for ozonated water in general medicine.

1 Stomatitis
2 Conditions after tooth extraction
3 Thrush (candidiasis)
4 Peridontitis
5 Carcinoma of the esophagus or stomach
6 Sub-acid or antacid gastritis
7 Colonic or ozone irrigation in colitis and
 all chronic intestinal inflammations
8 Rinsing the urinary bladder
9 Dentistry as a disinfectant

Their *The Use of Ozone in Medicine* is the standard German medicine ozone textbook and respected all over the world. Recent American developments have eclipsed some of the data. A big oxy-applause to the authors.

Clinical Ozone Body Washing

Erwin Dorsch is an extremely meticulous and creative German who is very skilled with his hands. He presently lives on his own sailboat—that he built from scratch—in Trinidad, and just opened his clinic there. He was the German translator and ozone engineer during the days of the Key Biscayne clinic. He went out on his own, selling and repairing many ozone generators during his time here in America. His crowning achievement was to take ozone and ozonated ultra-pure water to the next level.

The Aquacizer

Wanna take a ride?

Erwin created and built the 'Aquacizer,' originally called the Excalibur Rejuvenator. Unlike ozone body suits that worked by letting dry ozone gas soak into the skin, or ozonated steam cabinets that let ozone and steam fall on the body, his clinical machine was a marvelous piece of engineering. It sold for $35,000 dollars back in the '90s. The Aquacizer control head would take tap water and strip out all of the harmful fluorine, chlorine, or any other unnatural contaminants, and turn all this into 10 megohm ultra pure water. This ultra pure water was then heated and fed through an injector that thoroughly mixed it with 27 mcg/ml ozone gas. The water ozone mixture was sprayed out of hundreds of miniature nozzles inside the top of the clamshell shaped unit. Being in it was like bathing in an ozonated waterfall.

The patient would lie inside the clamshell with his or her head outside, away from the too-strong-to-breathe ozone on the inside. As in the smaller home units, the heat would open the pores and the ozonated water would draw the toxins out through them. At the same time, it filled the blood with oxygen species by soaking through to the under skin capillaries. People do not realize we breathe in and detoxify through the pores in our skin, and that if we can't, we die. That is why this method is so efficient, since the skin is like a tougher third lung.

Clinical units are different from the home units in that they actually ultra purify their water and mix the ozone into heated water (not steam) and do this *before* the water enters the chamber. The Aquacizer also has hundreds of nozzles spraying and treating the whole body at once.

Aquacizers are highly effective at total body purification, especially when it comes to skin and capillary purification. This Ozone Therapy method is best alternated with direct IV or Apheresis ozone. A few of these units are still in use, and a new incarnation of this is in the design stage. This unit, or one like it, should be immediately be put back into production.

Dentistry

HealOzone
Date: 08-07-2002; Publication: The Daily Mirror; UK, pp 6. Author: Richard Smith

"THE dreaded visit to the dentist for a filling could soon become pain-free. A revolutionary ozone gas can blitz decay in the affected tooth in just ten seconds. But more important - it doesn't hurt. Patient agony, needles

and drills may soon be consigned to history. The treatment is permanent and costs 45 pounds per tooth. A chain of dental clinics has just introduced it on trial.

"Dr Peter Murray, clinical director of dentistry with the firm pioneering the HealOzone treatment, believes it will be adopted worldwide. Dr Murray, who is based in Cardiff, said yesterday: "This system is a real Godsend for those people who have always had a fear of visiting the dentist. 'If we can identify the decay in time, there will be no need for them to experience drills. 'There are a lot of things that put people off going to the dentist - the feel and sound of the drill and the injections.

"'I believe HealOzone will replace the traditional tools of dentistry when it comes to treating decay - and eliminate the fear factor the public associate with dentists.' The process involves the dentist fitting an airtight rubber cap around the affected tooth and administering a gentle puff of ozone. Within 10 to 40 seconds the gas kills the soft decay-forming bacteria and transforms it into harmless hard decay. If the tooth is in a prominent position the dentist can insert a cosmetic white filling without the use of needles or drills.

"Office manager Sarah Wiltshire, 22, from Cardiff, Wales, England had the new treatment. Sarah said: 'I've always dreaded the dentist. I've had a couple of fillings before and I cannot stand the injections. But I couldn't believe how easy this treatment was. It only took a couple of minutes and I couldn't really feel anything.' "The procedure was developed by scientists at Belfast University, whose research showed that *just 10 seconds of the ozone treatment eliminated 99 percent of micro-organisms.*

"Professor Edward Lynch, principal investigator at the University's School of Clinical Dentistry, said: 'By trial and error, I discovered ozone eliminates decay in seconds. To be able to remove decay naturally is phenomenal and we are delighted.'

"Dentist's chain James Hull Associates have started offering the treatment at their branches in London, Cardiff and Birmingham. A British Dental Association spokesman said: 'This treatment is in its experimental stage at the moment.' Anyone considering any ozone procedure should talk about it with their dentist."

'Ozone' brand toothbrushes.

Since ozone's use in dentistry has been around since the turn of the century, one can wonder out loud why it is not taught in dental schools.

Ozone Insufflations

Ear Insufflation

The application of ozone infused into the ear is a fairly recent development. The science behind it is based on the idea that ozone/oxygen species will be absorbed through the small capillaries in the ear canal and enter the lymphatic system effectively. Patients report that they have gained great relief from swollen glands, and sore throats have been alleviated quickly. The sinuses may drain profusely. When combined with other ozone methods, this has been reported to be truly effective therapy for the head region.

Using a bottled oxygen medical grade generator, 10-to-30 ug/ml^3 ozone at no more than ¼ liter per minute, is bubbled through a bubble humidifier for five minutes before use, in order to fully oxygen charge the water. The ozone is then run through the humidifier and the humidified gas is delivered through tubing to the point of application. Caution is advised so that the ozone is not brought into the proximity of the eyes and nose. The ear may be moistened with pure water or DMSO to increase absorption. Hold the ear in question down and parallel to the floor. This strong medical grade ozone is heavier than air, and thereby falls away from the eyes and nose. The inner ear is irrigated with the humidified ozone/oxygen mixture.

The physician holds the output tube at the entrance to the ear. The output tubing is gently inserted part way into the ear cavity, allowing escape of some of the ozone so that no pressure builds up. Each ear is treated for about one minute upon the first application; two minutes for the second; and three minutes for the third. At the start this should not be continued for more than four days in a row. Crusty discharges are common. Irritation comes from over application. When you're through, you can drink the super–ozonated humidifier water while holding your nose, or apply it externally.

Can this burst my ear drum? Yes, if you rush it and don't allow the gas to escape. The essence of my teaching has always been to follow the instructions carefully as used by the many experienced physicians and not to be tempted to rush the treatment. In 14 years of this I have never met anyone who has had a burst eardrum, but there have been some causing themselves unnecessary discomfort from not following sound advice.

Remember, ozone detoxification is an exothermic reaction. When the ozone oxidizes a toxin or microbe it literally burns them up without flame, and burning generates heat. Keep the concentrations low and give them slow. Low and slow at first. Always take your time and approach this gradually.

Vaginal Insufflation

This is similar to rectal insufflation, except a vaginal cannula is used, and more ozone may be introduced into the vagina than into the rectum without a build up of pressure. The treatment may be used for up to five minutes at first, and many women have reported cures of yeast and other infections. Furthermore, it is historically effective in sexually transmitted diseases. It is not advised prior to menses, as it might increase the blood flow.

When the ozone generator is ready, the flow rate is set at .125-to-.5 liter per minute. The patient inserts a vaginal catheter and allows the humidified ozone to flow into the vagina, and the gas is allowed to escape the cavity. Unlike rectal insufflation, this can be administered for long periods of time; hours, in some cases. However, in certain conditions, Dr. Freibott comments that the growth of thrush can be aggravated temporarily with long applications. The reason being that ozone will accelerate the breakdown of carbohydrates into sugars which will feed the yeast. With repeated application it has removed this condition from many women. The advantage of this method is that the gas is absorbed into the uterus, fallopian tubes, nearby organs, and the abdominal cavity. Should this be used during Pregnancy? Most experts advise against it.

Penile Insufflation

Similar to vaginal is penile insufflation. A long thin blunt catheter is inserted into the penile opening and slowly and gently pushed in until resistance is met. A very slow flow of humidified medical ozone gas at very low pressure is used to irrigate the prostate area. Cancer patient Anthony Callela thought of this and reported to me that he rid himself of prostate cancer with it.

Rectal Insufflation

Used when injections are not desired, or as an adjunct to them. All rectal insufflation begins with a colonic or at least a bowel evacuation and enema to clear the bowel. No use wasting the gas on fecal matter. I was quite surprised when I got into this subject to find out that many (usually women) are conditioned to only have a bowel movement once every two days. We eat two or three times a day, and that means they are four drop shipments behind and no one talks about it! All that waste is being re-absorbed until it leaves. The way I heard about this chronic condition was when people started to report that their bowel movements were increasing to once or twice a day through oxygenation. Naturally, I asked them, "Do you mean you only went to the bathroom every two days, and thought it was normal?" I guess that's why oxygenating colon cleansers are top sellers.

Some wags who know little about science, medicine, or history claim it is "impossible to absorb oxygen through the colon." The colon is coated in miles of capillaries, and active oxygen easily passes into humidified tissue. Before syringes existed, doctors would commonly put drugs into the bloodstream through the exclusive use of suppositories. The drugs were absorbed easily, and oxygen is far smaller in size than drug molecules.

Single Catheter Method

A mixture of oxygen and ozone is slowly introduced into the rectum under very low pressure, which allows absorption through the portal vein of the rectum. This method is regarded as one of the safest forms of the ozone modalities. It is used by a significant number of people for a wide variety of diseases including cancer and HIV. Many have reported to me that rectal and systemic ozone is very effective for liver diseases of all types, including hepatitis, cirrhosis of the liver and so forth. In addition to the systemic effect, rectal ozone/oxygen insufflation is logically valuable in the treatment of colitis and proctitis.

The traditional rectal insufflation methods advocated letting ½ liter of medical grade 27mcg/ml₃ gas slowly flow into the rectal opening through an inserted blunt catheter for one minute, and then to hold it in for as long as possible. I worked daily applications of ozone into my routine by doing this method while brushing my teeth in the morning. This method proved hard to hold in—and painful for some people—because at first, ozone detoxifies so rapidly that it provokes a bowel peristaltic (cramping) reaction and immediate discharge. By holding it in, you are fighting yourself. If a cramp comes, it is far better to just let it all out in the toilet and start again. Some German doctors advocate using high concentrations (80-to-100 ug/ml) to begin with, then decreasing the concentrations to low levels to promote healing. The maximum amount of gas should not exceed 500 ml, which would be achieved at a flow rate of ½ liter per minute for one minute, or ¼ liter per minute for two minutes. Variations on this method use pre-charged ozone enema water or ozone filled bags as the supply. The drawback to using them is that the ozone concentration is constantly dropping with time and handling. The positive side is that bags are safer for people who get confused, since the only pressure created is when you squeeze them.

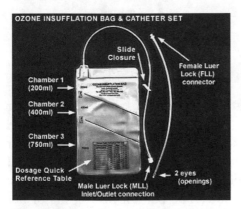

Ozone Services' ozone application bag for rectal insufflation

Whichever non-bag method is chosen, experts agree that one should *only* use ¼-to-½ liter per minute *pediatric* regulator regulated *low flow oxygen bottle pressure* to deliver the humidified gas.

A fellow in the industry who should have known better once experimented with adding a pump to the process and popped a hole in his colon like an over-inflated balloon. He wore a colostomy bag for a year. This is always a possibility if one puts in too much gas without letting the pressure out. I developed the Dual Catheter Method to overcome this liability.

Dual Catheter Method

I invented the following 'two catheter' method as a safer more efficient delivery alternative for myself. There is no pressure build up with this method. The medical grade humidified gas is released way up inside the colon from the end of the long catheter. The gas soaks through the colon walls into the miles of blood vessel capillaries surrounding it, and any waste gasses, liquids and particulates flow out of the short fat catheter so that pressure build-up is minimized. The patient should be lying flat, and the short catheter vent opening should be over a towel to catch the reduced oxygen and waste fluids. I've applied continuous ozone gas to myself in this manner for up to an hour at a time. It helps eliminate boredom if you meditate, read a book, or watch TV.

Two Catheter Method - Ozone Rectal Insufflation

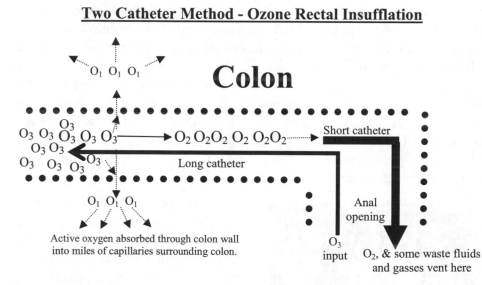

The patient is advised to first clean out the rectum by taking a colonic or high enema with ozonated water. The medical ozone generator is supplied oxygen from the *pediatric* regulator on the oxygen tank, and the generator output is connected to a bubble humidifier and ozone-resistant medical grade tubing going to a suitable long rectal catheter. The oxygen flow rate is set to deliver less than ½ liter per minute. The patient lies on his or her back and the long lubricated catheter and the short catheter are both inserted through the anal opening up to their collars. The long one goes halfway up the transverse colon.

When switched on, the ozone will quickly start to enter the rectum. The length of time spent applying the treatment varies according to the tolerance of the individual and the condition. Some people report that they can only tolerate it for 30 seconds to begin with. Others have reported that they can use it for up to an hour at a session. If any bowel discomfort arises, immediately discontinue the treatment for the time being. Be careful; patients do not have much feeling in the colon!

Gas can pool in an area, so it is useful to massage the abdomen whilst using this method. The gas is normally painless, and the rectal tubing is small and lubricated so that no discomfort occurs. There may be a slight sensation of warmth. Many have reported that they feel better as soon as treatment commences. Theoretically, using this method two or three times a day may easily surpass the amount of ozone delivered by autohemotherapy.

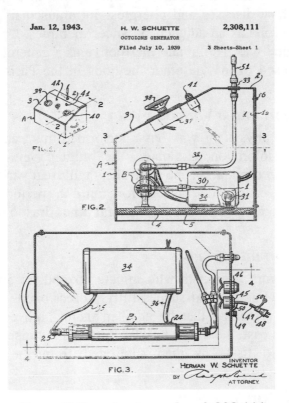

From U.S. patent number 2,308,111

"Parkes and Buckley have reported the satisfactory cure of chronic rheumatism, neuritic and neuralgic sciatica, varicose ulcer, anthrax, and diabetic gangrene. Malleterre has reported successful treatment of albuminuria, ankylosis, sciatica, prostatitis, gonorrhea, anthrax, and varicose ulcer. Hans reports Octozone extremely successful for sinusitis."

Ozone Injections and Drips

Minor Autohemotherapy

Minor, meaning only a small amount of blood, around 10cc, is removed from the patient. The treated blood is mixed with ozone and becomes a type of auto-vaccine that is derived from the ozone killed viruses and bacteria in the blood. The mixture is re-injected. Minor Autohemotherapy is believed to stimulate the immune system

Ozone Injection into a Muscle

A small amount of mixed ozone and oxygen (up to 10 ml.) is injected into the patient, usually into the buttocks. This method is commonly used to treat inflammatory diseases and allergies. Intramuscular injections have been used in Europe as an adjunct to cancer therapies. Dr. Paul Aubourg stated this method was painful and ineffective as a general therapy and so he commonly used ozone via other methods in the Parisian hospitals in 1938.

Ozone Injection into a Varicose Vein

I have spoken to physicians who have performed this treatment, saying it removes the discoloration, eliminates the bulging vein and abates the condition. However, not all ozone specialists will treat varicose veins. It is reported that it causes discomfort for quite a period of time after treatments, but so do stripping the veins and other drastic means of fixing them.

Ozone Injection into Tumors

Dr. Turska (Oregon naturopath who injected ozone for 50 years without incident) applied this method successfully. Direct ozone injection into a tumor is the only method that he advised using high concentrations of ozone for. In this case the ozone rapidly destroys the diseased tissue. The resultant necrosed tissue is then suctioned from the tumor site by aspiration.

Ozone Portal Vein Injection

Someone wrote to me and said, "My father is in the hospital with a failing liver. Are there any therapies that can help?"

Well, why would someone's liver fail? It would fail because it's constantly being bathed in poisons, so it's become overloaded and toxic and it's full of microbes due to low oxygen conditions around and inside it. In part, it's failing because bacteria are defecating in it.

We're talking here about cleaning the liver up. There's a problem if the poisons have been imbibed for so long that you've actually destroyed the liver. It's really tough for the body to grow a new one; we know of no cases where it has. It's not like you can go down to the body parts store and say, "Hey, give me a new liver," and they proceed to pull open the panel on your chest, rip the old one out and insert the new one! Well, actually they do that now, it's called liver transplant surgery, but I don't think that's really the way Nature intended for us to go. Naturopaths are of the opinion that the best therapy is to flood the liver and body with oxygen at a safe level where the body can stand the detoxification process. That's tougher when you have a liver problem because the liver is one of the organs of detoxification. If the liver has been compromised it must be approached slowly.

Portal Vein Protocol Used for 50 Years

I went to Mist, Oregon and documented and videotaped a doctor in action before his extensive knowledge was lost to us. The best method I have ever seen for the detoxification of the human liver was perfected by naturopath William Turska over 50 years of clinical practice. He would commonly **inject ozone into the portal vein** by going in through the anal opening.

I videotaped him performing this procedure. He first took a glass specula, an instrument that resembles the neck of a bottle about four inches in diameter and inserted it into the anal opening to hold it open. He then took an injection instrument that was in the shape of a pistol that included an on/off trigger and hooked it to an ozone machine. It had a long thin needle on it where the barrel would be. He found the needle insertion point by looking through the glass opening to position the needle.

He would locate the large interior vein pulse with his finger, or visually, and then insert the needle through the opening, right through the colon wall. Then he would inject ozone in short bursts directly into the portal vein over several minutes. New patients got shorter treatments and lower concentrations.

He explained that the portal vein is interesting in that it goes up from there and branches out like a fan all over the liver. Dr. Turska was so experienced he knew exactly where to insert every time. When the ozone is injected here it goes up into the liver, and then of course attacks anything that was in there bothering the liver. After several treatments the patients were healthy.

Dr. Turska (now deceased) showed me a before and after pictorial record of people with all kinds of diseases that quickly went away from the application of ozone into the portal vein. Since he passed away, it's going to be hard to find anybody anywhere that uses or knows of this method. I hope through this publication that interest in this procedure can be revived. It's the best liver treatment and/or general internal organ treatment I've ever seen.

One of the benefits from this method (or any injection method) is that the closer you can apply ozone to the problem area, the more of it that's going to get to where you want it. As ozone is introduced into the body its molecules rapidly break down every time it encounters a dead, weak, feeble, diseased, or dying cell or any toxic matter. Because it works on the first thing it encounters, if it is inserted too far from the problem area it gives up its effectiveness before it gets to the location of the problem. That's why we advocate one *flood* the body with oxygen, to get it to reach everywhere.

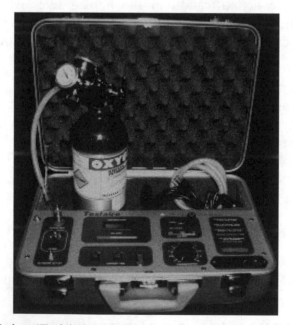

Modern 'Teslaire' Tesla Technology Octozone Generator.
Octazone - O_8 is a more powerful, and longer lasting, yet
milder form of ozone used by Naturopaths.

The Top Four
Most Effective Clinical Ozone Delivery Systems

After interviewing hundreds of doctors and thousands of Oxygen Therapy using patients worldwide, here are the four top general clinical ozone delivery methods, and my ranking of them, the most effective one listed last: Major Autohemotherapy, Direct IV, and Recirculatory Hemoperfusion (RHP) or Polyatomic Apheresis.

1. Major Autohemotherapy

The technique of Major Autohemotherapy was developed in the 1950s by the Germans. Undoubtedly they have been one of the leading authorities in

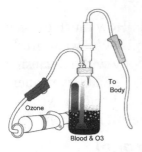

Medical Ozone Therapy. Major Autohemotherapy is something that's been practiced for about 50 years in Europe. There are probably more than 7,000 physicians in the world using this form of Ozone Therapy as you read this.

Autohemotherapy illustration from 15th Ozone World Congress 2001 IOA (International Ozone Association) London presentation of Renate Viebahn-Haensler, Iffezheim.

This method involves the withdrawal of approximately 250-to-600 ml (about half a quart) of blood and re-infusing it into the body after gently putting 27mcg/ml^3 of ozone into it. Ozone and oxygen are infused into the removed blood for several minutes. The ozone is gently swished around with the blood by hand, or you have a machine that does this automatically. It is gently moved around, not shaken hard because damage to blood cells will occur. When this process is completed, the once dark and now bright red ozonated blood is then reintroduced into the body by means of a vein in the form of an IV drip. The length of the procedure is from 20-to-30 minutes.

The drawback to its real effectiveness is that it is usually given only once or twice a week, because the patients can only afford that many treatments. If the doctors would switch to Direct IV, the patients would pay the same but triple their bang for their bucks.

By the time the ozonated blood gets back to the body there's little if any ozone left, but you have this sterile blood which is now full of

hydroperoxy radicals and other good free radicals which destroy bacteria and viruses and clean the body out. Also, you're creating a blood drug product. The ozonated blood and freshly killed microbes are dripped back into your bloodstream and you've just created vaccines for everything in your blood.

How does this procedure feel? If this process is handled skillfully it is barely felt, since a fine butterfly needle is used. After a while some may feel a sensation of fullness in the chest. This is usually a sign that the body has had enough for the moment as the blood is outgassing and causing detoxification in the lungs. Some people may cough for a short period of time after the treatment. This is only lasts for about 30 minutes. This is more of a mild irritation than anything else.

The main thing in Oxygen Therapy is the dose loading. How much oxygen can you safely get into the body with this method? We want fresh, energetic, clean active oxygen to safely get into the body without causing too much detoxification creating its own problems. Don't make the patient uncomfortable but build the effect of surrounding the bacteria, viruses, funguses, and pathogens with this oxygen so that it doesn't hurt them.

2. Ozonated Saline Drips (Parenteral Ozone)

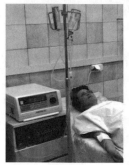

In 1924 British physicians were injecting peroxide and saline mixtures into Indian Ghurkas and curing the plague. Today the Russian medical establishment has a favorite delivery method of adding ozone to physiological saline solutions and putting the mixture on or into the body. They call it 'Parenteral Ozone,' and its use was developed in 1977 by Prof. Peretyagin and co-authors. Prof. Sergi P. Peretyagin is now the President of the Russian Association of Ozone. We met with Russian MDs at an IOA meeting in the early nineties where they gave us a short video of the then current Russian medical uses of ozone. The video included this saline technique and pictures of milk being ozonated for purity, and described tubes of ozone being instantly stuck into fresh gunshot wound holes during emergency room situations.

While developing these procedures the Russians also studied the kinetics of saturation and stability of ozone in normal saline. They developed a metrological system that allows the measurement of ozone concentrations not only in gas mixtures, but also in aqueous solutions. The Russians further say that as far as they know, "Similar metrological base in medical ozonising device does not exist in the world."

One Parenteral Ozone method is intravenous drop-by-drop infusion of ozonated physiological 0,9 percent sodium chloride solution and the use of possibly low, absolutely safe and therapeutically active ozone concentrations up to 10 000 mcg/L. This approach, through a long-lasting contact of medical ozone with sick human bodies, allows practitioners to achieve positive therapeutical effects by using considerably lower ozone concentrations in comparison with the Western school delivery methods. The safety and high efficiency of this ozone concentration range has been confirmed by a great number of experimental and clinical investigations. It is a simple and effective method.

Although to date, there have been no comparative effectiveness studies among the various ozone delivery methods, we do know that ozone's effectiveness is directly related to how much of it can be safely delivered intact inside the human or animal body. It stands to reason that this ozonated saline drip method is very safe, and effective, and also naturally limited by the amount of saline you can comfortably put into a patient's body. In the next two methods I discuss, the only limit to application is how much of the oxygen/ozone gas can be physically and comfortably absorbed by the blood and tissues. Therefore I would rank the next two methods higher in their ability to deliver a greater volume of oxygen/ozone gas into the body.

3. Direct IV Ozone

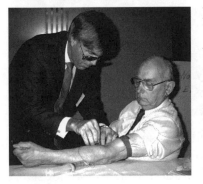

Lucas Boeve injects Wally Grotz with ozone at an IBOM meeting.

Popularized in America and Canada by my 1,500+ lectures that described the real world darkfield microscope verified results of Ozone Therapy during the Key Biscayne 'cowboy ozone house' days.

Discussion of this method is always met by the question, "You're gonna inject ozone, a gas, into a vein? Won't you kill somebody with an air bubble?" No. Ozone is only pure oxygen, completely soluble in blood, but air is 80 percent nitrogen which is insoluble in blood. We don't inject air; we inject oxygen and its more active form called ozone. If done slowly enough, there are no problems. The blood drinks up oxygen. Physicians and direct IV ozone users please note—best to obtain and use a syringe/infusion pump set at a slow steady 2cc. per minute loaded with sequential 10cc syringes instead of trusting the delivery to human hands and their attendant variables.

Ozone Does Not Cause Embolisms

From *Embolisms and their Formation* by Basil Wainwright—"To understand the process of a potential embolus (embolism) occurring in the bloodstream, one must appreciate the mechanisms involved. For example, an embolism in the blood can normally be caused by numerous processes such as surgery, severe chemistry imbalances, and accidents which can cause an artery or vein compression, dislodging damaged artery or vein tissue.

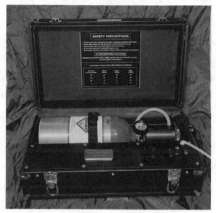

"Other causes of potential embolisms are fats, the introduction of air (nitrogen), amniotic fluid released into the bloodstream primarily during pregnancy, or other foreign material. *At no time has molecular oxygen ever been recorded to have caused an embolism*; simply because when calculated levels of molecular oxygen are introduced into the blood via the veins, immediate conversion takes place. The hydroxyls and peroxyls are immediately absorbed by the

High quality ozone water purifier

hemoglobin of the veinilar-deoxygenated blood. All veins except the pulmonary vein carry oxygenated blood from the tissues back to the vena cava for oxygenation.

"It is perfectly understandable that a doctor observing Apheresis [next section] would naturally fear the possibility of an embolism. Hopefully, the following explanation will eradicate those fears since it is impossible to create an embolism using Polyatomic Oxygen (O_3/O_4 ozone).

"The hemoglobin within each red blood cell is a special iron containing protein (globin), and is responsible for the transportation of oxygen throughout the body. The red pigment coloration (haem) obviously is responsible for the coloration of the red blood cells. There are typically six million red blood cells to every cubic milliliter of blood and each red blood cell contains 12 million iron atoms, it takes four iron atoms to transport 2.7 million oxygen molecules. This means that each milliliter of blood can transport 17 trillion, 100 billion oxygen atoms. In addition, if the patient through controlling dietary process has higher levels of iron, B complex, ascorbic acid, germanium and magnesium prior to treatment, then the *lymph and plasma* can also become significant oxygen carriers throughout the system.

"The complexity of the Iron atom (Fe) and its irrefutable scientific data and therapeutic results attest to the healing properties of polyatomic oxygen."

How Effective is Direct IV?

Direct intravenous injection is the process of injecting a specific quantity, purity, and concentration of ozone and oxygen gas into the body. In my experienced opinion, direct IV is the most aggressive, most active, next best therapy, second only to Recirculatory Hemoperfusion, also known as Polyatomic Apheresis. Got AIDS or Hepatitis C? Do not settle for less.

The direct IV method has also been successfully applied to the human body for over 50 years. The quantitatively superior results obtained in modern times through a more aggressive use of it have fueled the rapid spread of direct IV ozone knowledge and use. It is used whenever the Recirculatory Hemoperfusion methods are not available.

Direct IV was the most advanced and aggressive method around until Recirculatory Hemoperfusion came along, and it is superior to autohemotherapy because the IV method delivers a lot more ozone into the body. It is cheaper than autohemotherapy or Recirculatory Hemoperfusion, due to using less equipment.

Critics unjustly claim it is not as safe as autohemotherapy, and proponents say that such sentiments are rumor, and that such opinions are only being spread to scare users into buying expensive and less efficient autohemotherapy machines. Direct IV has always been safe in the clinics I visited, and whenever I have seen it used privately, but it was always used CORRECTLY.

Direct IV ozone is much more effective and less expensive to the patient, but the Germans and the Americans who learn from them are strangely reluctant to switch over to using it. This might be because of force of habit, or because of the investment in the autohemotherapy machines they already have. Some wits have proposed Direct IV might be purposely discouraged by the German industry and their adherents because practitioners can make more money through having to unnecessarily treat repeatedly due to using less effective autohemotherapy methods. I personally think people just stick with what they know, and it will take the next generation of Germans and their adherents to discover how superior direct IV ozone is. The present generation does not know what they're missing.

Common Protocol for Direct IV

The procedure entails very slowly infusing the ozone/oxygen gas into the blood to sterilize and purify the liquids and cells. The physician starts with pure dry medical grade bottled oxygen gas. The oxygen tank must have a pediatric regulator on it capable of regulating the flow rate of the oxygen into the machine at ½ liter per minute (lpm) or less. Due to the insoluble nitrogen, air is NEVER used to create ozone for use inside the body.

The generator creates a mixture of 2 percent ozone and 98 percent oxygen, which is equivalent to 27 mcg/ml^3. 27 mcg is the concentration proven to give the maximum amount of harmful microbe kill, with the least amount of cell disruption.

For safety, direct IVs are only given to patients, each lying completely flat before, during and for a while after treatment, so the ozone/oxygen is slowly distributed evenly throughout the body.

The only equipment needed is a number 25g winged infusion set (butterfly needle), some tape to hold the needle flat, a syringe, and the highest quality medical grade ozone generator available using only pure dry bottled oxygen as the source gas.

The injection rate will be 1cc per minute and applied twice daily. It is best to use 10cc syringes and to continually refill them as needed. This gas is to be injected intravenously. The key here is s...l...o...w... injections. If you rush it, you've got trouble. The physician has to give the blood time to absorb the ozone/oxygen gas.

While filling a syringe, normally one would pull out the plunger. In this instance just let the line pressure coming from the bottled oxygen tank push out the syringe plunger; that way no air will be drawn into the syringe. A ½ liter a minute oxygen flow rate, or less, fills up the syringe with the oxygen/ozone mixture. When filled the first time, the syringe is always purged once to be sure all air and contaminants are removed, and then refilled. Upon removal from the ozone generator filling port, the invisible clear ozone in the syringe must be treated as if it were water. Ozone is heavier than air and you don't want the ozone spilling out and air going in, so the open end of the syringe is always held upward.

Attach the tubing from the winged infusion set to the end of the syringe and remove the protective needle cover. Slowly push the syringe plunger slightly, until you can smell ozone coming from the needle. This is to purge any air from the tubing line.

Thus the purged needle is inserted into the vein of an arm pre-prepared with a standard injection tourniquet. If the patient is standing or sitting, he or she may have trouble with uneven blood pressure, so ALWAYS have the patient lying down before, during and after the injection. Do not skip this step and have every hundredth or so patient faint.

Upon insertion, with the blood being under pressure, a little of the patient's blood will move through the needle and into the small tubing. Notice the color of this first blood during each successive therapy session. If the blood is dark brown, black, or dark purple in color, this indicates the patient's current state of health is quite poor, no matter how he or she 'feels'. Practitioners are strongly advised to apply this method only after they have received thorough training.

Continuing the protocol, the blood just grabs the ozone as it goes by and then circulates it everywhere, but if the ozone is applied too fast, it might irritate the vein of any malnourished patients who do not get enough natural antioxidants in their diets. Medical ozone is just active pure oxygen. The blood loves oxygen. There is no danger in this method when used correctly.

How much ozone can be safely given? The first time try 10cc. The second time try 15cc. The next time, 20cc, and so forth. Each patient is very different as to his or her individual tolerance level of toxicity. What you are looking for with this slowly increasing dosage method is the patient's individual 'full' level. When the patient begins to get a chest or throat tickle or cough, you stop immediately because they're full. If you continue injecting during this session, the patient will soon start coughing, and probably cough uncomfortably for an extended period of time. Sometimes a coughing cycle can be stopped by doing ozone ear insufflation.

I always have used the analogy of filling up the gas tank in your car. You pump the gas in and when it's full, if you keep pumping it in the gas runs down the side of the car. The lungs are the oxygen overflow mechanism for the blood. When the bloodstream is full, the blood out-gasses into the lungs, and the oxygen-ozone sub species 'run down' the inside of the lungs, causing rapid lung pollution detoxification, heat, and possible slight temporary edema. All the patient knows is that he or she can't stop coughing if you do not quickly stop the procedure at the first sign of this.

Managed properly, this is not a drawback but analogous to having a direct indicator of that patient's particular capacity to absorb the therapy on that day. Historically, patients with clean blood, usually young vegetarian athletes, can absorb the most gas because their antioxidant systems are working quite well. I saw one such person take 6, 60cc syringes of 42 ug/ml^3 ozone once without effect. At the opposite end of the curve sometimes we see very plugged-up, toxic people with black blood also soak it up like a sponge. In their case, there is no out-gas, since the toxins use all of the active oxygen up before it reaches the lungs. Most everyone else is in the middle, and they usually start topping out at 25cc at first.

Individual results will always vary, and the majority of people I've seen get better from taking direct IV ozone had HIV, cancer, multiple sclerosis, arthritis, STD, chronic fatigue, heart disease, or arthritis. This has been my personal observation from years of experience, interviews, and research with the actual people that I have met backing that statement up in clinics worldwide. Unlike many other writers, this is not theory or hearsay. Some other authors merely review the literature and sum it up for you. I collate the original discoveries and deliver the goods from the inside.

The Main Advantages of Direct IV:

- A precise dosage is administered.
- Direct IV is inexpensive and requires fewer applications.
- Direct IV produces better results than autohemotherapy, especially in lung cancer, allergies, and AIDS.
- It consistently removes unwanted antibodies from the blood stream.
- It eliminates allergic components which contribute to cancer formation and virtually all other degenerative diseases, vascular disease, immune disease, environmental illness and allergies.
- The effect is more consistent than autohemotherapy.
- When the ozone is injected into the blood stream, it immediately starts to react with any type of oxidizable substrate, especially lipids of the cell membrane. Lipid peroxidation products of ozone include alkaoxyl and peroxyl radicals, singlet oxygen, hydrogen peroxide (peroxide burst is the mechanism by which viruses, bacteria, and fungi are killed by macrophages and microphages) oxonides, carbonyls, alkanes and alkenes.
- People with AIDS are seeing doctors using this method and they are reporting getting better and returning to normal lives.

New Diagnostic Procedure

Dirty blood equals toxicity and is one of the two major causes of disease, as I explained earlier. A patient with dirty fluids is diseased; or, if left untreated, disease will certainly follow. As the number of treatments progresses day after day, the first blood will get lighter and lighter, brighter, and brighter, eventually turning pure clean natural bright cherry red if the therapy is continued. Look at the back cover of this book for comparison purposes. I have seen blood turn from death black, to brown, to purple to beautiful bright life red in just six weeks through using direct IV alone. Right here I have just revealed the most important and inexpensive objective low-tech whole-body diagnostic procedure available anywhere. Anyone with sight can now gauge the level of garbage in his or her own fluids, organs, and cells by just looking at the color.

Part of the problem is that everyone's blood is filthy. Healthcare workers that give or take blood look at diseased blood all the time in this over-polluted society, so they tell you it's 'normal' for venous blood to look almost black. Sorry, all blood used to be red; bright cherry red. And if it were, we would all be healthy instead of facing the longest running chronic and other disease epidemics in history.

I Meet Ozone, the King

In the early days I wanted to understand ozone completely by living through it so that I could explain it correctly to people. I went to Key Biscayne and gave myself direct IV ozone twice a day for one week while daily looking at my blood through a darkfield microscope. When I first injected I could see my individual blood cells all perfectly lined up. But by the third day, my blood looked like it was a field of mud from the oxidized pollution now thrown into solution. I had previously done a lot of peroxide cleansing, but nothing is as strong a detoxifier as ozone; it's The Breath of God. It was burning up the rest of the pollution and toxicity that the other cleanses I spent so much time on had missed! And I didn't have any major disease. I merely had the occasional flu and a little bit of arthritis, and a general feeling of exhaustion.

A Drawback to Direct IV

I interviewed a man that was formerly a male prostitute. He developed AIDS. Naturally, he had a great deal to overcome besides low-self esteem. When the usual drugs didn't make him better, he decided to try ozone. He taught himself how to inject it, and started slowly injecting using multiple 10cc syringes of 27-mcg ozone/oxygen via a syringe pump. [It is recommended that you use a syringe pump during IV work to remove a source of human error.] He was cruising along and feeling great, and over a few weeks he had worked himself up to 15-to-20 syringes full, or 150-to-200cc of ozone per day. He never got the cough indicating he was full of ozone, but he did get the chest tightness and was smart enough to stop putting any more into him for that day.

Then one day he hit his first detox plateau. Suddenly he had shingles breaking out all over his skin. (A form of the herpes virus.) This caused him concern enough to go the emergency room one night at three a.m. The breakout was the classic detox response to a too aggressive direct IV ozone usage.

Our bodies take every illness we have had and suppress them long enough to inactivate the illness cause by covering it up with dried hardened mucopolysaccharides, or mucous. The body is 'walling it off.' Once it is kept away from the rest of the body somewhere and the symptoms stop, we say we are cured. Maybe, but we're not clean yet. I'm describing classic drug use for symptom suppression without fixing the real problems. In this manner we are still carrying around the remnants of our old childhood diseases.

The ozone comes along, burns up the mucous patches, and out pop the old dead or dormant pathogens. They still cause the same symptoms as when they were fresh, but with all that ozone/sub species floating around they are burnt up instantly. When detoxing, we go through all our previous disease symptoms in reverse order. This is why. Just like peeling an onion.

The man in our example did have a warning sign that he was pushing his therapy too fast. For a few days before this, every time he put the butterfly needle in his vein, no blood readily came up the clear tube attached to the needle. And he was extremely tired. He thought he wasn't getting the needle into his vein. Actually this lack of normal pre-injection backward up-the-needle flow indicated his blood was so full of thrown out and oxidized debris that it was getting way too thick. He was tired because so many toxins were stalling in his thickening blood and lymph. The garbage couldn't leave easily, and less oxygen was getting through the blood fluid traffic jam and into the cells —so he was low on energy.

The classic solution at this point is to lay off the ozone injections for a bit, while keeping some oral oxygen supplements going (enough to keep the toxins in solution) and including mild exercise (to keep the fluids moving) and colonics (to keep the drains open) and lots of water (to continue slowly diluting and cleaning it all out). Once the blood and fluids return to a normal clearer thinner consistency, the ozone therapy is resumed at a lower level. His reaction could have been diarrhea, sweats, swelling, itchiness, pain, rashes, or whatever the patient's weakest link is. For him, it was a temporary re-occurrence of the symptoms of his old shingles.

Direct IV is the least expensive heavy-duty ozone delivery method. It is continued until all symptoms go away and the blood remains a bright cherry red. Assume a minimum of six-eight weeks of daily treatment while eating lots of natural food full of antioxidants. A drawback of this method is feeling like a pin cushion and then not wanting to continue, especially if mistakes are made that make you sore.

The good news is that this detox stuff is only temporary and entirely avoidable in the first place by *going slower*. However, the fact that it happens always proves the therapy really is working and burning up the toxins as advertised. No need to go this fast and be uncomfortable. The next therapy I will describe (RHP) does not have this problem.

The Dormant 'AIDS Therapy Institute'
Ian St. Claire, John Page, Townsend Hoopes, Alex Bradford, and Associates started a research company to get Direct IV ozone through the FDA process but switched to ozonated saline when the old guard at the FDA became fearful and incorrectly thought IV ozone might cause embolisms. Money was raised, 7 animal studies were completed at the U of VA and Cornell, etc., and in 1992 they filed Investigative New Drug Status. But ozone was stopped short again by its old nemesis *human ego*. $8 Million short, they fell apart before entering Phase 1 human testing.

"Queen Mum is Staying Young with Ozone Jabs"
05-13-01 Sunday Mirror, UK, by Terry O'Hanlon, pp 11

"The Queen Mother is trying to fend off the effects of old age by having 'ozone injections'. The regular jabs increase oxygen levels in the 100 year old's blood and the treatment's supporters say it removes toxins and rejuvenates the whole body.

"She let slip the treatment to friends when they commented how good she looked. A Clarence House spokesman would not comment on the Queen Mother's treatment, but a palace insider said: "'Everyone says how remarkable the Queen Mother is for her age. Perhaps this is the reason.'"

If ozone injections were good enough for the Queen Mum,
Aren't they good enough for you?

---

4. Extracorporeal Recirculatory Hemoperfusion, RHP, or 'Polyatomic Apheresis'

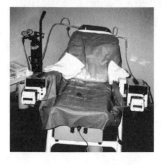

What I want ultimately is all the doctors using this method! Finally the answer to "So if flooding the body is the ultimate goal, how do we safely and comfortably detoxify the whole bloodstream of the body completely and all at once?" My 'all time favorite' Oxygen Therapy is Recirculatory Hemoperfusion, RHP, (I coined this phrase) or Polyatomic Apheresis Oxygen Ozone Therapy. Think of it as blood/ozone dialysis.

This is the most advanced emerging technology within medicine today. It uses expensive specialized equipment and has to be performed by a medical professional. Pure molecular oxygen—O_2 is the prime carrier and ozone as O_3, O_4, and O_8 is added to it. This mixture is continually added to and circulated through the bloodstream of a patient by means of highly sophisticated dialysis type method. The patient's blood is mixed with ozone (polyatomic oxygen) in an extra corporeal (outside the body) closed loop. In a continuous fashion the blood ozone mixture is being re-infused into the patient. The blood only leaves the body momentarily before coming back in completely sterilized and oxygenated.

The beauty of this method is that after pumping blood out of one arm, it is then ozonated and filtered *outside* of the body.

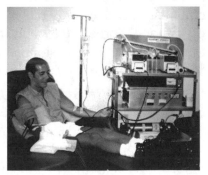

The first time I saw Recirculatory blood perfusion in operation on a human, I was suddenly amazed by two immediately very striking visual facts. The profound importance of these two simple facts burst into my consciousness like lightning. I was looking right at the answer of how to quickly prove to any scoffers that these therapies work!

The diseased blood continually coming out of the patient and flowing into the blood/ozone-mixing chamber was very dark, almost black. However (and here is the whole point of this book) the sterile ozonated blood coming out of the other end of the chamber and going back into the patient was now bright cherry red! The blood went from being a filthy disease carrier to becoming perfectly clean. The immediate visual proof.

Inside the far end of the clear glass mixing chamber, just under where the newly sterilized bright red blood passes through the filter and then leaves to go back into the patient, there was a steadily growing pile of pale powdery substances. As this pile of ex-blood garbage got bigger and bigger, I realized I was looking straight at the reason why all the other Oxygen Therapy delivery systems are doomed to pale in comparison to this one. I realized that the flocculated piles of burnt up toxins and microbes were the only things holding back ozone's true potential. All other delivery systems mix active oxygen and blood *inside* the body—so they are always sending the pile of newly oxidized sludge back through the already stressed organs. This creates the cleansing reactions that always limits how fast you can give someone the active forms of oxygen.

By having the ozone/blood interface reaction *outside* the body, there are a number of benefits to the patient:

➤ We do not force the liver, the kidney, and the lymph system to process the dead microbes and carcasses and inactivated toxins that ozone creates. By not having the body deposit these things into the fluids for removal, irritating cleansing reactions are minimized. Patient comfort and willingness to continue with the therapy 'long term' is increased substantially.

➤ Because the efficiency of this ozone delivery method is optimized, this results in each treatment being quantitatively more effective than standard ozone delivery methods, which by themselves are already proven to be way ahead of most so-called 'normal' medical treatments here in America.

➤ The amount of ozone that this method puts into the blood is substantially greater than in any other method, necessitating far fewer treatments.

➤ This means, in the long run and compared to any other method, instead of having to take the usual six months or perhaps years of Ozone Therapy, even worst case scenario patients get significant results within just weeks!

And the top reason for using this method? It works, it works quickly, and it is the least expensive therapy on the planet when compared to the good it does you! People (or their insurance companies) commonly spend far more for questionable results from non-Ozone Therapies.

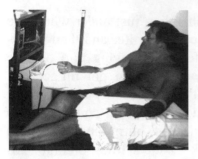

Recirculatory Hemoperfusion is the present cutting edge and the mainstream future of the heavy-duty clinical Oxygen Therapies. Owing to its complexity, a Recirculatory Hemoperfusion setup is definitely not a home unit; yet I think Recirculatory Hemoperfusion is a golden answer to treating most disease. The whole slow-down with other Oxygen Therapies comes about because we have to limit the dose increases due to the unwanted detoxification effects created when you start oxidizing someone's toxins rapidly. This unnecessary extra detox work makes already stressed organs work unnecessarily hard, and any serious patient pursuing aggressive therapy will feel uncomfortable.

The polyatomic active oxygen is shown to react by ionization, oxidation and by restructuring the cells' protein structures, thereby eradicating any diseased cells. It destroys bacterial, fungal, protozoal, parasitic and other microorganisms in the bloodstream. It removes viruses and prions. Ozone and oxygen reaches all the organs and eventually perfuses into all body tissue, including the bones. It also crosses the blood brain barrier very effectively. Do it long enough, and nothing can hide from it. Nothing. The laws of physics will apply.

Common Protocol for Recirculatory Hemoperfusion

Recirculatory Hemoperfusion therapy is such a highly effective treatment that it must be administered by trained personnel and monitored so that it does not induce any unwarranted side effects in the patient. Prior to the treatment, the patient is subject to extensive blood sampling and laboratory work. This blood work is also conducted after the treatments in order to depict the changing status of the patients' blood.

The patient must have good veins. With the patient lying in a reclined chair two catheters or blood ports are inserted, one into the vein in one arm and another into the vein in the other arm (legs, feet or the groin area may also be used). Blood is continually withdrawn out of one arm and then flows down an elevated 'dimpled glass' inside a hollow tube filled with ozone. The dimpled staircase  gently turbulates the blood in the presence of ozone. Picture a series of waterfalls picking up oxygen from the air. These 'waterfalls' stretch the surface tension of the blood fluid and create a good low pressure mix of the blood and ozone. The sludge separates out of the fluid at the bottom of the tube and is filtered out. The filter used is 0.09 microns. The remaining clean sterile bright red blood is then pumped back into the bloodstream in the other arm. A one hour treatment is superior to *any* other ozone delivery method.

Experimental pushing-the-envelope private trials had a few patients successfully ozonating, filtering and recirculating their blood this way for more than eight hours! Imagine the volume of blood, lymph, and organ purification obtained in an eight hour treatment.

The continual circulation and filtering will take out a majority of the blood garbage; and if the patient stays hooked up for at least three hours, every single bit of body blood is totally ozonated and purified. A typical daily treatment to start with can last from 45-to-90 minutes. The therapy allows the patient to resume a normal work schedule after a treatment. Treatments can be performed on a daily or 'every other day' basis until the desired effects are achieved. The procedure provides a safe and effective treatment for a wide range of diseases.

It's an amazingly effective treatment. I interviewed one doctor who said a literally dying unconscious AIDS patient was brought in on a stretcher. They said, "Well, he's almost dead. It doesn't matter much at this point, so let's hook him up and see what happens." After a few hours of Recirculatory Hemoperfusion treatment the patient got up, dusted himself off and walked home!

Summary

In more than 14 years of visiting ozone clinics around the world, I continually witnessed wonderful effects taking place as old toxin, microbe, parasite, and cancer-cell-laden brown, dark purple or black blood came out of the diseased bodies receiving daily ozone treatment. I witnessed the diseased blood turning a bright cherry red over the course of a few weeks. Bright cherry red is the color of health, free from the old toxins and microbes and parasites and cancer cells that have been oxidized away by the Ozone and/or other Oxygen Therapies. The removal of their inactivated oxidized remains by the body is a bit irritating, and sometimes causes the classic harmless temporary cleansing reactions. To minimize cleansing reactions, I advocate the use of the recirculatory ozone blood washing delivery methods whenever possible.

The Commercialization of Recirculatory Ozone

In order to interface the blood with ozone, one publicly traded U.S. company called Medizone International (Symbol: MZEI) proposed using a hollow fiber bundle interface, and another, Polyatomic Apheresis International (Basil Wainwright's PAI Company) uses the elevated dimpled hollow tube filled with ozone.

Why don't we hear more about this superior method of treating people? Very, very good question. Authority ignores ozone or has prevented USA research under the guise of 'protecting' the public. That's one of the reasons why I wrote and why you bought this book, the strange unavailability of information from traditional trusted sources.

Basil Earle Wainwright and His Polyatomic Apheresis,

Basil Earle Wainwright is another 'colorful' chapter in the annals of ozone history. Basil is a British research   scientist and inventor who has worked at a host of ways to solve several of the most daunting problems encountered by all the traditional methods of

I interviewed Basil many times while visiting his lab and research facility in Florida.

delivering ozone to the human body. Many investors lost a lot of money over the years as Basil's various operations struggled to survive. He would often take in investment money and it would quickly be eaten up by expenses before any returns could be realized. Sometimes he managed the money, and sometimes he let other people manage the money and build the machines so he could concentrate on research. Often those chosen as fiduciaries were mistakes. The result was that often people's feelings were hurt due to promises left unfulfilled.

Basil often had to do whatever he thought was necessary to keep going. Over the years many have told me sad stories about how they invested everything they had in his various businesses. After all, like in the 'cowboy clinic' days of Key Biscayne, weren't they seeing amazing cures right before their eyes? Usually they were drawn in with thoughts of making millions, or some joined in for humanitarian reasons, but all of them had one thing in common. They put up a lot of money and then it was gone. Others would send in machines for updates or repairs and never see them again, or get an inferior one in return instead. These stories were personally reported to me several times over the years by several of the people involved.

Naturally Basil was hounded by dissatisfied authorities and newshounds because he was right in their face challenging their medical status quo, and also because people that had lost money were seeking redress by going to the same authorities. But it's not that simple a situation to judge. Some of the authorities really were out to keep things just the way they are, and conversely, along the way his ozone generators were being used by desperate people who were having remarkable success.

 Basil was off and on a political prisoner, first being jailed by the British as a fraud as the British press vilified him, and then by the Americans. He was also vilified by the local Florida press and then put in jail in Florida for four years without being convicted, and all his motions were ignored by the court.

And now he is sought by the Kenyans, including of course, more unhappy investors and ex-employees, and the usual press vilification is going on. But despite conditions that would have caused most to walk away from the whole thing, he perseveres and keeps working at perfecting his technologies. The man has an honestly deserved persecution complex. Later on I'll show you the results of the *positive* aspects of his work.

Don't fall for red herrings. His history, personality, character, and the politics surrounding him have **nothing** at all to do with whether or not the *science* of Ozone Therapy is valid. Obviously Ozone Therapy is valid, and obviously Basil is very intelligent and has personally been both a blessing and enigma in our industry and to those who seek him out. In my opinion, I have never seen someone do so much good for the downtrodden and ill, while also causing so much chaos with those who buy from or invest in him.

I know all about both sides of Basil's stories having personally spoken with most of the actors involved as the stories unfolded over many years. Politics is a slippery slope, and Basil is English, so I'll use an example from British folklore. The Sheriff of Nottingham and the Noblemen thought Robin Hood and his band of merry men were crooks, and on the other hand, the poor suffering people loved Robin for passing out the Nobleman's' money. Was Robin a hero or a thief? Depends who you ask. The hitch always comes in the details of the execution. The question is always, is one operating from integrity and honesty?

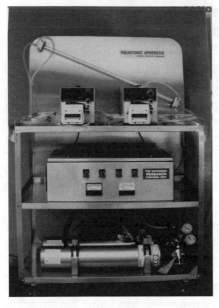

From Basil's Website

"On March 14th, 1990, in Miami, Florida, Basil Wainwright [Note: following the lead of the other ozone experts over the past 100 years and in this instant case] converted his first HIV+ patient to PCR undetectable. Following this major medical breakthrough, it was only a matter of time before the FDA, with weapons drawn, stormed and ransacked Basil's Research facilities in Florida. The United States authorities seized and confiscated his extensive research data, results, and equipment, and Basil was thrown into jail for four years without trial, on the pretext of *practicing medicine without a license*. During this term in a U.S. prison, one of the police told Basil *'he could walk'* if he would just go on national television and denounce the efficacy of Medical Ozone. Altogether, six recorded attempts were made on his life and only through certain international scientific pressure groups and Christian organizations was Basil ever released.

"The Kenya Government and Elders of the Pentecostal Church in Kenya invited Basil Earle Wainwright to establish his research facilities and utilize his proven *Polyatomic Apheresis Technologies and Polyatomic Oxygen Therapy* in Kenya, in early 1996.

"In the last six years Basil Earle Wainwright has made Kenya his home [no longer]. He has been able to carry out his research and has repeatedly recorded staggering results. From a 24-patient-sample carrying the HIV/AIDS retrovirus, *Polyatomic Oxygen Therapy* achieved 24 conversions ~ 19 PCR undetectable conversions and 5 HIV+ to PCR undetectable conversions with P 24 antibody negative. Already, more than fifty-eight (58) fully-documented, scientifically irrefutable HIV+ to PCR undetectable conversions have been achieved and verified by (Pre–Polymerase Chain Reaction) PCR analysis. [Since increased] The PCR tests indicated 'undetectable' Viral loads and included in many cases, blood plasma P 24 marker anti-body negative, amongst many other medical conditions *miraculously* treated using *Polyatomic Oxygen Therapy*.

"Today, Basil's fight to expose his technologies has not eased. Many forms of intimidation or harassment continue to be deployed to bar any access to these technologies and therapies to the general public around the world. This is easily achieved through apparently benign bureaucratic measures or sheer ignorance, to more aggressive, nefarious, corrupt and sinister manipulations. The fact is that he still lives with the fear that some day someone will take his life. Basil feels that he is still being constantly attacked and frustrated by 'the establishment.' A typical example was when Basil Wainwright invited Winnie Mandela to visit his Research facility in Nairobi and meet the HIV reversed patients herself. This caused such a great media hue and outcry, instigated by the Women's Association and the Medical Board, that within 24-hours Basil Wainwright was once again smeared scandalously across the front page of the national Standard Newspaper as ... '*a con man, crook and fugitive.*'"

Kenyan Headlines

"1350 FAKE CLAIMS FRUSTRATE KENYA'S HOPE FOR AN AIDS CURE"

Date: 10-20-1998;
Publication: Africa News Service

"Anyone would want to find a cure for the dreaded AIDS. But claim after claim have turned out to be mere attempts to make a quick buck." said the Kenyan newspaper.

The Kenya Medical Association banned Basil's treatment in July of 1998, strangely suddenly pronouncing it had "no scientific basis" without any investigation. Basil was declared a charlatan in the media. "The saga reached a high pitch when Winnie Mandela, estranged wife of South African President Nelson Mandela, visited Kenya the same month. She toured Basil's clinic in the posh Nairobi suburb of Karen and declared the treatment a big success. In her characteristic flamboyant and defiant fashion, she took issue with western scientists who, according to her, did not believe anything good could come out of Africa."

"Mrs. Mandela's rebuff caught the Kenya Medical Association with their pants down but galvanized it into further action. The Kenya Medical Association went to great lengths to condemn Basil and his treatment, and Kenya's health minister Jackson Kalweo declared the treatment banned, warning that his ministry would *close down any clinic that dared administer it.*"

I have to ask one more time, WHY? Ozone is completely safe when applied properly, so what is the real reason to stop the research when people using it are testing negative for diseases?

This standard operating procedure of press vilification was exactly the same authority-plus-media led tactic used to stop all ozone medical research in Australia, and Florida before that. The paper even falsely claimed Basil was stopped by "popular outcry," reminiscent of the old American newspaper barons creating the news by writing tomorrow's headlines to make something happen later. Why are they so afraid of Basil's research that they will send *every* media outlet and bought-off politician after anyone doing research? Isn't it obvious, in order to *prevent* any research - because somebody knows it works.

Here's how Basil described what happened:

"Dear Ed.
Despite the phenomenal successes achieved in Kenya, treating numerous diseases from HIV, Cancer, and Cardiovascular disease to name but a few, I was forced to leave, due to the rampant corruption and cash money brought in by a courier (whose identity is known) on behalf of (name of large multinational drug company withheld). Even $ 217 million USD given specifically for HIV research and to support PWA's has totally evaporated with all the politicians denying knowing where the money has gone. This is not a single occurrence but it goes on all the time."

He told me he was able to keep treating the sick by 'paying off the locals' to not enforce the politicized national bans, but that could only last so long. So what happened? The sick got tired of having their bodies used as pawns by the politicians when they were obviously getting well from the Ozone Therapy. More headlines emerged:

AIDS PATIENTS IN KENYA SUE OVER
BAN ON OZONE THERAPY
Date: 02-02-2000;
Publication: Africa News Service
Nairobi (The East African)

"Ten HIV-positive people have taken the Kenyan government to court for banning the controversial Polyatomic Apheresis treatment practiced by the self-styled Aids 'doctor', Basil Wainwright. The 10 say the ban is an infringement on their constitutional right to effective treatment of their condition.

"Led by the chairman of the Kenya Aids Society, Mr. Joe Muriuki, they want the ban on the procedure lifted and Mr. Wainwright allowed to practice. They have named Kenya's attorney-general, the minister for health and the Pharmacy and Poisons Board (PPB) as respondents. They claim that the therapy is one of the most effective treatments for Aids. Kenyan health authorities banned polyatomic Oxygen Therapy on September 7, 1998, through legal notice number 128 signed by the then Minister for Health, Mr. Jackson Kalweo. The minister outlawed 'the manufacture, sale, advertisement or possession' of polyatomic oxygen, also known as ozone, which is used by Mr. Wainwright to allegedly 'cleanse' a patient's blood of HIV.

"On his part, Mr. Wainwright accuses the health ministry of "deliberately" failing to give his 'polyatomic apheresis' technology a chance to 'prove its efficacy.' Speaking to *The East African* in his office last Wednesday, however, Kenya's director of medical services, Dr Richard Muga, said that Wainwright has been contravening the ban and illegally 'treating' patients at a secret clinic in Nyali, Mombassa.

"At the time of the 1998 ban, Wainwright was practicing from a clinic in Karen, a Nairobi suburb. 'It would seem that Wainwright secretly converted a residential house into an illegal clinic, without a license from the medical board as required by the law,' Dr Muga said, adding that the ministry had been unaware of the facility. A raid carried out by ministry officials and police a week ago on the Nyali premises found a well organized five-bed 'ward', including a nurse. The nurse confirmed that they had been 'treating' Aids patients using the outlawed procedure. No patients were however found admitted in the 'clinic' at the time."

What's the Bottom Line?

Away from politics and morality, the reality is that you can't outlaw the laws of physics no matter how hard you try. Anaerobic viruses *still* cannot live when surrounded by active forms of oxygen like ozone which is harmless when used correctly, no matter how many learned pronouncements to the contrary are made by those in high places.

Basil Wainwright among others is continuing to reverse a multitude of diseases, including HIV/AIDS, practically on a daily basis, using his own Recirculatory Hemoperfusion technology he first called 'Polyatomic Apheresis.' 'poly atomic' = many atoms of ozone—as in O_3, O_4, and O_8, and 'Apheresis'—as in *putting something into the blood.*

Back at the polyatomic website, Basil explains that his devices "synthesize the combinant ratios of stabilized concentrations of O_2, O_3, O_4, and O_8 oxygen molecules, which is achieved by passing pure medical grade oxygen (O_2), through an electrical corona discharge which, at specific frequencies and Hertz settings, creates the stabilized forms of recombinant oxygen atoms. Hence, Polyatomic Oxygen is created. There are recorded combinations of oxygen in excess of O_{64}. The device injects the Polyatomic Oxygen into the bloodstream in a precisely (digitally controlled) measured dosage and flow rate time to body weight and other medical therapy treatment parameters, for circulation through the blood throughout the patient's veins, arteries and capillaries. This is amazing, considering that adults have more than 96,000 kilometers of blood vessels, and the heart completely circulates our entire blood volume every 60 seconds.

"Polyatomic Oxygen Therapy and Polyatomic Apheresis Technologies do not pretend to be the ultimate panacea for all infection and medical conditions from which mankind suffers. It is, however, a complete disease cellular purification platform, a dramatic advance and a highly viable alternative medical therapy and health care option which can complement other forms of conventional medicine.

(Photos from Basil's video.)

It is capable of and has proven to eliminate all known bacterial, fungicidal, protozoa and virucidal infections,

Hopeful Kenyans lined up for ozone treatments whenever Basil's bright blue Polyatomic Ozone bus would pull into their village.

including the HIV/AIDS retrovirus, Ebola, cardiovascular disease, all kinds of cancers and tumors, Hepatitis 'A', 'B', 'C', 'D', 'E' and 'F', MS, CMV, EB, Parkinson's, Alzheimer's disease Tuberculosis, Pneumonia, Highland Malaria, amongst a host of other infections and diseases in in~vivo human applications."

Claims of 635 Irrefutable PCR Negative AIDS Cures?

March 5, 2002, In a telephone conversation with Basil Wainwright, in Nairobi, Kenya, on the current results of his highly efficient Polyatomic Apheresis treatments.—"Ed, we now have 635 Irrefutable AIDS Cures, and at one of our satellite clinics out in one of the small towns where there is no PCR testing available, we counted 400 more as very probable." Part of the list Basil sent me is reproduced next for your inspection.

Name	Condition	Test Results 1 (Viral load)	Test Results 2 (Viral Load)
Ann Guantai	HIV	482,000 copies/ml	53,000 copies/ml
Cathrine Hunter	HEP C / CFS	693,000 copies/ml	Undetectable
Christopher Kariuki	HIV	86,493 copies/ml	16,000 copies/ml
Clement Mhango	HIV	701,000 copies/ml	Undetectable
Daniel Okondi	HIV	460 copies/ml	Undetectable
David Oloinyeyei	HIV	492,000 copies/ml	82,000 copies/ml
Dekhana Kodzo	HIV	651,580 copies/ml	Undetectable
Eleni Peyioti	Cancer / HEP C	369,000 copies/ml	Undetectable, C/Clear
Elijah Kariuki	HIV	604,000 copies/ml	Undetectable
Grace Wekesa	HIV	362,000 copies/ml	29,000 copies/ml
Ibrahim Paranisat	HIV	530,000 copies/ml	15,000 copies/ml
Irina Bednina	HIV	740,000 copies/ml	Undetectable
Jackness Muthany	HIV	354,000 copies/ml	Negative
Jean Mathiodekas	HIV	263,000 copies/ml	Undetectable
John Odhiambo	HIV	102,186 copies/ml	83,831 copies/ml
Joseph Kipkenda	HIV	302,000 copies/ml	23,000 copies/ml
Jovita Awour	HIV	51,000 copies/ml	6,000 copies/ml
Louis McVicker	HIV	220,000 copies/ml	56,000 copies/ml
Margaret Makimei	HIV	160,000 copies/ml	1,320 copies/ml
Martha Kamau	HIV	280,000 copies/ml	Undetectable
Mary Kisira	HIV	204,000 copies/ml	Undetectable
Mary's Baby (Phylis	HIV	31,663 copies/ml	8,000 copies/ml
Micah Kigen	HIV	294,000 copies/ml	150,000 copies/ml
Murathi Elisabeth	HIV	48,227 copies/ml	Undetectable
Ochieng John	HIV	350,000 copies/ml	Undetectable
Onyango Samuel	HIV	492,914 copies/ml	86,041 copies/ml
Patel Indudhai	Prostate/ Diabet	490,000 copies/ml	Undetectable
Rose Naigah	HIV	4,764 copies/ml	Undetectable
Rosemary Orioki	HIV	1,341 copies/ml	Undetectable
Salome Kamau	HIV	484,000 copies/ml	26,000 copies/ml
Sarah Hicks	EB / CMV / HEP	692,000 copies/ml	Undetectable
Shompa E. Simaloi	HIV	692,000 copies/ml	10,230 copies/ml
Soki Chantal	HIV	329,452 copies/ml	29,000 copies/ml
Stephen Makimei	HIV	104,000 copies/ml	16,000 copies/ml

Sample page in list indicating 635+ Kenyans tested as cured through use
of Recirculatory Hemoperfusion of Ozone Therapy as provided by
WDDS Polyatomic Apheresis. (Supplied by Basil Wainwright.)

In interviews with others it was reported to me that many actual test results did come back as PCR negative, and people did dramatically get up from their deathbeds and have other similar amazing results. I was also told five of those listed on the above results chart page as 'good examples' were too far gone, and naturally died. One never had a test, and another remained inconclusive, and one was a doctor, not a patient.

The science of Ozone Therapy needs no such shenanigans! Ozone can rightfully stand upon its own true historical record of successes without any investor-pleasing aggressive 'embellishments.' Especially if the treatment is continued long enough. Any untruths in data set everything in our industry back by creating unwarranted suspicion fueling opposition. I am personally disgusted whenever I have to bring up these incidents, focus on Basil's character instead of science, waste 'sound bites,' and risk scaring people away from valuable therapies. However, truth is king here.

The poorest-responders are generally those who fail to adhere (by choice or circumstance) to sound nutritional regimes. Live food diets and proper hydration, minerals, and oils are crucial in enhancing the transportation of active molecular oxygen and related sub-species throughout the body.

Cure for AIDS Patented in 2000

Basil has reported his new *state of the art* treatment centers are covered by 17 separate World Patent Applications.

United States Patent 6,027,688, Wainwright, February 22, 2000

Apparatus and method for *inactivation of human immunodeficiency virus*. Abstract. An apparatus and method for the inactivation of infectious organisms such as viruses, bacteria, fungi and protozoa, and especially for the inactivation of human immunodeficiency virus in proteinaceous material such as blood and blood products, without adversely affecting the normal physiological activity of the material, by contacting it for a time interval of only about 16 seconds with an ozone-oxygen mixture having an ozone concentration of only about 27 .mcg/ml. Inventors: Wainwright; Basil E. Assignee: Polyatomic Apheresis, Ltd. Appl. No.: 237713, Filed: May 2, 1994.

Basil Location Update

Basil telephoned someone who had an email sent to me (through a series of third parties) just as this book went to press. Since the below letter, he's back on the internet: www.polyatomicoxygen.com.

"Dear Ed.

"Despite the phenomenal successes achieved in Kenya, treating numerous diseases, from HIV and Cancer to Cardiovascular disease to name but a few, I was forced to leave, due to the rampant corruption and cash money brought in by a courier....

"The government I am working with at this time has already analysed test results, and has given full governmental clearance and authorisation for this

Basil Wainwright
(From Kenyan video.)

technology to be used throughout Asia. (This is not approval for a research programme but the full go ahead for setting up clinics using this technology throughout Asia.) The Government officials are so concerned for my well being, they have provided me with armed guards (bristling with guns).

"Although I am not allowed because of security reasons (James Bond stuff) to let anyone know where I am at this time, we recently learnt that the Kenyan government has offered rewards of 6 million KSH for any information which will enable them to seize me.

All my love and kind regards, Basil"

[Others rebut that although he is brilliant, he also uses charitable causes as shields while avoiding creditors whom he could have paid but didn't.]

"Dear Ed.   Somewhere in Asia, [undisclosed location] 09/03/02

"Recently we have made the single biggest breakthrough in Medical history.

"Following incredibly successful meetings yesterday, the following countries have officially agreed and committed that Polyatomic Apheresis and its interrelated Technologies can be used extensively in their countries; Malaysia, Singapore, Brunei, Thailand, Indonesia, Laos, Myanmar Cambodia, and Vietnam.

"You should be the one who can use this breakthrough to help the suffering people of the world.

"The necessary financial input to do this is required right now. I mean right now. No longer can I or the dying people of the world continue with promises. GOD has given us the total Autonomy to do his work. Everyday, 3.9 million people die from man created diseases most of them unnecessarily. In the final equation do we say to GOD, we did our best? Did we really?

All my love and kind regards, Basil Earle Wainwright"

With the authorities and creditors nipping at his heels, Basil is still pulling magic out of his hat. He just sent me this:

"You will see in the photograph one of the most advanced blood purification systems in the world."

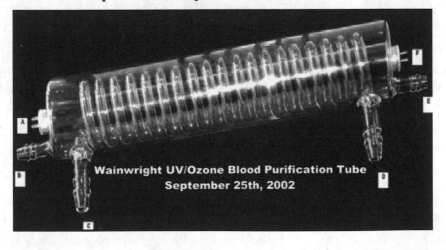

Wainwright UV/Ozone Blood Purification Tube
September 25th, 2002

A) INTERNAL UV TRANSMISSION UNIT   B) R-600 THERMAL CONTROL FLUID INPUT
C) BLOOD INPUT   D) BLOOD OUTPUT
E) R-600 THERMAL CONTROL FLUID OUTPUT   F) UV INTERNAL TRANSMISSION UNIT

"THIS UNIQUE SYSTEM CAN EFFECTIVELY BREAK DOWN ALL THE TOXIC COMPONENTS IN THE BLOOD INTO HARMLESS ABSORBENT NON-TOXIC COMPONENTS, WITHOUT APPLICATIONS OF IONIC FLUIDS OR ANY OTHER SEPARATION PROCESSES."

"These unique blood purification units will also eliminate cellular mechanical shear, which is caused by forcing blood cells through a multitude of hollow fibre semi permeable membranes. Not shown in the photograph are two ferrite clamps providing alternating electromagnetic field polarization, breaking down all the nitrogenous components within the blood by destabilization of their subatomic structures.

"All renal and peritoneal dialysers (without exception) cause significant amounts of cellular trauma upon the blood entering the membranes. As can be seen from the picture, the blood is merely recirculated through the capillary coil without trauma, and the pre-supersaturated oxygenated blood, is then subjected to four incrementally spaced UV frequencies in the nanometer range.

"These four frequencies are spaced at 60 millisecond separation, and convert the pre-supersaturated oxygenated blood into Polyatomic Oxygen within the blood itself (O2, O3, O4, O8). Providing a very effective antiviral, antibacterial, antifungal & antiprotozoic elimination process. This unique blood purification process also provides a thermal temperature control for the blood flowing through this new purification cellular system. (World patents pending)"

This new ozone configuration takes blood from the patient that has been pre-flooded with oxygen, and the oxygen charged blood is passed through a pulsing electromagnetic field while being hit with specific ozone-producing wavelengths of ultraviolet light. The premise is to turn the oxygen already in the blood into ozone, instead of putting ozone into the blood.

Doctors are already running blood over UV sterilizing lamps in single tiny tubes, and I have always taught that the good effects they were getting were due a little bit of oxygen in the blood turning into ozone, and that what they had was really a minor Ozone Therapy, but most of the doctors using this method never realized this.

Basil's new design is an improvement over these traditional blood UV therapies due to the larger volume of ozone created in the blood, but without trials I have no way of knowing if it is better than the standard recirculatory Hemoperfusion using the dimpled tubes. It would be very interesting if one was to put the patient's blood through *both* an RHP and this device before returning it to the patient. We eagerly await the results of actual field usage.

Other Recirculatory Hemoperfusion

I have spoken with a few Doctors over the years who said they have seen European extracorporeal blood ozonation and recirculation machines in the past. In addition, Peter Jovan, one of Basil Wainwright's ex-investors has become director of Aakaash Hospital in Coimbatore, India, where I have been appointed 'Senior Consultant.' He told me he was in Russia to buy some ozone generators for the hospital, and one of the Russian doctors showed him a book written 10 years ago. It had pictures of a patented Recirculatory Hemoperfusion unit the doctor had built within it.

I asked my Russian contacts about this book and got the following reply: *"Prof. Peretyagin's book is available only in Russian, the title is* **'Ozonated Extracorporeal Blood Circulation, Experimental Grounds and Results of Clinical Application.'** *The authors are: Prof. Gennadij A. Boyarinov MD, head of the department for anesthesiology, reanimation and transfusiology of the Nizhny Novgorod Military-Medical Institute and Prof. Vladimir V. Sokolov MD, head physician of the Nizhny Novgorod Cardio-Surgical Center. This method of extracorporeal ozonated Hemoperfusion was developed by the Russian school of Academician Prof. Korolyov about 20 years ago and has been* **successfully used primarily in cardio surgery during a heart operation** *by means of heart-lung machine (oxygenator). Ozone concentrations used for this purpose range from 80 to 150 mcg/L (0,08 to 0,15 mcg/ml) while the treatment lasts for about 1,5 - 3 hours and* **the whole patient's blood has been exposed to ozone**, *the* **results are remarkable.**

Recirculatory Hemoperfusion Near El Paso, Texas

Closer to home, Dr. Alberto Martinez offers Recirculatory Hemoperfusion and other ozone treatments in Juarez, Mexico, just across the bridge from El Paso, Texas. (See Resources) I have seen a PCR negative HIV lab test result from one of his patients being treated with the Recirculatory Hemoperfusion.

My Conclusions

With the knowledge we already have about the effectiveness of Recirculatory Hemoperfusion, and the fact that European and Russian concerns have known about similar methods for years, it is amazing to me that up to this point in time no one has seriously implemented, commercialized, and promoted this 'best of the best' ozone delivery system. It will change the face of medicine on earth if it ever gets honest money behind it and it is marketed correctly. Naturally it will gain wide acceptance. Let's offer ozone with all the existing heart-lung machines.

Combining Various Treatments

What I meant by *flooding the body* safely with activated oxygen should by now be clear to you. Through the development of modern equipment many Oxygen/Ozone Therapies can be used at home and in the clinic with ease. In my opinion many of the supplements can be used sequentially during the day. For example, you can be under a doctor's care and getting serious blood ozonation from him or her, and still drink lots of ozonated or oxygenized water, breathe correctly, slightly ozonate your room air, take an oxygen/mineral/enzyme/amino supplement like Hydroxygen or OxyMune, and clean your colon and oxygenate with Homozon, and possibly use ozone rectal insufflation and ozone saunas and/or dilute peroxide bathing. Of course one never starts out with all this at once.

The resulting totally high oxygen tissue and fluid saturation would necessitate eating the correct natural diet, lots of minerals, antioxidants, and bowl cleansing. Since individual results and responses will always vary, your competent active oxygen trained doctor will always be able to advise you far better than a book. Time and experience, along with common sense, will show you what is best for your own particular needs, but always go slow for safety.

Doctor James Boyce

Dr. Boyce successfully used a program in Bay St. Louis, Mississippi incorporating IV ozone and a little autohemotherapy. He also had his patients soak in a bathtub full of water with a gallon of 35 percent food grade hydrogen peroxide in it. He gave them oxygen supplements to drink, he ozonated their air, and put them in a hyperbaric chamber right after injections. Of everyone not using Recirculatory Hemoperfusion, he obtained the best results. He was incorporated multi Oxy-Therapies into a program whereby he was saturating the patients' bodies with various forms of oxygen; and guess what?—His patients were getting well. His combination therapies turned 118 people HIV negative.

I visited his clinic and talked to some of the people using his therapies. In the case of all his patients, it is very important to note that all their secondary diseases had gone away. Forget 'cure.' The fact of all secondary diseases disappearing is reason enough to use ozone.

For some, it had been five years since they had been treated and turned negative (on the Western Blot and p24 tests currently available at that time). But I could tell just by looking at them they were obviously not sickly and facing imminent death, and this was years later. The former

Boyce patients I met were healthy and happy. They were all restaurant and antique store owners near New Orleans, and socially afraid of being known as even 'former' AIDS patients in their local communities.

Dr. Boyce spent five years in prison on trumped up charges stemming from his clinic successfully turning those 118 people HIV negative (PCR didn't exist yet) in the early to mid '90s. He never made claims, he knew better, but the government put him in jail for claiming he could successfully treat diseases with ozone. He was also told he could get out of jail if he would denounce ozone. He didn't.

Cancer and Body Maintenance

Gary wrote to me saying. "I have had ozone do some real fantastic things with cancer. I have a question about long-term remission and ozone maintenance after the cancer is gone. One friend is having a reoccurrence after six months of remission. She is now doing two rectal insufflations and two vaginal insufflations a day for ovary and rectal cancer."

We Do Not Treat Disease

Well, as far as this question goes, the whole point of Oxygen Therapies is to remove the underlying cause of the disease. *We don't actually treat any diseases in Oxygen Therapies. What we simply do is restore the body to the natural balance it was in before it got sick,* before the oxygen declined in the environment, and before the food, water, and air became empty of oxygen, and our diets became empty of important minerals. So what we're talking about is putting oxygen back into the body. When you put the missing oxygen and ionic minerals back into the body, amazing things happen. The body's immune system goes, "Hey! This is what I've been looking for all along!" The immune system takes the oxygen and applies it to bacteria, viruses, funguses, pathogens, toxins, and other chronic imbalances.

Understanding Problems That May Arise During Therapy

Although ozone is extremely safe, nothing is perfect. Ozone increases the effects of drugs and supplements owing to its' cleaning of the cells, which leads to increased cell absorption. The dosage of these medications may need to be reduced. If anything becomes uncomfortable (indicating too rapid a detox), reduce your oxy-dosages accordingly.

Ozone should never be used in the presence of ether. A combination of the two will cause the compound to explode.

There were no harmful side effects when ozone was used correctly by the vast majority of people that I have interviewed. I haven't met anybody who has personally been hurt *when they followed instructions*. It is very rare. And any side effects we tracked down had usually resulted from operator error or incorrect procedure. When used correctly, these are the safest medical therapies on the planet, bar none. Why? Because they're the most natural. We're only putting oxygen into the body!

Someone might do too much of the direct IV too frequently and increase the dosage too quickly for his or her particular body—this is *not the fault of the protocol, since by doing so the protocol was not followed.* In such a case, the ozone will of course go through the body very quickly. We assume this too rapid oxygen overload could cause toxins to emerge, or junk food remnants and plaque to fall off the arterial walls. This might, for example, temporarily chemically or physically block the blood flow to a portion of the optic nerve, and the sight might dim. However as we have seen before, after a couple of days of a resumed lesser therapy the body dissolves the plaque and then the sight returns. One person I knew had that happen to him. He was an AIDS patient, and one day he had trouble seeing. It lasted for three days. He kept getting a reduced amount of the Ozone Therapy and the blindness went away and he could see again, and far *better* than before!

This and all other seeming side-effect phenomena experienced are 99.99999 percent harmless, since these reactions always disappear as the patient continues to do the therapy, which soon un-blocks the problems it created by a too rapid detox. Again, if this happens *the patient is not following the safety protocols*. By slowly going back and reapplying the ozone at a reduced level, one usually eventually breaks up the log-jam 'log by log' and ends up cleaner than when starting out.

On the other hand, I heard of another person who was blind to start with received the Ozone Therapy and then was able to see again. So it goes both ways. It must be used *correctly*.

If someone in the media should seek to discredit this therapy and protect the status quo, he will ignore the multiple proven benefits, and tomorrow's tabloid headlines are going to misleadingly and misquotingly say, "Man writes about ozone cure, people go blind!," all the while ignoring the 250,000 people who die from regular healthcare misadventures every year. I never say it will cure anything; they are lying if they say I do.

If you have a considerable amount of toxicity, it is essential to follow the protocols correctly and have a competent health care professional guide you through the process.

When doing concentrated therapies, your body might start to mimic all of your previous diseases, and any problems you've had, in reverse order. As the layers of toxicity are exposed, oxidized, and swept away and floated out of the body, ozone will reach deeper and deeper into the cells.

This is what happens. At the site of every cell in the body, when finally enough oxygen reaches that cell, layer upon layer upon layer of stored hardened mucopolysaccharide (mucous) toxicity is dissolved. As each layer is dissolved, the leftover hidden remnants of any particular ailment concealed beneath it are exposed. Pollutants in your body that come out may be the drugs you took as a child or college student, the dead hosts of viruses and bacteria from all the diseases you've had, and the active viruses and bacteria that you're unearthing and deactivating as you sweep away the garbage. Out it goes, layer by layer by layer, attached to oxygen molecule by molecule, pulling it out, atom, by atom, by atom. This takes time.

As this process goes on you may develop a fever, you may get a rash, and you may get a temporary runny nose, or a headache, because all this irritating old stuff is stored in the cells.

Almost every problem that crops up is ultimately owing to over-zealous regulatory agencies preventing research, teaching, or clinical trials of these wondrous and extremely beneficial internationally applied therapies here in America. For years, this negative repression had created an ozone underground with a siege mentality where desperate people rush into things, and where physicians have to combine their training with informal educational avenues while trying to learn what they can about ozone and quietly go about applying it in their practices. Thousands of people are out there doing this stuff every day. If Oxygen Therapies were completely accepted and openly taught in the medical schools, I never would have had to write this book, especially this clinical section, in order to try and keep everyone safe.

Never Stop an Oxygen Therapy Abruptly

If the man in the example above had stopped the therapy abruptly, he might still be blind. Never get into a detox mode and slam on the brakes by stopping abruptly. Once the toxins are unearthed and floating in solution, keep it all flowing out. If one gets a reaction, cut back the dosage to half and then half again, if needed, but NEVER stop completely. If one stops completely all the garbage and toxins thrown out into the fluids could stop leaving and settle into new places, causing new problems.

The theory is that whatever bacteria or other cellular garbage that was impairing his optic nerves or other sight mechanisms were being washed out; but too rapidly and so they log jammed somewhere, causing the blindness.

The problem with an eye doctor or any other U.S. trained healer is that he or she usually has zero training in detoxification processes. When confronted with a patient who has a detox reaction that is severe, these doctors point at the temporary problem and yell "Damage!"—Instead of waiting for it to go away with proper continued detoxification procedures.

They usually at that point report the healer using detoxification processes to the nearest authority because of a lack of understanding of this 'process oriented' therapy. Unfortunately many of the patients have little understanding of the process as well, for owing to this lack of understanding they might give up the process midstream out of fear, and then be stuck with something that may last a long time. If they continued getting the active oxygen at a reduced rate, combined with colonics and other elimination organ cleansings, any detoxification problem might go away in just a few days. Indeed, this disappearing of that which is NOT permanent damage is what we see all the time, but it is usually just a case of 'the sniffles' or something a lot less frightening than lessening sight.

Let's put all of this negative talk into perspective. A few people use it incorrectly and of course they have trouble. This is one of the safest therapies in existence. Compare it to the statistics of side effects and death coming out of what the vast number of people reading this book consider to be normal medical care here in America.

Are Ozone Therapies Really Safe?

There are more than **3,000 medical references** in the German, Russian, Italian, and Cuban literature showing ozone's use throughout **50 years of application** to humans by way of **millions of dosages**. The International Ozone Association and the ozone machine manufacturers report more than 7,000 MDs in Europe using medical ozone safely and effectively, some for more than 40 years.

I have interviewed many people with cancer sent home to die, who then used proper Ozone Therapies and ended up still perfectly healthy five years later. **BUT they stuck to a full protocol—getting it daily, in the right dosages, and the right concentrations, and combined it with other significant oxygen and other modalities**. People who are too far gone, or have never tried it, or only just 'dabbled' in it, or who went to a well meaning but under-trained ozone therapist—these are the only ones who end up being the nay-sayers. To be fair, always keep in mind that whenever you are examining the question of ozone's safety you must compare Ozone Therapy's statistics to the statistics in the American medical literature. Our nursing homes, hospitals, and cemeteries are full of the results of business as usual without ozone.

Go ahead and query anyone who is disrespecting ozone. *Ask* them, did you work up to using at least 150cc (not the starting dosage) of 27-42 mcg/ml3 concentration strength of only pure medical ozone gas? **Was it applied once or twice a day, every single day, for at least six weeks?** Was the ozone delivered by IV or better? If anyone has concluded it's dangerous, or ineffective, they're doing it wrong!

Of the many successful ozone using people that I have interviewed—and written or spoken about—99 percent of them have received ozone in the known proper ways. And none were hurt. Those who use ozone correctly continue to come back for more, because they live the benefits within their own bodies.

Are Oxygen Therapies Compatible With...?

Part of your using these aforementioned solutions properly is knowing how to use them with other things. I'm continually asked, "Mr. McCabe, are Oxygen Therapies compatible with (they name a drug or supplement)?" I ask, "While you're doing this other therapy, are you breathing?" The point is that when administered correctly, Oxygen Therapies are generally as safe as breathing. If they cause discomfort, somebody's doing something wrong.

Active oxygen supplements and treatments are usually potent oxidizers that react with everything except healthy tissue, and that's exactly what they are supposed to do. For safety and effectiveness I always advocate people take their Oxygen Therapy separate from any other medication.

Also, Oxygen Therapies may eventually clean you out so well that you do not need as much of the other things you were taking because they will be absorbed easier, so notice your dosages and how they affect you. If you are going to continue to use both orthodox and oxidative therapies, choose a practitioner who understands both sides of the fence.

One of the Most Important Messages in this Book

If there is a single message I want every doctor reading this book to understand about the <u>use</u> of these active oxygen therapies in removing diseases and toxins, it is contained within the two following sections entitled:

"The Clinical Ozone Methods"

and

"Mechanics of O₃ in Medicine"

The Clinical Ozone Methods
The Collected Wisdom of 50+ Years of Clinical Ozone Gas Usage

Ozone Consumer Guidelines and Application Methods

The first thing to keep in mind is that not all ozone treatment is the same, and the effectiveness of any ozone treatment increases with the number of times it is given per day or week, the strength of the concentrations used, the quantities applied, and the delivery methods used. For example, 50cc of your ozonated blood reinjected (Minor Autohemotherapy) into you in a clinic every other week is nowhere near as effective as drinking ozonated water at home every day. Quantity, concentration, and frequency are the keys. The aim is to safely and comfortably *flood the body* with oxygen by *slowly* building it up as you slowly detoxify.

General guidelines: For best results during the treatment phase, ozone (O_8 and O_4 are 'smoother' than O_3) is applied once or twice daily, or perhaps every other day, in concentrations varying from 1-to-80 micrograms per cubic milliliter (mcg/ml^3), but 27 mcg/ml^3 is generally best for maximum viral kill and minimum hemolysis (blood cell injury).

As much quantity as can be safely and comfortably absorbed by the body is applied. You can tell if they have had enough (or too much) when the patient's throat gets irritated by blood outgassing lung-irritating oxidative sub species into the lungs. Stop just before this point, every patient has with a different tolerance. This treatment is cycled and continued for as many days or as long as it takes, or until the problems go away. Mild diseases may take a few treatments; chronic ones, several months.

The lower concentrations and quantities of ozone will aid healing, and stimulate the immune system slightly, but are usually ignored in favor of the real power of medical ozone which is found to be generally centered on daily applications of 27 mcg/ml^3 concentrations for internal work. Higher concentrations are used for external body work. The upper range tops out at around 75 mcg/ml^3, and beyond that it becomes controversial. These high concentrations are never allowed to enter the lungs, which are too sensitive for anything other than concentrations averaging around normal air levels of ozone or slightly higher.

If you are a seriously ill patient, please only seek out a competently trained professional, one experienced in ozone, to guide you.

I maintain that it is incumbent upon every therapist desiring to treat seriously ill people with ozone to study the methods presented through this book, and especially to study and thoroughly understand this next paper I wrote on *The Mechanics of Ozone* for the IBOM physicians in 1993.

Ozone and the therapies to be used around it have many subtleties, and I have experienced many well-meaning MDs overlook them because they already know so much about medicine. I have seen unfortunate results from these "Just tell me a little bit, I already know the rest," types of viewpoints being held.

The doctor and the patient should never assume that ozone is "just another drug to be injected," and that a tiny bit of ozone once or twice a week is the same as a lot, and nor should one assume, "It doesn't matter what you eat." That way no one has to watch in disbelief when a patient does not respond completely after everyone had heard so many good reports about Ozone Therapy.

Without experience or training, when it comes to ozone a physician may have little idea what it is really all about. I implore every ozone therapist and doctor contemplating the use of ozone to grasp the subject thoroughly.

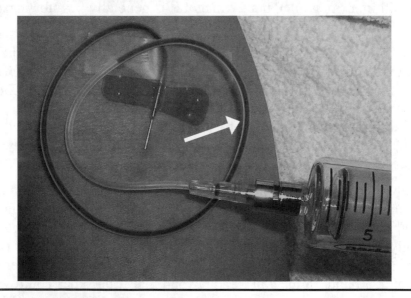

THE WHOLE POINT OF THIS BOOK IS THE OVERALL BODY *OXYGEN LEVEL* AND RESULTING *PURITY* OF THE BODY FLUIDS – EASILY INDICATED BY THE *COLOR* OF FRESH DRAWN BLOOD.

DISEASED PEOPLE HAVE FILTHY POLLUTED *VERY DARK* BLOOD.

I HAVE SEEN *WEEKS* OF DAILY OZONE INJECTIONS CLEAN PATIENT BLOOD SO WELL IT RETURNED TO NATURAL *BRIGHT RED*!

A few of the points in the following section have already been raised. The purpose of this repetition is to allow the article to stand on its own and for easy photocopying and distribution.

MECHANICS OF O₃ IN MEDICINE
FIELD FINDINGS PRESENTED BY ED MCCABE AT THE
1993 INTERNATIONAL BIO-OXIDATIVE MEDICINE PHYSICIANS' CONFERENCE.

Ever hear this one? *"I know someone who said they 'got ozone,' and they said it doesn't work."* Or, incredibly, as a well meaning but ozone-inexperienced doctor once told me: *"I gave them the shot (of ozone) but they still died anyway."* Why doesn't ozone work in every case? And if it is used correctly, why do some AIDS patients using ozone turn HIV negative for a little while, then go back to positive? And can we improve on ozone delivery methods?

Even among all the ozone doctors, only a few rare souls really understand the 'back-away-and-look-at-the-whole-thing concept' that I have advocated for years. The concept is simple: The human body is 66 percent water. Inside a 150 pound man is about 100 pounds of water needing stcrilization, and it has been measured in these proportions:

Percentage of water making up tissues, organs, fluids and bone

Brain 75%	**Heart 75%**	**Lungs 86%**	**Muscle 75%**	**Liver 85%**
Kidney 83%	**Bone 22%**	**Blood 83%**	**Saliva 95%**	**Perspiration 95%**

Some are unfortunately so conditioned that they think of ozone as just 'another drug.' It would be useful to take a minute and forget the idea of putting a little drug in the body to bring about changes. Instead, look at modern naturopathic and medical ozone delivery systems as having to face the giant task of cleaning up a 100+ pound bag of dirty water. To understand the concept, instead of thinking like a doctor, think like an engineer approaching a mechanical water purification problem.

Away from the human body, how would you physically completely detoxify and completely purify 100 pounds of dirty disease-laden wastewater? Most municipalities and Naval vessels do this by micro-bubbling a lot of strong ozone through drinking and waste water for a long time while also running it through filters. This would rid our 100 pounds of water of any bacteria, viruses, funguses, pathogens, or any other toxic contamination. This is easily always true for wastewater, but, since we have nerves that sense discomfort and pain, the human body needs more delicate methods applied to it that dispense the active oxygen slowly over many days. We're a lot more than just a bag of dirty fluids.

How are you going to completely purify 100 pounds of body water? By injecting only 10cc of 1.5 mcg/ml³ (very low concentration) ozone into a muscle once a week? Of course not, yet some have tried this, failed, and then outrageously published their failure in the literature while declaring ozone to be a failed Therapy. That scant and ineffective method may be useful to create a killed virus preparation vaccine, and it may bring about minor blood chemistry changes. But complete purification? No. You must safely and properly flood the body with singlet oxygen—long enough to purify it and yet at a rate slow enough so that the body filters (eliminative organs) won't clog from the detoxification. This is how to achieve optimum purification and healing using ozone.

Our typical 150 pound man with 100 pounds of water has approximately 12 pints of blood in him. His blood is 83 percent water, so in our case, therefore, he's got about 10 pounds of water in his blood. Even if you completely purify the bloodstream, you are only purifying 10 percent of the whole-body 'dirty water' problem! The blood circulates through the body 12-to-20 times per minute, but what about cleaning the lymph, the other half of the flowing fluid system?

From an AMA "Today's Health" article, December, 1964, by J.D. Ratcliff— "The Lymph ... As vital as the main bloodstream, the intricate and all but invisible lymphatic network... It is one of the world's rivers of mystery —sluggish, largely unmapped, many miles long. The Lymphatic system has puzzled physiologists since early Greek times...our health, even our lives depend upon how well this complex system functions.

"In contrast to the bloodstream, which follows a swift flowing closed circuit from arteries to capillaries to veins and then back to arteries, the lymphatic system flows slowly in a single direction back to the heart. Its initial rivulets, microscopic in dimension, originate in intercellular spaces. [Picking up the cellular wastes for transport out.] Fluid gathered here passes through ever-enlarging ducts until it reaches the lower neck [subclavian] region, where it empties into veins leading to the heart to be mixed back into the blood. [Note: It takes 24-hours for the lymph to completely circulate through the body once. Five-minute or sporadic treatments won't come close to reaching all the way in and cleaning the fluid volumes in the intracellular spaces and lymph.]

"Much of the mystery surrounding the lymphatic system traces to the fact that most of its ducts are so fragile that they are invisible—the smallest have walls of only 'one cell' thickness. And the fluid they carry is ordinarily almost as clear as water [let's hope!]. Moreover, at the touch of a probe, all but the largest lymphatic vessels collapse, as they do at death. Exploring such a gossamer stream has called for supreme ingenuity. In many respects the body is like a vast swamp. Its trillions of fluid bathed cells live an aquatic life. The lymphatic network provides an all important drainage system."

So when you're up to your neck in alligators (facing terminal disease) its hard to remember that your original objective was to drain the 'vast body swamp' by purifying *all* the fluids, beyond the blood, and into the lymph, inside the gossamer passageways, through the organs, inside the bones and cartilage, within the cell walls and the cell fluids, and including the aquatic life surrounding the DNA. How are you going to completely purify *all* of it?

We can contrast the 100 pounds of human body water needing cleaning in our 'vast swamp' example with autohemotherapy's only cleaning a scant pint or more of blood three times a week.. In contrast to this, and more to the point, pioneers in the U.S. inject several syringes of oxygen/ozone into the bloodstream every day for six plus weeks while addressing diet, colon cleansing, and lymphasizing. The best of all worlds except one.

The problem remaining with autohemotherapy, and with any other in the body ozone delivery method except RHP, is that after the ozone is applied the newly oxidized toxins and newly dead microbes and disrupted viruses are *still in the body* and still have to be sent through the already stressed organs of elimination. That's why heavy-duty uses of the *in-the-body* ozone methods use up body resources and energy, and may cause unpleasant detoxification reactions.

The present common methods like autohemotherapy do work, but can we improve upon them? Are these methods enough? Also, blood and lymph eventually carry oxygen to every cell, but what about being sure we reach in and purify all the hidden backwaters of the body swamp? Fortunately, there are three devices in existence answering the call of approaching total purification:

'Aquacizer' Ozone Water Bath and Steam Cabinets —The patient reclines naked inside a whole body clamshell chamber with their head protruding outside

the chamber, while hundreds of micro nozzles spray warm and hopefully 'ultrapure' water homogenized with pure oxygen/ozone onto the body. The skin capillaries dilate and discharge body toxins outside the body and into the 'very hungry' 10 megohm ultra-pure water washing the toxins away, while the pores open and absorb the ozone and oxygen sub-species into the skin, lymph, and bloodstream at the same time. Dilute peroxide bathing and ozone saunas fall into this category to a lesser degree.

Hyperbaric Chambers —New, soft portable units roll up like a tent. Clinical units are large and expensive. Patients lie or sit in a pressurized oxygen atmosphere. The oxygen pressure (around 2.3 atmospheres) on the whole body harmlessly forces stable oxygen through the skin and deep into the most hidden body cavities and bone marrow. This treatment is most efficient when employed right after the patient's blood is filled with an ozone injection or two.

Recirculatory Ozone Hemoperfusion, or 'Polyatomic Apheresis.'
This has to be the most advanced medical ozone delivery system to date. The patient lies in a reclined chair while blood is continually withdrawn out of one arm, ozonated at slight pressure, filtered, and pumped sterilized and clean back into the other arm. A simple one hour treatment with this method alone is vastly superior to any other ozone delivery method. Experimental trials conducted privately had a few patients successfully ozonating and recirculating their blood

for more than *eight hours* in one treatment! Imagine the volume of blood, lymph, and organ purification obtained in an eight hour treatment. That's how to get to all the hidden backwaters. Only in this RHP system are the toxins and dead microbes filtered away OUTSIDE of the body and not sent through the weak and over-stressed organs of elimination! No more heavy cleansing reactions.

'Mr. Oxygen's'™ Holy Grail

Now even beyond that, imagine the results possible if we first give someone this Recirculatory Hemoperfusion or Polyatomic Apheresis Ozone, and then put them in an Aquacizer type ultra-pure ozone warm water wash, and immediately afterwards while still freshly flooded with oxygen we put them into a hyperbaric chamber where they are suitably re-*flooded* with oxygen. We follow all this with colon cleansing, and all the while we daily and continually apply the rest of my *Crown Jewels of Health*; lots of pure oxygenated water, correct diet, nano-ionic minerals, amino and enzyme support, exercise, physical cleansing, emotional and spiritual cleansing and repair, and lots of non-judgmental loving acceptance. 'Mr. Oxygen's'™ Holy Grail of Ozone Therapy. Total purification and repair inside and out, going as far as body treatments can take you. This has never been done. I hope someday to have enough funding to put this all together for you in an official clinical setting to be able to prove it over and over with proper documentation. It will be expensive to do it correctly, but this will set a new standard for healthcare and will be totally worth it.

Why Are Some Positive Blood Test Results Premature?

A little ozone given continuously can clean anybody's blood up so well that an immediate blood test will show a state of improved health. But, as I have been explaining, unless all of the little gossamer lymph vessels and ducts and the bone marrow and intercellular spaces are cleaned up, we're not finished. After a few days of the not-yet-clean lymph slowly recirculating once a day and picking up hiding microbes and re-mixing with the blood, a person could test positive for some virus again only a few weeks after being taken off the 'too-little' ozone.

For example, a patient with freshly ozonated blood may test on an HIV PCR test that she only has one viral particle in two million, or she may even test virus 'undetectable.' Upon hearing the great results, such a sense of euphoria (or lack of finances) sets in that the patient goes away and drops the therapy. Six months later they re-test, and surprise! A measurable viral load has returned, and to some undesirable level. The ozone just wasn't applied often enough or long enough to clean *all* the backwaters. If there is one point about ozone delivery systems that I wish to get across everywhere to everybody, this is it:

To become, and remain, totally PCR or any other test negative,

All of the patient's body fluids must be totally purified

<u>Russia</u>

1,300 medical ozone machines (600 are Russian Medozons units).

1,950 ozone MDs.

1,300 ozone nurses.

1,100 medical organizations using ozone.

660 are large state, regional, or city hospitals that have departments for Ozone Therapy.

440 private clinics or surgeries or medical centers specialize only in Ozone Therapy or provide ozone treatments.

40 state medical institutes, universities and academies teach Ozone Therapy to doctors.

200 Ozone doctoral dissertations.

Russia - State Diagnostic Center Employs Ozone Therapy

Modern Russian Medical Ozone Generator 'Medozons BM.' Manufactured at Former Defense Industry Plant.

Some models offer precise dosage metering.

Chapter 18

Oxygen, Light, & Sound

At the cosmic level, science speaks of the planets being 'falling bodies' because our planet and all of us are falling or whirling through space, circling the sun within a gravity field. At the microcosmic level, everything in life is dancing atoms composed of electrons circling around neutrons and protons.

The sounds we hear come from something vibrating within a frequency range our ears can hear. But just because we don't hear something doesn't mean there aren't other 'sounds' around us such as infrasound, ultrasound, and other vibrating phenomena such as radio waves. You don't see these waves, but we know they exist because your car shakes when you pull up next to some soon to be deaf homeboy's car radio at the traffic light.

Because they are in constant motion, the atoms of all creation also vibrate or 'sing' microscopically as they move at super high speeds. Some people with extended sensory abilities have reported actually hearing some of these sounds that most of us can't hear. On some level, if we could hear it, life is God playing a beautiful symphony while we live inside it. Since everything is vibrating, if we had the right technology we could actually measure and record the collective 'key-note' given off by any object.

In fact, there already is a new discipline combining art and science that explores the harmonies and melodies given off by the sequencing of everybody's DNA amino acids. From MIT: *"Abstract Artist John Dunn and biologist Mary Anne Clark have collaborated on the sonification of protein data to produce the audio CD, "Life Music" The authors describe the process by which this collaboration merges scientific knowledge and artistic expression to produce soundscapes from these basic building blocks of life that may be encountered as esthetic experiences, as scientific inquiry, or both."* An artistic use of the science and scientific use of the art can be found at http://algoart.com under 'DNA Music.'

Your whole body could be said to manifest a certain natural aggregate keynote composed of everything that is you giving off all the vibrations that are blending like a symphony inside you. Individually, each organ, like your liver, reveals its own separate personal natural healthy tone. If you drink too much the liver cells stress and if you continue drinking the liver becomes diseased. When diseased, its natural collective cellular 'music' is altered and the once harmonious vibrational 'music' of the liver becomes unnatural and disharmonious. The average doctor or scientist will look at this condition from a different mental viewpoint and use different terminology based upon chemical measurements to describe this condition. Your health insurance carrier will recognize their descriptive 'cirrhosis of the liver' chemical terminology and reimburse them accordingly.

Serious disease is discordance to such a degree that it may disrupt everything, and could end up stopping the body symphony permanently. Harmonious existence allows full oxygenation, and full oxygenation promotes harmonious existence.

Wouldn't it be great if we could restore the harmony of the whole body organ orchestra? There have been many healing therapies advanced based upon these non-chemical science facts, but they have not been widely accepted. They usually were not perfected enough, or, if they *did* work, they didn't fit well into the currently acceptable social and political world-view. Whatever the reason, the result has been that many promising or even successful therapies based upon altering our inner waves and atomic structures somehow have been lost to history.

In the main, these therapies have all relied upon the body's amazing ability to absorb harmonious electrons or even the mysterious life energy itself out of any number of sources—as long as they are reasonably harmonious with the body.

Direct application of electricity, electric fields, magnetism, magnetic fields, radio frequencies, light, color, sound, tones, music, prayer, spiritual and other energies to the body have all been applied with the hope of giving us 'more energy' or 'healing energies.' The ultimate 'energy therapies' would be able to work on everyone, and probably require *very* exact subtle energies in order to produce what most earthlings would call repeatable 'miracles.'

Beyond our present day understanding of chemistry, electricity, and magnetics, our society as a whole has barely explored the potential of super subtle exotic areas like gravity engineering, or the use of light, or sound, or other vibrating energies. I have always known intuitively that some of the best therapies, once perfected, would end up using combinations of light, sound, and perhaps even gravity. Here's an example of the potential of sound that we don't understand.

Once, an inverted rare brass Tibetan 'singing bowl' was chosen for me and placed upon my head like a helmet and struck. The low loud ringing tone went deep inside me. Hours later, and in a different location, while doing something else, the same exact loud sound spontaneously came OUT of me from deep inside! It was just as real and crisp and present as the first 'boooooooooong.' I was suitably impressed. Stories also exist of the old Mayan Priests striking a large cymbal or gong (called the 'Sun Disc') in order to heal, levitate, and transport objects. There is much available in these areas that we know little about, and I advocate more exploration.

In the area of using light as therapy, I admire the work of Indian physician, scientist, engineer, civil reformer, editor, aviator, scholar, metaphysician, inventor, and color therapy researcher Dinshah P. Ghadiali (1873-1966). From a web biography of Dinshah: "This man was extraordinary in many ways and pursued a wide range of investigative inquiry; scientific and otherwise, in his long, but persecuted lifetime wherein the AMA branded him–*without any investigation* whatsoever–and his color therapy as worthless and fraudulent. He was truly a Renaissance man in every sense of the word. His greatest legacy is a simple, but enormously successful system of light therapy which he had labeled '*Spectro-Chrome*'.

"Dinshah P. Ghadiali originally used five colored glass slides (glass plates) along with an ordinary incandescent light bulb for his light source to apply the Spectro-Chrome therapy. Dinshah experimented with higher wattage bulbs but found that a 60-watt incandescent light bulb or even a *flashlight* worked just as effectively as a 2,000-watt bulb! High power lighting wasn't necessary for Spectro-Chrome therapy to work. The colored light energy applied in Spectro-Chrome is intended to buttress and enhance the color frequency spectrum of the human aura to achieve results, and the refinements of that Etheric energy matrix require no bombardment of high intensity photons to yield results." As I always say, precise gentle frequencies harmonious to human biosystems work the best. And they must also be politically and socially acceptable.

One of the reasons we have not seen many more successes in these areas of electric, magnetic, sound and light therapies is that our bodies have built-in survival safeguards that resist force. Our body's survival mechanisms have a prime directive to maintain our present existence just the way it is—by preserving the inner status quo, since that has already proven to be the safest average course. Obviously it must be, because our body has survived up to this point by relying upon it.

So, Nature sees to it that, if possible, our bodies wisely and stubbornly resist our inner energies being affected by most forms of outside energy being directed at us. For example, if we had no such built in protection, all the TV, radio, military, CB, cellular, police, and taxicab transmissions would have already scrambled our brains.

Often, the first instinct is to just increase the power on an electro-therapy-healing device in order to overcome the body's inner resistance to change. But that often causes the body to resist even more, in a rebound type effect. To create healing, the body must be gently coaxed into harmony, like that first shy kiss.

In terms of employing new therapies, that means the exact resonant harmonic frequencies and power densities of light, or sound, or electrons must be found that permit the healer and his therapy access to the controlling mechanisms of our inner atomic worlds. The right key, at the right pressure, inserted into the right lock.

Another therapeutic route showing promise is to first discover the exact keynote frequency of the disease itself, and then invert it and feed it back onto the disease. In this world of duality, exact opposites cancel each other out.

In order to work well, precise therapeutic frequencies and harmonics and energy forms at the right power levels must be so harmonious that the body does not resist them. The therapy must also produce the desired results, and be repeatable. The energy coming from the top 'energy healers' and the best devices all have these same characteristics in common.

I have been to Expos all over the world displaying cutting edge therapies while lecturing at them weekend after weekend over the past 14 years. While attending, I have had my aura photographed many times, and I have seen, been bathed in, and experienced many electric, magnetic, sound, music, and light therapies. Numerous people want to believe that such systems will work, and have parted with much cash to get them. In my opinion, a few systems have some effect, but on the whole the results usually are not very dramatic or reproducible. In most cases wonderful promises of how they would transform my life overnight did not materialize even with repeated usage.

Many Expo attendees now have a house full of these devices collecting dust. Some might say these people were cheated. I say they were expressing their preference for the type of things they want to have in their lives—and encouraging their further development by funding future research into these ignored areas.

When we were kids we probably all pictured some mythical ray gun in our minds that we would simply shine on an injury that would immediately make it heal, as in Star Trek's 'dermal regenerator,' so we are naturally drawn to these technologies. Perhaps we possess a common ancient subconscious memory of some place or time where such devices were commonplace.

Our favorite Oxygen Therapies work on so many people because they donate the sun's energy in the form of harmonious electrons to the tissues and cells. Some of these new therapies also deliver electrons, but that's not the whole story. There are subtle differences that are crucial to a therapy being successful.

Low Level Pulsed Laser Light Therapy

Signs that a present day electro-therapy has approached being able to repeatedly manipulate healing control mechanisms through the use of light have appeared in Low Level Lasers. Lasers are different from regular light, in that the charged light particles or photons are all lined up, or coherent, and in a beam instead of scattering in every direction. Evidence is showing that the body can respond well to low level polarized light particles following each other in a stream.

Low level pulsing laser light emitting diodes gently employing specific frequencies and power densities have already appeared in industrial and veterinary applications. Experimental use of Laser Therapy for healing, especially when combined with appropriate Oxygen Therapies, has already produced many positive testimonials. Please refer to the Evidence section of this book for examples.

The folks at Low Level Lasers, Inc., and 2035, Inc., are the makers of the Rotary Multiplex and Q-1000 series of Lasers. They were kind enough to supply me with enough information so I could determine that current research, clinical experimentation, and published scientific papers on Low Level Laser Therapy indicate the following:

➤ Unlike high-powered cutting lasers, Low Level Lasers Therapy is gentle and completely safe and non-toxic, as safe as sitting in a room with a light on.

➤ The most beneficial laser wavelengths have been identified through years of trial and error.

➢ To prevent the body from setting up resistance to the therapy, the Low Level Laser Therapy must be gentle and it must maintain a standard frequency and power density. The specifics of doing this are proprietary, but that's why battery driven lasers and lasers that plug into 110 volts don't produce the same results. Unevenly fluctuating power supplied by batteries and unstable power supplies prevent the body from recognizing its varying energy patterns as beneficial.

➢ Osmosis transfer of essential nutrients cannot occur across the cellular membrane of a depolarized cell. It is believed that one of the mechanisms of action of the Low Level Laser Therapy devices, as in Oxygen Therapies, is the donating of charges via coherent photons passing through the skin, which increases the polarity of cell membranes, allowing increased osmosis transfer of nutrition to occur.

[Note: Photons are electromagnetic radiation particles, and electrons are elementary particles with a negative electric charge. The body commonly uses both of these tiny charged particles, when delivered by sunlight or oxygen.

Some propose that both photons and electrons are simply made up of slightly differently combined little known 'fine particles,' 'no mass' particles, or 'gravity particles' that are called by different names. A gravity particle is 1/80th the size of the already tiny electron.]

➢ The proper type and amount of delivered laser energy causes the cells to relax muscles, decrease swelling, and produce more of their own beta endorphin pain killers. Also, Low Level Laser Therapy reduces seven inflammatory enzymes by 75 percent increases two healing enzymes by 75 percent, balances the cell sodium/potassium pumps, and increases collagen, fibroblast, and osteoblast production so bones will heal quicker. Perhaps most importantly of all, Low Level Laser Therapy probably does this by re-energizing any de-polarized cell membranes that result from sickness or injury, thus allowing osmosis of essential nutrients into the injured cell and increasing the growth of healthy cells.

➢ In *Low Level Laser Therapy,* by Turner and Hode, Low Level Laser Therapy has demonstrated an ability to re-energize injured cell membranes and increase the mitochondrial energy by 150 percent. Picture photons being sprayed on, and re-filling, a cell depleted of charged particles with more charged particles. Once charged up, the fluids start moving in and out of the cell again.

The outer layers of a cell, as well as the cell itself are just energy. A bunch of positively charged protons, neutrally charged neutrons, and negatively charged electrons compose the outer layers of the cell. Modern studies show that the sense receptors on the skin of the cell membranes are the true controllers of what the cell does, not the DNA in the nucleus. In fact, according to what the membranes 'senses,' the DNA adapts accordingly. This is the opposite of what most people were taught.

When these charges comprising the composition of the cell membranes get out of balance, the cell is in a state of dis-ease. Electrons are transferred, lost, and gained very freely on the cell membranes, and the cellular membrane or 'wall's' cleanliness and balance determine how efficiently it can transfer food, oxygen and minerals, etc., in–and also how efficiently it can transfer the waste out. How well the cells do this is of prime importance in creating dis-ease or health.

This transfer of nutrients and wastes is called osmosis. Osmosis is a basic biological principle and is one of the methods involved in the transport of all nutrients across cell membranes. However, nothing can be transported across a depolarized (uncharged) membrane. Depolarization happens because of injury or disease of the cell. Computer controlled Low Level Laser Therapy is an effective method for re-polarizing the cell membranes and reestablishing electro-balance. —From my research (there are over 2000 LLLT successful studies) and the Low Level Lasers, Inc., and 2035 Inc. literature.)

Dr. Prakash, of Pune, India, reported at the 2002 meeting of The International Society of Lasers in Surgery and Medicine, that he cured 60 out of 64 heart diseased patients with LLLT. He used just the laser—no other treatment. This is a first in LLLT, and this announcement has raised many eyebrows!

Low Level Laser Theory

All the clues are there but let's put all this together. Remember my "all bodies vibrate and vibration gives off music" concept? The doctors who already commonly measure this body energy with MRI waves prove that this is a valid statement. Again, from the Low Level Lasers literature: "Energy comes off the body very predictably as demonstrated in diagnostic tests such as EKG (electrocardiogram), EEG (electroencephalogram) CAT scans, MRI (magnetic resonance imagery), Kinesiology, and various dermal screening devices." The *music* or energy radiation from vibration is there, and the body gives it off and absorbs it.

So how do we get the right kind of subtle energy, music, electrons, photons, or 'charged particles' into the body? Like so many other mysteries being solved for you in this book, we look to Nature first.

In order to produce this 'right kind' of laser, it was necessary to study how healthy people stay healthy and heal themselves when needed. It was learned that they accomplished this feat by controlling their own frequencies via very subtle energy. Low Level Lasers use computerized controls to maintain exact frequencies and power densities that emulate and harmonize with the body's own frequencies.

It has been demonstrated experimentally that this new method of low level laser application correctly 'resonates with the body,' and donates charged particles that allow the body to restore energy and polarity to its cell membranes.

Low Level Laser Therapy has recently received limited FDA approval for pain control and for specific uses such as carpal tunnel. There are two manufactures of industrial and veterinary lasers that may be utilized safely for humans. Low Level Lasers, Inc., manufactures industrial Low Level Lasers called 'The Resonator' and the 'Rotary Multiplex.' 2035, Inc. manufactures two new industrial lasers called the 'Q-100' and 'Q-1000.' These industrial and veterinary lasers are available through Oxygen America. (See resources.)

I believe that Low Level Pulsed Laser Light Therapy at harmonious frequencies and proper power levels delivers many charged particles via photonic energy that the body will accept in the same way that it accepts electrons from Oxygen Therapies, or light from the sun. Naturally, the two Nature-based therapies would work well together. Ultimately they clean and move the Lymphatic fluid. The next section on Lymph will further tie all this together.

Chapter 19

Lymphology

Although they are as pervasive and as important as your blood system, you don't hear much about the lymph vessels and the lymph fluid. This is probably because the lymph vessels are clear, and collapse easily, so they don't attract as much notice. Some opine that information about this other half of the circulation story has been suppressed, since understanding it makes healing too easy.

Your blood system is only half the story of your circulation. Your heart pumps your blood through a system of vessels and then through tiny capillaries in order to carry food, minerals, and oxygen out to all the cells. When they get to the end of the line, the blood cells stay in the capillaries, discharge their cargo into the lymph, enter the veins, and finally return to the heart, lungs, and digestive system for recharging.

Where the blood system ends, the lymphatic system begins. The lymph system is a pool of fluid surrounding the blood capillaries all over the body. It picks up the minerals, fats, vitamins, sugars (glucose) and oxygen. Water and blood proteins (albumin, globulin, and fibrinogen) also seep out of the blood system into the lymph. Everything in the body is floating in this fluid.

Lymph fluid has a dual function. First it floats nourishment and oxygen out to the cells, and then, continuing to flow, it carries away our waste. There are two ways our waste leaves the spaces around the cells. Excess water and cell wastes are absorbed by the capillaries and carried away by the veins, but blood proteins, hormones, and debris are slowly carried away by the lymphatic system.

This is how our cells are supposed to be kept clean, but the system only works well when it's moving and when there is enough *active* oxygen in our fluids in the first place, and when we have enough fluid. Low oxygen equals low energy and means less cleaning going on. Less cellular cleaning further lowers the energy because everything gets coated with debris, mucous, and fungus. The lymph system recirculates much slower than our blood.

When finished with its duties, the lymph slowly flows up to the base of the neck through a series of filtering lymph nodes and one way valves. At this point, the lymph fluid is dumped into the subclavian vein-which flows into the heart where the lymph mixes with the blood again. This completes the fluid circle of life within you.

Injuries, Swelling, and Lymph

Whenever a cell is injured and torn open, plasma proteins leak out. The inner potassium and outer sodium also mix and both lose their charge. These inner cell proteins don't belong outside the cells, so they irritate the body. In defensive response to their irritation, the body swells up at the site of injury. This immobilizes its lymph fluid locally and prevents all this from spreading further. That's why a twisted ankle or a black eye swells up. Cells are torn open. But the swelling creates a lymph fluid non-flow stagnancy which creates further swelling and pain which increasingly slows and blocks both the inflow of healing oxygen and the outflow of debris. At this point the lymph fluid 'stalls.'

Dr. Samuel West, the premier Lymphology expert, taught me all of this, and one of the examples of this process is observed when a person eats a food that is not good for them. Their body reacts as if it was slightly poisoned. This causes the pores in their blood vessel walls to enlarge, and this in turn allows plasma proteins—that normally are kept in the blood vessels—to infiltrate into the lymph fluid. The body does this to try and 'water down,' or dilute the concentration of the offending undigested food particles in the bloodstream—by dumping some of them into the lymph pool. The way you know which foods are causing this imbalance is to notice which foods make you very thirsty immediately after eating them. Junk food does this a lot.

An extreme life-threatening example of this lymph contamination via vessel escaping plasma proteins is called 'going into shock.' When a body goes into shock, the same pores in the blood vessel walls open up wider—causing more blood plasma proteins to flow out of the vessels and contaminate the lymph. The irritation created by the trapped dirty water and plasma proteins cause tissues to swell and the lymph to stall. Stalled lymph means fresh lymph is not going out to the cells. So, you may be breathing but there might not be any oxygen delivery or waste removal!

All the cells just sit there, helpless, needing the lymph to bring them lots of fresh oxygen and water all the time. When the Lymph stalls, from injury, poison, emotional distress, or disease, the oxygen delivery slows and blood proteins cannot be removed from the spaces around the cells. This can cause death within a few hours! **All of a sudden disease names mean very little, as similar preconditions prevail in all cases.**

Stalled lymph also means no waste gases, or debris can ever *leave* the cells, because nothing is carrying away the cellular byproducts, or 'excrement.' This suffocation and toxin build up cycle explains how someone can die from shock. It's really cell suffocation and poisoning.

The solution, of course, is to get the cells charged by getting the full amount of active oxygen and sun energy into the lymph fluid, and to keep this well-oxygenated fluid *moving* around the cells. Hyperbaric oxygen therapy uses pressure to force fresh oxygen into stalled lymph. Oxygen Therapies and exercise also bring in the oxygen, as well as lymphasizing, deep breathing, the correct use of low level lasers, and other similar body harmonious energy-delivery systems.

Lymphasizing

The essence of Dr. West's teachings to me were that if we could just get the lymph fluid moving again, we could enable the body to rapidly heal many injuries and other conditions. We need to help the body bring healing oxygen and minerals to the cells, and to flush out the cellular waste.

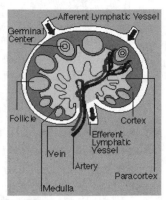

Dr. West and I are in agreement. After exhaustive research, he independently concluded the following: "*Diseases stem from the accumulation of cellular debris, toxins, and especially excess fluid in the spaces around the body's cells. This prevents the cell from getting enough oxygen to fuel its functions. This condition comes about when there is both inadequate lymphatic circulation and particularly when there is inadequate intake of oxygen through the lungs. A cell is quite like a candle flame whose light gets dimmer as oxygen is depleted, and eventually it goes out.*"

Dr. West demonstrated that deep breathing both activates the lymphatic system that flushes the body of trapped proteins and fluids, while the fresh oxygen turns on the sodium-potassium cell pumps that provide electrical energy for the functioning of the entire body. Dr. West simply stated that, "If we have oxygen poor blood, we just die…so to be a constantly shallow breather is suicide."

'Lymphasizing' is the little known art of speeding up the movement of the lymph. Speeding it up means more oxygen gets to the cells, and more waste is removed. We all know what that means.

This is why exercise is always mentioned as important. Even just walking pumps the lymph collected in the soft pads in the center of the soles of the feet back up the legs. Jumping up and down, rebounding on a trampoline, and other devices techniques and methods all are designed to get that lymph flowing again.

The 'Chi Machine' movement is patterned after a fish swimming. The back and forth or side to side movement of the legs therefore flexes the spine, and this motion pumps the lymph fluid around.

Jonathan Hayes takes a relaxing air-swim 'ride' on a Chi Machine.

Strong cellular electrical polarities cause fluids to move across the cellular membranes, and moving the lymph enables the delivery of fresh oxygen to the cells. Dr. West proved that the electrical potentials radiating out of our personal electrical bio-fields could be distributed to other people, and to specific areas on our own bodies. He advocated the use of imagery and coupled with specific hand movements; pressure, massage, and stroking to accomplish this. By increasing the cellular bio-electrical potentials, and moving the hands over the patient's body in specific patterns—such as stroking toward the neck—the lymph flow is increased.

Of course, flooding the body with oxygen detoxifies and at the same time donates many charged particle electrons to the fluids and cell membranes. In turn the body uses these to raise its electrical potentials to the correct levels. Healthy strong polarities move the lymph. Because the re-established strong polarities foster movement that allows the oxygen to clean up the fluid and osmose into the cells again, the swelling soon lessens and the injury repairs quickly, if it can.

Loss of Charge, Aging, and Pollution

I have long taught that many of our challenges stem from pollution of all sorts eventually compromising the polarities of the cells. The strong repelling forces in the atoms are what keep the DNA coils apart, and pollution can contaminate the atomic forces. Our DNA coils shrink when we get older, and elderly people get shorter, partly because the atoms making up the DNA coil and cells are losing their strong opposing charges through interfering toxins prematurely neutralizing them. As a result, millions of our atoms can gradually 'pack tighter,' and we 'shrink.'

"A normal cell's electrical potential can be measured at around 6,800 angstrom. A cancer cell has a charge of only 1,800 Angstrom units, and when the cell's potential drops to 1,300 or below, that's when microbes can enter the cell."–Dr. Mona Harrison, MD.

In our atmosphere, more and more oxygen atoms find it harder to turn into ozone as their nuclei become polluted with fine particle radiation. Once the oxygen atoms collect extra protons from pollution they are isotopic and the oxygen isotopes can no longer turn into ozone.

It's all part of the large malaise caused by pollution covering and mixing with everything and turning it all charge neutral, 'average,' gray, and without any distinction. Healthy life here is based upon the dynamic tension involved in our existing between two strong fully charged energetic opposite polarities. Spiritual explorer Joseph Campbell, author of *The Power of Myth* series, described life for humans as our existing between the awesome beauty of creation—and the absolute terror of existence.

Chapter 20

Examples of How Oxygen Therapies Affect Specific Diseases

So the therapies do what history says they do, they're safe, they're inexpensive, and we know in general how they work. Now let's look at disease in general, the specific diseases they have worked on, and the rationale behind why they did work.

Why Are Some of Us Sick?

OK, so you're sitting watching TV or maybe reading the latest edition of *The Organic Carrot* and wondering why you—or your parents or children or friends or relatives—are sick. Oh, we don't really think we're 'sick,' it's just that we *keep getting* the flu, or have this rash or pain or infection that is always there, or, even after age 25 we still have 'adolescent skin.' And every few weeks our nose starts running, and our athlete's foot keeps coming back. We can't shake our halitosis, or we never have the energy we should have, and our mate has that pesky 'yeast thing' and recurring herpes outbreaks.

I'm a journalist, analyst, editorial writer and medical historian, not a doctor, so please don't construe any of this book as medical advice, but consider the following scenario as a possibility.

You are at war. The most insidious form of evil imaginable has invaded you. The enemy has quietly moved into your personal territory and set up headquarters. They continually send out scouts and raiding parties that take over new turf, new parts of your body. The newly established invaders have connected psionic and physiological 'taps' onto your cellular metabolic energy systems and are drawing off your chemistry and Life Force and are diverting and perverting it to their own use. They actually force you to create more of them. Yep, you have a cold. Or it's a bacterial, or viral, or yeast, or parasitic infection.

When so many of us constantly have so many symptoms, no one knows what health is anymore—because the average keeps dropping! Everyone joins the comfort of futility and says, "Well, it's just one of those things we have to put up with during our lifetimes." My Dad was a great one for saying, "Eddie, it's just the way it is." As if I should resign myself to it! Luckily for you my Irish had not been beaten down like his was, and so here I am advocating a mass revolution in health and production consciousness.

There is one thing we can say about all of the pathogens that invade you and cause you to suffer through a disease—they are foreign to you! Because you have a low oxygen condition in your cells, they take hold and start eating parts of you. Then they reproduce and grow a colony also excreting toxins that make you sick, or sometimes give you a fever, or sore throat, or arthritis, or cancer, or AIDS, or Epstein-Barr/Chronic Fatigue Syndrome, or Herpes, or Candida, or a cold, etc. Since you no longer get enough oxygen from breathing, or from your food, or your water, they can grow in you because of the low oxygen level in your cells.

Sure, we live, but only because our bodies were created as marvels of adaptation. We are quickly finding out that there are limits to the amount of self-imposed abuse and partying that our bodies can endure

And as for our environment, we used to think that the earth was a giant sponge that could soak up anything we threw away, kind of like magic. *Well folks, the sponge is finally full* and we've run out of that kind of magic. The choice now is between limping along to age 60, or living full happy lives up to 120 years of age and beyond. Within our little range of lifestyle choices we make the decision between these two extremes many times every day. Visit any hospital and you'll see how most have chosen.

Aakaash Ozone Hospital
Coimbatore, Tamil Nadu, India.
Ed McCabe is a Senior Consultant.

Anthrax, Botulism, and
Other Bio-Warfare Diseases

Let's talk candidly about these and other infectious bio-terror agents....
Night after night we were told a new disease, or terrorist bio attack could
be right next door, or coming soon. Should you *cower in fear*, and hide in
your homes, shunning all human contact? Or is there something positive
you can do *right now*?

**Anthrax comes from anaerobic spore forming bacteria. Botulism
comes from anaerobic spore forming bacteria. Smallpox comes from
an anaerobic virus. All the other plagues are from anaerobic bacteria
or viruses as well.**

I have said this over and over publicly for 14 years, on over 1,500 radio
and TV shows and speaking platforms. Thousands or perhaps million of
active oxygen using patients (and their doctors) have proven for more than
50 years that: "Primitive anaerobic life forms and toxins—such as bacteria
and viruses—can't exist in our bodies when surrounded consistently by
active forms of oxygen!" You do the math. Even if some terrorist was
spraying non-living poisons at us, continually having a high tissue-oxygen
level ready to oxidize them if possible is your first and best edge.

Why isn't everyone using these therapies? Why isn't it the first answer
announced on TV when the disasters strike? Because no one with money
owns patents on it, so there is no incentive to tell you about it, so it rarely
makes the news. In my own well-researched and very experienced opinion
which I have been expressing all these years, if you clean your body out
with active oxygen, and then you keep your fluids loaded with oxygen
daily, *you* are the one who has a *far better* chance of surviving disease
exposure than *anyone* else! People who get sustained ozone Recirculatory
Hemoperfusion will be the first to recover if stricken.

Keep in mind that *any* treatment you end up with will be enhanced if you
oxygenate your body with active forms of oxygen. But, to achieve
maximum effectiveness, the detoxification must be completed; and the
Oxygen or Ozone Therapies must be given safely, consistently, and
correctly for months at a time before any possible exposure. Here's a good
reason why you should act now: "U.S. national smallpox vaccine stocks
are sufficient to immunize only six-to-seven million persons." —CDC
Website.

In my opinion, anyone will be far better off if they start supplementing with active oxygen now. Playing catch-up after being infected is tough — and at best like playing Russian roulette. No therapy is perfect. Don't wait.

Summary of Published Anthrax Literature

As an example, I'll focus on the recent scares from Anthrax. Anthrax is a facultative anaerobe, if surrounded by active forms of oxygen it cannot survive. As you must realize by now, ozone gas is a highly aggressive pure oxygen molecule comprised of three or more oxygen atoms, and non-toxic when used correctly. Ultimately, it is harmless oxygen. Ozone is commonly employed as a sterilizer. **Anthrax spores 'soften' and are easier to eliminate in a humid environment. The greater the ozone concentration, the more rapidly ozone eliminates spores**. We can employ these variables efficiently by coupling proper relative humidity with ozone concentrations greater than were contemplated in the following successful quoted studies. Doing so will result in more rapid and greater spore destruction efficiencies.

Ozone has already been proven effective against spores. The Journal of Applied Bacteriology published *Inactivation of Bacillus Spores by Gaseous Ozone*, which states, "The exponential decrease in (the number of surviving spores) varie[s] roughly in inverse proportion to the ozone concentration." [The more ozone, the more dead bugs.]

Ozone is commonly used for sanitation, as reported in: *Efficacy of Ozone fumigation for Laboratory animal sanitation*, Keio University School of Medicine, Tokyo, Japan. This also revealed that ozone eliminates Bacillus spores.

Another study ...*Ozone resistance of Bacillus spores*..., also from Keio University School of Medicine, Tokyo, Japan, shows Ozone might be effective for routine sterilization; but the strength of ozone concentrations used, contact time and relative humidity determine how effective ozone is at killing spores.

Another study, this one from Department of Food Science, North Carolina State University, and published in Poultry Science, described ozone as a disinfectant of Bacillus against poisonous formaldehyde. It stated that a low concentration of "Ozone (1.52-to-1.65 percent by weight) resulted in bacterial reductions of greater than 4 log10 and fungal reductions of greater than 4 log10." The study adds "Ozone may be used as a disinfectant against selected microorganisms."

Another study, published in the Journal of Microbiology and Biotechnology and performed at the University of Tennessee's Department of Microbiology proved, again, that ozone (created by uniform glow discharge plasma) "provided antimicrobial active species to surfaces and workpieces [workpieces, same as packages and mail] at room temperature as judged by viable plate counts" and "(ozone) reduced log numbers of bacteria, Staphylococcus aureas and Escherichia coli, and endospores from Bacillus...." Again, the inactivation was dose-dependent. Each of these referenced studies, already impressive enough, would have shown spectacular results proving ozone's sporicidal and sterilization abilities if in each case greater concentrations of ozone and more humidity were used.

Another study, *Inactivation of Bacillus subtilis spores ...during ozonation*, by the Swiss Federal Institute for Environmental Science and Technology, found typical sporicidal activity by ozone in water.

Another study on sterilizing surfaces with ozonated water was published in Applied Environmental Microbiology, *An evaluation of the sporicidal activity of ozone*, showed spores of Bacillus and Clostridium suspended in water were inactivated after 10 minutes of ozone exposure.

A specific Anthrax destruct study was privately commissioned at the College of Food Science in Shanghi to research ozone's effect on Anthrax. Please note the study conclusively proved that ozone destroys Anthrax in water, *leaving no harmful residue.*

Wally Grotz also sent me a stack of studies showing that hydrogen peroxide also destroys Anthrax.

During the height of the Anthrax scare we communicated all this information to the higher officials in the Post Office; they sent us to the CDC. We got right to the top managers of the CDC's emergency task force, and to the front line people below them. In an effort to save lives we communicated several times, sent faxes, and spoke to everyone we could find. They did not act on the knowledge that we gave them.

West Nile Virus

In early September of 2002, 850 cases, 43 deaths (CNN), and there is no quick test to screen donated blood or transplants. Mosquitoes breeding in stagnant water left in flower pots, old tires holding water, and mud puddles after the rain are the latest source of plague to challenge us. They tell us since West Nile Virus is carried by mosquitoes we should spray poison everywhere. Should we destroy beauty and the food chain and poison all the poor birds by saturating the food chain with poison? Should we paint our permeable skin with (and absorb) side-effect causing DEET? There is a better way. Just like *all* the other viruses, the **West Nile virus is still *anaerobic***, and if you haven't skipped ahead, by now you must know what that means.

Transfusions, transplants, and mosquitoes inject the viruses into us, the viruses travel through our fluids and breed in the brain area. The brain swells with the microbe waste toxins and humans are said to die from encephalitis. What if your body and fluids were full of oxygen and active oxygen sub species, how could the *anaerobic* virus even travel, let alone settle in and breed? Small oxygen molecules easily pass through the blood brain barrier. And no one has to *ever* get sick from a blood transfusion anymore. **No virus or microbe can survive in living or bagged blood that has had enough ozone bubbled through it!** They just showed bags of dirty blood on TV in a West Nile virus news story. The blood was filthy in brownish-purple color just like on the back cover of this book. NATO and the Canadian Defense Forces advocate purifying all stored blood with ozone. I have been telling everyone I could about this for *14 years*. Why aren't we purifying our blood yet? Before there is more suffering, immediately send everyone you know who works in health care several copies of this book! Especially send them to the ones that brag of using hip 'complimentary' treatments while continuing to ignore Nature's already proven and safe Ozone Therapies.

Please recall that Jack Stewart in New Zealand told us how all his prize race horses were cured of encephalitis by SEO oxygen supplementation, and Dr. Mayer injected ozone into children's spinal and brain fluids at Miami Children's Hospital for 50 years while curing bacterial and viral meningitis. How is this new encephalitis disease transmission vector any different? See: Akey DH, Walton DE. Liquid phase study of ozone inactivation of Venezuelan equine encephalitis virus. *Applied Environmental Microbiology* 1985; 50:882

Smallpox

Due to smallpox still being yet another *anaerobic* viral pathogen (like West Nile), smallpox has the same vulnerability to ozone and the other Oxygen Therapies. Because Smallpox rapidly spreads infesting all warm and wet tissues, it may require much higher levels of active oxygen applied very quickly to overcome it, especially after fever and blistering have set in.

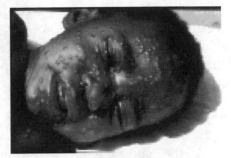

Imagine how fast and how many viruses bred inside this poor fellow.

From the Centers for Disease Control:
"Smallpox is believed to have originated over 3,000 years ago. In earlier years, epidemics of this devastating disease swept across continents killing large populations and literally changing the course of history. Attacking royalty and peasants alike, 30 percent of those infected died and the survivors were left with scars and often blinded.

"You can only get it from people [generally within six feet of you.] The greatest risk of infection occurs among household members and close contacts of persons with smallpox, especially those with prolonged face-to-face exposure. There are no symptoms for about 12-to-14 days. During this 'incubation period,' which could range from 7-to-17 days, one feels fine and cannot infect others.[The viruses are multiplying rapidly, but have not pushed out to the surface yet.]

"The first symptoms are much like the flu: sudden onset of fever, tiredness, severe back pain, and sometimes stomach pain and vomiting. This lasts 2-to-3 days, and at this time people become highly contagious to others. When the fever begins to drop and the people begin to feel better, a rash develops in the nose and mouth. The rash becomes bumps that spread to the face, hands, and forearms, and then spread to the trunk and legs. The bumps have been even found on the palms of the hands and on the soles of the feet. All of the bumps evolve at the same rate, and are most dense on the face and extremities.

"If the vaccine were to be given to everyone in the United States, it is estimated that 350-to-500 people would die from the vaccine. That is about one or two deaths for each million doses of smallpox vaccine administered." That is supposed to make us feel better? No problem? That is unless your number is up, and YOU are number one or two.

May I point out the obvious? There *is* a harmless available proven treatment you don't have to play Russian roulette with. This treatment, when maintained at proper levels, will completely surround any pathogens that cannot live in it and quickly oxidize them. This treatment is harmless to our normal human cells, has no side effects, and has never killed or hurt anyone when used correctly. It's a little thing we like to call 'Ozone Therapy.' Have you heard of it?

> **At the end of the day, it all comes down to the one simple remaining fact.** *All* **infectious warfare agents are composed of toxins or anaerobic bacteria, microbes, and viruses that** *cannot resist* **being destroyed when concentrated active forms of nature's oxygen and ozone surround them. This is always the same pesky law of physics that just won't go away. It is so simple, yet some who refuse to look still cannot see the obvious. When TV says "There is no cure"...**

Arthritis

One man of 38 years that I met had arthritis. It caused his hands and feet to completely curl over; and he couldn't walk or drive a car. After several weeks of ozone treatment his hands uncurled and his kids yelled, "Daddy look! You're moving your toes!" In the U.S. 75,000 children have rheumatoid arthritis.

What's arthritis? Well, according to Rheumatoid Disease Foundation, arthritis is probably from an amoebae/microbe living in the synovial fluid (that's the joint fluid) excreting calciferous waste matter into the fluid where it 'cements' into the joints, and creates friction due to piled up toxicity. Pain, microbial invasion, swelling, deformed joints; it's all down to low oxygen toxicity. You take away the toxicity, and if the body uses its oxygen to burn up the microbes and hydrogen, carbon, nitrogen, and sulfur that composes your problems, why shouldn't the joint work again? Your body can work on any specific area with Oxygen Therapies—if there *is* enough oxygen, *and* there's something left to repair.

Cancer

In a survey of 79 oncologists from McGill University Cancer Center in Canada, 64 said they would <u>not</u> consent to treatment with Cisplatin, a common chemotherapy drug, while 58 oncologists said *they would <u>reject</u>*

all the current trials being carried out by their establishment! The reason: "The ineffectiveness of chemotherapy and its unacceptable degree of toxicity."—Phillip Day, *Cancer: Why We're Still Dying to Know the Truth*

Cancer Cases

Colon/Rectal cancer	95,600
Lung cancer	171,500
Prostate cancer	184,500
Breast cancer	189,300

And these figures were from 1998. The World Health Organization says that up to one-third of all cancers of the colon, breast, kidney and digestive tract are attributable to too much weight gain and too little exercise.

Warburg's Prime Cause of Cancer Simplified

Adapted from two-time Nobel Prize winner (respiratory enzymes) Otto Warburg's:
The Prime Cause and Prevention of Cancer and his other papers summarized by Ed McCabe.

"Because no cancer cell exists, the respiration of which is intact, it cannot be disputed that cancer could be prevented if the respiration of the body cells would be kept intact."

[Everything in quotes in this Warburg section is from Warburg himself, and everything else is my summation of his work.]

There are many secondary causes of cancer such as poisons, microbes, radiation, and foreign objects in the body, viruses, retroviruses, and other injuries.

But there is only one prime cause of cancer. The only thing all cancers have in common is that something took away the cell's ability to breathe, either mechanically, chemically, or energetically. The prime cause of all cancers is impaired cellular respiration.

Cells are constantly dividing. When any newly formed (embryonic) cell is denied 35 percent or more of the oxygen it needs, its tiny breathing mechanisms and respiratory enzymes are no longer saturated with oxygen. When the oxygen transferring enzymes in the cell are no longer saturated with oxygen, the cell is damaged severely—since cell respiration decreases irreversibly. Respiratory enzymes not saturated in oxygen end up as though you burned up your car engine by running it without oil. This can happen within two rounds of cell division under low oxygen conditions. As respiration falls, the cells struggle but can't keep up the high-energy production levels normally sustained by converting oxygen into ATP energy.

As the cell's damaged breathing (its high energy producing complex) fails, the cell loses all of its higher functions and de-evolves, or 'de-differentiates' into a simple plant type cell. Under proper magnification cancer cells often are seen as green, and look just like a plant. Why does Nature convert the damaged human cells into 'plant type' cells? Because that's the only evolutionary option left to a human cell under low oxygen conditions that allow it to maintain life. Plant cells use fermentation, a much simpler and inefficient form of energy creation common to simple organisms. Fermentation is a simple process that converts body sugars (glucose) into a weak form of ATP energy and produces a number of lactic acid byproducts. But fermentation merely buys time because it is so inefficient that it only leaves enough energy for the cell to grow and grow and grow. The damaged cells can no longer be special or individualized (differentiated) with special functions. Think of this cancer process as a defensive mechanism of life trying to continue temporarily in the form of a plant type cell, since the human form is no longer working correctly.

If circumstances keep denying these severely damaged cells access to 35 percent or more of the oxygen they need, they keep growing, but incorrectly. That's why we have to remove the external cancer causing agents. The cells know they are lacking in energy, so in a valiant attempt to create more energy they ferment more and more while trying to catch up. And when it comes time for them to reproduce, the damaged cells make exact copies of their own damaged selves which in turn make more copies of more fermenting cells. I just described cancer. As one definition put it, *"For cancer formation there is necessary not only an irreversible damaging of the respiration but also an increase in the fermentation."*

If the host body is suffering from radiation or other poisons, is low on oxygen and doesn't have many active respiration enzymes, the unnatural conditions allow these acidic fermenting cells to spread. Their unregulated growth and increasing amounts of acidic waste products crowd out and choke off the oxygen from normal cells, and they eventually take over.

Is there a latency period? Not exactly. The slow takeover may show up again many years later, after the respiratory enzyme producing cells no longer saturated with oxygen are damaged. These acidic cells cause the nearby cells to keep turning into fermenting cancer cells, and 'latency' is merely the time it takes for the cancer cells to silently grow and grow as they take over. Due to the lack of oxygen, the cancer was always there. *"The most important fact in this field is that there is no physical or chemical agent with which the fermentation of cells in the body can be increased directly: for increasing fermentation, a long time and many cell divisions are always necessary."* And, *"The mysterious latency period of the production of cancer is, therefore, nothing more than the time in which the fermentation increases after a damaging of the respiration."*

This is why it is useless (except as part of the proposed repair) to only cut out a cancer without reversing the underlying causes. Without enough oxygen, the normal cells will just keep mutating into acidic fermenting cells. In the very earliest stages of embryonic development normal body cells ferment for a bit, but quickly evolve; the fermentation mechanism drops away and is replaced by normal oxygen respiration only. But if the oxygen is kept low during their development, they have to keep fermenting to survive. When this happens they mature as fermenting cells! You can remove small tumors; however, the surgery may release these fermenting cells into the body. If the body is not clean and saturated with oxygen and enzymes, the fermenting cells may find new homes someplace else.

> *"There would be no cancers if there were no fermentation of normal body cells."*

Unlike oxygen using cells, fermenting cells have, relatively speaking, incomplete digestion. They excrete mostly lactic acid and depending upon where they are in the body, other actual metabolic toxins and amines that freeze up muscles and cause other damage like further blocking cellular respirations. The elimination of toxins is another burdensome drain on the body's resources. To summarize, anything that causes cancer:

➤ **Somehow lowers the body's ability to carry oxygen to the cells.**
➤ **Damages the respiratory enzyme transport of the oxygen into the cells.**
➤ **Damages the reactions in the cell using the oxygen to make energy.**

There are many secondary causes of cancer. A good deal of money has been wasted funding research on chasing these tangents. The prime cause is known, and we need to focus in on the disruption of cellular respiration via many sources.

Retroviruses can thrive in low oxygen conditions attacking the body and disrupting cell respiration. We know viruses and bacteria can thrive in us when cellular oxygen levels are too low. We also know that the low oxygen levels could be made worse because of the continual piling up in the body of pollution and foreign substances.

Another respiration related problem could be the body defensively growing aberrant starved-for-oxygen capillary networks around foreign objects like implants. (Sadly, many women called me after getting breast cancer from breast implants).

Poisons or X-rays can also damage the cell's respiratory enzymes. *"Carcinogenesis by X-rays is obviously nothing else than destruction of respiration by elimination of the respiring grana."* You can kill cancer cells with chemotherapy or radiation because cancer cells are weaker. Normal cells appear, at the treatment time, to survive these poisons because they are stronger than mutated cancer cells, but the poisons have also damaged their respiration. And we know what that means in time: *"...the descendants of the surviving normal cells may in the course of the latent period compensate the respiration decrease by the fermentation increase and thence become cancer cells."*

Budwig Agrees with Warburg

Dr. Johanna Budwig also states her microscope work proves cancer cells are oxygen deficient. Cancer cells show a high level of fat in the cytoplasmic space and in the nucleus. And cancer cells exhibit many forms.

The reason the cancer cells are oxygen deficient is the lack of essential fatty acids (remember Udo's research?) to provide the needed transport of oxygen to and into the cells. Without EFA's the plasma membrane which makes up the cell wall changes, destroying its ability to correctly regulate the flow of nutrients, waste products and toxins into and out of the cell. Inappropriate fats which would otherwise be controlled or excluded are able to gather within the cytochrome space within the cell—further choking off normal cell function. Without sufficient oxygen the cell has to adopt another means of supplying itself with energy or die and so it reverts to a more primitive glucose digesting system which requires little or no oxygen.

On Repairing the Damage

Is it possible to fix the problems? Indeed, I have interviewed hundreds of newly converted Oxygen Therapy using healthy people who were once cancerous. Many of them had doctors who had sent them home 'to die' because there was nothing they could do. Standing there talking with these bright-eyed healthy survivors, one by one for over 12 years—yes, this experience has convinced me that the problem can be fixed for the vast majority who correctly follow the protocols.

Warburg postulated that because young cancer cells with partially damaged respiration live in the body almost aerobically, inhibition of any further growth on their part using fermentation should be possible by repairing their damaged respiration before the fermentation gets locked into a mature cell.

Three *pre*-conditions exist for this proposed repair. Like planting a garden to raise healthy cells in, you have to clean up and prepare the soil before you plant.

1. "All growing body cells are **SATURATED** with oxygen." Warburg emphasizes again, ...**all body cells [must] be SATURATED** with oxygen.... So grab your favorite Oxygen Therapy and get at it. As I have said for years, "Flood the body with oxygen, properly and safely."

2. External cancer-causing agents are kept away, at least during the treatment. Stop smoking, drinking, using dope, breathing poorly, and eating processed, hydrogenated, and unhealthy food; also avoid being overly X-rayed, working around toxins, and try to resolve whatever causes depression, guilt and anger. Start exercising.

3. Safely flood the body with (replace the damaged and missing) active groups of respiratory enzymes. As Warburg said, "These enzymes are harmless and should be increasingly added to food, in the greatest amount, even forever." Be careful—not too much iron.

I took the Warburg research data and discussed the ingredients with others knowledgeable about modern supplements. Here is the exact ingredient list I came up with. I would like to get this list turned into a supplement someday.

Mr. Oxygen™ Suggested Respiratory Enzyme 'Warburg' Formula
(Warburg's Active Groups of Replacement Respiratory Apo-enzymes)

➤ Chlorophyll
➤ Iron salts like Ferric Fructose, Iron Fructose
➤ Riboflavin (B2). Important for body growth and red cell production. There is no known toxicity to riboflavin. Because riboflavin is a water-soluble vitamin, excess amounts are excreted by the body in the urine.
➤ Nicotinamide (B3), (Niacin). Niacin in high dosages gives a redness flush from increased circulation, and may cause itching.
➤ Pantothenic Acid (B5)
➤ Cyanocobalamin (B12) required for phase one detoxification of chemicals in the liver
➤ Cytohemin (plant iron groups)–IV injections OK d-amino-leuvlinic acid (precursor to oxygen transferring hemins).

So far, the closest product we found specifically designed to answer Warburg's admonitions is 'Oxy Plus' which the label bills as Warburg's Oxygen Booster. It focuses on chlorophyll and vitamins that are to help the body produce new red blood cells. OxyMune and Oxy-Moxy both feature oxygen uptake ingredients as well, Oxy-Moxy is made from sea vegetables.

The body will attempt to normalize the metabolisms of cancer cells by using the extra active groups of enzymes we give it. The body's attempt to normalize the undifferentiated cancer cells typically results in their elimination. This it's expected that the growth of metastases can be inhibited with the enzymes. If the respiration of the body cells can be kept intact by adding enough enzymes, cancer could be prevented

Keep doing these three things—oxygen boosting, carcinogen removal, and enzyme replacement—and the body should halt any further growth of fermenting cells. Then one exercises patience and just waits for the old damaged cells to die out and be digested. This may take months. Bromelain is an excellent enzyme supplement for digesting this old dead cell protein.

"These proposals are in no way utopian. On the contrary, they may be realized by everybody, everywhere at any hour... The prevention of cancer requires no government help and no extra money." (My favorite Warburg quote.) What if you don't have cancer, but don't want to develop it, either? Don't worry, you're still covered by Warburg.

> A nurse who works in medical research called me up and said "It's so simple. I don't know why I never thought of it. When we're working with cell cultures in the lab, if we want cells to mutate, we turn down the oxygen. To stop them, we turn the oxygen back up."

Warburg Stated, "To Prevent Cancer":

➤ *Keep the speed of the bloodstream so high the venous blood still contains sufficient oxygen.* You could add seaweed extracts like Dulse to your food to boost the metabolic rate. You could drink lots of clean water, exercise regularly, especially by rebounding, and using a Chi Machine along with the Hsin Ten far infrared light emitting Hothouse.

In its early stages, and quickly coming over the horizon, is Low Level Laser Light Therapy as well.

> *Keep a high concentration of hemoglobin in the blood.*

> *Always add the active groups of respiratory enzymes to the food and keep increasing them if a pre-cancerous state has already developed.* (Check the OxygenAmerica.com website for the best current combo.)

> *Exclude external carcinogens rigorously.* Live and work where it's clean.

The National Cancer Problem

According to a major cancer society, 175,000 women had breast cancer nationwide in 2001. Why is there so much cancer? Low available oxygen at the cellular level, and pollution, combined with processed, unnatural food creating the perfect acid Ph condition in the body which allows cancer to bloom. Remember, keeping the oxygen up, and the respiratory enzymes intact, are *the* keys to the prevention and repair of cancer. We know sufficient oxygen is chronically missing from our food and environment. But what about the enzymes you assume are in your food? After all, eating 'fresh live foods' means that the enzymes are all supposed to be in there. Aren't they? Maybe not. Watch out; even if you try to shop correctly at the supermarket you can unknowingly be shortchanged. Here are a few principles and quotes from a typical anti-cancer 'Stop Cancer' website.

Cooking destroys all enzymes. Unfortunately, cooking any food at temperatures above about 116 degrees Fahrenheit kills all enzymes. All canned or bottled foods contain no enzymes because they are cooked before being processed. Most fresh-frozen vegetables also generally have no enzymes because they are usually dipped in hot water before freezing.

Even the raw vegetables and fruits you eat may be enzyme-deficient! Raw vegetables and fruits can be excellent natural sources of enzymes. Unfortunately, they contain no enzymes when they are picked 'green' (often the case in supermarkets because they have to be transported over long distances). Enzymes can only develop when the plant ripens in the soil. Irradiating food or treating it with preservatives can also destroy enzymes.

And now consider that collective groups of fermenting cancer cells—like colonies of bacteria—will protect themselves as best they can from your own immune system because they are trying to survive. "Cancer cells hide themselves under a thick coat of adhesive fibrin, a coat that is some 15 times more thick than the fibrin over normal cells. The thickened coat hides away their suspicious markings, including their antigens, from the body's immune defenders." And under low oxygen conditions, "The cancer cells with their sticky coating can adhere to tissues where they congregate and multiply. To throw the body's immune cells further off track, the cancerous cells may slough off their antigens. The immune cells immediately attack these harmless proteins but leave the cancerous cells unharmed. It is a type of warfare that could make a military general envious.

The cancer cells grow because of the absence or inadequate presence of enzymes that are capable of stripping the fibrin away from the individual cancer cells. Adequate enzyme activity can lay bare their antigens and so pave the way for their destruction by the body's immune cells."

Echoing Warburg's admonishment, "*These enzymes are harmless and should be added to food, in the greatest amount, even forever.*" The anti cancer site goes on to say. "The more cancer cells the body produces, the more enzymes that are required."

Apart from all the people that I interviewed who no longer have cancer because they *saturated* their bodies with oxygen during clinical Ozone Therapy, is there any recent proof? Try this one out on a skeptic:

1980 Aug 22nd 'Science' Vol. 209:931-933, a U.S. peer reviewed scientific journal, published: *Ozone Selectively Inhibits Human Cancer Cell Growth*. They got their results using only .3-to-.5 ppm of ozone. Imagine what would have happen if they used the standard dosage of 2 percent (27 mcg)! The ozone dose they used was way too low.

"Since Warburg's discovery, this difference in respiration has remained the most fundamental and, some say, only physiological difference consistently found between normal and abnormal cells. Using cell culture studies, I decided to examine the differential responses of normal and cancer cells to *changes* in the oxygen environment. The results that I found were rather remarkable. I found that 'normal' O_2 tension actually maximized the growth of the cancer tissue, and that **high O_2 tensions** were **lethal** to cancer tissue, 95 percent being *very* toxic, whereas, in general, normal tissues were not harmed by high oxygen tensions. Indeed, some normal tissues require high O_2 tensions.

"It does seem to demonstrate the possibility that if the O_2 tension in cancer tissues can be elevated, then the cancer tissue may be killed selectively, since it seems that the cancer cells are incapable of handling the O_2 in a high O_2 environment."—J.B. Kizer: Biochemist/Physicist, Gungnir Research, Portsmith, OH.

Cancer Treatment Example

During another of my weekly conversations with Dr. Boyce, while his clinic was in full swing, he was telling me of a man who visited his clinic saying he had squamous cell carcinoma. In other words—cancer of the ear. He had a big hole in his ear and he looked awful. He asked Dr. Boyce to treat him.

This person had been to the Mayo clinic where they told him that he would have to have his ear cut off, as there was nothing they could do. He went to Mexico. They also said, "Well, you're going to have to have that ear cut off. There's nothing we can do for you, either." With all of the alternative clinics in Mexico, you would think they would be able to do something.

Fortunately he went to Dr. Boyce. Upon looking at his blood reports Dr. Boyce noticed the man hadn't mentioned that he had leukemia. He held it back because he figured they might not take him into the clinic if he had leukemia (since it is supposed to be incurable). They started treating him with the Medical Ozone Therapy, the Hyperbaric Oxygen Therapy, the peroxide intravenous drip, the chelation, the peroxide bathing, all the things I discuss. The quote the attending physician delivered to me was: "Lymphatic case is 80 percent cured. No more squamous cells in the ear. We don't have to amputate."

That was the report even though everybody he went to had seriously informed him that he was going to have to have that ear amputated. Two months after getting Oxygen/Ozone and Chlorophyll Paste Therapy, his ear was completely healed and completely filled in. Completely! I saw the before and after pictures. Dr. Boyce used a chlorophyll-based salve externally to encourage the new cell growth that filled in the hole.

The clinic had slowly increased combinations of Oxygen Therapy protocols and included just about all of them. Dr. Boyce flooded the body with oxygen. He then removed the man's bodily waste products with colonics, and with DMSO, a solvent and sulfur supplement. He then proceeded to rebuild the man's immune system with vitamins and minerals; and by giving his body all the building blocks it needed. Dr. Boyce made it impossible for the aberrant cells to remain there. The patient was able to keep his health and his dignity.

As we go to press there are many positive testimonials being ascribed to a new super absorbable form of a liquid *ionic* cesium supplement being used by cancer patients. When used alongside Oxygen Therapies, cesium appears to be very potent. Unlike other cesium supplements, it is so active that doctors using it tell me it must always be taken along with ionic minerals and ionic electrolytes. Eniva's Ionic Cesium is postulated to rapidly alkalinize the body, and an alkaline pH of 8 is deadly to cancer. Another mechanism of action occurring with Ionic Cesium use is that it is reported to block glucose sugar uptake by the cancer cells when they need to ferment glucose for food. The cancer can literally starve to death. Ionic minerals are 10,000 times smaller than colloids, so they go everywhere quickly and are absorbed in the body immediately.

Cesium and Cancer

As I have explained, Otto Warburg was awarded two Nobel prizes for his lifelong research showing that cancer is caused by damaged cell respiration due to lack of oxygen at the cellular level. According to Warburg, damaged cell respiration causes fermentation, resulting in low pH (acidity) at the cellular level.

In 1984 Keith Brewer, Ph.D. (Physics) translated Warburg's theories into a practical, cost efficient treatment for cancer. Brewer successfully treated 30 patients with various forms of cancer, using Nature's most alkaline mineral, cesium. The results of Brewer's work? *All 30 patients survived.*

Dr. Warburg, in his Nobel Prize winning paper, illustrated that in the absence of oxygen the cell reverts to a primal nutritional program to nourish itself by converting glucose through the process of fermentation. Fermentation creates extra lactic acid. The lactic acid produced by fermentation lowers the cell pH (acid/alkaline balance), and makes the body too acid—which destroys the ability of DNA and RNA to control cell division further enabling the cancer cells to wildly multiply. The *lactic acid simultaneously causes severe local pain* as it destroys cell enzymes. The cancer appears as a rapidly growing external cell covering with a core of dead cells. Oxygen oxidizes lactic acid and dead cells.

Cesium, a naturally occurring alkaline element, the *most* alkaline one, has been shown to change the cancer cell in two ways:

1. Cesium limits the cellular uptake of glucose, which starves the cancer cell and reduces fermentation.

2. The normal pH of the blood is 7.35-to-7.45. Cancer cells cannot exist in a pH of 8.0 or more. *Cesium raises the cancer cell pH to approximately 8.0.* This alkaline state along with oxygen can neutralize the weak lactic acid and reportedly has stopped pain within 12-to-24 hours. Because an alkaline pH range of 8.0 is a deadly environment for the cancer cell, cancer cells have died in previous cancer patients within a few days. After their demise, the remains of the dead cancer cells can be absorbed and eliminated by the body. As I have shown you, the only way the body can eliminate wastes is by having plenty of oxygen to combine with the waste.

The Work and Findings of H.E. Sartori

Certain foods contain biologically active compounds and/or ingredients, i.e., vitamins, inorganic salts, organic compounds, essential fatty acids, minerals, and chelating agents, which may either precipitate or prevent cancer development.

The presence of Cesium (Cs+) or Rb+ *in the adjacent fluids* of the tumor cell is believed to raise the pH of the cancer cell, causing cell growth and reproduction to cease, which results in reduction of life span of the cancer cell. The *normal healthy cells do not uptake the Cesium, only the cancer cells can,* so the healthy pH of the normal cells is maintained. The introduction of the alkaline pH through these alkali salts may also neutralize the acidic and toxic material within the cancer cell.

Treatment was performed on 50 patients at Life Science Universal Medical Center Clinics in Rockville, MD and in Washington, DC. All patients were terminal subjects with generalized metastatic disease. Forty-seven of the 50 patients studied had received maximum modalities of treatment, i.e., surgery, radiation, and various chemotherapies, before the metabolic Cs-treatment was initiated. [These previously absorbed toxic system challenges probably dampened the results.] Three patients were comatose and 14 of the patients were considered terminal.

The diet during treatment consisted mainly of whole grains, vegetables, linoleic acid rich oils (linseed, walnut, soy, wheat germ) and other supplemental food. To increase efficiency of the treatment and improve the circulation and oxygenation, the patients received the chelating agent EDTA, dimethyl sulfoxide (DMSO) and also a combination of vitamin K and Mg salts.

An overall 50 percent recovery from cancer by the Cs-therapy (cesium) was determined in the 50 patients treated. One of the most striking effects of the treatment was the *disappearance of pain* in all patients within one-to-three days after initiating Cs-therapy. The results demonstrate the rate of effectiveness of CsCl [alkali salt cesium chloride] in cancer therapy.

How The Body Selectively Uses Cesium To Destroy Cancer

Dr. Brewer demonstrated that both Cs+ and Rb+ can enter cancer cells and embryonic cells —but not normal adult cells. While the amount of Na+ (sodium) is greater in cancer cells, *the amount of calcium in cancer cells is only about one percent of that in normal cells*. The great *reduction in the amount of calcium found in cancer cells* is due to the *higher acid content in the cancer cell*. The *lactic acid eats away at the alkaline calcium* in the body's attempt to neutralize itself. [Did I just hear a great rationale for taking Coral Calcium?]

This would explain why osteoporosis and cancer go hand-in-hand and often reside in the same body.

When Cs+ or Rb+ enters the cancer cells, the pH increases from as low as 5.5 to over pH 7.0. *At a pH of 7.6, the cancer cell division will stop. At a pH of 8.0-to-8.5 the life span of the cancer cell is considerably shortened* (to only hours).

The Studies of A. Keith Brewer, Ph.D.

There are four steps to changing a normal healthy cell into a cancer cell:

1. Disturbance of the cell membrane by carcinogens or energy [or a buildup of toxins or other injury or parasites]:

Glucose can still enter the cell [contributing to fermentation turning the cell into a veritable plant] but oxygen cannot. The cell thus becomes anaerobic. Another way of disturbing the cell membrane surface is by means of radiation (X-rays, alpha-, beta-, or gamma-rays, UV and other), which prevents oxygen from entering the cell. Through glucose, K, Rb and Cs may still enter the cell. Through lack of oxygen, any metabolism in the cells has to proceed anaerobically. Both chemical and physical factors, including emotional stresses, can cause or precipitate oxidative damage of the cell membrane from free radicals and other related activated species. These are not only the primary instigators of cancer but virtually all degenerative illnesses including; allergies and auto-immune diseases, multiple sclerosis, rheumatoid diseases, immune suppression syndromes, most endocrine diseases, including diabetes, hypothyroidism, adrenal insufficiency, and many others.

2. In the absence of oxygen, glucose undergoes fermentation to lactic acid. This causes the cell pH to drop from between 7.3-to-7.2 down to 7 and later to a more acidic 6.5; in more advanced stages of cancer and in metastases the pH drops to 6.0 and even 5.7.

3. DNA and RNA in an acidic medium lose positive and negative radical sequencing. In addition, the nucleic acids and amino acids entering, and those within the cell, are altered [mutate].

4. Loss of control mechanism and spread of cancer cells. In an acidic medium, the various cell enzymes are changed in structure and function. As a consequence, enzymatic processes become ineffective, the cell completely loses its control mechanisms, and chromosomal aberrations may occur.

The German scientist, Hans Nieper, has also shown that cesium chloride is effective in the management of most tumors, that is, of advanced brochogenic carcinoma with bone metastisation. Cesium therapy seems to be the treatment of choice for these types of cancers, even with only relatively minor changes in lifestyle.

To enhance any cancer program, including one based on Cesium therapy, one must incorporate the missing items listed in my *Crown Jewels of Health* section of this book for the utmost success.

The *Crown Jewels* are all factors that are equally if not more important than any selective therapy. Most people would rather die than change their diet, and often do. You especially don't need your artificial sugary drinks and snacks. Natural foods grown and prepared correctly are delicious. The therapies in this book may clean you out during your present emergency, but without a radical change in lifestyle and eating habits it is unlikely that a patient will remain cancer or disease free. If you don't change what got you there, it will keep taking you there.

The greatest challenge with this approach to cancer treatment is not so much a problem of getting rid of the cancer, but of *having the patient take responsibility for his or her health*. This includes continuing to follow the instructions from your competent health practitioner and maintaining the necessary changes in diet and lifestyle for the rest of ones life. Otherwise, if not cancer, almost unavoidably another degenerative disease will develop.

The treatment of cancer [and possibly the prevention of cancer] by cesium is a very practical and intelligent one. It is inexpensive and non-toxic over unlimited time. Cesium treatment is one of the most far-reaching discoveries in the field of cancer therapies. [Cesium information collated by my friend Dr. Darryl Wolfe of the Wolfe Clinic in Canada.] The most absorbable cesium supplement I know of is Eniva's ionic cesium. *Oxygen America* carries it, and they also offer access to the full range of other innovative Eniva products like ionic germanium, gold, copper, silver, and other special ionic mineral blends. Supplements in ionic form are proving to be super-absorbable and therefore delivering the greatest benefit.

Note: All treatments for any chronic illness mentioned in this book should be closely monitored by a competent practitioner, hopefully one skilled in the oxidative modalities.

The above text is based on the research studies and findings of the following:

Otto Warburg, M.D. – Nobel Prize Laureate (1931)
H.E. Sartori, M.D. *Cancer 1985 – Orweillian or Utopian*
The work of Keith Brewer, Ph.D
The Curious Man: The Life and Works of Dr. Hans Nieper

Many naturopaths I spoke with say the best way to dissolve dead cancer cells is with a good enzyme like the Bromelain in pineapples. With that in mind, I can't help but wonder what would happen if somehow all cancer patients started taking ozone injections, supplementing their natural antioxidant rich live food meals with humic shale minerals, coral calcium, Bromelain enzymes, ionic Cesium, DMSO and MSM while drinking plenty of water, being surrounded with love, and exercising. Hmmmm….

Injecting Cancer Tumors with Ozone

You must have noticed I keep saying cancer cells are anaerobic and cannot survive when surrounded by high concentrations of ozone. You have seen a picture in this book showing ozone instantly eating the dead organic rubber cells of a surgical glove, so now put the two concepts together.

Here's visual proof your doctor can try him/herself. I have interviewed ozone technicians that have injected syringe fulls of ozone directly inside cancerous tumors over the course of a few days. At first there is a resisting back pressure owing to the large hardened tissue mass, but soon the cancer cells start to melt and liquefy at the injection's ozone discharge point radiating outward from the center of the tumor. The technicians reported that the same syringe that inserted the ozone, as it is empty and still inside the tumor, can in a little while be used to suck out the liquefied (yellow-green or grey) freshly dead cancer cells by simply slowly backing out the plunger.

Kit Carson, the old time film cowboy had a big tumor on his throat that disappeared in a few days, shrinking from the inside out after getting a few injections of ozone. The old cowboy had gone to the now defunct Key Biscayne cowboy clinic, and one of the technicians described to me how he was treated in detail. This one cancer fact and practitioner tip alone is definitely worth looking into a great deal further and could alleviate major amounts of suffering when ozone tumor injections are adopted as a standard treatment at the first sign of a tumor.

Candida, Allergies, Asthma, Fatigue

Laurie contacted me and said, "I have ongoing Candida, fatigue, and major year-round allergies. Now I've developed a mild case of asthma. I'd like to try Oxygen/Ozone Therapy but I'm hearing from folks that with any sort of asthma you want to stay away from ozone."

I don't know what folks she's been talking to. She's probably referring to some of the research-lazy associations that are out there printing all these lies about ozone and confusing the public and telling them that medical grade ozone, made from pure medical grade oxygen, is the same thing as air pollution. This is blatantly not true.

Granted, you should stay away from air pollution whether you have asthma or not. However, let's look at her problems. Fatigue. Why is she tired? Because she has insufficient oxygen and insufficient fresh fluids in her body. Her cells are not alive like they're supposed to be because they're full of toxicity. How do we know this? Because she's got candida. Candida is a yeast. A yeast that's anaerobic. A yeast that cannot live if it's surrounded by oxygen. If she has it that proves she definitely does not have enough oxygen at the cellular levels. That's why so many people with candida tell me they have used rectal insufflation of ozone, or they take a liquid oxygen supplement, Homozon, or some other oxygen supplements like OxyMune or Hydroxygen.

They all report to me that their Candida goes away after they take in enough oxygen over a long enough period of time at a safe level, hopefully, working with some sort of a physician or naturopath. Laurie's also got major year-round allergies. She's full of toxicity.

Well, at least she's normal. Normal meaning one of our people who lack knowledge about oxygen, which sadly at present is the majority of us.

Almost everybody who's taken an antibiotic ends up being full of yeast within two years because the antibiotics kill most of the beneficial bacterial flora in the body, within the colon and intestines where this bacteria is needed to digest our food and hold back the anaerobes. Did you know that your fecal matter is half bacteria? At least half of it is just aerobic bacteria. That's how necessary aerobes are in the intestines. Digestive bacteria love oxygen because they are aerobic.

However, the disease bacteria in your intestinal flora are anaerobic. They harm the body and they live on fermentation. That's why they make you crave sugar. If you have a yeast infection, you will crave sugar. That's the yeast trying to get you to feed them. Yeast love sugar because they are trying to ferment. It's an anaerobic disease.

To be accurate, there are harmful intestinal bacteria that are aerobic and they can live in air but not in *active oxygen*. I'm speaking of aerobic pathogens such as Salmonella, Shigella, E Coli, etc., but the colony ratios of the good ones to the bad ones are what active oxygen straightens out.

Then she got asthma. Well, of course she did. She was toxic from all this lack of oxygen, the yeast, the candida, and the chronic fatigue. The toxicity built up to the point where the body was irritated and so low on oxygen that fatigue, asthma, allergies and whatever her particular weakest link is, just broke out in the body because of the lack of oxygen and the resulting lack of cleanliness at the cellular level.

Common Colds & Flu

Everyone I know using active oxygen supplements or practicing deep breathing never gets a cold or flu! Anaerobic bacteria and viruses cannot live in active oxygen. Individual results will vary, but people who are already sick and only then start taking an active oxygen supplement usually report they can cut the duration of the cold or fever way down.

Heart

Heart Attack Causes—They May Not Be What You Think!

What is a heart attack? It's one of the top two causes of death. If we look at the work of one of the physicians in Europe who had the coroner's job, we'll discover some striking information. Among other things, he would examine people's bodies after they had died from heart attacks. In performing hundreds of autopsies, he made a startling observation. The leading cause of heart attacks is supposed to be coronary occlusions. In other words, blocked blood vessels. Yet he found that the vast majority of people who died from a heart attack did *not* have any blockage in their arteries.

This puzzled him. He asked himself, "Why are these people having heart attacks?" To find out he performed many experiments on animals. What he concluded was that **the cause of heart attacks is a lack of oxygen in the heart muscles**. When the muscles in the heart do not get sufficient oxygen, they try to 'beat,' and instead, they cramp. When they cramp, they scar over. As that microscopic scar slowly repeatedly increases in size, to the size of a pinhead, it then locks up the muscle and you have a heart attack. The cause of heart attacks is not enough oxygen in the heart muscles!

With an amazing 247,000 American women dying from heart attacks each year, compared to 41,000 dying from breast cancer and 42,000 from lung cancer, this is worthy of examination, ladies.

Interesting side note. Over the years people have reported to me that when they used one of the Oxygen Therapies at a high dosage level for long enough, all the scar tissue on their bodies disappeared. As I said, the body it is programmed to fix itself and all it needs are the missing ingredients. Why wouldn't these principals apply to the heart muscle scars?

Ozone and Ischemic Cerebrovascular Disease

A study regarding the ability of ozone to improve the health status of older adults suffering from ischemic cerebrovascular disease was carried out at the Geriatric Complex of the Salvador Allende Hospital in Havana. A group of 120 was chosen for the study.

Extensive physical, neurological (including CT scans and EEGs), multidimensional psychological and psychomotor tests were given before and after treatment. Special attention was devoted to the patients' mental condition, their ability to participate in daily activities, their ability to administer their own medications, and their social interaction with friends and family. Patients were then classified into three standardized groups according to their symptoms: There were 48 (40 percent) in an 'acute' phase of the disease, 42 (35 percent) in an 'ancient' phase, and 25 percent were classified as being in the 'chronic' phase of the disease. Treatment consisted of 15 sessions of Ozone Therapy given through rectal insufflation over a period of three weeks.

The results were impressive. The mental condition of all acute phase patients participating in the study improved by the end of therapy, along with 91 percent of those in the ancient phase and 67 percent in the chronic phase. By the same token, the medical condition of all acute phase patients improved, while improvement took place in 67 percent of those in the ancient phase, and in 47 percent of the those in the chronic phase of the disease. Post-therapy tests revealed that the subject's ability to participate in daily life situations improved among all of the acute phase patients, in 95 percent in the ancient phase, and for 80 percent in the chronic phase.

The Salvador Allende Hospital researchers concluded:

➢ Ozone Therapy, in the manner and doses applied in the study, produced significant improvement in the group of patients with cerebrovascular disease of the ischemic type, being more effective the sooner the therapy begins.
➢ The initial clinical state improved in 88 percent of the patients treated, obtaining better results in those in the acute phase.
➢ In the multidimensional evaluation, all parameters measured improved, especially daily life activities.
➢ No adverse reactions were reported during the treatment.

These results have led the physicians at the Salvador Allende Hospital to routinely use Ozone Therapy on all patients who enter the facility for the treatment of ischemic cerebrovascular disease.

Angina (A Personal Observation)

Nathaniel Altman wrote: "In Cuba, Ozone Therapy has become a routine treatment for patients suffering from angina and heart attacks. During my visit there in January, 1994 to gather material for *Oxygen Healing Therapies*, I was introduced to the 80 year old mother of Manuel Gomez, the cofounder and former director of the Department of Ozone. Three years before, Dona Lilia had such severe angina that she could barely get from one room to the other without pain and shortness of breath. After much resistance (she is a very stubborn woman), her son finally persuaded her to undergo three weeks of daily Major Autohemotherapy, during which the angina completely disappeared. Although Dona Lilia hasn't changed her lifestyle ('I don't like taking long walks') she has tremendous energy and her angina symptoms have never returned."

Heart Attack Suggestions

Many heart attacks occur because of lack of vitamin E, lack of oxygen, lack of water, and low Co-Q10 in the heart muscles. It is believed that most heart attacks can be helped by these decisive actions:

➢ *A person who fears a stroke coming on should drink water <u>instantly</u> as soon as he or she feels weakness on one side.*

➢ *Immediately starting deep coughing before you pass out. Coughing pumps the heart and forces deep breathing that oxygenates.*

➢ *Immediately taking cayenne pepper under the tongue, which quickly opens up the capillary blood flows. Many carry cayenne capsules around to help themselves or others.*

➢ *Immediately laying the sufferer on his or her back, grabbing the ankles and raising the legs up, and then swinging the legs back and forth, left to right repeatedly, just like a swimming fish pumping the lymph and spinal fluid and re-oxygenating it.*

Here's an area for future study. I have witnessed several people damaged from a stroke tell me that they eventually regained some or most lost arm functions by using correct applications of either oxygen supplements or ozone, etc. It has been postulated that the active oxygen dissolved the blockages and allowed the blood to flow into a deprived area again.

Lungs, Kidneys, Liver, Blood

They're all filters. Many advocate that you can safely and slowly clean them out with active oxygen and re-build them with ionic minerals!

Pancreatitis & Diabetes

On the pancreas you have these sites which are called receptor sites. In other words, something must come along and fit into this specific site. It's a lock and a key kind of thing. What if all these receptor site 'openings' on the pancreas itself become covered up with bacteria or virus colonies, or some kind of fungus? Don't you think these chemical receptor sites will no longer be able to recognize the lock and the key?

"Well, we'd better give you some insulin," say the doctors. Insulin is given to force chemical reactions, and it eventually burns these receptor sites out so they never work again. And as you get older, you have to take more and more insulin, because you keep burning the sites more and more often. And, of course, you never solve the lack of oxygen problem in the first place, so the anaerobic bacteria really start to thrive and toxically plug up the works which impairs organ function while confusing the healthy cells. Doctors really don't know what to do because they would have to admit that all of the body water is filthy, and that is not in their training.

"Diabetes can be caused by e-coli in the pancreas" –Dr. Mona Harrison, Global Health Trax lecture 2001. Since e-coli are anaerobic, and can't live in active oxygen, how would you treat diabetes? All I know is that any diabetic I met who started using an oxygen supplementing substance soon stopped the progressive decline that usually occurs dead in its tracks. And most told me they felt great, had more energy, and all their friends were commenting on how much better they looked. I personally believe no one ever had to lose their limbs if only the truth of these therapies were known so they could be widely used.

Sickle Cell Anemia

My wife Leeda had a debilitating form of chronic anemia her whole life. School kids called her 'snow-white' because she was so pale. Many times I would watch her climb up a single flight of stairs and have to sit down from heart-pounding, pain in the arms, legs, and joints and feeling completely exhausted. It hurt her to breathe, to walk, to talk, to eat and to sleep. She says she felt like a lead weight. Of course she would, she was not getting enough oxygen into her cells.

She got her first ozone shot. She immediately felt better. The constant bone pain, chest pain, limb pain, joint pain, breathlessness and extreme fatigue subsided somewhat, she started breathing better. She *felt* after all these years in and out of the doctor's office and hospital (that did nothing for her) that she was finally onto something. She started doing her own ozone insufflations. She repeatedly breathed ozone in the air, drank it in water, and even put it in her ears. She can now *run* up the stairs, and in her whole life she had never before been able to do this. The first time this happened, she didn't realize she had done it until she was already at the top of the stairs. The pain is all gone, and the rapid heart beat has stopped. Nobody can tell her ozone doesn't work. This experience has taught me first hand how much people with any type of anemia suffer on a daily basis, and that it is all unnecessary.

Sickle Cell & Ozone Therapy

From Welcomat Newspaper. (The Philadelphia Weekly: News, Arts & Opinion) Volume XXIV, No. 26, January 11, 1995

"It feels like somebody's stabbing you with a knife," says Dorothy Simmons Hardy of Philadelphia. "It's a constant, sharp, achy pain. It never lets up. The pain is real severe, and it's real constant."

"Simmons Hardy is one of about 80,000 people in the United States who suffer from sickle cell anemia. When the oxygen level in her blood drops too low, her blood cannot flow normally. The pain usually hits her legs, but sometimes she has arm and stomach pain. Frequent hospital stays make it hard for her to hold a job. She almost died twice during pregnancy because of sickle-cell complications, including a blood clot on the lung.

"NIH spends about $70 million a year to study sickle cell anemia. Statistically, relatively few people suffer from this disease, compared to other ailments. But American scientists say a disproportionate amount of energy is focused on sickle cell because they consider it a great medical challenge: In 1954, Linus Pauling won a Nobel Prize for identifying sickle cell anemia as a genetic defect, making it the first genetic defect found in humans.

"…Philadelphia businessman and ozone proponent Jim Caplan has convinced doctors in Cuba to treat sickle cell anemia with ozone, much as the Germans are treating diabetics.

"At the moment, the best American doctors can do is put their sickle-cell patients on antibiotics to stave off infections while the blood circulation is limited, and give them painkillers to help them ride out the painful crises. But **Cuban doctors have documented that in their research, the use of ozone cut the length and the severity of painful episodes in half.** So far NIH grant reviewers have opted not to put research money toward this treatment.

"Too little research, too little profit.

"What if scientists came up with a stunning medical breakthrough that might be able to help doctors treat a variety of diseases, but nobody had the chance to get rich from it?" −End of article.

Fifty-five adult sickle cell anemia patients were studied by Dr. Sylvia Mendez, et al., each suffering from a painful crisis. Notice they used only the less effective rectal ozone insufflation ozone delivery method (only *once* every 14 days!), yet it still took only *half the time* to resolve the crisis compared to controls. Also, *skin ulcers, which are common among sickle cell patients, completely disappeared* among the patients receiving ozone.

When blood oxygen levels drop, the blood cells lose their normal elasticity and shape and then shrink, or get 'sickle-shaped' due to the substitution of glutamic acid by valine in the amino acid chain. This reduces the ability of the blood cells to carry and deliver oxygen. The resulting lack of oxygen produces the painful crisis; infarction, abdominal and/or muscular pain, ulcers, etc.

Notice: When the blood oxygen levels drop, the blood cells sickle, and the people go into the painful crisis. Duuuh! Why not keep the blood oxygen levels up any way you can, and then where will the crisis come from?

Research shows medical ozone significantly raises the blood and tissue oxygen levels, it increases the rate and capacity of oxygen absorption by the blood cells, it prevents blood cell clumping, and it increases membrane permeability and flexibility while scavenging free radicals. Together, all these properties offer the sickle cell sufferer a potent and safe arsenal. Imagine the results available when the ozone or oxygen supplements are taken more often than they were in the Cuban study. It turned out the research showed one ozone application every two weeks was enough to prevent the crisis from starting. Of course, I always advocate daily active oxygen supplementation for everybody anyway.

Dr. Gomez designed the Cuban study and then went to Italy, defected, and moved to Spain. Due to his defection, back in Cuba he became a 'non-person,' and many of his plans and their funding that would have supplied ozone generators to rural clinics all dried up. Despite the politics, the results of this simple low cost Ozone Therapy were so impressive that the Cuban Ministry of Public Health (MINSAP) later approved Ozone Therapy as a standard treatment for sickle cell anemia throughout the country.

In America, Jim Caplan first tried to get the study done at the Philadelphia Children's Hospital. They strung him along for years while seeming 'concerned,' to his face, but occasionally he heard anti-Semite comments, "Don't your people have their *own* disease?" from the black doctors he was trying to help. Sufferers be damned, their pride wanted a black man to be first to solve the problem, and Jim was white. They never did the study. That's how the Cubans ended up doing it when Jim had to take it elsewhere. Influential blacks and the Congressional Black Caucus were no help either, they all ignored without any response the studies he sent them, and this included people with the disease in their own families. Jim relates these human weaknesses as being a 'very sad' state of affairs.

> **The only thing stopping our widespread use of ozone is human ego.**

Then, in a major event for us, on March 25th, 1992, James A. Caplan's Philadelphia company CAPMED received **'Orphan Drug' Designation** from the U.S. FDA for research into the use of ozone/oxygen for the treatment of Sickle Cell Anemia. "Congratulations on obtaining your orphan drug designation." The way is therefore now paved to get ozone approved for sickle cell, but we need people to step up and supply the funding.

Incidentally, black people have the best shot at getting ozone FDA approved in America. This is due to their being perceived as a segment of society that's fatal to ignore the wishes of–when they get their organizations together. Sickle cell and AIDS have hit the black communities hardest of all. Imagine if the black communities get behind this book and demand access to the therapies. Now imagine what would happen if they did so while holding all the foreign studies in their hands that have already proven safety and effectiveness on humans! Black movie stars and politicians could then give safe non-sectarian and non-political humanitarian aid back to their communities by opening ozone clinics where the 'The Man's' liquor stores are now.

Such an organized effort of combined political will would be able to circumvent any delaying tactics thrown up by the guardians of the status quo, including both the misinformed doctors and politicians alike. They will start with "First you must test small animals for a few years," and then "O.K., now test large animals for a few more." Of course this would waste another 10 or 15 years. Medizone and The Canadian Armed Forces already did $2 million worth of animal studies proving safety and efficacy.

On the personal immediate need level, it is entirely possible that since ozone already enjoys FDA 'orphan drug' status, someone with enough clout could get an individual 'compassionate use' permit from the FDA. This would allow them federal personal ozone usage and also shield them and their doctor from any political outfall.

Stomach Ulcers

People drinking some form of Oxygen Therapy supplement have been reporting to me for years that their ulcers went away. Recently the Center for Disease Control reported that "Over 25 million Americans will suffer from an ulcer at some point during their lifetime. The good news is that most ulcers are caused by an infection with the bacterium, *Helicobacter pylori*, and can be cured in about two weeks with antibiotics." But antibiotics kill the friendly aerobic bacteria necessary for digestion, taking out the good and the bad bacteria. Active oxygen used as an alternative only kills the bacteria causing problems, the anaerobes.

The Internet newsletter *Med Tech 1 Beat* said: "Acid-reducing medications can make ulcers feel better, but will not cure the infection." The best part about this news: if bacteria cause ulcers, why can't active oxygen forms cure them, and permanently? I have met many people who told me their ulcers went away from drinking some oxygen supplement, even something as simple as a few weeks of freshly ozonated water. This is great news if you are one of the 25 million Americans, men and women, who will suffer from an ulcer in their lifetime. Ulcers send one million people to the hospital each year. And the bacteria can eventually lead to gastric cancer, the second most common form of cancer worldwide. H. pylori infection is detected with an easy blood or saliva test (see www.salv.com for one example.)

Viral Illnesses & AIDS

There are 900,000 Americans are living with HIV. The CDC says one third do not know they are infected. Worldwide, an estimated 36 million people are infected with HIV, and 22 million have died. I have seen—and doctors have reported to me—many victims of AIDS returning to health using Oxygen Therapies due to one simple fact. The laws of physics have not changed. All the bacteria and viruses associated with AIDS cannot live when surrounded by active forms of oxygen, and oxygen is harmless.

Kenya Ground Zero, Twelve Die Per Minute

A frustrated ozone technician wrote me from The Wainwright Polyatomic Apheresis Clinic in Kenya: "We forget, just in Kenya, at least 12 of God's children die from AIDS each and every minute of the day. Can you live with that thought? We collectively are allowing this to happen. We have the technology at our disposal (Oxygen Therapies), yet we do nothing to expand it on a global scale. We allow ourselves the excuses of why we can't—but there is really no excuse, is there Ed?"

"The world needs to act decisively now or AIDS could spread to countries that have so far avoided the worst of the disease," Dr. Peter Piot, the head of UNAIDS, told reporters in Cape Town, South Africa, on the 20th anniversary of the first report of AIDS.

I saw this poster in northwestern Miami after Leeda and I got lost getting off the turnpike. "8 out of every 10 people getting AIDS are black." We asked for directions, and the cops told us, "You are two very white people. Lock your doors, and do not get out of your car for any reason, even if bumped." AIDS is the background of the urban economic deprivation and crime scene. Those affected should have access to the real solutions.

AIDS History and Comment

More than 22 million people have died of AIDS and another 36 million have been infected with HIV since that first official report some 20 years ago by the U.S. Center for Disease Control. The report came in a nine paragraph article about the curious deaths of five gay men. "Looking back, no one could have imagined the extent of the problem," observed Dr. Peter Piot, head of UNAIDS, yet he and other global health officials fear that the widespread disease is still in its early stages, and therefore could spread to countries that so far have avoided the worst of it.

"When you look particularly at Asia, at Western Africa, at Eastern Europe, it is clear that we are really at the very early phases of the spread of HIV," Piot explained. So far more than 70 percent of those with the virus that causes AIDS remain in sub-Saharan Africa, the world's poorest region. Yet global health officials note that India, as one example, with one billion people, could suffer the spread of the disease as rapidly as it has throughout South Africa, where 11 percent of the country's 43 million people are infected.

Yet in the 20 years since that first report an estimated 58 million people have contracted HIV, some 22 million of them fatally. Efforts to develop a vaccine persist, yet a cure is still only a dream for the infected. "This is now, without any doubt, the largest epidemic in human history, and we are certainly not at the end of it," he concludes. "The devastating pandemic of AIDS has proven how quickly a disease can spread globally . . . . teaching the world a lesson in what can happen when governments react too slowly."

U.N. Secretary-general Kofi Annan has asked wealthy countries to contribute from $7-to-10 billion annually in the effort to help prevent and treat AIDS in the developing world, with some half of the money to be earmarked to fight AIDS in Africa. Bill Gates gave millions to India. Will ANY of it be spent on the proven Oxygen Therapies?

"Particularly," Piot elaborated, "We must overcome the poor healthcare infrastructure in many of the worst-infected countries, the refusal of many to get tested for HIV, the costs associated with the care of those infected. and the high price of the AIDS drugs."

"Unfortunately," Piot points out, "the primary focus on the price of antiretroviral drugs oversimplifies the problem. [It actually does more than that. By focusing all discussion on drug price, the public agenda is steered away from discovering and implementing the Oxygen Therapy solutions.]

More than 25 million of the 36 million African region people are infected with HIV. In 2000, 2.4 million people died in the region. Botswana has the highest rate of infection–35.8 percent. Life expectancy has been cut from 69 down to 44 years. And these are just reported cases! African villages aren't big on infrastructure. The questions is, with all this infection going on around and within America, call it AIDS, or whatever, how can we remain in denial and do nothing about the answers available in this book?

Modern Plague

The 1918 influenza pandemic swept through the United States Army and spread to slaughter more young adults than did World War I. Has the situation changed? No. Disease is rampant on a worldwide scale. Many hundreds of millions suffer from every disease imaginable. To add injury to insult, AIDS, the new slow spreading disease complex on the block, has rapidly moved into the number one position as a mercenary killer. It respects no one, has a rapacious appetite, and quietly infiltrates the lives of people, leaving its insidious mark.

"The unrelenting spread of **AIDS kills 300 people every hour.** AIDS is no longer a disease, it is a disaster," reports Peter Walker, director of disaster policy for the International federation of Red Cross and Red Crescent societies. The mayhem that what we call AIDS produces is beyond comprehension. Yet the evidence of ozone's efficacy in the treatment of this merciless disease is staring political bodies in the face. Let's look closer at the evidence.

HIV Virus Can't Live in Oxygen!

Scientifically speaking, the Syracuse University study proved what the ozone experts all know, that HIV virus and any other bacteria and viruses can't live in active oxygen. The human body and its' blood components do quite well with ozone at the proper levels! There's your answer. Maybe you don't believe me. "How could it be so simple?" you ask.

Dr. Bernard Poiesz and associates from the State University of New York Research Hospital in Syracuse repeated an ozone study 15 times. They put ozone through a hollow fiber delivery system, interfaced it with factor 8 blood infected with the HIV virus, and achieved a 97-to-100 percent cure ratio. **The ozone was non-toxic to normal healthy blood components. The ozone was deadly to viruses, and the ozone was completely safe.** It's that simple. HIV was surrounded with active oxygen (in this case in the form of ozone). If HIV and all other viruses and bacteria can't exist long in active O_1 oxygen, and you surround them with active forms of O_1 oxygen, what happens? It's so logical, isn't it? The scientific documentation proving this is all over the table.

I wonder why this announcement of the way to stop anyone ever getting an infection from a blood transfusion—and possibly the end of AIDS itself —wasn't headlined or even mentioned in your daily newspaper? I announced this study back in 1988, in my first *Oxygen Therapies* book.

Syracuse University proved ozone can completely *inactivate blood viruses 97-to-100 percent of the time*, and yet not harm the blood. Dr. Poiesz's studies were published in the peer review Journal of The American Society of Hematology.

By the way, the earliest known use of Ozone Therapy specifically on AIDS was in the '80s by Dr. George Freibott, ND, as also published in my *O₃ vs. AIDS* book.

I have witnessed innumerable AIDS and other patients receiving ozone infusion therapy. When they start out, just like cancer patients, their blood is filthy, diseased, and so empty of oxygen that it is almost black in color. Ozone was put into them for a few weeks and their blood turned back to a bright cherry red color. It became fresh, clean, and full of life giving oxygen.

Besides the reversal cases from Africa using Hydroxygen or peroxide in the front of this book, I have seen people sero-convert to HIV negative; and, even more importantly, lose all secondary infections from use of ozone. Individual cases will always vary. But they stuck to a FULL protocol: getting IV at the right dosage daily, at the right concentrations, and combining it with other significant modalities. People who have never tried it or only just 'dabbled' in it end up being the only nay-sayers. Those who use it correctly continue to come back for more-because they live the benefits within their own bodies.

T-Cells

It is essential to understand the following if someone embarks upon a course of Ozone Therapy and has AIDS. If someone with HIV is tracking his or her T-cell count and that T-cell count has been steadily declining over time, more and more lymphocytes have been eaten up by the AIDS complex. Stem cells that make T-cells have become infected, and the T-cells become infected as well. When Ozone Therapy is applied, what's going to happen to the T-cell count? If the T-cells are infected, zoom, down their numbers go, right? Because ozone attacks only infected cells.

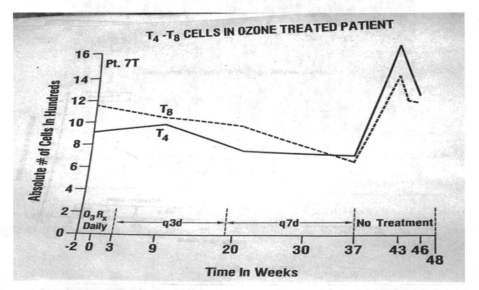

You worry, have I lost my T-cells? You start thinking, is Ozone Therapy a mistake? Has it destroyed my immune system? No. Good ol' benign Ozone Therapy is doing its job. It is spotting infected cells and it is killing the feeble, weak, diseased and dying anaerobes that are plugging up the system.

The diseased non-functional cells are only in the way. The no-good infected T-cells are cellular trash. They're evil bio-factories stealing your life energy to replicate more parasitic viruses while taking up the spaces where your body should be noticing that it lacks enough immune cells and then signaling itself to supply more new ones.

But the infected cells are only there simply because they must survive in an environment with a lack of proper oxygen levels. So, if the defective cells are automatically eliminated, your T-cell count will drop, but the test gave a falsely high count to begin with, as it counted all the *sick* cells *along with* the *healthy* ones. After a cleanout of sick cells that leaves only the healthy ones in place, the numbers are finally accurate. You see your *true* numbers after a cleanout.

So during the course of therapy the count dips and then rises, and after so many immune cells are no longer needed for a disease that has abated, the count will drop into the normal range. You'll probably freak out if you take this therapy and you or your healthcare provider don't understand what's *really* happening. Being informed will enable you to stick with the Oxygen/Ozone Therapy.

In time, if it can, the body replaces all these missing T-cells with stronger, healthy ones. Most report their T-cell counts go up after ozone. If the cells that make the T-cells are destroyed by infection, then you can still take hope because I have met people who came to ozone after falling to *zero* T-cells! Their T-cells are gone, but amazingly they reported to me they use ozone as a replacement for their immune system!

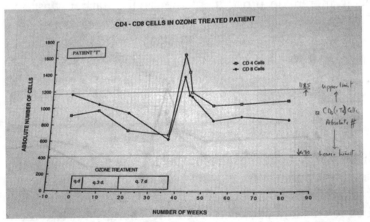

Flood Your Body With Oxygen, Therapy For Our Polluted World

They said they have to continue to get bi-weekly ozone booster treatments to use as a supplement to, or outright replacement for, their missing disease destroyed immune systems.

The people I interviewed who used the correct protocols generally saw a doubling and tripling of the T-cell counts within a matter of months. But keep in mind these were usually people who had something left to repair and started the therapy with greater than 300 T-cells.

Starting with less than somewhere around 300 T-cells has historically meant that it will take a longer time to work on you when you have AIDS. And if you're less than 300 plus T-cells, and you've taken AZT, DDI or put some other pharmaceutical thing in there, poisoning the cells further; it's going to take even longer for the body to clean it all out and turn itself around. So there's no guarantee with any therapy, even ozone, but you can see the function and direction of the process here.

Some of the drugs people take have often forced sort of a rushed *artificially induced* production of immature immune system cells. This apparently causes T-cells *numbers* to rise, but the T-cells themselves do not work well for lack of maturity. At least that's what was explained to me, so be watchful in this area. So when you take Ozone Therapy the unhealthy useless T-cells giving phony high counts are removed as well.

The drop of the T-cells is actually showing you that the ozone is working well, doing exactly what it is supposed to do—removing useless, dead, defective and infected T-cells!

Virus Counts Go Up

Another interesting phenomenon we often see is the *virus counts* seem to go *up* for a while in the people when they first get on the Ozone Therapy. The same former male prostitute I mentioned earlier was on an 'AIDS cocktail' of drugs for six months. He was a physical mess and the drugs and lifestyle were killing him. He'd had pneumonia three times, shingles twice, candida, thrush five times, sinus infections six times, a bladder infection, genital warts for four years, anal fistulae constantly draining for four years, night sweats twice a month, a rash, anxiety, depression, dementia to the point of being imprisoned and sent to a mental institution, and low energy with slow recovery after exercise.

Although his virus count test came back 'undetectable,' because of the drugs, his T-cell count at one point was only 7. He was using cigarettes and pot and felt so sick he decided to stop taking the drugs. At this point he switched to substantial direct IV Ozone Therapy for two months. Using a syringe loaded infusion pump he slowly injected a total of 210, 10cc sized syringes of 30 mcg/ml and sometimes 70 mcg/ml strength ozone directly into his bloodstream during that time.

In a classic detoxification response, he had the major healing crisis, during which time he coughed up yellow mucous and his body broke out in shingles and a rash all over for a few days. The dead husks of all his freshly killed microbes were causing uncomfortable temporary symptoms on their way out. This is why Recirculatory Auto-Hemoperfusion is better than direct IV; all the symptom causing dead husks and oxidized toxins are filtered out **outside** the body. Anyway, to his amazement, after his healing crisis all his above listed diseases and symptoms he suffered with for so many years simply went away!

Of course, to allow his Oxygen Therapy to work, he stopped smoking pot, started taking Hydroxygen, and began eating fresh foods and spirulina, while supplementing his meals with vitamins and minerals.

Here's the interesting part. After only two months on ozone his PCR 'virus' count went from '0' before ozone to 'over 500,000.' What the heck was going on? The virus doesn't replicate that fast. He gets on the ozone and starts healing his secondary diseases, and feeling great, but his "HIV gene fragment count extrapolated by a mathematical equation" shoots straight up. The problem was once again, the test. The test was taken in the middle of his healing crisis, while he was still in the middle of getting daily ozone injections inactivating microbes and stirring everything up. The immediate tell-tale indicator of what was *really* going on was his blood, it was still *black*. If the blood is black, you're *not* there yet. His blood was still full of exiting, oxidized garbage.

Just as it was supposed to the PCR test dutifully counted all the dead gene fragments, maybe of the HIV virus and maybe not, that ozone had dug up from deep in the tissues while killing them. He had so many problems it was taking the body a long time to clean out, and the debris was being counted. The body was still dumping the fragments into the blood solution for elimination, exactly where the test would find them. A PCR 'too high to count' number of dead "viruses" was actually a great sign in this case! It showed a lot of detoxification going on. Only much later, after detox and all the dross has settled down and been flushed out, will the test be more reliable. But the doctors won't realize this. In a classic response they will look at a detox and assume ozone causes HIV to worsen. They must be taught.

Remember, a well oxygenated body throws *all* dead bacteria and viral debris up out of the background and into solution, and the gene fragment presence PCR tests count live *and dead* particles. The virus count numbers might artificially *seem* to jump way up when you first test immediately after a course of Ozone Therapy because you're still washing out piles of fragments.

Depending upon how sick the patient was when starting on ozone, one may get cleaned out the first time around. If there's still a problem, one may need to start over and do another six week course of injections after a period of rest and rebuilding. Historical evidence shows each time it gets easier and shorter.

Some get the therapy and feel so good that they leave it way too soon. If you're not real sick, you might start feeling better right away from being oxygenated again, and your blood may test negative quickly. But you must know how and when to interpret the tests. Keep in mind the tests given *after* you have been away from the deeply cleansing Oxygen Therapy for a while are the true picture of your status. The above two typical false test results are why you cannot judge the Oxygen Therapies with immediate tests. Think long term for testing. In the immediate term after the healing crisis ask, "Do I feel any better?"

Another thing, if you drink or use dope, forget everything I am saying, because you will only get palliative benefits. Substance abuse negates Oxygen Therapies by using up the oxygen. Period.

Still smoking? How many people with AIDS or cancer smoke?

"Blood is a vehicle for delivering oxygen and nutrients to our body's tissues and organs. Without it they die. Our blood vessels (circulatory system) are the piping highways in which our blood flows. The inside of each healthy blood vessel is coated with a thin teflon like layer of cells that ensure smooth blood flow. Carbon monoxide from smoking or second-hand smoke damages this important layer of cells, allowing fats and plaque to stick to vessel walls. Nicotine then performs a double whammy of sorts.

"First, each time new nicotine arrives in our brain in causes a release of adrenaline which in turn immediately releases stored fats into our blood. Yes, the extra food we smokers eat during our big meals is converted to fat and stored, and then pumped back into our blood with each new puff. It's how we were able to skip meals and it's what causes many of us to experience low blood sugar levels when trying to quit. In fact, many of the symptoms of withdrawal - like an inability to concentrate - are due to nicotine no longer feeding us while we continue to skip meals

"These heavy blasts of stored fats, being released by nicotine, stick to vessel walls previously damaged by carbon monoxide. It only gets worse. We've recently learned that nicotine itself, inside our vessels, somehow causes the growth of new blood vessels (vascularization) that then provide a rich supply of oxygen and nutrients to the fats and plaques that have stuck to our vessel walls. This internal nicotine vascularization (vessels within vessels) hardens a smoker's blood arteries and veins and further accelerates their narrowing and clogging.

"We each have a rough sense of the damage we've done to our lungs but what degree of clogging has already occurred in our blood vessels? How long do we have before our coronary arteries - that supply life giving oxygen and nutrients to our heart muscle - become 100 percent clogged? It's called a heart attack and the portion of the heart muscle serviced by the artery will quickly die. How long do we have before our carotid arteries - supplying life giving oxygen and nutrients to our brain - become 100 percent clogged? It's called a stroke and the portion of the brain serviced by the artery will quickly die. The damage being done isn't just to the vessels supplying blood to our heart and brain, it's occurring - to one degree or another - inside *every* vessel in your body. It effects everything from blood vessels associated with hearing to the skin's blood supply that shows itself in wrinkles and early aging." – From *Quit Now*.

For many years the tobacco companies had to report the ingredients in cigarettes to the feds, who promptly locked them in a safe so no one could see them. Lawsuits have brought out some of the truth. Here's a highpoint list of those addiction satisfying poisonous gasses smokers are sucking down that compromise their immune systems.

Tobacco Fact Sheets – Cigarettes – What's in them?

Imagine a road being coated with tar, a similar process happens to the lungs and wind pipes of smokers because of the tar in cigarettes. I remember seeing gross photos of black gooey shriveled smoker's lungs in high school science class. Do you think repeatedly taking these poisons into your body every day might be perpetuating any of your health problems?

Some of the Chemical Poisons and Gasses in Tobacco Smoke

- Acetone found in paint stripper
- Insecticide residues e.g. DDT found in insecticides
- Naphthylamine*
- Phenol
- Toluidine*
- Butane found in lighter fluid
- Methanol found in rocket fuel
- Polonium~210*
- Arsenic found in white ant poison
- Potassium~40*
- Toluene found in industrial solvent
- Sulphur dioxide
- Nitrogen dioxide
- Cadmium found in car batteries
- Urethane*
- Benzpyrene*
- Pyrene*
- Vinyl chloride*
- Dibenzacridine*
- Naphthalene* found in mothballs
- Hydrogen cyanide used in gas chambers for execution purposes
- Dimethylnitrosamine
- Formaldehyde used to preserve body tissue
- Ammonia found in floor cleaner

*Known carcinogens (cancer causing substances)

Courtesy the South Australian Smoking and Health Project.

Tobacco the Truth is Out There: Prevention activities for the middle school, Book 2, 1997. And, Commonwealth Department of Health & Family Services. *Smoking - What you're really smoking* pamphlet.

Wheezing is not glamorous. Too many pitiful Souls have called me up wheezing, barely able to speak between gasps for air, and begging for help. They all once fell for the lie that smoking is 'glamorous.' And kids, becoming a smoker is not a rite of passage into adulthood, that's just a sales trick playing on your strong adolescent need for acceptance and rebellion. The corporations want you hooked for life, and don't care if they induce you to turn yourself into a craving fool that keeps slavishly giving your money to the pushers. The pushers all sleep well, telling themselves that if you're stupid enough to smoke or use dope then you deserve what you get.

I know the Surgeon General said cigarettes are more addictive than Heroin, but at some point you must wake up. Reason must overcome addiction. If you're addicted, life is giving you yet another opportunity to overcome addiction. I have seen Chinese herbs that worked well with oxygen to get smokers through withdrawal effortlessly.

Does HIV Cause AIDS?

It's been a bit of a balancing act writing this book for readers on both sides of this AIDS causation fence. Up to this point I wrote for the general population and media employees with all their assumptions formed by majority opinion. To help them I must. But the Emperor may have no clothes. Ultimately, it's a wash anyway because the Oxygen Therapy solution stays the same either way, no matter what the cause.

The PCR test is often used, but the government and the manufacturers say it is not definitive. The Center for Disease Control in Atlanta, Georgia, stated that AIDS is defined as when you test positive on the Western Blot, and ELISA diagnostic tests. [By the way, there are 62 factors other than HIV that can cause false positive readings on the antibody tests.] The CDC also says that in order to have AIDS, you must also have at least 1 out of 29 diseases (It used to be you had to have 7 out of 26): 'secondary diseases' they call them.

Some things to keep in mind in their criteria: There has never been a scientific study proving that HIV caused AIDS. It was simply announced that it was the '*probable*' cause of AIDS. Two weeks later the 'probable' was dropped and all the newspapers said, "HIV, the cause of AIDS," every time they ran a story.

At the most, according to the up to date advocacy groups around the front line in America, they now call HIV either the 'promoter' of the disease or a 'co-factor or a component' of the disease: not the actual cause of the disease. It's all in the literature.

"No one had ever proven that HIV causes AIDS." —Kary B. Mullis, Nobel Prize Winner in Chemistry, 1993. He invented the PCR test.

HIV grows harmoniously with the cells it infects. *"The failure to kill T-cells even under optimal conditions is the Achilles' heel of the supposed AIDS virus."* —Duesberg. *AIDS Acquired by Drug Consumption.*

"As a scientist who has studied AIDS for 16 years, I have determined that AIDS has little to do with science and is not even primarily a medical issue. AIDS is a sociological phenomenon held together by fear, creating a kind of medical McCarthyism that has transgressed and collapsed all the rules of science, and has imposed a brew of belief and pseudoscience on a vulnerable public." —Dr. David Rasnick, Designer of Protease Inhibitors, SPIN magazine, June 1997.

The point of my explaining all this to my readers is to raise enough doubt so that you don't prematurely end your life if you have been told you have

AIDS. How many sick people are being convinced they will die and they must take poisonous drugs by doctors and family that are full of loving compassion and possibly absolutely wrong? Do your own research.

Whether HIV causes AIDS or not, we can still use the readily available PCR testing to measure gene fragments as we continue with a course of Oxygen Therapy and afterwards. The PCR test may or may not indicate HIV accurately, but it's a great general indicator of how clean you're getting the blood and tissues. If ozone is burning up gene fragments and microbes including HIV, then in my opinion—an opinion earned the hard way, by interviewing thousands of Oxygen Therapy using patients—we can bet you're probably eliminating ALL other harmful bacteria and viruses as well, and that the real cause of your malady is among them.

As the PCR count goes up and down, so are the levels of the live infectious agents and dead fragments of all the other infectious materials as well. Ozone and the other active oxygen species do not discriminate between live or dead infectious agents, they go after them *all*. Naming the disease or the infecting microbe isn't necessary to the treatment because the treatment stays the same.

Nobody dies from AIDS. They actually die from suffocation at the cellular level, usually after taking lots of poison. What is the quickest way to end suffocation? Give them oxygen. What's the only way the body can detox? By using oxygen to oxidize toxins. But there must be enough life left in the body for it to have enough time to work and have a way to repair itself.

> *"Several Aids, Hepatitis C, and other viral sufferers reported to me they were willing to use rectal insufflation or portable ozone steam sauna applications, but not the intense Direct IV or RHP plus total body saturation and dietary and lifestyle work that is always necessary with such serious diseases. They didn't want the bother, and they wasted valuable time in denial while the viruses multiplied. One thought being a vegetarian would somehow kill the viruses? True, they all felt somewhat better with the minor ozone therapies, but it wasn't until they were confronted with disappointment in their actual test results or persistant symptoms that they really listened to me, got serious, and did Direct IV or RHP at least every other day for several weeks. That's the true point after which they saw the tests and symptoms change. How deep into this you have to go depends upon how sick you are when you start the journey."* —Author.

Some AIDS Case Histories
A Letter from Dr. Frank Shallenberger

August 3, 1992
Mr. Ed McCabe

❶ RE: Brief Summary of Four Cases of AIDS Treatment with Intravenous Ozone Therapy.

Dear Mr. McCabe,

Four male patients were treated with two weeks of daily ozone infusions each lasting one hour. These patients were also eating whole fresh foods, taking nutritional supplements, abstaining form recreational drugs, and practicing safe sex.

Patient #1 - Diffuse Kaposi's Sarcoma for nine months.

Patient #2 - Chronic Diarrhea, Chronic Cryptosporidium Intestinalis, Chronic Fatigue, Weight Loss.

Patient #3 - Chronic Fatigue. Chronic Thrush.

Patient #4 - Chronic Fatigue

In addition, a fifth patient was treated with an eight day course of similar therapy who presented with chronic rash, chronic fatigue, and neuro leukodystrophy related AIDS.

Summary of Results

We now have a four-month follow-up on the first four patients. All of them are in complete clinical remission, including the patient with Kaposi's.

The fifth patient with neuro-leukodystrophy is now 50 percent improved one month after treatment.

No significant improvement in the laboratory abnormalities associated with AIDS has been noted to date. No complications were observed.

Comments:

This therapy is not only safe, but also easy to administer and inexpensive. In view of the results, it is imperative that it be further explored.

More Documentation:

"So these people had the classical preliminary results of having their secondary infections go away. Some of their results were reported in two weeks only; the minimum is six weeks. We had a four-month follow up on the first four patients. All of them are in complete clinical remission, including the patient with kaposi's. The fifth patient with neuro leukodystrophy was 50 percent improved one month after treatment. Because it was only two weeks into it, he saw no significant improvement in the laboratory abnormalities. In other words, the blood hadn't changed that much; it won't in two weeks. No complications observed. This therapy is not only safe, but also easy to administer and inexpensive. In view of the results, it is imperative that it be further explored."

Sincerely, Dr. Frank Shallenberger, MD, ND

Ozone Paper

Dr. Shallenberger also published this paper: *Selective compartmental dominance: an explanation for a noninfectious, multifactoral etiology for acquired immune deficiency syndrome (AIDS), and a rationale for Ozone Therapy and other immune modulating therapies.* It was published in the peer reviewed journal Medical Hypotheses 1998 Jan: 50(1):67-80 and it is available on the net at Medline.

In his publication Dr. Shallenberger demonstrates "a substantial improvement in the immune systems of all the patients treated, despite the fact that each one had a very advanced disease, and none were on anti-viral therapy. He also wrote of bypassing HIV theory and explained that lots of people may have AIDS because of a chronic feedback loop in the immune system which is turned on by chronic 'high-dose antigenic challenge' (living badly). This results in chronic suppression of the cellular immune compartment, and he suggests a rational course of nontoxic Ozone Therapy can potentially reverse cases in the earlier stages.

Dr. Shallenberger is being modest and politically reserved. I know he wrote that because he's repeatedly seen ozone do exactly that with his own eyes. His work parallels (from a physical viewpoint) my own informal thinking that indicates ultimately AIDS starts with an error in T-cell production triggered by errors within the emotions and mind of the patient. I have learned that most disease manifests on the inner energy fields before making its way out to the physical body.

Perhaps the person has been programmed by the group consciousness going around that promotes 'AIDS will kill you,' and the person also has a genetic weakness in the T cells, and also is sensitive to emotional distress. This setup opens the way for succumbing to infections that match the inner invisible energy imbalances. Such a person, when told they "have AIDS," may turn their back on life, and start taking poisons that slowly destroys him or her.

I was recently told by someone in 'the scene' that there are actually underground gay 'parties' and videos around where AIDS infected people kill themselves. Lunacy, error, voyeurism, and ego all wrapped up in fear. Can you see how dangerous withholding the information in this book can be?

Woman Had AIDS for Eight Years

Dr. Mayer in Miami was one of the people who first had a patient turn HIV negative from ozone injections. I had her medical lab reports. Her name is Lisa, and she had AIDS for about eight years. She had herpes, shingles, chronic fatigue, dipsy, and viral candida. She started ozone in December 1985. Her chronic fatigue and dipsy then tested negative. Her candida left after 10 treatments of ozone, and her herpes went after 21 treatments of ozone. Her energy level was a 2, and then her energy level soared to 10. She began thinking clearly and she started reading again. She's no longer concerned about having a life-threatening illness. She was trying the ozone for the HIV, yet her herpes, shingles, chronic fatigue, dipsy, and candida all went away.

On December 1, 1991, at an Australian lecture I held up blood tests showing that first she was positive for HIV, as per Western Blot and ELISA tests: SmithKline Beecham Laboratories. And then showing her HIV negative, as per Western Blot, and ELISA with a re-testing of each one.

Lisa had AIDS for eight years and then in December of 1991 she went 'non-reactive,' was re-tested three times on the Western blot test and three times on ELISA and came up HIV negative again. This is a documented case I reported back then of someone turning HIV negative.

Two Medical Studies of Women with HIV

Two girls came out of a study. I had their medical blood records. They were HIV positive. They had AIDS and a lot of the secondary symptoms. They went to a clinic where they had ozone injected into them every single day for 60 days. They did other things as well. They looked at their diet. They had fruit juices. They turned away from anything that would harm them. They started out positive for AIDS, and their HIV virus and P24 count was positive. And when checked after the ozone, the new blood medical record from a standard laboratory said 'HIV negative.' The P24 counts were negative. In how long? Just 60 days. Do you see this in the news? No. I sent a press release to all the major evening news shows with videotape of these women standing up in front of a crowd saying, "Yes, this is my record and this is what happened to me." But it was ignored.

HIV Infected Man

I had the medical records of a man from New York City. In April, 1989 he was HIV positive. He had fever, night sweats, malaise, muscle aches, and swollen glands. It was recorded on 8/6/91 that he's completely asymptomatic. In other words, all these diseases are gone. He's HIV negative on the Western Blot. I'll accept that this is a huge success for Oxygen Therapies.

AIDS Sufferers Helped Each Other, Co-op Formed

I traveled to San Francisco to videotape a number of men who formed an ozone self-treatment co-op in the early '90s. Some had read my book but they had not consulted with me beforehand. None of them had the money to get an $8,000 German generator, so they all chipped in to get one. They bought a medical ozone machine from Germany, rented a little room, passed out keys, and taught each other how to do rectal insufflation. There were about a 160 men in the Bay area that treated themselves with ozone. One told me that he is PCR negative now. He showed me his laboratory proof (see below), including the technician's disbelief that anyone with AIDS could test negative. He would keep testing, and the technician would keep writing on the tests 'PCR Negative. Technician requests additional testing!'

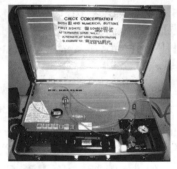

He explained how he had started self-treating in the co-op with everyone else, but found the group atmosphere less than healing—so he got his own machine. He gave himself rectal ozone weekly for about a year and a half. He came up PCR negative, which means no trace of any of the seven nucleotides that form the AIDS virus could be found in his body. He was delighted with that.

Many of the people trying ozone in the group (and I am sad to report I have seen this phenomenon over and over) started gossiping among themselves and venting all their anger at getting sick, and how life was 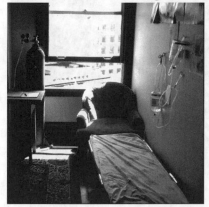 unfair, and in general let all their negativity come out. They told each other how it was impossible to cure AIDS saying, "This won't work, but I'm trying it, sort of." Naturally, negativity and halfhearted attempts accompanied the worst results.

The co-op people only used the minor therapy of rectal insufflation (because that's all they knew about) after studying Dr. Carpendale's preliminary San Francisco Veterans Hospital work. Besides using a minor Ozone Therapy in a situation that really called for a heavy duty Oxygen Therapy, most only took the ozone occasionally, so I knew when I heard this later that the results would be erratic. And, rather than stick to the known protocol, they kept clubbing, getting high, drinking, picking up strangers and re-infecting themselves randomly. They wanted the party and the cure. You can't have both. When they didn't all get immediate cures, they started disrespecting ozone and stopped taking the treatments. Their egos got in the way, they chose the party over healing, and they took themselves out.

Four of the men said that within three-to-six weeks of doing the ozone on a daily basis their T-cell count first stabilized and then doubled. Almost immediately they had an increase in energy. They all felt fantastic right away. Their night sweats stopped immediately. All their secondary infections completely went away. So these men were greatly encouraged. These are the ones I videotaped.

And they were not doing anything else special. They were not taking the antioxidants that most doctors say should be given with Ozone Therapy to ensure that the body has enough of the antioxidants in it. They were not eating the live foods that most experts say helps the immune system work well. They were mostly living their lives as they were before, but they also were not foolish as the others were.

They did the ozone consistently and they didn't feel sick anymore. Many of them still tested positive for the HIV virus, but they weren't under a doctor's care, and they weren't doing the strict protocols necessary for the best results either. They're just living their lives and running around. Yet, even under those conditions, one of them who was eating a more natural diet and has become PCR negative after doing ozone long enough.

strument Three Part fferential	% of Leucocytes		Absolute No. Cells/cu.-mm.	
	Patient	Normal Range	Patient	Normal Range
Lymphocytes	31	(25-40)	2542	1,050-4,160
Monocytes	7	(0-8)		
Granulocytes	62	(50-77)		

Technologist Initials: GK PCR: NEGATIVE TECHNICIAN ASKS FOR ADDITIONAL SAMPLE 8/90

Doctor Reports

We also had the case of Dr. Pierre R. Steiner, who retired in the United States. He had been using ozone in his practice for more than 40 years in Switzerland. He has nothing but rave reviews for the use of ozone. Yet we always find the protagonists. Let's compare Dr. Steiner's viewpoint—learned from 40 years of ozone experience—with that of one of our country's most educated doctors from a big hospital in a major city.

During the picture taking and video shooting of the former AIDS patients whose pictures I have just shown you above, a woman came along and said the chief doctor of one of the local San Francisco hospitals—who obviously should have known better—told her, "Oxygen never leaves the bloodstream, so don't bother using it." Another doctor said to me. "Yes, ozone works fine in the blood, but how do you get it into where the virus is hiding?" Viruses can hide in the central nervous system. These doctors just don't know. They need to be taught.

We know oxygen and the active oxygen byproducts go everywhere in the body. Oxygen passes the blood brain barrier and it goes to every single cell because all cells need oxygen constantly and the body is set up to constantly deliver it to them. Oxygen is carried everywhere; so, for a doctor to assume that the ozone and active sub-species just stay in the bloodstream is lazy thinking. It goes *everywhere*. It goes out of the blood vessels, into the *lymph*, bathes the cells and goes into the cells. After it cleans and invigorates the cells, the oxygen combines with wastes and gets picked up by the lymph fluid as it comes out of the cells again attached to cellular wastes. After it enters the lymph channels it re-mixes with the blood. It washes, purifies and feeds the cells along the way. Well, all this doctor had to say after my explanation was. "Now the problem is, oh, er-hum."

Another doctor in Florida told me that he and his friend were using the same protocol. His therapy was to give the patient the ozone using just the old autohemotherapy method as the Germans do it. But then, immediately following the infusion, the patient was put into a hyperbaric chamber. The oxygen in the chamber, under high atmospheric pressure, was being forced through the skin of the patient lying in the tank. The effect of this is to drive the ozone deep into the tissues and deep into the central nervous system where the virus is hiding.

Here is the amazing thing. These men in San Francisco were doing basically home-brew rectal insufflation for a year and a half and getting great results. That's what's so nice about ozone, use just a little bit, and sometimes you might still get benefit. But this doctor (who like most, make me promise not to give out their name) combined Ozone Therapy with Hyperbaric Oxygen Therapy, and reported to me that he brought more than 200 people from HIV positive to HIV negative within around 60 days each. And that's phenomenal. Think about it again: This man figured this out all by himself and all it took was 60 days for results!

Doctor Treated HIV—In 30 DAYS using Ozone and DMSO

I was doing a series of lectures and radio and TV appearances in Florida. I came across some more doctors who had been using ozone and also were figuring things out for themselves. One reported, "The thing about ozone is, we know it works. We know it kills the virus. We know it kills bacteria and fungus. If people have cancer, they take ozone. Now, the problem with AIDS is that the virus hides in the central nervous system and hides deep in the tissue and bones. Now that's why it's hard to get with the ozone."

I asked him "What can we do to speed this up?" He thought about it and said. "I know, I'll give my patients DMSO, which is a natural solvent. I give them that intravenously on the same days that I give them the Ozone Therapy, so wouldn't it make sense that the ozone would be carried across all the cell membranes very quickly by the DMSO?" We further speculated that the IV method might not be necessary, since DMSO goes everywhere. He was having very good success using this method of treatment.

This doctor had worked out his own protocol where he was giving people DMSO. In the first day they would get three grams per kilogram of body weight. And he would give it to them orally, intravenously, or anally. That was for the first eight hours, then he would reduce it a half a gram each eight hours until he came down to one gram, and then he'd keep them at the one gram level. He said, however, "You have to do blood checking along the way. The blood has to be monitored to be sure not too much hemolysis (blood damage) was occurring." He said that using the combination of ozone with anal application is pretty much the only thing he used, and he still got results.

In another state, a therapist said he had enabled six people to turn from HIV positive to HIV negative in thirty days. So, the more I went around interviewing, the more it seemed that the treatment time period was shorter in which people reported how the ozone was able to completely eradicate the virus. It depended upon how sick they were to start with.

Most of this happened in the early '90s. Because of a raid I no longer have these documents, just the memory of them, so please don't ask for them. It's just that these things were common in my life then, and it did happen.

I remember the first TV show I did while saying I had seen AIDS disappear. A very angry young man got up from the audience and laced into me in front of everyone saying, "I have friends with AIDS, how dare you go around spreading false hope!" Right there the very people I was trying to help were calling me a liar. The alternative would be to not tell everyone what I have seen, so I could be safe.

Which would be worse, to let the unnecessary suffering continue, or be the object of scorn for trying to tell you the scientific fact that anaerobic bacteria and viruses can't live in active oxygen? Somebody give me 15 million dollars, a fair playing field, protection, a clinic, and sympathetic co-operative doctors. We'll end up with all the up to date proof you want, especially with the early stage cases of all the major diseases.

Dr. Freibott and I were talking about how in this country we have Ozone Therapy that's being used in a growing professional situation. All over the country doctors are using it in their offices, and we also have desperate people treating themselves. Faced with death, these people try it and they tell their friends, "Well, I don't care, I'm going to treat myself. I'm going to save my life. I have that right—it works." So we have this Ozone/Oxygen Therapy very strongly endorsed by a large and growing group of professionals and lay people without a PR firm behind them.

Compare and contrast that with the drug therapies routinely approved by the FDA that have no testing at all, have no history, are of dubious value, and come complete with damaging side effects, and yet are instantly all over the news. What an odd situation we have going on in the United States when doctors witness cures and are afraid they'll go to jail if they tell anyone. Of course no American ozone studies are funded or published, and that's the way some like it.

FDA Advisers Tied to Industry

By USA TODAY'S Dennis Cauchon

"'More than half of the experts hired to advise the government on the safety and effectiveness of medicine have financial relationships with the pharmaceutical companies that will be helped or hurt by their decisions,' a USA TODAY study found.

"These experts are hired to advise the Food and Drug Administration on which medicines should be approved for sale, what the warning labels should say and how studies of drugs should be designed.

"The experts are supposed to be independent, but USA TODAY found that 54 percent of the time, they have a direct financial interest in the drug or topic they are asked to evaluate. These conflicts include helping a pharmaceutical company develop a medicine, then serving on an FDA advisory committee that judges the drug.

"The conflicts typically include stock ownership, consulting fees or research grants. Federal law generally prohibits the FDA from using experts with financial conflicts of interest, but the FDA has *waived* the restriction more than 800 times since 1998.

"These pharmaceutical experts, about 300 on 18 advisory committees, make decisions that affect the health of millions of Americans and billions of dollars in drugs sales. With few exceptions, the FDA follows the committees' advice.

"The FDA reveals when financial conflicts exist, but it has kept details secret since 1992, so it is not possible to determine the amount of money or the drug company involved."

Should You Blindly Trust the Status Quo With Your Disease?

Guess what the *third* leading cause of death is? According to one of their own, it's not cancer or heart disease. Dr. Barbara Starfield of the Johns Hopkins School of Hygiene and Public Health describes how the U.S. health care system may contribute to poor health. Her well-researched article was published in the Journal of the American Medical Association

(JAMA). It has been called "the best article ever written in the published literature documenting the tragedy of the traditional medical paradigm."

Doctors Are the Third Leading Cause of Death in the U.S., Causing 250,000 Deaths Every Year

ALL THESE ARE DEATHS CAUSED PER YEAR

7,000	Medication errors in hospitals
12,000	Unnecessary surgery
20,000	Other errors in hospitals
80,000	Infections in hospitals
106,000	Non-error, negative effects of drugs

These total 250,000 deaths per year from iatrogenic (doctoring) causes, and don't include outside of hospital events! The problems are so prevalent that lawyers are aggressively looking for malpractice suits with TV advertising. Their own statistics prove very clearly that the system is just not working. It is broken and is in desperate need of repair.

My hat is definitely off to the brave, hard working compassionate healthcare workers and physicians who daily slug it out through so much suffering while handicapped by a lack of oxygen knowledge. Imagine the courage needed to keep going to work in this profession like they do, knowing what's wrong and feeling powerless to change it. But they show up to work anyway, because if they didn't show up it would be worse. We owe them a big oxy-thank you as well. I would simply like to see their talents better utilized than by repeatedly chasing avoidable episodes.

Reportedly 87 percent of the population will retire with less than $12,000 per year income! This is below the poverty level. And that's when the typical 'old age' and 'side-effects of medication' diseases strike. Right now, with our expensive and ineffective present health care system and all its faults, 90 percent of all bankruptcies are from medical bills. This tells us that instead of our enjoying the sunset years of quiet reflection, we work our whole lives only to have it all taken away at the end by the equivalent of institutionalized plunder. This plunder obviously has risen to the level of criminality.

Everybody thinks it's just him or her who had to suffer from these 'sudden illness' tragedies as part of an elderly minority. Nothing changes because the retirees are then too poor, too tired and all of their time and energy is taken up managing the unnecessary emergency crisis. And then they're gone, only to be replaced by the next customer in line. Then they too, before they realize everything is *not* as advertised, are also gone and being replaced. After reading this book, I hope you agree that the majority of this mess is unnecessary and avoidable through implementing the widespread application of Oxygen Therapies.

As it turns out, to no one's surprise, this mess is fueled by a big money game. Every politician, incumbent or challenging, from either party, has in their number three or four donator slot the medical-industrial complex as 'contributor.' The politicians do what they are paid to do by the biggest business around. A rare few Souls do what's right. The system might be too big to easily change because so many in power ignore the plight of their brothers and sisters and live off their suffering somehow; but if we all use and demand Oxygen Therapies, we might be able to move things toward the light.

Look how the new prolific TV ads for slightly changed and renamed old drugs have driven so many to demand them every time they see their doctors. They ads don't even say what they do or what they're for! The population is so trained that they demand them anyway. You got a new drug doc? If everyone else is doing it I must need it too, so hand it over!

Drug sales have skyrocketed. The ads must contain some new powerful subliminals because after I watch them even I have the uneasy feeling that there is something wrong with me and I must need this shiny new drug surrounded with all the nice people in the ad, even though I have no idea at all what it does or what it's for.

Our poor doctors have to cave in to the demand to keep their customers. What would happen if everyone seeing their doctor (*you!*) demanded that their doctor research Oxygen and Ozone Therapies? And while you're at it, why not leave a copy (or a few cases) of this book with your healer so they have all the official study references at their fingertips? By putting the major English language references (to date) all in this one place, I have made it real easy for them to find the time to research the literature.

Chapter 21

Mr. Oxygen's™ Crown Jewels of Health™

As I told you, and will keep telling you, every problem has a solution. Since this is a dualistic universe, our lives exist within the dynamic tension of opposites where nothing exists without its exact opposite as well. You have a problem? There has to be a solution, somewhere. But, the trick is, you may not find it where you're looking. Open minds win.

Disease does not come from a lack of medicine.

As I set out to find in the introduction, here are the most common worldwide mass tested practical and easily attainable winning solutions. The best of the best, as promised. Our 'Holy Grail' of what causes the most health.

Internationally, these methods consistently gave the best real world results. My tallying of the results was conducted without any theory, or political, commercial, or scientific agendas involved. The only drawback for some is that most of it was necessarily delivered to me verbally. But if you look hard enough, you can find lots of proof backing up every single thing I tell you. I am only one man, with only so much time in the day. Next time around, maybe someone who wants to pay the world back for his or her own good fortune will grant me millions of dollars and several specialized support staffs to write and catalogue the information in detail.

Too many think their answer lies in chasing the ever-changing latest 'fantastic' product pumped up by the "fastest growing sales organization ever." I have watched these fads come and go since the '50s. Unfortunately, once someone is loyal to a particular treatment or particular healer they sometimes ignore the truth of what I have to say. The neighborhood and national 'alternative' healers are sometimes the most closed minded. You would think the opposite would be true. They have their kingdoms and they are enjoying fame. Their egos are jealous if someone else gets attention, so they close their ears. That is exactly how it was when I started.

I have watched them initially think what I teach is just one more fad like the 16 they have already been through. But what I teach is the bottom line, and the beginning, and the end. It doesn't get more 'basic' than this.

Not too surprisingly to me, it turned out that t he top solutions of our time for our present health problems logically follow the old naturopathic principals. "You are a living, breathing part of Nature, so the more highly oxygenated, mineralized, hydrated, and clean you get, both externally and internally, the healthier you usually become over time." A bag of dirty fluid (you) works better when cleaned out and well fed.

This is not medical advice and individual results will vary, but in my experienced opinion I have seen most people being healed of the most diseases the most often when they use these methods. Since they are the best of the best, I call them my:

'Mr. Oxygen's Crown Jewels of Health'™

> **Oxygen**
> **Minerals**
> **Water**
> **Enzymes**
> **Colon**
> **Light, Sound & Other Energy**
> **Emotional, Mental and Spiritual Balance**

1.   **OXYGEN.** (Air, active oxygen, and ozone). Health always starts with lots of fresh active cellular oxygen, the body cleanser and immune booster. Again, doubters hold your breath.

Trying anything else before getting your cellular oxygen levels back up is wasting your money and time. You can try active oxygen supplements and treatments for a while to see what can quickly and inexpensively be repaired first by only this simple expedient.

You live here, and you may look pretty, but since you do live here your fluids have been—and are continuing to become—increasingly filthy inside. Don't be embarrassed; everybody's fluids and tissues are. You don't have to take my word for it. Look at what's in your own blood under a darkfield microscope. You will then easily prove this to yourself.

Burn the garbage up with oxygen, clean it out with oxygen and water, and rebuild yourself with oxygen and minerals! Pump up the oxygen so the bugs die, the toxins flush out, and the background pH changes. That gives the anaerobic disease causing germs and bugs nowhere to live, and no garbage to live on, and there are no toxins to harm you. What's left then to stop you from getting your energy and health back? Hint: The rest of this list. And consider using any of the products I mentioned.

2. MINERALS. (Food). You must have an abundant daily supply of the basic building blocks of plant softened and ionic full spectrum major and trace minerals. Ionic minerals are 10,000 times smaller than colloidal minerals, and due to their tiny size they quickly go everywhere when taken internally or externally. This is a whole new area of healing using tiny bits of negatively charged substances. In addition, Ionic Silver, Ionic Germanium, Ionic Co-Q_{10} and Ionic Vitamin C are also the perfect partners for your oxygenation efforts.

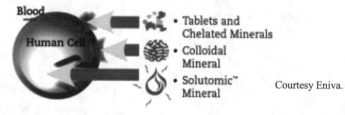

Courtesy Eniva.

Minerals are as important to life as oxygen is. Catalysts for the 5,000 chemical reactions that occur in the body every second, they transmit messages between the cells. They maintain the sub-alkaline balance of the blood and skin and carry oxygen's electrons around. Mere chelated minerals are tiny rocks and are not very absorbable into the body. The ionic minerals you need already come pre-softened from plants. Without the proper minerals being available, few of your vitamins will work, there would be no inter-cellular communication, and *no* healing would occur. First you clean it all up with the oxygen, and now you've got to re-build it all starting with the minerals.

Let me make it real simple why this is so important. When we eat we chew our food. You chew apples. You chew sandwiches. You chew steak. Chewing chops it all up coats it with enzymes and it all goes down your gullet. Then your stomach acid hits it and works on it. Then it's alkalinized and passed on down the line. The food bolus continues to pass through this process until, in the end, the once food is now all bleached, broken down, and reduced back into the elemental minerals it started out as in the soil where Nature and the animals packaged it for you as plants and meat.

Eating food is ultimately a *mineral delivery system*. Eating and digesting are the processes of creating minerals from food. The plant softened minerals are then absorbed into the body, redistributed, and that's why you are solid. When you were little, you didn't have a large volume of minerals making up your little body. Now look how many you have. You grew big and solid because of the minerals that you absorbed by eating along the way.

So if your pancreas is bad, your stomach is bad, your heart is bad, your arteries are bad, your lymph is bad, your head is bad, your joints are bad, whatever, your body needs to repair it. You have to give it the building blocks it's already made of, don't you? Don't you have to give your body these same building blocks needed for it to build with? You can't give the body a drug and lots of hope and concern and tell it to go build or repair itself! Drugs don't give you the building block minerals. Drugs don't clean you out. They leave the problems in place and alter and hide the symptoms, but by taking this lazy route you'll pay more, a lot more, later.

If I tell you, 'Go build a house,' but don't let you have any construction materials, or I'm only going to give you a couple of bricks, but not all of the bricks, how can you build a house? The house is your metaphorical body and immune system. That is the precise situation we are all in when we eat these days, because many of our national farm soils are used up. All the high mineral contents have been gone from our food since the '30s.

These mineral elements we're discussing are what your body is actually made of, and therefore exactly what it must have, at all times—all of them—in order to repair and re-build itself. Without all of the 78 or so available minerals and trace minerals softened and packaged by Nature into a tiny absorbable form, you can't build or repair your body-house.

You continually need the base element building blocks contained in food; especially organic food grown in mineralized soils. I'm sure everybody has eaten a carrot at one time that has been really bitter. That carrot was bitter because it was grown on de-mineralized soil, and probably picked early and rushed along so the next crop could be put in, long before the soil minerals could be absorbed fully and turned into sweet natural sugars by the plant. Carrots grown on mineralized soil are very, very sweet. They have everything in them they're supposed to have.

Because our BIG Farming soils became 'played out' of almost all their natural minerals back in the thirties, you can't get enough oxygen and minerals solely from your food. Sorry, not even if it does display the expensive 'organic' label. Many studies have confirmed that the food on our table today no longer gives us the nutrients our bodies require, including oxygen, minerals, potassium, sulfur, MSM, and antioxidants.

In 1926, Dr. William A. Albrecht, the head of the agricultural department at the University of Missouri, tested 100 bushels of Kansas wheat for mineral content. In 1968, 42 years later he tested wheat from the same farm. To get the *same* amount of mineral content, it took 1,000 bushels!

A recent soil analysis conducted in 1996 found that less than 20 minerals are still present. To compare, in 1940 American soil contained 60 minerals.

> **You have to now eat 75 bowls - a bathtub sized salad today - to equal the nutritional value of one bowl 60 years ago. It is this lack of nutrients in our foods that makes it imperative for us to supplement oxygen and minerals top of our diet!**

"Today foods are grown in mineral depleted soils, which yield mineral-deficient plants, which when eaten always result in mineral deficient people and animals. Because of the deficiency in the soil, it now takes 75 bowls of spinach *to equal the iron that was in one bowl* of spinach 50 years ago! Could you eat eight loaves of bread? That's what it would take to get the nutrition from the wheat of years gone by. No wonder people are ill and mineral starved. Could you eat a bathtub of salad to get the few nutrients out of it? Of course not! Now you can understand why the vast majority of Americans are overweight [from junk food] and literally starving to death."—Shari Karp in *Royal Reporter*

And, what you think that word 'organic' means is quite different from what the law says it has to mean. The mega-corporation food people know this, and during interviews they speak of marketing to 'your *perceptions*,' rather than displaying full truth and producing real nutritious food. Mom and Pop don't own the farm anymore, kids. You have to supplement.

Ancient Sea Vegetables Condensed Into Minerals

Your daily food, on every table (or in the bag on your lap while you sit at the drive-thru) should naturally have in it all the minerals that Nature designed you to run on. To make up the difference you can take supplemental liquid or powdered full spectrum minerals along with you in your pocket, and always add them to whatever you eat. At least you'll stand a much better chance this way, even if you tell yourself you can't eat right. For minerals I start with OxyMune and Hydroxygen Plus and their easily absorbable ionic mineral ingredients. Then I boost my overall mineral content up to what I really need with a high

concentration blend of plant derived humic shale 'nature softened' colloidal minerals and traces called MinRaSol, produced since 1926 from the only original Clark mine, in Utah. Also check out the GHT line of ionic sea minerals. Fast food addicts, full minerals are #1 what's missing.

> Even when you eat 85 percent raw foods you may still have to supplement — just to come out even.

High Grade Marine Coral Calcium

There is a further supplemental source of minerals that also warrant our attention. Nature normally uses calcium (when we can get enough of it in the right form along with dietary magnesium) to endeavor to keep our cellular pH slightly alkaline, and this allows maximum blood oxygen absorption. Calcium and oxygen in our bodies act as absorbers, absorbing the deadly carbolic and lactic acid wastes left over from the process of living. If we adhered to my 'Mr. Oxygen's Crown Jewels of Health' precepts and had enough nature softened absorbable calcium and other minerals in the right form on a daily basis, we would probably never need the use of many of these last minute emergency therapies I write about.

Ancient ocean vegetable plant life is mined from deep under in the Utah desert in the form of humic shale deposits and sold as MinRaSol and other mineral supplements, and today marine mineral deposits are collected from ravines below the coral reefs surrounding the island of Okinawa. These mineral rich coral components are swept from seabeds, bottled, and sold as 'coral calcium' supplements. Doctors in Russia, Germany and India regularly prescribe coral calcium supplementation.

Natural dietary calcium is vital to oxygen transport. As far as the name goes, don't let the 'calcium' name fool you, these coral deposits are not just calcium. They work so well as a supplement because they are highly absorbable, being 'pre-softened' by nature just like the humic shale. They have 75 minerals in them and that also sort of puts them in the same league as the humic shale colloidal mineral supplements. All life came out of the sea, and it all started out feeding on the volcano melted and blended minerals at the bottom of the food chain, and these basic to life minerals and traces are exactly what is missing in our food and lives today.

The commercial coral calcium blends with the highest mineral content and the highest nutritional/health value also have a high natural magnesium and Vitamin D content. Magnesium and Vitamin D are the keys to making calcium absorbable. Some grades washed up on nearby beaches are also sold as 'coral calcium' but they have lost most of their value due to sun and rain washing the nutrients away over time. So if you go this route be careful to get the high grade coral calcium with lots of natural magnesium. Our bodies need a ratio of two magnesium to one calcium.

Robert 'Bob' Barefoot is my counterpart living in the world of coral supplementation and his infomercials for 'Coral Calcium Supreme' are all over the TV. When we were conversing recently I was pleasantly surprised to find him using the phrase "Flood your Body with oxygen" over and over as we spoke. When I told him "Do you realize you keep saying the name of my new book over and over?" He said, "No, what's the name of your new book?" Upon hearing that we were both using this 'Flood your body with oxygen' phrase, and that we were obviously on parallel courses to educate the American people, we had a good laugh.

Note to the agribusiness folks and the ex-hippies running their 'organic' research departments. You already know we're designed to run on pure natural food, air, and water, but your dilemma is how to get there from here. Easy. You're lost in detail. The crop feed waters and soils should be oxygenated (peroxide is extremely cheap in bulk), and the soils laced with volcanic or humic shale (dirt is dirt cheap), or perhaps sea minerals (seawater is cheap). Then grow the crops while recordings of birds singing are being played at them in the morning as you oxy-foliar feed them. In other words, inexpensively put the missing ingredients back into the mix so it's closer to the way it was before all this messing with Nature started. That's how to realize your original dream of transforming our food supply into something healthy. Just like people, properly oxygenated and mineralized plants will grow faster, bigger, and stronger. They'll pick up more soil minerals, and become disease-resistant and they will taste better.

Bottom line, taking these inexpensive measures will make your bosses and shareholders happy by earning you superior natural crops producing more yields per acre, and saving you lots of money on unneeded pesticides and fertilizers. Besides your economic incentives to do this, you could really live your dream and honestly and proudly call your produce 'organic' without having to fudge on labels and induce politicians. Oh, by the way, the people eating your food will also suffer less disease, and humanity will have a higher collective intelligence and consciousnesses. Not bad, eh?

When I speak of 'minerals' as a crown jewel, I refer to natural food as their delivery system also delivering amino acids, enzymes, and co-enzymes. Co-enzyme Q_{10}, for example, is a co-enzyme found mostly around heart muscle because your heart is beating constantly. You need to transfer oxygen's electrons real quick to keep the muscles going all the time in there. Funny how no-one gets 'heart cancer' with all that oxygenating Co-Q$_{10}$ around the heart tissue. Vitamins are included as part of the mineral crown jewel. And nobody has enough natural vegetable derived potassium or MSM.

My friend Billy put off going to the dentist for a long time, and when he did, the dentist told him that half his jaw had to be removed. He went home and started taking minerals, and several months later he strolled back into the dentist's office and suggested he be X-rayed again. The dentist's jaw dropped in astonishment when the X-ray pictures revealed his jaw had grown back. "Why that's impossible!" the dentist exclaimed. His assistant was also amazed. "And yet, there it is," smiled Billy.

3. WATER. Next you have to continually move the oxygen and minerals through your body. Lots of good tasting very clean water, is the major cleanser and transport mechanism for everything; our oxygen, hydrogen, vitamins, minerals and enzymes. To oxygenate, mineralize, enzyme-ize, pump up, burn up, purify, and to flush it away, it's all gotta have a carrier or vehicle with plenty of room for everything to float in. Nature has given us clear clean oxygenated water. I cannot over-emphasize how important this step is. Flush all the dross out with lots of clear, energetic, natural oxygenated water!

> "The cell is immortal. It is merely the fluid in which it floats that degenerates."
> Dr. Alexis Carrel, 1912 Nobel Prize winner.

We have to constantly change our water, but usually we neglect to do so. Dr. Batmanghelidj teaches us that water is second in importance only to oxygen. I agree, because water is how the oxygen gets around.

Remember the fun you had as a kid bouncing on the bed or on a trampoline? Well now they call it rebounding. Rebounding is a wonderful way to move watery lymph fluids carrying oxygen while vertically breaking the gravitic lines of force surrounding us by using your body as the armature of a natural energy generator. Exercising and jumping up and down pumps your lymph fluids around so they can get fresh oxygen. I suggest you use the best personal rebounder (trampoline), David Hall's Cellerciser, the one with the tapered springs. While we're on the subject of

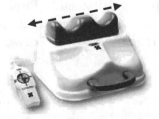

pumping your watery fluids or lymph, the use of a Chi Machine is also a good fluid pumper.

The Chi Machine lets you relax and helps move your oxygen. The Hsin-Ten 'Chi Machine' moves your legs like a fish swimming while you lie on the floor. While you relax, the side to side 'swimming fish' leg movement mechanically pumps your watery spinal and lymph fluids around to oxygenate them (and you) nicely—especially when you're supplementing with active oxygen. The same company sells  a semicircular solid-state $1/3^{rd}$ circle far infrared light emitter called the FIR 'Hothouse.' We lent ours to a friend's husband who had lung cancer. They called to say that by his lying under it, his pain completely stopped the first day! Far infrared rays are not stopped by the atmosphere and penetrate the body and increase blood and lymph flow, and therefore the oxygen flow. Pain comes from stored lactic acid when there is not enough oxygen and water.

Try drinking lots of oxygenated or ozonated water, and exercise to move it around into and throughout all the little capillaries and lymph ducts.

4. ENZYMES. The purifying and dissolving energetic scavengers and catalysts that, along with oxygen, remove waste, and dead, weak, diseased, and dying old cells, as well as undigested food particles. Undigested food particles are the main source of food feeding microbes living in your body.

Why do you think the largest number of selections on the shelves in the pharmacy area are stomach acidifiers? Everybody eats cooked food. Heat and the stupid radiation they want to use on food destroy enzymes. As we get older, no one in our society can digest his or her food properly. This is because of processed foods, a lack of oxygen, minerals, and not enough natural vegetable sodium to make the stomach's hydrochloric acid. The food you grew up eating has largely been devoid of the many natural enzymes you needed to digest it. Nature puts them in, man takes them out.

The fewer the enzymes the more undigested the food, and the more undigested the food, the more the undigested food particles end up in our blood for microbes to feast upon. The older we get, the more microbe colonies grow, the more waste coming from them piles up, the more the waste gets in our fluids, and the poorer the cell functions become as the garbage is still constantly piling up, and piling up, and piling up in the cells in charge of digestion. Once the clean blush of youth is sufficiently sullied by enough undigested garbage piling up, digestion and stomach acid production start their relentless slow decline unto your premature mechanical failure. Taking extra enzymes at mealtime AND between meals is when I really see them go to work. You should consider a full blend of good enzymes.

Case in point. I stopped all exercise and supplements for a few months to see what would happen if I just ate a variety of raw food salads all day, some cooked meat (all tailored for my own blood type), very few carbs, and of course, a very tiny bit of real world cheating. In other words, I created the exact 'health diet' that most educated working people with a sedentary lifestyle would be living, minus any drugs, alcohol, caffeine, dairy, or sodas. The All-American perception of a health diet without supplements, and supplied by the restaurants everyone goes to.

Darkfield microscope examination of my blood later revealed I was back on the road to slow deterioration. I suddenly had lowered digestion again, sticky platelets, and the beginning of yeast, protoplasts, plaque, crystals, spicules, thrombocytes, and erythrocyte aggregations. I was again taught I obviously needed the supplementation I teach about. The starting to emerge blood impediments were, of course, due to the missing nutrition of our 'pretty' but low quality oxygen, enzyme, and mineral deficient food supply. I definitely needed my oxygen, mineral and enzyme supplements back. I would have fared better on raw vegetable juices, meats, whole dairy and eggs, but that would have been an unfair test of 'averageness' as many never have the time or opportunity to juice or eat like this when they primarily eat in restaurants.

5. A CLEAN COLON. If it comes in, it has to go out! What if you go through all the steps of oxygenating, mineralizing, enzyme-izing, and hydrating, and you finally start moving the accumulated garbage into your bodily fluids for transport to the organs of elimination? Or, even if you don't do any of that, what's going to happen if any of the usual waste arrives at the colon and the colon is partially or mostly closed for business? Where is it all going to go? Back up into the organs and skin. You must clean the clear mucous that plugs up your colon wall and blood capillaries in order to stop automatic re-absorption of your trash. Without a clean colon it would be like bringing your trash out to the curb, letting it rot, and then bringing it right back inside.

Here's the order you are plugging up (and failing) in as the body attempts to neutralize the poisonous acids building up in order to maintain a healthy slightly alkaline pH in the face of being toxicity overwhelmed:

The first thing the body uses to fight acidity is getting more oxygen the only way it knows how, through breathing harder so it can push more CO_2 out of the lungs. How many people get winded and pant easily? When there isn't enough oxygen, minerals, and sulfur/MSM, excess carbon dioxide and lactic acid collects, so the intra-cellular amino acids are used up trying to buffer the acids. Lymph and saliva try to neutralize and dilute them but they each thicken as you dehydrate — lowering their efficiency.

Next, your high pH electrolytes; calcium, magnesium, sodium and potassium are used up binding salt acids, and then your skin, urinary tract, colon and respiratory system overload trying to filter them out. Then blood plasma changes while loading with bicarbonate in an attempt to neutralize the increasing acidity. If the low oxygen and water conditions persist and no change in diet and elimination are forthcoming, then the bones, teeth, joints and muscles will be robbed of their calcium, magnesium, sodium, and potassium reserves causing severe mineral deficiency, and when all this fails because the acidic mucoid sludge continues to block everything and pile up, than the body can no longer neutralize the acids. In a final attempt to survive, the body pushes the excess acids and toxins out to be stored in the peripheral vital areas of the skin and extremities.

Diseases Related to Emergency Peripheral Toxin Storage

Acid and toxins in the wrist—carpel tunnel
Acid and toxins in the knees—osteoarthritis
Acid and toxins in the feet and toes—gout
Acid and toxins in the skin— dermatitis and eczema
Acid and toxins in the joints and fingers—rheumatoid arthritis
Acid and toxins in the tissues—fibromyalgia, chronic fatigue,
and degenerative disease, etc., etc,. etc.

As the last straw, when these areas are filled, the vital organs will start filling up and you get the resulting end cancer, diabetes, heart disease and serious arthritis. —From Crusador magazine article, *What in the Cell Is going On?* - Gary Tunsky

Don't reabsorb the garbage because your colon is plugged up at the cell surfaces. How can you drain your swamp if the drains are plugged with mucous? Oxygenate, drink lots of water, clean your colon regularly. There are lots of cleansers out there like Homozon. Try fasting one day a week. Give your digestive system a rest.

6. LIGHT, SOUND, & OTHER ENERGY. (New to our Crown Jewels) Energy; like sunlight, birds chirping, water flowing, and being surrounded with peace and love, are all essentially and ultimately natural beneficial energy delivery systems. We perceive them when we feel the warmth, listen to the melodies, and feel the warm love energy pour out of Mom's hand as she checks our forehead for a fever. We also seek to acquire more unconditional warm love for ourselves through touching and being with others. The reflection we feel of God's love and acceptance for us is the powerful drive behind the urge to procreate.

More and more we see energy healers around. Magnets, Complex Magnetic Units, light, radio waves, acoustic, and other energy devices are hitting the market at a record pace. Their successes, if any, all come down to this. Our bodies are wonderfully adaptive; they can draw missing healing energy from just about any source.

Sick people draw energy off healthy ones. Gurus, politicians, actors, musicians, and sports heroes draw energy off crowds and give it back, hopefully in equal shares, and without ego involvement. Life is all energy exchanges. Every conversation and especially coupling exchange energy.

Natural healers have the ability to radiate energy, and many of us have heard of the Tesla coil, Multi–Wave Oscillator, Violet Ray, and Rife Ray Beam machines used in healing starting in the first part of the last century. Today we have other energy emitters using light, sound, and other energies. The ones that work by nullification will emit the exact equal and opposite energy of the imbalance or disease agent and be able to nullify them. Conversely, success can come from application and absorption of enough of the perfect additive natural healing energy that can strengthen and boost the keynote energy of the person or organ being treated up to a state where it is so purely just itself energetically and so highly charged that no imperfection can exist within it. Miracles are in this category. The miracle energy comes from *somewhere*.

7. EMOTIONAL, MENTAL, & SPIRITUAL BALANCE. The emotional, mental, and spiritual non-physical components of disease and toxicity are too often ignored. When I say that all disease is ultimately caused by toxicity, I'm also referring to any spiritual or mental or emotional pollution stored within you. If you are out of harmony with who and what you really are, at some point your electrical and energetic fields are warped. Along with that, your mental process also won't be quite right. You will then be drawn to do, think, associate with, absorb, and consume things not good for you.

Too many are angry too often. He's mad at her, and she's mad at him; and round and round it goes. I once interviewed a clairvoyant and asked what he saw as he observed the downtown Jacksonville business lunch crowd. He replied "Anger, lots and lots of anger." Anger negates love; love is what keeps us alive. The less love, the less health.

Your health and the electrical energy in your body depend upon your emotions, your thoughts, and your way of being (free will choices). And if you have any guilt, conflict, or negativity in your life, it's automatically depressing your immune system. Nothing is worth that. If something is making you feel that way, solve it. Leave that situation or those people, or change your attitude to one of loving acceptance that whatever is happening is God's will for you. We only meet the rest of God. Life has a way of providing —if you are sincere about making things better. If you did something that makes you guilty, go back and ask for forgiveness. If the object of your guilt is no longer around, ask anyway.

We came here to learn by doing how to harmoniously use our free will in stressful situations. Angry, guilty, lustful, greedy, vain, or overly attached, if your emotions and thoughts are out of whack, at some point you are not producing the correct immune chemicals within your body because things are out of harmony. You can stop living with guilt and disease because of having made the wrong choices that you thought were unavoidable. They weren't, but you were lost in some illusion at the time. The bad choices you made then were the only ones available due to your lack of *full knowledge* at that moment. So although you free will chose yourself into a pickle, it's not your fault; there's no guilt necessary if you are now sincere about change. Now do something right instead to make up for it.

> *"There are two paths you can go by,*
> *but in the long run,*
> *you still have time to change the road you're on."*
> —Led Zeppelin

It has been my experience that if one cleans out and properly oxygenates for long enough, the emotions tend to smooth, and the thinking tends to become clearer. Then one might be enabled to make fewer and fewer bad choices. When you think clearer you no longer feel comfortable with self and other person abuse and judgment, so that frees up wasted energy in our body temple that can be used to repair and rejuvenate instead.

The immune system runs best when you are happy, when you're satisfied with life, when you're full of love, and when you're giving love to other people. It also runs best when you're surrounded with warm, loving relationships and people who never criticize you, and when your environment is harmonious, clean, unpolluted, and you're guilt free and at peace. Real easy to find here, right? That always gets a laugh at lectures.

Funny how everyone thirsts so bad for these things and laughs at how infrequently we get them. That's the beauty of the Human Spirit's ability to *overcome*. We tend to forget in the heat of the moment that we get the good things by giving them. A firm hand is good, but must always be tempered with compassion.

Happiness. This is the ideal, the ultimate goal. What we all work toward. Actually, I understand the score-keepers of life only require you to do the best you can. That's not so hard is it? Just be aware, interestingly enough, that the closer you come to this goal the happier you are and the healthier you are. Studies prove this. I have witnessed this. I see it in the lives of others. So if you have a spiritual connection, use it. If you aren't spiritual in any way, apply your personal choice of a moral code and live up to it; or just be nice, helpful, and forgiving. The way you feel—whether or not you're guilty, angry, or feel worthy about your life—changes your immune system. Negatives lower the electricity in your immune system.

Here's a good healthy code you can borrow from Mother Theresa. When I first read it, I kept thinking how cool it was that she and I had ultimately come to the same conclusions independent of each other. Well, it made sense, these are ultimately the only logical choices:

People are often unreasonable, illogical, and self-centered.

...Forgive them, anyway.

If you are kind, people may accuse you of selfish, ulterior motives.

...Be kind, anyway.

If you are successful, you will win some false friends and some true enemies.

...Succeed, anyway.

If you are honest and frank, people may cheat you.

...Be honest and frank, anyway.

What you spend years building, someone could destroy overnight.

...Build, anyway.

If you find serenity and happiness, they may be jealous.

...Be happy, anyway.

The good you do today, people will often forget tomorrow.

...Do good, anyway.

Give the world the best you have, and it may never be enough.

...Give the world the best you've got, anyway.

You see, in the final analysis it is between you and God.

...It was never between you and them, anyway!

While engaging in a holding action against the daily onslaught of stress and toxins by oxygenating, etc., you can also engage in positive evolution and character building. Daily prayer, meditation, or spiritual exercises of some sort are highly advised for you. This is not idle speculation, but demonstrated fact.

I have personally seen a rib bone healed *instantly* and verified by before and after X-rays through my Chiropractor friend Larry's "merging his consciousness with the perfection of God that allows no imperfection to exist," as he explained to me. He had broken his rib after falling while we were water-skiing. He was curled up on the floor of my boat in pain. Later he X-rayed the break, then meditated and "merged with perfection," re-X-rayed it, and you could see where it had instantly healed. I found out about all this when the next day after the accident he jogged up to my house slapping his side all excited and saying "Ed! Look!" Later I saw the X-rays proving it for myself.

Dipping into the Life Force daily with our consciousness and intent charges our creativity and immune defenses with this same energy. Larry meditated regularly and was willing to completely surrender to the *All That Is*. We are truly wondrous creations. Take the chance, try it.

JUMP

"You've got to jump off cliffs everyday,
and build your wings on the way down." ~Ray Bradbury~

Here's the point of all these jewels. I have witnessed each and every one of my Crown Jewels alone work miracles in many people. During the countless interviews I have personally conducted I have been shown or heard of astonishing, amazing, so called 'miracle' results appearing consistently and quickly in people who have properly applied only <u>ONE</u> of these jewels. Perhaps they applied only oxygen supplementation, or only ionic minerals, or only enzymes, or simply lots of clean water, or just colon cleansing, or just spiritual re-integration practices. Each one has proven to be capable of being a 'power-house' solution ALL BY ITSELF.

NOW IMAGINE WHAT WOULD HAPPEN TO YOU IF EVERY DAY YOU CONSISTENTLY **COMBINE ALL SEVEN** CROWN JEWELS INTO EVERY ASPECT OF YOUR LIFE!

WOW!

If you don't have this entire set of jewels in your crown, you won't be able to get the full benefit possible from any Oxygen Therapy. Oxygen therapies are definitely not another pill that you take to try and buy your way out of your problems with.

It doesn't work in the traditional way of going to a doctor where you throw money out and say, "I don't care what it costs. Fix me." What this costs is *your* time and effort and the use of *your* common sense over a long period of time. Your health is something that *you* are responsible for.

For all that, Oxygen Therapies are still the rock that the house of health is built upon. But it all works together as a whole. Your immune system needs the entire set of crown jewels of health: oxygen, minerals, enzymes, clean oxygenated water, a clean colon, harmonious light and sound, and that precious elusive emotional, mental, and spiritual balance. Get these seven things straight before you do anything else! You just might be amazed at the difference in your health. Individual results will always vary due to many factors, especially free will and karma, but everyone else I interviewed who uses the jewels wears the crown proudly.

Part Three

The Evidence

Anyone who poses as a concerned consumer health watchdog while scaring people away from Ozone Therapy does not really care one bit about the suffering. What they are doing is closing their minds to innovation and protecting the status quo—at your expense.

As you read this, Ozone Therapy is already in use in thousands of clinics worldwide. So why would anyone who really cares about the mass suffering ignore it unless they have another agenda? More than 50 years of historical evidence and practical results prove that there is no real need for 'waiting' for more studies while people suffer. We know how it works, we know how to use it, and every healer should know all about it and be able to freely use it safely as they see fit and the situation demands.

Chapter 22

Testimonials

This section includes only a handful of the testimonials available worldwide. What I'm doing is reporting to you interviews I've held with people, or showing you testimonials that have been sent to me. As a journalist I have talked to thousands of people concerning Oxygen Therapies, to hundreds of doctors using it and to scores of clinics around the world. I've been told endless stories. For example, I've heard over and over and over that someone has been sent home to die with cancer and told to put their affairs in order. They found out about ozone or some other Oxygen Therapy, they got it somewhere, wherever they could, whenever they could find it where it's not being suppressed. It was done correctly and they came home well.

I have video taped many of them. They showed me their medical records. They sit there and smile at everybody and say, "It worked for me, and it's harmless. Why not try it?" It's a grass roots thing. We are not hearing about this from the authorities. It's word of mouth, and has mostly been underground, but with the current influence of the Internet it is now easier to access all kinds of information about these therapies.

In all cases the number of testimonials shown was an arbitrary decision. At first I created a much larger book than the one you're reading and removed many testimonials to get to a book of this size. Please do not assume that one area having more testimonials indicates any one product or service is better than any other.

Individual results will always vary!

The following are the stories of real people who had real experiences, wrote them up, and sent them in. They are grouped according to the specific therapy employed. Experts try to dismiss testimonials because some lack independent verification.

You have to ask yourself, can all these people who don't know each other and are concurrently having the same experiences while using different Oxygen Therapies all be mistaken, or liars?

Bottled Oxygen

Cluster Headaches

W. L., Australia—"I am 51 years old and I have suffered from horrendous cluster headaches for about 30 years. These headaches are equal to the intensity of migraines but they come in waves of three or four a day for one or two months, then they disappear. The treatment for about two years was an ergot constrictor that eventually put me in hospital with a stimulated heart attack. The treatment involved shutting down the heart! A specialist advised that the treatment would have to end or I would face serious complications. It was a wild card, but he suggested intense oxygen uptake at the onset of the headaches and he had no idea if it would work. During the next headache attack (which actually occurred whilst I was still in hospital with the stimulated heart attack) I received a high dose of oxygen and the headache stopped within five minutes! The ergot treatment would take hours and you had to take it before the headache started. This was a catch 22, since you never knew when to take the drug and whether it really did prevent a future headache. I have since purchased two oxygen bottles and I self-administer at the first sign of a headache. Not only does it offer an instant fix, I think it actually breaks the cycle and the headaches do not take hold for the duration that they used to. I wish the treatment had been around 30 years ago instead of the dangerous constrictor treatment."

Hyperbaric Oxygen

Czechoslovakian Dr. Oldrich Capek—wrote to me about some of his clinical results using hyperbarics. Here are his comments.

"Here is just part of my experience with Hyperbaric Oxygen and Hyperbaric Ozone Therapy.

"I started to experiment with Oxygen/Ozone Therapies in 1993, after a visit by (Czech-born) Dr. Lilly Holemar from Vancouver, who 'lobbied' me enthusiastically on behalf of ozone and chelation therapies. At that time I worked together with my colleague, in a very successful alternative-medical practice, in the historical center of Prague, CZ. My experience with Oxygen/Ozone Therapies followed a visit by Saul Pressman of Vancouver (who gave me a book written by Renate Viebahn) and after contact with Czech-born Zdenek Den Rasplicka, also from Vancouver, who gave me a book written by Ed McCabe!

"Here I have only time-limited experience; nevertheless some of my observations are very interesting (exciting). First, it really works extremely good in most cases of headaches and migraines. One needs from one to six applications in general, in order to completely avert the problem. Very often even one single application brings instant relief.

"Then in elderly people (surprisingly now I have only female ones), with brain atrophy and related behavioral and memory disorders it very quickly restores available brain functions, circulation and oxygenation of the brain. Typically I have been using pressure of 3-to-4 PSI, with oxygen supply up to 1(one) liter per minute. (One has to count on multiplication of oxygen concentration under the pressure, so I did not dare to use higher concentrations so far). Typical time of application is 45-to-60 minutes. I have observed that those old ladies, who were already 'behind the glass', first became more interactive, more curious, more demanding and finally quite naughty with the care-taking family. Particularly, I consider this symptom as a good return of brain functions, since those patients virtually 'come back' and are an integral part of the family, demanding care and attention after a long period of silence.

"Finally, I would like to mention one case: A young woman with constant headaches and dizziness, suffering for several months after her traffic accident—where she had had a (cranial) head injury (without brain concussion)—was treated in a Hyperbaric Oxygen Chamber. After five sessions most of her pains and dizziness disappeared. Now, after seven sessions she is completely symptom free and without any medication. It is important to note that she had tried most of the latest diagnostics (CT, MRI, EEG) with no [negative] findings at all and that previously she was using a number of different painkillers and anti-migraine drugs without any effect."

Oxygenated Waters

Oxy-Water

Consumer Testimonials

Mike Marcolini, Active Healthstyles—"Having weight-trained myself for nearly 20 years and now training other athletes, I am very suspicious of anything that promises immediate and tangible results. My mottoes to my students are: 'Too much hard work is never enough,' and 'Nothing is immediate,' it has to be part of a program.

"Except for this: OXY-WATER is IMMEDIATE and TANGIBLE.

"My concept is simple. Weights make you strong and running shoes and hills put the finishing touches on things.

"I notice a slight difference when doing aerobics and a HUGE DIFFERENCE in HEART RATE and RECOVERY TIME when performing heavy-duty white muscle fiber sets.

"My routine is to drink a third of the bottle 15 minutes before and continue throughout the workout. The best results have come from two asthmatic lifters who no longer need their puffers during sets. IT WORKS!"

Michael Yessis, Professor Emeritus, CSUF President, Sports Training, Inc.—"I am writing to you in regard to some of the negative comments that I have heard from several pseudo-experts as well as (about) what has been written in some magazines regarding improving the concentrations of oxygen in water and in the body. I am in full agreement that everyone is entitled to one's own opinion; but opinions should be based on some knowledge, not strictly on one's thoughts. Individuals who state that it is impossible to put greater concentrations of oxygen in water have not presented one iota of substantiation for their comments. Meanwhile there are studies and proven technologies showing that additional oxygen can not only be dissolved in water but is absorbed into the body to show even greater oxygen saturation in the blood.

"If these pseudo-experts were aware of the many studies done with oxygen they would know it is rare to find an individual with 100 percent blood oxygen saturation. By increasing oxygen saturation in the blood you can improve performance and have faster and more effective recovery after exertion. If these individuals would use products such as Oxy-Water which contain oxygenized water, they, too, would be able to experience the benefits of extra oxygen in the body. This is especially true in situations where the body is 'down,' as for example, after heavy mental or physical stress, during fatigue or, in general, when the body has lowered concentrations of oxygen or perhaps from fighting infections.

"There are also studies showing that there is now less oxygen concentration in the air, especially in the large cities because of the increasing amount of pollution in the air and less oxygen being introduced into the atmosphere by plants. Therefore the need for supplemental oxygen is greater today than ever before! By supplying greater oxygen to the body, we can make up for what the body is not getting through normal breathing or because of other medical or environmental problems.

"I should also call your attention to what the Soviets did with oxygen and their athletes. They had drinks called oxygen cocktails into which oxygen was introduced that resulted in up to 7-to-8 percent greater saturation of oxygen in the blood. The pseudo-experts say that Russian technology and research is inferior to ours and therefore their studies cannot be considered valid. This is a great way to escape facing reality, since we have not even come close to approaching the technology developed by the Russians to improve athletic performance and to help the body recover after heavy workouts.

"The bottom line is that oxygen can be absorbed into the body from water that has been 'oxygenized' and even oxygenated if there is a binding agent to hold the oxygen in the oxygenated process. The greater concentrations of oxygen do many great things for the body, too many to enumerate here. The pseudo-experts sit back and criticize instead of opening their eyes, ears and minds to technologies so that they can learn about the latest advances that may be unknown to them. When these experts substantiate their claims with solid data and not just opinion, I will listen to them. But for now they should be ignored."

John Schaeffer, mfs-ssc-spn
Author: _Winning Factor_ _Book_ _of Fitness_—"Just a note about Oxy-Water. I first started using your product about six months ago. I really feel Oxy-Water is a product everyone should use. I train harder and recover faster since using Oxy-Water. As a World Champion powerlifter, trainer and author of _The Winning Factor_, I've seen about every product on the market; some good and some just hype. I was so impressed with Oxy-Water, I have all the athletes I train using it. Since I introduced it in my fitness center I can barely keep it in stock! Oxy-Water is AWESOME!"

Asthma

Rich Kopchak—"I suffer from asthma. I normally use my inhaler two or three times daily. I tried Oxy-Water because I like good tasting water. To my surprise I found that Oxy-Water not only tasted great, but also significantly improved my asthmatic condition. When I drink Oxy-Water I feel as if I have had a great boost of air to my system. I have discovered that when I drink two-to-three bottles of Oxy-Water each day, I don't need to use my inhaler. This product is a miracle in a bottle.—I feel anyone with asthma should try this product. I don't know if they'll have the same results as I did, but when you have asthma anything that can help improve your condition is worth its weight in gold. Thank you for Oxy-Water."

Hardcore Muscle & Fitness

Stephen Puzyk—"As you probably know, most physical fitness enthusiasts are set on their regiment of supplements, protein powders and drinks. Therefore, under these conditions, new products don't stand a chance of making an impact.

"However, I must say that Oxy-Water is able to break that barrier and our sales show it.

"We currently sell 10 cases per week and are looking forward to the one-liter bottles.

"Another item that you may find interesting is that we have had customers that belong to other gyms buy Oxy-Water from our pro shop. Thank you for a wonderful product."

Hurst-Jensen, Amateur Cycling Team

Trent D. Lundberg—"As captain for the Hurst-Jensen cycling team, I wanted to write and update you on our squad's recent string of victories during the month of June. Since bringing Oxy-Water on board as the official water supplier for Hurst-Jensen in mid-May, the team has garnered five victories on both road and mountain bike circuits. Most notable among these wins was the prestigious Newsweek 24-hours of Canaan relay race on June 7th and 8th in West Virginia. Hurst-Jensen fielded a team of five riders, including one woman and bested more than 100 other squads to bring home a terrific win despite grueling conditions such as snow, sleet, rain, and stiff competition. Fortunately, our team had an unfair advantage: Oxy-Water!

"From the beginning of this partnership, the riders on the team were intrigued by Oxy-Water and its claims of more stamina, faster recovery times, and less fatigue. Articles have been written about the secret methods of Russian and German athletes, mentioning 'oxygen cocktails' that the athletes would consume prior to, during and after competition to give them a competitive edge. These drinks were unavailable in the United States until we discovered your terrific product and we cannot thank you enough for supplying us throughout the 1998 racing season. Quite simply, Oxy-Water works. Riders have told me about their legs feeling fresh after many days of consecutive competition, lowered heart rates both at rest as well as during racing, and also increased metabolism, all due to Oxy-Water. Training schedules have remained consistent, diets haven't changed...the only difference we can attribute it to is the water!"

Overweight Asthmatic Hockey Goalie

S. West, Ohio, U.S.—"Dear Oxy-Water, I am writing this letter to let you know how much I believe in your product. My first experience with Oxy-Water was back in February of this year. My 14 year old son plays hockey on a travel-team here in Columbus, Ohio. There are days when they have intense practices that end with lots of sprint skating back and forth, utilizing the full length of the ice. The goalie for his team is somewhat over weight and has never been able to participate in these sprint skating sessions without having to stop and gasp for breath. He would get so out of breath that he would have to be helped off the ice by his teammates. There were occasions when he would have to use his inhaler for an asthma attack brought on by this exertion or even make himself sick trying to complete the exercise the way his teammates could.

"The manager of the team brought Oxy-Water for the boys to try out at one practice; they had some before and after the practice. During this practice the overweight kid (after drinking your Oxy-Water) skated his heart out and never had to stop once for his inhaler or to take a breath. He finished the whole practice and all the sprints for the first time and skated off the ice on his own. His teammates were so excited for him and so impressed with his achievement that they actually had an Oxy-Water party after practice with the remaining bottles, waving them in the air like trophies. After witnessing this experience, I took some Oxy-Water home, which I consumed before and after my own workouts. I have noticed a great difference in the energy I have during exercising and I have considerably less aches and pains when I drink your Oxy-Water. Thank You."

Aqua Rush

Brain Aneurysm

F. B., U.S.—"On the evening of August 17 in Kenai, Alaska, my wife's youngest sister, Sophie, suffered a ruptured brain aneurysm. Within a few hours she was on a MEDIVAC flight to Harborview Trauma Center in Seattle. After a very difficult night, the medical staff had stabilized Sophie but her condition required surgery. We were told that her chances of survival were 50/50 at best. On Wednesday the 19th, Sophie came through a difficult but successful surgery. What a relief! However, the doctors were guardedly optimistic because they were expecting Sophie to enter into a phase they described as having 'brain spasms' which could also be life threatening and would require treatment that could very possibly cause a stroke. The purpose of the treatment was to maintain blood flow to the brain through blood vessels that were constricting as a result of the spasms. The spasms set in with a vengeance on Sunday the 23rd and Sophie was back on the respirator and sedated, but by Wednesday afternoon she had awakened and was responding and on Thursday the 27th her respirator was removed. We checked with the doctors and nurses and they had no problem with her using bottled water; so, from here on out, everything Sophie drinks either is, or has, Aqua Rush in it.

"To shorten this story, Friday she is sitting up. Saturday she walks across the room with a walker. Sunday the 30th she is walking down the hall with a walker. Thursday, September 3rd, Sophie walks on a plane with my wife and flies back to Anchorage to complete her therapy! We were told at the outset to expect months of hospitalization and therapy. I asked the doctor if oxygenated water could have contributed to Sophie's speedy recovery. He balked, but could not give me any other reason. Even the medical staff was surprised at Sophie's progress. The completion of her therapy in Anchorage lasted less than two weeks and she was sent home. I believe Aqua Rush oxygenated water was an answer to prayer and was responsible for Sophie's remarkable recovery."

Brain Surgery/Head Injuries

J. H. N., U.S.—"On July 30, my 16 year old daughter Abby was injured severely in a car wreck. She sustained major head injuries and had emergency brain surgery. I insisted to the nursing staff that Abby be allowed to drink only the oxygenated water that she had been drinking for the past two months. They agreed that it may be beneficial for her brain condition while in the intensive care facility. They were amazed when Abby would take just a couple sips and the monitor for blood saturation levels would increase from 90-to-98-to-99 percent in seconds and stay that way for five-to-six minutes. We feel that the water contributed to her quick recovery and release from the hospital. She was up and walking in 24-hours and released from the hospital in four days. As of this date, 10/1/98, she is still drinking the water and the neurosurgeon predicts that at her present rate of recovery, she should be back to 98 percent normal in a year."

CFIDS

Richard, WA, U.S.—"In 1982 I contracted (CFIDS). If anyone out there has it, they will recognize the abbreviation and the near hopelessness of this malady. Because of the lack of energy and endurance I have spent most of my time in bed. My average day would be to get up in the morning, have two cups of coffee and breakfast, then right back to bed. Then be in that bed for most of the day. However, since I have been drinking Aqua Rush (three liters a day) I get up as usual and very seldom have to go right back to bed. In fact most of the daylight hours I'm able to stay up. There is a side point also; my wife has been bleeding from one kidney for over one year. One week after drinking the water her bleeding slowed down to about half. Thank you very much, Aqua Rush for providing this valuable product."

Hydrogen Peroxide

Acute Allergies

Theresa Hayes, Minnesota, U.S.—"I developed acute allergic reactions to almost all airborne fumes and pollutants. I would get headaches, coughing, sneezing, difficulty breathing, burning and itchy watery eyes and fatigue. I learned about Oxygen Therapies and I started using H_2O_2 orally. I used an eyedropper to put H_2O_2 into cellophane capsules. I did this for about two months and my allergies pretty much disappeared; except for cats, but that greatly improved too. I took the H_2O_2 about three-to-five times per day. I also noted that I had much more energy and I wasn't getting sick with a cold as I usually would every couple of weeks or so. I did develop a stomach intolerance for the H_2O_2."

Emphysema

Darlene Glenn, California, U.S.—"My uncle has emphysema. We use the peroxide gel for everything. He can't breathe. We put it on his chest and back. He said that for the first time he only lost air two times instead of five or six times that night. He felt a difference in his body in five minutes. Sadly, he never stuck with it."

Candida

John Bayford, England—"After reading your last book about two years ago, I use hydrogen peroxide with most of my Candida patients. I find it really gets things moving and patients swear by it."

Candidiasis

Jerry Bergemeyer, Arizona, U.S.—"I was fortunate enough to have a booth next to Ed's at a Preparedness Expo in Salt Lake City in 1993. I bought his book after doctors diagnosed me with chronic fatigue syndrome. Thanks to that fateful meeting, I was saved much mental travail. I still have that original paperback on Oxygen Therapies and still use the advice it contains. May God continue to bless you Ed, as you have blessed us with your compassionate service. Humankind is the better for your being."

Chronic Fatigue

Chris Savage, Australia—"In October 1989 I was training at the Queensland police Service Academy in Oxley, Queensland, Australia, when I was injected with a vaccine containing Hepatitis B Virus, formaldehyde, aluminum and mercury. It was midday, and after lunch our squad was put through a physical training test which included a 2.4km run followed by an 'obstacle' course. After that I was exhausted and went home. But then I noticed I was still exhausted hours and even days later. In fact, 14 days later I was still so tired that I stayed in bed and could barely find the energy to get up and go to the bathroom. I felt that I was going down and down and knew where that led to. So I dragged myself out and saw a medical doctor who was unable to assist. I then went to a naturopath who advised that I take four grams of vitamin C every hour for four hours. This helped tremendously and I was able to resume training at the academy. I thought then that I was cured. However, my energy and health was not the same as before. I was sleeping a lot and my energy was down 50 percent.

"For years I looked but couldn't find anything or anyone to help me and my life was an absolute mess, which it had previously not been. Then in December, 1994, a friend gave me a video of a public address by Ed McCabe at Brisbane, in Queensland, Australia in 1992. As a result I then started drinking dilute hydrogen peroxide and basically got 'stuck on' Oxygen Therapies. I am so happy because my health is better than even before the vaccine and my life has been changed for the better too. I now literally thank God for giving us Ed McCabe. It is through his courage and conviction that my life was saved. Thank you, Ed."

Colds

Farrell Caesar, Illinois, U.S.—"I have been taking H_2O_2 for one year. I haven't had a cold or flu in that time. I take 15 drops in water every other day. I eat orange slices before and after to avoid the taste. Thank you, Ed. Great Book."

Cold/Flu/Acne

Linda Dale, Florida, U.S.—"I started having bad colds once a month and real bad hacking coughs and also breakouts with acne on my face when I'd feel the colds coming on. A cold lasted for two weeks but it just never felt like it went away completely and then would come back twice as strong. This went on for four months. I once had throat itching real bad, and localized breakout on the exterior skin where the itching was internal. I knew it was from the bugs I had. I went to my MD and she laughed at me when I suggested they were related in any way. I was shocked that this was so obvious and she didn't want to hear of any relations! So I threw out the three prescriptions that she wrote me without even taking a culture, and referred to Ed McCabe's first book. I started with 30 drops in a quart of water. For three days I gradually increased to a level I was comfortable with. I did this for three weeks to be sure to kill whatever it was and it worked! I used aloe and cranberry juice to disguise the weird flavor and I will do it again in a heartbeat if I feel sick again. It is a blessing to know other options other than being a guinea pig! Thank you Mr. McCabe, you're a gift from a higher entity."

Colon Cancer

Valdy Jensen, South Africa—"My friend Bob Whall was diagnosed and operated on for cancer, lower colon—almost rectal, early stages. Re-growth after the operation. Radium and chemotherapy followed, but was advised after maximum doses that he still indicated positive. Finally took my advice and commenced from three drops, three times per day to twenty-eight drops three times per day. After some eight weeks, he showed clear and the doctor was amazed at the change and asked what had taken place. Only then was he told. Bob has remained on low dosage since then and remains clear now four years on."

Hepatitis C

"W. L. Williams, Oregon, U.S.—"I received a blood transfusion in 1974 while in the Marine Corps. I did not know I had Hep C until 1990. I am taking 35 percent hydrogen peroxide and my liver is less than 5 percent destroyed. My doctor told me if I did not take Interferon my liver would be destroyed and I would be dead by the year 2000. It's mid 2001, and I am well and working two jobs. Thanks."

Hepatitis C/CFID/Stealth Virus/Polycythemia Vera

A. M. B., MD, Virginia, U.S.—"I have hepatitis C which failed interferon. I subsequently used IV peroxide treatments alternating with chelation therapy and high dose vitamin C. I have been negative by PCR since 1995. My liver bx (biopsy) stabilized with just slight signs of inflammation. I have had intermittent long-standing bouts of chronic fatigue. The last one was very serious. It responded to UBI, autohemotherapy and IV peroxide with intermittent high dose Vitamin C infusions. Because of my P. Vera I ran out of veins. Prior to having a medi-port placed, I used a tanning booth to get a base tan and then I covered myself with a tanning accelerator which dilates skin capillaries. Then I covered myself with three percent peroxide and got in the tanning booth. This treatment was very helpful. Then I got medi-port and did the other treatments. I think oxidative therapies also stabilized my P. Vera."

High Cholesterol/Skin cancer/Rheumatoid Arthritis/Age Spots

Al Myers, California, U.S.—"When I started three drops at three times a day, I had no idea that it was going to do all it has done for me. At seven drops I went in for blood work. My cholesterol was back in the normal range. I started mixing one part H_2O_2 and four parts water and applying it to an age spot on my face. It scabbed up and came off. The arthritis took a while, but it got better all the time. Sometimes a small bit of discomfort is about all there is."

Lung Infection

Marion Kirbo, Texas, U.S.—"Couldn't walk across living room without resting. I was treated with two rounds of antibiotics. Still sick. Dr. decided to hospitalize to administer IV drips of antibiotics but I just couldn't stand it anymore. Someone told me about Vladimir Risov, MD, in Austin Texas. He treated me with H_2O_2 IV and I began to feel better during the treatment. Went to mall for short visit afterward."

Lung Cancer

D. Endersby, Canada—"Bronc was diagnosed with lung cancer in December 97 and was told there wasn't anything they would recommend doing; he's 80 years old. He was given information about Bruce Mailer from Stettler, Alberta, taking an H_2O_2 treatment and extending his life an extra 10 years (to date he's still alive). This treatment involved taking up to 25 drops in a glass of water three times a day until cancer was stopped, then maintaining 20 drops once per day for healthy living. Bronc went to Dr. Mike Gammon on May 9th, 2000, he told him how he started out at three drops three times per day and increased by one drop per glass every day until reaching the max of 25 drops three times per day. Mike took the X-ray and compared it to the December X-ray and the tumor had shrunk by a third. He gained eight pounds and feels much better than he has in five years. Mike says it would appear it is killing the tumor and healing the lung. Hope it works for someone else. I lost my dad in 1985 to lung cancer."

Lymphoma/Paralysis

J. D. Reilly, Iowa, U.S.—"In 94, I was diagnosed with lymphoma. It was accompanied by partial paralysis from the waist down, due to a spinal tumor. After chemotherapy knocked out my immune system and I contracted a mysterious immune deficiency, they finally decided it was pneumonia. Two people in the same day turned me to your book. I immediately started drinking Oxy-Toddy, Dr. Wallach's H_2O_2/Aloe drink. Along with a change of diet and new ways of thinking, I somehow managed to clear myself of all the cancer within six weeks without the final five of my chemo sessions. My doctor was amazed and told me to just keep up whatever I was doing."

Reversing the Aging Process

Gene Gerrard, Hawaii—"Reversing the aging process with 35 percent H_2O_2 baths three times each week for two years. Using one pint of 35 percent food grade hydrogen peroxide in half a tub of rather hot water and soaking for 30 minutes at least three times each week for two years. Among the many benefits, the most significant was a change of hair color from white to a shade lighter than the original color. See my website for more: www.oxygenheal.gq.nu Peroxide bathing has made it possible for me at 76 years young to do things you would not believe. Go to www.acrobaticshawaii.gq.nu for a picture of me balancing two adult women at the same time."

Shingles

R. Turner, Ohio, U.S.—"My six year old daughter developed a painful rash on her abdomen. I recently learned that hydrogen peroxide can kill viruses. I treated the rash twice with hydrogen peroxide and then took her to the doctor. The doctor diagnosed shingles and said it will blister and be painful before healing within three weeks. Additional hydrogen peroxide allowed my daughter's viral rash to heal within five days."

Skin Cancer

Kelvin Currie, Australia—"Total success. Slightly scratch surface, add drop of H_2O_2 (neat). Hey, presto! One-to-three days and treatments. Gone!"

Ulcers/Hemorrhoids/Feet

Tom Farr, Washington, U.S.—"I received your book and read it 10 years ago. I drank H_2O_2 diluted for four months and the symptoms went away. I think if I had kept going my asthma symptoms would have been improved also. I still gargle with it at three percent. Soaked my sister's feet in three percent for an hour and her feet didn't smell bad again. She is 51 years old. Her feet haven't smelled bad since and that was six years ago. Thank you."

Wisdom Tooth Infection

Annette, New York, U.S.—"I have a wisdom tooth growing sideways which causes food to get stuck between the tooth and the gum, creating an infection. I started rinsing that side of my mouth with hydrogen peroxide mouthwash and it cleans it right up so that I don't have to have surgery to remove the tooth."

Peroxide and Emus

Doug France, Sue France, Kiya, Sareena, Sam, Lisa, Emma-Jo, Asher Rain France, Australia—"Hello Ed, I don't know how to say thanks enough and show the fullness of the feelings attached to the words, since I took the plunge, breathed deeply and sent the email publicly asking for help earlier I've had emails from all over.

"Asher's is a really nasty little critter seeing as that it has metastasized, according to the oncologists. I'm onto the TGA people in Canberra about importing Laetrile, very hard to get a permit for a special access drug. In the meantime I've been loaned your book; you've convinced me, I put some food grade peroxide into the water on the property and all the livestock (emus) passed tons of worms and began to gloss up their feathers, put on weight and gained vigor. What can I say—thanx, thanx and for a change thanx. Dr. S. Pressman has contacted me with directions to a Professor here in Australia and I've just shot a quick email roughly outlining the problems. Here's hoping. Again, Ed, Thank you for your book and your help. God Bless you."

Peroxide and Pools

Donna, Missouri, U.S.—"I have been putting six cups of 35 percent in my 18' by 24' backyard swimming pool each Sunday night instead of the 'pool shock'. Every bird now knows where to bathe and drink. My pool is always clear and clean. The algae have not been a problem, as it filters or floats to the top for easy skimming. This amount is working fine and the neighborhood now knows (quietly) where to swim for what ails ya!"

Koch Therapy

Courtesy, Viktor Goncharov ND, Ph.D.—Complete disappearance of arthritis and hay fever in two consecutive cases (two young people age 23 and 26). A positive effect could be seen within a few hours.

Absolutely mind-boggling case of the disappearance of a uterine fibroma in 10 days. The case was pretty unusual as the patient was a young lady of 25. A little bit premature for fibroma. Nevertheless her family doctor was very much amazed.

Another complete case of relief from hepatic colic occurred within seconds after the patient was unsuccessfully treated by the physician who failed using all kinds of muscle relaxants. The patient, an old lady, experienced an unusual surge of energy and joy besides relief.

Another case of long-standing ulcerative colitis. No more cramps and no blood discharge. (Case is not over yet) As for any acute condition, the Koch treatment is a treatment of choice.

Stabilized Electrolytes of Oxygen

Genesis 1000

Testimonials collected while I toured New Zealand with Bruce Smeaton, previous owner of Oxy-Genesis in New Zealand.

Genesis 1000 is sold as Oxy Genesis in the United States.

Asthma

An asthmatic wrote to Bruce and reported that she was able to decrease her medications by 75 percent per day.

Bruce reports that contrary to some reports, these products do not release large amounts of oxygen into the bodies of users. I previously verified this with peroxide test strips.

Stable forms of sodium chlorite aren't that good as water purifiers, as they don't break down in the water glass. Less stable forms that break down in the glass would purify water.

Of 1,120 chemist shops (pharmacies) in New Zealand, 500 carried Genesis 1000. That's an example of how popular an Oxygen Therapy can be!

Dr. John Muntz, Ph.D. ND DO—"In the almost seven years I have been using (SEO) in my practice, I have dispensed it to several hundred patients with not one report of any ill or untoward effect. At this time I have seen it work in a variety of ways. I know that (SEO) has a significant contribution to make to the field of medicine and I'll be happy to cooperate in any way I can to bring it about.

Dr. Muntz continued:

Acne/Weight Problems/Headaches

"In two weeks, G1000 and the proper diet removed headaches, lethargy and nausea from a gentleman. He brought his two overweight daughters in, and they were helped dramatically by it. One of his students had an overweight acne ridden friend. She started taking 10 drops of G1000 twice a day, changed her diet and she lost not only weight but her acne as well.

Asthma

"Mrs. M. V., a severe asthmatic in New Zealand, wrote—'I have been able to cut down on my medication since taking Genesis 1000 and after two weeks cut my asthma medication—the steroid inhaler—down to 1/4 of the previous dosage. I also haven't had the Prednisone tablets since I've been on G1000. Sometimes I can manage on half the quantity of Ventolin tablet too. Thank you for your fine product.'

Flu/Infected Ear/Runny Nose

"A three month old boy had infected ears, runny nose and other flu symptoms. The mother said she didn't want the child to take antibiotics because 'that causes more problems than it solves.' The diet was already good, so she started giving him G1000, three diluted drops three times a day. The day after beginning treatment, the symptoms began to disappear. He was quickly much better.

Giardia/Water Pathogens

"When 20-to-30 drops of G1000 are left to stand in a liter of questionable water for 24-hours, the manufacturers state it will completely neutralize any harmful pathogens in the water, including Giardia.

Insect Bites

"An 11 year old boy with yearly infected insect bites that made his abdomen look like it was covered with sores started taking five drops of G1000 three times a day along with a homeopathic remedy. In two days the inflammation was gone and in four days the sores had disappeared.

Mumps

"A 14 year old girl with mumps had all her symptoms disappear in two days. The same happened to a flu patient with swollen glands for four weeks. The symptoms were gone in two days.

Rash/Vomiting

"A 10 month old baby boy wouldn't sleep, cried often to the point of screaming. He wouldn't eat and if food was forced upon him he would vomit it up. He was covered with a rash from head to foot. His skin was raw! A dietary change was tried. He was calmer for two days and then everything was the same as before. Since he was so small, I then suggested three drops in a bottle of water three times a day. Two weeks later the mother brought him back. The rash was gone; the boy had been happy and contented for over a week."

Dr. Muntz sums up—"I am convinced this is one of the most versatile products we have ever seen in the natural health field."

Dr. David Holden, N.D. Dip. Bio Chem.—"My asthma patients report energy increases within the first two weeks of treatment. For many this accompanied noticeable increases in their peak flow rates. Other practitioners notice these, as well as other dramatic improvements, in cases of eczema and psoriasis. I know an Auckland practitioner who eliminated a uterine tumor with G1000 and other oxygen metabolism enhancers. In three weeks her white cell count was back to normal. Undoubtedly, the areas of highest efficacy are in the treatment of herpes, candida, influenza, low grade chest infections, asthma, sinus and eczema."

Dr Carol Taylor—After many of her patients started getting results with Genesis 1000, they said they no longer wanted to continue taking the usual steroid-based drugs, so she began a controlled, double blind study on asthma and emphysema patients. Thus began her oxygen clinic work at Greenlane Hospital, New Zealand.

External Applications

Anecdotal and clinical reports tell of external applications of 1-to-10 dilutions (1 part SEO to 10 parts of water) being successful on psoriasis, eczema, solar keratosis, herpes, and acne.

Sports Results

Canoeist

Warren Thompson, A New Zealand competitive canoeist, states he shaved 13 seconds off his best time on a 1,000 meter course. He also can also perform repetitive sprints with only seconds in between each attempt, without going into oxygen debt.

Weightlifting Champion

Power lifting Champion Steven King reports the results he got from G1000 "have been nothing short of spectacular." He has increased his repetitions without undue fatigue or stress and boosted his recovery rate by at least 50 percent.

World Class Marathon Runner

New Zealand professional Road Runner John Bowden can now run 40 extra miles per week. He went from 120-to-140 up to 180 miles a week. His heart rate also returns to normal in half the time."

Note: No claims of 'cures' are being made. These are just examples of the actual reports we are getting.

Aerobic 07

Down's Syndrome

On July 25, 1989, I had a phone conversation with Ed Goodloe, the President of Aerobic Life Products, and an SEO manufacturer. He told me the story of a baby who had Down's syndrome (a Mongoloid child). They started giving this baby the Aerobic 07 drops and the doctors are saying that with this rate of change back to normalcy, this child should be totally normal in about a month or so. Apparently, the doctors' feel that one of the causes of this Down's syndrome is an oxygen shortage to the brain and there were certain cells in the brain that just were not burning the glucose. With the addition of the extra oxygen through the Aerobic 07, the cells have started burning glucose again. Glucose is like the gasoline in a cell, if the cell were a car. The minerals you take in through your colon with your food would be the electrical charge. The oxygen is the spark that ignites it—it burns the glucose.

Fruit Acid, Mineral, Enzyme, & Amino Oxygenating Supplements

As you will see below, this inexpensive class of products is being reported as very safe and powerful when used correctly on a wide range of ailments. They are designed primarily as overall system balancers and long-term gradual oxygenators. If you have a very serious disease like AIDS they are wonderful compliments to Ozone Therapy.

Hydroxygen Plus

Allergies/Asthma

L. D, Texas, U.S.—"My son has had problems with allergies since he was eight months old. I gave him the H+ because my mom was told about it, and tried it for problems she was having. When she read to me what it does for the body and how it would help him with his asthma, I tried it. It has made a world of difference in my son. He is breathing better, no more coughing in the middle of the nights, no choking up and having to sit up on pillows to sleep. I only have him on five drops twice a day.

"When I received my package in the mail, I read it all. In it, it said you could put it in the animal's water to keep it fresh. Well we have two cats, male and female. The female had started to get a lump in the area of her milk gland and it began to grow. Well, having worked for vets for years, I thought the worst: cancer. So since we used the same medications on animals that humans take, I decided to try some H+ on the cat. I put five drops in her water, several times a day. She began to sleep a lot, but that's the only side effect I noticed. After two weeks the lump was almost gone. Now it can hardly be felt. I have cut back on the amount now to one drop daily. She is acting like a normal six month old kitten again. This stuff is amazing."

Cerebral Palsy

P. R., Georgia, U.S.—"I started my son on Hydroxygen Plus because he has CP and so many health problems. Well let me tell you, it works wonders for him. He has three types of uncontrollable seizures and with the three medications he is on, Depakote, Phenobarbital, and Klonopin, he is usually quite lethargic. Well his therapists have noticed some amazing improvements in him. He seems to have 'perked up. His comprehension has improved dramatically, whereas before he didn't understand you, now he's much more alert and responsive. Where he was lax before, he is now in control and stronger in his movements."

Cervical Dysplasia

D. S., Tennessee, U.S.—"Four months ago I was diagnosed with cervical dysplasia or pre cancerous cells of the cervix. In late December my friend told me about Hydroxygen Plus. I started on it in early January. In March, I returned to the doctor for another pap smear. The dysplasia was completely gone. My Pap smear result was normal. I feel so blessed to have learned of Hydroxygen Plus before needing surgery. My niece sister-in law both had the same thing and they both had to have surgery."

Congestive Heart Failure/Asthma/Weight

D. H., Texas, U.S.—"I went to the hospital for Congestive Heart Failure. Three doctors said that the right side of my heart was damaged and sent me to a heart specialist. I started Hydroxygen on the 1st of March. May 5th, I went to the Heart Specialist. He had taken an Echo Cardiogram and an EKG and said; 'Why did the doctors send you to me, there is nothing wrong with your heart!' I can only attribute that to God and Hydroxygen. I have lost 35 pounds in two months on the product. As an asthmatic, I can also testify that I breathe deeper and better than I have in years."

Diabetes

J. W., Florida, U.S.—"This is one POWERFUL product for insulin dependent diabetics, or at least for me it has been! I heal faster than I have ever healed; and not only consuming it but topically on cuts and scrapes it's amazing. My tolerances on my high and low blood sugars have been widened tremendously. My eyesight is crystal clear . . . and I can go on and on. I basically have never felt better in all my 44 years and I have a ton of new energy and stamina in my workouts, I recuperate faster with no muscle soreness to speak of! I am just flat in AWE of this stuff. I'll take it (Hydroxygen/Silica Plus) for as long as I am alive. These products are the Rolls Royce products of the industry bar none! I love GHT and have finally found a 'home!' I was so fortunate to have found GHT that I will be forever in debt to my sponsor! Thanks GHT for your excellent products and your one-of-a-kind compensation plan. I love this company."

Giving Birth

Michelle Garrod, England—"During the birth of my beautiful son Zak, my doctor kept asking me if I wanted to take anything to ease the pain. I said that I was fine with Hydroxygen Plus, which I had with me. It helped me cope better with the birth. The hospital staff was just amazed. One nurse wanted to try it after the birth."

Menopause/Arthritis/Carpal Tunnel/Wrinkles

B. C., North Carolina, U.S.—"I feel like I am 17 again with vitality energy and an overall sense of wellness. I am sleeping much better which allows me to get up more rested and ready for the day. I have had no more 'hot flashes' or other symptoms of menopause. My arthritis and carpal tunnel hurts no more. My mind is sharp and alert. I breathe much easier and deep (I am a smoker) and can sing again since I have more air! I have noticed my wrinkles, cellulite, and hemorrhoids are disappearing and my skin is much improved. My digestion problems have vanished and I can eat what I want without paying dearly. I truly have been blessed."

Multiple Organ Shutdown/Near Death

L. M., California, U.S.—"On March 18, 2000, I was sent home from the hospital. I was in bed immobile and in pain, so much so that I would literally holler when I was moved at all. I was being given morphine to control the pain but it wasn't working too well. Within the next two weeks, I was back in the hospital two more times. On the 1st of April, my doctor came to the house and said, 'She is already in multiple organ shutdown and will last a few days or weeks at the very most.' I was out in hospice care in preparation for the end of my life. My diagnosis was as follows: methicillin resistant, Staphylococcus aureus (MRSA). The medical profession once had no known cure for this disease! It normally starts in the lower respiratory tract where it becomes an airborne system. This is a super bug and usually becomes fatal. I was also diagnosed with congestive heart (arrhythmia), osteoporosis, severe asthma, emphysema, thrombophlebitis (causing the legs from the knee down to be a dark red and purple), and retention of fluids (causing swelling) and of course painful arthritis. I also have several allergies, but in light of the above feel that they are not worth mentioning.

"At the time, I was on 12 different medications, 24-hour oxygen and a nebulizer every four hours. The first day of September, I started on Hydroxygen (a little too heavy I'll admit but what did I have to lose?) Two months later to the day, I was kicked out of hospice as being too well. At this same time, we ordered culture samples to be taken on the MRSA and the medical profession did not believe what they were seeing, so they took three culture samples to make sure that the MRSA was totally gone. I was using Hydroxygen Plus and Silica Plus. Also, I was on the vitamins and minerals faithfully. I went down only to six medications and have gradually improved. I was put into home health and one month later to the day I was discharged as being too well. I had come to the place where I only had two medications. I am up and around with minimal pain and resuming my life.

"In three months, I had come from death to healthy and I owe it all to a deep-seated faith, Global Health Trax, a faithful missionary Bob Minut and a loving daughter and son-in-law. I need not tell you that I will be on these products for the rest of my life."

Stroke/Leaky Heart

E. S., U.S.—"I use the Hydroxygen Plus drops. I had a stroke a month ago and already I am beginning to recover faster than I have ever recovered from other serious illness before. Also, I discovered that I have a hole in my heart. It will need surgery. But in the meantime I have been blessed with feeling more alive and alert since the Hydroxygen Plus drops. The hole in my heart leaks the blue un-oxygenated blood into the chamber of the fresh oxygenated blood; thus, 80 percent oxygen is reaching my vital organs. With the Hydroxygen Plus drops, I feel more alive. What a gift, and a blessing it is for my precious pets (my children) and for myself as well."

Systemic Lupus/Fibromyalgia/Pleurisy/Asthma/Chronic Fatigue Syndrome/Deep Vein Thrombosis

Z. S., Kansas, U.S.—"I have been sick since 1984 with systemic Lupus, Fibromyalgia, Asthma, Chronic Fatigue Syndrome, and deep vein Thrombosis. I was in the hospital several times every year, my chest being torn apart because my lungs had collapsed. I was on masses of drugs, at times spending almost $1,000 a month on medications. I had to use a walker, circulation aides, leg braces and leg wraps to walk. I was scheduled to have a double amputation. Then finally a friend told me about Hydroxygen Plus and Colostrum FM and sent them to me just two and a half weeks before my scheduled double leg amputation.

"I got the product and the first day I felt it work in my body! I was amazed! Within three days I no longer used my braces, no more walker, all pain was GONE! For the first time in 17 years, I no longer use my inhaler. For the first time since 1984, I was PAIN FREE. I had a sonogram and more than 50 percent of my clots are gone! And I will be keeping both of my legs. I walk like I am 20 years old again; my energy is the greatest it has been in my entire life."

Hydroxygen Plus and Animals

No Belief Factor or 'Placebo Effect' Here!

Horses

Experiment

An anonymous owner who runs his Quarter horses in New York State and Louisiana decided to try an experiment to see if he could get more usefulness out of his older stock by running them longer. He gave 15 horses each 10-to-12 drops of Hydroxygen per bucket of water. The Hydroxygen doesn't necessarily increase their speed, but their endurance is definitely increased and their recovery times are much shorter. They aren't as winded after their races. His doctor reports they are winning again.

Lung Infection

K. V. H., U.S.—"My dad has an awesome story about his horse! His Arabian show horse had a lung infection and the vets were really worried about him. They gave the horse the Hydroxygen Plus and Silica drops and he healed up wonderfully. The vets were amazed because usually it takes a year to get over it. The other amazing part of the story is that two other horses at the same place had the lung infection too but it was awful . . . they died because of the lung infection. We are so glad that King of Queens is doing great!"

Dogs

Coat/Skin

S.W., U.S.—"We raise Quarter Horses and pure bred Rat Terriers. I have one female Rat Terrier that was having skin problems. She had bumps and itchy places on her back. When I started using Hydroxygen Plus about two months ago, I started putting about three drops of it in her water every day and her coat and skin are beautiful again. We both love the products! We haven't tried it on the horses yet. I imagine it would take a bigger dose to work on an animal five times bigger than a human."

Coat Thinning/Bald Spots/Rashes/Sores/Fungi/Anal Gland/Infection/Skin/Pain

M. B., President, NET WINN—"I am the proud owner of a malamute wolf dog. He was, up until about two years ago, the most beautiful dog I'd ever seen. He was so beautiful that when we'd walk down the street, people would cross over, go out of their way to touch him, even get down on their knees to hug him.

"But then something went wrong: About two years ago his hair started thinning. At first I didn't worry as his coat was so thick. But then it started getting too thin, with bald spots, then rashes, then sores and a mysterious ear fungus. In and out of the vet's office, wearing me down emotionally and financially. It was getting to where I had to choose between his medical needs and my own food. Then something worse happened. His anal glands filled up with bacteria on the inside. He couldn't poop. It was a medical emergency, major surgery.

"Then it happened again about six months later. Then the anal gland problem, a third time. And then about a month ago his skin problem was so severe that his belly and inner thighs turned bright red and on fire; he was hot, not just warm and he was in pain.

"One month later and Wolf now has hair on the belly, fluffy coat coming back. So far, rashes are gone; sores are gone, definite hair re-growth. Wolf's only been on the product one month. NO visible skin problems. I hope that Wolf and I never have to live without Hydroxygen."

Coughing/Tonsils

Z. S., U.S.—"I have a little Pekinese, Remi, (Short for Rembrandt because he's one of a kind.) Well, every time he would drink water he would hack and cough terribly. I took him to the Vet and they kept giving me pills to give him and said his tonsils were always getting infected. Well the pills never worked, so I put three drops of Hydroxygen Plus in his water tank and he no longer coughs at all! Now he doesn't have to have his tonsils out. I didn't want to put him through unnecessary surgery because I love him so much. What a relief."

Heartworms

J. N., U.S.—"I give my dog Hydroxygen Plus every day twice a day. Eighteen months ago he was diagnosed with Stage 4 Heartworms and was practically given a death sentence. The vet said that my dog's heart was so compromised by the Heartworms and that it was at such a late stage that he didn't hold much hope for survival. However, I insisted that they do everything possible to save him. His heart had swollen to four times its normal size and his lungs were almost completely full of fluid because his heart couldn't do the proper job. Needless to say, he also wasn't getting proper oxygen to any part of his body. His eyes were cloudy and his coat was dull. He was sluggish and a really unhappy dog.

"I was already taking Hydroxygen myself. I decided that there was no harm in giving it to him, as well. I started putting it in his drinking water and also giving it to him directly via a dropper. I mixed 24 drops with eight ounces of water and give him a dropper full each morning and each evening.

"Clyde had the regular treatment for heartworms. However, he also got his Hydroxygen every day, twice a day. And when I took him for his follow up appointment after one month the vet couldn't believe it was the same dog. His eyes were clear, his coat was shiny, he had loads of energy and most importantly, his heart had returned to its normal size and appeared healthy. He didn't even have to go on the heart pills that the vet thought the dog would have to take if he survived the heartworm treatment."

Lockjaw

Pug Lady, U.S.—We've been using the Hydroxygen for our dogs for about three months. We had a male neutered and his jaws became locked; he couldn't open them an inch. We didn't notice right away because he was managing to suck in his food and water. But I noticed that he seemed to be barking with his mouth closed and then discovered his problem. Took him to vet, could find no reason for this. Everything else was normal. The vet gave him a couple shots, but there was no improvement after four days more. So, on suggestion of a friend who introduced the product to us, we gave him one drop of Hydroxygen in a syringe of water. The next morning, he ran in and jumped on the bed and licked me! He was already able to get his mouth open about two inches. We continued with a couple of drops per day, and he returned to normal in a week. This was after three weeks of no improvement."

Parasites

K. H., U.S.—"I have all my animals on the Hydroxygen Plus and the silica Plus. We have not had any serious health issues with my animals and hope not to. I can say that they are far healthier and parasite free; no need to worm them any longer, the Hydroxygen does the trick."

Water

G.G., U.S.—"I have had several dog owners tell me that after putting several drops in the dog's bowl, they preferred the Hydroxygen Plus water over regular. One of them ceased to drink outdoors and waited to come inside to drink out of his 'better' water. One would wait patiently for the drops to be added, not drinking until they had been added."

Cats

Chemical Poisoning/ Paralysis

P. S., U.S.—"One afternoon we came home and he was sitting on the floor paralyzed and not able to even open his eyes. He was completely limp. We discovered that he had walked through some chemicals that had been spilled on the carpet in a spare room that we have and had evidently licked it off his feet. The lady who had owned our house had many chemicals and cleaners that we had thrown away, but somehow had missed a powder that had been spilled in the bedroom on the carpet, a very strong medication that should have killed the cat and would have if we hadn't found him very soon after it happened. He was only a few weeks old, too. Tom opened his mouth and held it with his finger in it to squirt Hydroxygen Plus down his throat. At first, he did not even respond. The next time he did this, only a few minutes later because I was almost hysterical over my cat. Starbuck shuddered a little, so we knew at least he was responding. He was not able to move at all that night and not until the next day did he begin to wobble around a little. We continued to do this for him and to put the Hydroxygen in his water and food as always, adding more than usual, though, and after a few days, he came out of the groggy, doped up look and now he is a very large, though rather spoiled and silly, beautifully fluffy, temperamental member of our family! Thanks to Hydroxygen Plus."

Leukemia

D. J., U.S.—"The cat had leukemia and looked like it had one or two days to live. It was completely emaciated and lifeless and hair was falling out. But when Dan saw the condition of his cat he said, "What the heck. The cat's gonna die anyway. I'll give it a try." He doesn't just put a few drops in the water; he squeezes the bottle so a great big stream of Hydroxygen hits the cat's drinking water. His wife also administered some Hydroxygen directly into the cat's mouth in the beginning stages. It only took about three days and the cat was back to normal. The cat is in vibrant health and the coat of hair is beautiful. I saw the cat for myself."

Seizures

Monique, U.S.—"One month ago, my sweet two year old kitty, Sid, got deathly ill. I found him on a Sunday morning convulsing violently for four minutes straight. He then tried to get up and his little white paws folded in under him and he couldn't walk. I rushed him to the emergency room in tears where I found out he had had a grand mal seizure. They said I could either keep him there overnight, or take him home and see if he had another one. To avoid a big vet bill, I took him home. One hour later, He went into another major seizure and hit his head on the coffee table. I tried holding him down, but he was so violent that he spun on the ground like a tornado and spun all the way around the TV room and was soon unconscious.

"That night I woke up every few hours to Sid having another seizure. I thought for sure it was the last night I would spend with my sweet baby. Monday morning I rushed him to my family vet where they ran every test they could on him. Nothing showed up. Both my vet and the emergency vet suggested I take Sid to a neurologist, but the MRI would cost 1,000 dollars! I could not afford this at all and ended up spending hundreds on just the test and keeping Sid at the vet overnight for three days. Within a 48-hour period, Sid had 10 grand mal seizures. He looked utterly hopeless. When I picked Sid up from the vet, he could hardly walk. He weighed four pounds and the vet said I should put him to sleep. I decided to wait and take him home with his eight different medications. I prayed for a miracle when out of the blue my friend called me. He told me about Hydroxygen Plus. I was rushed the bottle and received it the next day. I gave Sid one drop, three times a day, and then gradually moved to six drops a day. Within the first moment I gave Sid the H+, he became a little livelier and played with a rock outside. Within a week he could walk again without falling over! It's been about three weeks since Sid has been on the H+ and he is like his old self again! He can walk, jump, and play! My vet was astounded! We still thought he had a neurological disease, but the H+ has relieved all his symptoms and I'm sure it will make his life much longer! He hasn't had one seizure since being on H+ and he has lots of energy (almost too much energy!) This stuff honestly works."

Hydroxygen Plus and Plants

B. C., U.S.—"My houseplants are looking so bright and healthy after adding Hydroxygen Plus to the water. We set our Christmas tree outside on the patio, still in its water container and I started adding the drops to the water. It is now March and the tree is still beautifully green and the needles are soft and alive."

L. D., U.S.—"I love Hydroxygen Plus for my cut flowers. They stay fresh and beautiful twice as long and my container plants get a drop or two in their water every couple of weeks. Good Luck bamboo is so popular now and I put a drop in the container every month. My plants and flowers always receive so many compliments from friends and family."

OxyMune

OxyMune is so new we only have testimonials from the beta test program, but I fully expect it to equal Hydroxygen's performance record in time due to similar chemistries and special proprietary formulation. OxyMune has an enhancement—intended to boost oxygen uptake in the body—that I have not seen in any other product, so keep your eyes on this rising star.

Athlete's Foot/Warts/ Sinuses/Lack of Energy/Snoring

F.G., Westchester, U.S.—"Six months ago, I complained constantly of lack of energy. A friend introduced me to OxyMune Results are amazing. Tons of energy and mental alertness. Additionally, my prior monthly colds have vanished. I have used the product topically on everything from athlete's foot to warts with equally amazing and swift results. When diluted not more than 10 drops to one ounce of water and inhaled, my sinuses and snoring are greatly improved and my wife loves the good nights."

Bumpy, Itchy Finger

B. M., Long Island, U.S.—"About fifteen years ago I started to have an itchy finger and also a bumpy effect similar to a wart. I went to a dermatologist who ordered a biopsy. The test results found it was attributed to stagnant water, such as in a fishpond or tank. It happened in the springtime, so he froze it off and gave me medication for the itching. It was okay for a while. For the next eight-to-ten years it periodically reoccurred. So I would renew my prescription and continue to use it when the itching and bumpy area would return. The skin would become very rough to the touch. After hearing about the marvels of OxyMune, I decided to give it a try. I applied it two-to-three times a day to the affected area and in about three weeks the itch and bumpy area had disappeared. My skin is very smooth again."

Full of Energy

J. K. Brooklyn Park, U.S.—"My introduction to OxyMune was three drops before going biking in eighty degrees; I rode hard for one hour. A workout like that normally would have left me winded and tired. However, I felt light and full of energy. OxyMune is now my most valued daily supplement. I love OxyMune. A word of caution. Two friends of mine insisted on ignoring the directions on the label and took OxyMune before going to bed, their extra energy kept them awake for hours."

Lack of Energy

M. N., Connecticut, U.S.—"For a couple of years, on a nearly constant basis, I had been feeling weak, tired and energy-deficient. My close friends used to remark how drawn I looked. Then about nine months ago, my cousin discovered OxyMune and began to sing its praises. She gave me a bottle to try. Almost immediately, I began to feel a major improvement. My strength and energy returned. In addition, most significantly, my skin has been wonderfully transformed. It is fresh, clear and wrinkle free. I get unsolicited compliments from friends and colleagues. Acquaintances express surprise that I'm not in my late '30s (I'm 50+). While in Macy's and Bloomingdales, strangers have asked my advice on make-up. Now I believe I am what I consume."

Noticeable Health Benefits

C. C., Tobin Farms West, & B. T., Tobin Farms Velvet Antler, U.S.— "For the past several months, Tobin Farms West in San Diego and Tobin Farms Velvet Antler in Maine have been using OxyMune in our foundational program. The wonderful oxygenating effects of this product have impressed us. Due to the high level of toxicity in our modern day world, we believe oxygen supplementation is desperately needed by humans and animals to maintain optimum health. The noticeable health benefits that we have experienced as a result of using this supplement have convinced us that OxyMune is a very valuable tool that we will continue to use in our line of foundational health products."

Sinus Infection/Coughing/Headaches/Skin Blemish

Mary Smith, New York City, U.S.—"With considerable skepticism I began taking OxyMune one year ago. Being prone toward getting allergy-driven sinus infections each season that frequently required prescribed antibiotics, my initial reactions to taking OxyMune were early-morning coughing, excessive excretion in my eyes and nose and headaches.

"At the time I began taking OxyMune, I was entering a new period in my life that involved considerable stress. With a lifetime regime of physical exercise, I considerably increased my daily exercise activity to help contain the newfound stress while gradually increasing the OxyMune daily dosage. I also increased my daily water intake, drinking at least eight glasses of water per day—not an easy task. The coughing, the eyes and nose excretion, and headaches all began to abate. One year later, I am elated to report that I have not had a sinus infection or any other illness! The increased physical strength, the improved skin tone and the additional energy that I have gained from the combination of OxyMune daily dosage and daily exercise are nothing short of miraculous! Although still under considerable stress, I am sleeping soundly—another wonderful benefit of OxyMune. Out of curiosity, I put OxyMune directly on a sun-related skin blemish located on my left hand. One month later, the sizable blemish disappeared! I am now attending to another skin blemish located on my right hand. Finally, my bowel movements are now not only daily, but frequently twice per day—another benefit from OxyMune. The skeptic has become convinced that OxyMune is a powerful product that obviously works!"

This testimonial was from the beta test program, and now let's see what she has to say after taking it for a year and a half:

Date: Saturday, September 28, 2002 3:15 PM

Thanks for forwarding your new website and all the information.

I have been taking OxyMune for 1-1/2 year and am happy to say that I haven't been sick since. I am getting slightly anxious because my supply is almost depleted. Can you tell me how I can get more OxyMune?

Sports Stamina

R. R., Wisconsin, U.S.—"OxyMune …. Just one word … WOW! This supplement is responsible for so many changes in my life. I though it was just supposed to increase oxygen concentration and with increased oxygen, increase my energy. If that was the only thing it did I would be happy, but that's only the start. Not only did my energy skyrocket. I sleep like a baby. My sleep is deep and relaxing. I wake ready to take on the day. With all the new energy and deeper sleep, I don't get sick. As a consultant, my mornings come early and my nights are long, but now I'm still able to find the energy to workout. Have you ever seen athletes breathing in oxygen on the sidelines during sporting events? With OxyMune, it seems like your hooked up to the oxygen tank while competing. I play tennis competitively. When I'm deep into the match, I now have the confidence to dig deep and find the energy to win."

Oxygenating Powders

Homozon

Bowel Problems

V. M., Creston, U.S.—"I am 77 years old and I have been using Homozon for over two years. I know I just couldn't be without it. I have had bowel problems from childhood but since using Homozon I've never taken any medication. My health and strength are amazing. I can't say enough for Homozon.

"At this time, I counted my fall as a bit of real bad luck but it turned out to be a truly fortunate event. If I hadn't slipped down those darned steps, I wouldn't have met the Freibotts. My recovery was little short of remarkable. And Homozon was responsible all the way."

Cancer

M.M., New York, U.S.—"Larry's left eye was removed because of a cancerous tumor growing on his optic nerve. The doctors told him that he would probably lose his other eye because the cancer would probably spread to it. Well, the doctors were right, the cancer was already growing on the other optic nerve and they wanted to remove that eye right away before the cancer spread to the brain. Larry came to me as a last resort. Within three weeks of treatment the cancer was gone. When Larry went back to the doctors they were amazed. They never saw cancer of the optic nerve disappear without eye damage. They wanted to know what he did and Larry did not tell them. They said whatever you are doing, keep doing it. This was all done with a colon cleanse, change of diet and Homozon."

Chemicals

Chris, U.S.—"We have been healthy and doing better daily. The Homozon has made the difference between night and day with me. I started working again in February at the truck stop and I have energy of my own. Lyle and I had the hardest struggle of our relationship when we quit chemicals and the Homozon has made the difference between living and dying for us. My mental health got questionable and it was very frightening for me. I'm working full time now. Thank you. God Bless."

Cleansing

T. S. L., U.S.—"Someone, I don't remember who, gave me a can and said try this. I tried it and was amazed. I've done extensive fasting, know my body very well and know how to oxygenate with health."

Eczema, Dermatitis, Acne

Marcus, U.S.—"I have had a history of eczema, dermatitis, acne and other minor skin disorders from six months of age until my present age of 37 years. Though various treatments such as cortisone, coal tar, worked at first, I later became resistant to their effects and had worse skin flare-ups because of them. Then I discovered your ozone products: Homozon, Sonozone and Cutazone. Homozon keeps my digestive system clean by flushing and purifying it. As a result, clear skin with no major eruptions. Sanozone is super topical treatment for surface infections and rash flare-ups. I have given several friends Sanozone for their skin problems including one with a dense rash over his whole upper torso including neck and arms, giving him immediate relief. Thank you so much for making your products available to me and others and helping us become aware of a much better and safer alternative healing system."

Irritable Bowel Syndrome

(Anon), U.S.—"About age five, my son developed symptoms which have been described as irritable bowel syndrome. This is a digestive irregularity swinging back and forth between extreme constipation and extreme diarrhea. In my son's case the constipation periods lasted much longer than the diarrhea and became so severe that the stool was so large and so hard that he lost control of his bowel on a daily basis, having movements at night in bed or in his clothes during the day. I began giving him Homozon, usually half a teaspoon twice daily. The only other change was that I would insist on his ingesting a small amount of vegetable at least once a day. An improvement in the bowel movement and consistency was observed in the first day. This Homozon dose was continued for at least two weeks at which point his digestion appeared fairly normal. Whenever we had a relapse over the subsequent year, we would give ½ teaspoon of Homozon once or twice daily and the digestion would always improve. He continued in this way with less apparent need for the Homozon until a little over a year ago when he was admitted to the hospital for an appendectomy. A wound infection developed after the surgery so that the incision had to be reopened and cleaned daily for a couple of weeks until it began to heal. At this point, a granulation developed under the skin at one end of the incision which prevented final healing. We applied a solution of Sanozone with a small amount of lemon juice to the wound several times a day. This cleared up the granulation and the incision finally healed well with no further complications."

Pleurisy

(Anon), U.S.—"I would like to say that since taking the Homozon my immune system has gotten so strong that I have not come down with Pleurisy. This is something I would come down with two or three times a year. Over all health for my wife and I has improved dramatically."

MSM

Brain Cancer/ Blood Cancer

John Lallo, California, U.S.—"At the age of 58, John Lallo's life started to change almost overnight. John could no longer remember the simple things of everyday life, forgetting everything, from saying his prayers to turning off the gas stove. He complained constantly about massive headaches every day. He was bumping into walls, having his legs involuntary collapse, falling down and loosing control of his bowels. The news was very bad. John was diagnosed with 3rd stage osteosarcoma (cancer of the of the brain) and immediate brain surgery was needed to remove a two inch diameter tumor pressing on his optic nerve causing double vision and progressive loss of the use of his legs, speech and other brain functions.

"John's surgery was completed and his tumor was removed, but it left John with all his brain functions operating abnormally. His doctor told him that he now had stage four osteosarcoma and that the cancer had spread into his blood. John was given 6-to-10 months to live with radiation and even less without it. At this point in time, John could no longer walk or drive. As he was about to die, John was then told to 'Get your affairs in order.'

"The family was devastated and had no hope in sight until their son was told by a person on his UPS route of a man who had been successful in teaching people how to turn around all types of cancer, hepatitis and HIV. John's wife made the call and John and his family found out about the amazing world of Oxygen Therapies and how they have been known to effectively kill cancer cells.

"John immediately started taking MSM (organic sulfur), an increasing dosage of Hydroxygen Plus, changing his diet to an alkaline based diet and drinking lots of clean water everyday. In two weeks time John *drove* to his doctor and *walked* into his office for a check up. To his doctor's amazement, John's swelling in his brain had been reduced by 90 percent and his speech, his vision and his walking had returned to almost normal. Remember this was only two weeks later! His doctors could not believe his progress and they were very mad at John for driving a car and not taken his medication. Nevertheless, they told John, 'Whatever you're doing, keep doing it.' They still wanted to give him radiation because he had a particularly rapid growing aggressive type of cancer. John refused the radiation and told them that he was going to be just fine and that he was going to beat this cancer faster than anything that they have ever seen.

John religiously continued taking the MSM and the Hydroxygen Plus. He also continued his strict diet and drank lots of pure water every day.

"Six weeks after his surgery John went back to his doctor for a MRI to see where the cancer was at that point. To his doctors' complete amazement, they found absolutely no swelling in the brain. All brain functions had returned to normal and there was NO CANCER at all. His incredulous doctors could not begin to figure out how this was possible, or if it was even real.

"Six months later John has received a clean bill of health and has no trace of any cancer. He is back to work building houses and his doctors told him that he should have been dead a month ago and they are still in complete disbelief. John and his family are happy and healthy and telling the world about the amazing healing powers of Oxygen Therapies, diet, water and the power of the mind." A truly amazing story!—as related to the author by witness: Keith Ranch, Planet of Health.

Folks, I have repeatedly been hearing these stories concerning all the Oxygen Therapies for 14 years. Individual results will always vary. No one claims users will always get these results or not, but such blatantly verified positive results—even if they only happen 'now and then' (a definite understatement)—should lead to immediate inquiry by everyone.

Diabetic Neuropathy

O. T., U.S.—"I had suffered from Bi-lateral diabetic neuropathy. My lower legs and feet were severely discolored with edema, constant major pain and almost no circulation, also affected by post polio. These conditions were getting worse over the years with no hope in sight. A friend of mine told me about the amazing benefits of MSM. I am very happy to say that I noticed a great improvement in the coloration of my feet and legs and I can move so much better. I am feeling great all the time now, thanks to MSM."

Scars

R. D., California, U.S.—"My 18-½" long and 1" wide scar on my left thigh is almost invisible thanks to my applications of MSM lotion on it. This is seven years after my operation for my broken femur. I have been taking MSM crystals twice a day for four years now: my health has improved tremendously and so has the health of my mother age eighty-seven and father age eighty-one. Believe this, both your hair and nails will grow and improve within one month."

Ozone

Air Ozonation

(Too many thousands to list. Ask any ozone air purifier salesperson. They all have them.)

Headaches

Grace Janacek, U.S.—"Mr. McCabe, Just wanted to thank you for your tireless work on ozone awareness. I was recently laid off from my job with a major airline due to the events of the last six weeks: (Sooo, I thought I'd find out why this little machine I bought had taken away my headaches…"

Pneumonia

D. B., Hong Kong—"An ozone machine (for the ambient air) was placed into his hospital room after brain tumor surgery and has been used 24/7 since. The doctors are amazed that he has not contracted pneumonia but he gives credit to the ozone. So simple."

Ozone Bagging

Crushed Little Toe

S. S. Denver, U.S.—"A friend crushed her little toe doing aerobics.

1st She self-treated. No results.

2nd One week after event goes to Dr. and gets pain medications. Told to stay off her feet.

3rd Four weeks now and on vacation goes to hospital because pain is so bad can't walk; they give her a supportive boot and more pain medication.

4th Back home and now toe is almost purple and green and brown; goes to the Dr. and he states that it looks like amputating is the only solution.

5th Now she is ready to listen to me. I wrap her foot up in a baggy, elevate her foot and stick a hose with ozone going into the bag, one hour and 20 minutes later we had a pink toe. Still pain but now she saw proof and repeated this over a period of one week and saved her toe."

Ozone Injections

AIDS

R. K., Florida, U.S.—"I am currently using ozone to treat AIDS. I have been using ozone IVs for two weeks now and my T Cells count has gone from 22-to-154 and the viral load has come down from 80,000-to-57. I am using 70MG/Ml. I also am using rectal insufflation to kill what I believe is candida. I see it coming out of me after I hold the ozone in me for one hour. I also do herbal colon cleaning and dry brush massage. Therapy ongoing."

AIDS. HIV/ Hepatitis B & C

R. T., Florida, U.S.—"I had Hepatitis C real bad. Viral load was almost six million along with three million viral load on HIV. I also had only 79 T-Cells. Couldn't hold down water. Doctor told me to get my house in order, that I would be dead within a month or less. I met Dr. Reyes from Life Extension after Frank Sedarnee was killed. I used ozone (The breath of God). Also lots of other stuff. For the past five years I don't show any antibodies for Hepatitis B and C. The Vanderbilt medical school and doctors in France are looking at my blood. I am the first person who does not show any antibodies for the virus as though I never had it. And my liver is normal. I tell them that my Lord Jesus Christ healed me along with the breath of God. When he heals you he does it all the way. My last T-cell count is 379 and my viral load is 954, not far from non-detectable."

Candidiasis

J. S., Seattle, U.S.—"I had severe systemic candidiasis from taking antibiotics for several years straight. I tried diet drugs, and natural remedies to no avail. Out of complete frustration and exhaustion I went to a clinic in Kent, WA, and got ozone treatments by pulling out the blood and mixing ozone gas and letting it, via IV, flow back into my body. I had 10 treatments over two months. It was miraculous how I healed and I am still free of candidiasis two years later. I even have a fairly 'loose' diet."

Cancer/Hodgkin's Lymphoma

J. F., Australia—"I have been affected personally and socially by the information that Ed has brought forward. I have studied all the information and downloaded the entire content of the website and have implemented the strategies that he talks about. Anyone not open minded enough to realize that our lives are governed by bias will end up diseased of the mind and not the body. The quality of one's life can be measured by the quality of one's thoughts. LOOK into this; this is so complicated because it is so simple and in that, also, hard to believe. I wish you luck as I have been saved and the information I have used in this material I believe has saved my brother's life from stage 2 Hodgkin's Lymphoma to the neck. Thanks Ed, You're a God Damn legend and are brave. I am sending out your message to hundreds daily in my presentations.

Thanks Ed, You saved my brothers life."

Hepatitis C

P. A., Oregon, U.S.—"Story not completed yet. But will be done in August. Ozone treatment, blood removed, mixed, re-injected. Have been feeling great; liver pain had decreased dramatically. I'm also being treated with acupuncture and Chinese herbs."

Multiple Sclerosis

Janet Beniglio was in a wheelchair with MS. I videotaped her after she had been getting a series of ozone injections and she said she started bicycling around town.

Stroke

Ozone Therapy in the treatment of a possible cerebral vascular accident, compound with aggressive LUL Pneumonia and secondary Pre-existing Ocular Nystagmus and genital Herpes.

C. K., Canada—"The patient is a 58 year old gentleman who had been in reasonably good health and was traveling for three days in a motor vehicle when he experienced a sudden change in consciousness that resulted in him becoming disorientated, uncomprehending and finally unconscious and incontinent of urine and feces. On initial examination his statistics are: height 6'4", weight 280 lbs., with a history of smoking for 35 plus years. His vital signs were: Blood Pressure 122/45, HR-120 irregular, RR-36 regular and labored and a rectal temperature of 104.6F. His pupils were 3mm equal and reacting and he was unconscious with no response. He had been incontinent of urine and feces. With persistent tactile stimulation he uttered incomprehensible sounds. A complete neurological assessment could not be conducted due to the experience of the practitioner and the inability of the patient to follow simple commands. The patient had been unconscious for almost five hours.

"After the patient was cleaned up, intravenous Octazone treatment was implemented, using a 23-gauge butterfly IV into the (L) antecubital vein. The initial injection of 100 percent Octazone was infused at a rate of 10cc per minute for a total volume of 50cc, at which time involuntary coughing was noted and the infusion discontinued.

"Over the next few hours, the patient's vital signs were measured and recorded, remaining within the same parameters as previously noted in his record. Within five hours of receiving the first Octazone injection, the patient became oriented to time, place and person and woke up wondering where he was, with no memory of what had happened. A second Octazone was given at this time within an eight hour period.

"As soon as the patient regained consciousness, Homozon treatment was implemented on an hourly basis. The patient was given one heaping TBSP every hour for eight hours; then every two-to-three hours thereafter for six days. IV Octazone injections were given on a daily basis, with a maximum saturation dose of 360cc on the seventh day of treatment.

"The patient continued to improve rapidly and was able to tolerate food by the third day of treatment. He was placed on a completely organic light vegetarian diet and the following nutritional supplements were included: High potency garlic (6 caps daily); cayenne pepper 130,000 heat (3 caps daily) floradix liquid multivitamins (2 ½ oz. Doses daily) dandelion root extract (15cc daily); colloidal silver (1 tsp., 3 times daily); vitamin C, (2000 mg, 3 times daily) acidophilus (4 caps, daily); and to treat the ongoing lung congestion-Breath Easy tea (2-to-3 cups daily); mullein/hyssop/myrrh extract (15cc daily); Zingetuss Expectorant cough syrup and electrolyte water. Other fluids given were ozonated water mixed with fresh organic lemon juice.

"This case in particular was a profound recovery experience for all not only because of the apparent complete recovery with any deficits from the stroke, but because of complete absence of the bilateral pneumonia which my father had been treating prior to his departure to the U.S. It was also discovered during observation that my father's 'wandering eyeball' had centered itself and he'd discovered a significant improvement in his vision. He only just recently revealed to me that he had also been suffering form genital herpes and none of these problems has returned or recurred to this day. He continues to use Homozon daily to maintain his health.

"Ed, I just wanted to say that I've been a believer, follower, educator and practitioner of Oxygen Therapy since 1991. Your work has been impeccable, profound, encouraging and courageous—well, put it this way: there aren't the words to describe the respect and admiration I have for you and for what you've done to advance Oxygen Therapy internationally.

"From 1995 to the present I have spent a lot of time traveling and continue to spread the news of Oxygen Therapy in a more fluid way. I've become a traveling Florence Nightingale. I've discovered in my travels and everywhere I go people suffering needlessly when I know they don't have to. I guess it's deep in my heart to help people."

Terminal Emphysema

J. A., New York, U.S.—"Evelyn was terminally ill with emphysema according to the pulmonary specialists and confined to a wheel chair when I heard about ozone treatment in Manhattan. After about four IV ozone treatments, we put away the wheelchair, as she could not walk to the clinic. I enrolled her in a lung rehabilitation program at Lawrence Hospital in Westchester where she started to lift weights and exercise on a treadmill. I couldn't believe my eyes at her progress, but she fell, fractured her elbow and was given general anesthesia. I [heard] from the ER physician that they would only use local anesthesia. She died from pneumonia. The hospital wrote to me that on the morning of the operation they got her approved for general anesthesia."

Ozone Rectal Insufflation

Glaucoma

A. Yepes, Mexico—"Twenty sessions of Magnet therapy and ozone by rectal insufflation 200ml at 46 micorgrams concentration, were enough to recover right vision. Also, the blood pressure was regulated and a tremendous trigeminous pain was relieved. I was the medical doctor using ozone and electromagnetic field."

Lyme Disease

Kathleen Swan, Oregon, U.S.—"My daughter contracted Lyme disease from a deer tick on a Colorado vacation. Blood lab testing proved this condition. Six weeks of antibiotics could NOT begin to touch the pain my daughter was in and they said, 'We don't know what to tell you to do except to have her take the Tylenol w/Codeine.'

"She could not walk for more than half a block without having to sit down from exhaustion and pain. Dr. Juergen Buche, N.D., suggested we try Ozone Therapies. Rectally insufflated 21 days, 21 days off, 21 more days.

"The second day into the first set of treatments Rebekka was out of pain, the third day she able to stand up 'straight' again instead of walking like a cripple with arthritis and each subsequent day found her stronger with the Lyme spirochete being annihilated from her body because of the ozone. It has been an entire year now almost to the day. Had we not had ozone available as treatment protocol, she could be wheelchair ridden with severe damage to her heart. Please folks. Take seriously to heart this testimony. It works. Period."

"P.S. I met a lady whose Multiple Sclerosis, although it had reached a point wherein she was bedridden, incontinent, unable to stand and going downhill fast (in her early thirties); she, too, utilized ozone and nutritional supplements (via liquid form). Just two months ago, after a brief phone conversation, she didn't have time to talk that day because she was in the process of helping her husband pack their home for a move to their brand new one. OZONE THERAPIES WORK!

Ozone Steam Cabinets

Arthritis

S.T, Courtesy, The Finchley Clinic, England—"I have had arthritis for the best part of 20 years, along with two replacement knees and a restructured hand. I know that there is no cure. But as my arthritis is spreading gradually up my spine, they don't do replacements for that; my hips and foot; I had little to lose by giving it a try. I have tried most of everything else. After just three sessions of Crystal and Ozone Therapy and a mineral supplement which you suggested I am almost pain free for the first time in my life. I am actually able to cope with public transport for the first time in years. Just to prove it is not the power of my imagination, wishful thinking or anything else, let me tell you that the first two of your treatments did not include my right hip. While other parts of my body are now pain free, the pain in my hip only improved when it was included during the third session. With the removal of constant pain and the debilitating effect this has, not to mention the depression that goes with it, I feel not just better but rejuvenated. This I believe is due to the ozone treatment, which you use in conjunction with Crystal Therapy. My skin is smoother, old operation scars fading and my system seems to be regenerating itself if that is possible."

Cervical Cancer

Kim, Courtesy, The Finchley Clinic, England—"When I was in South Africa in April, I decided to go for all my normal check ups as I know the doctors and felt more confident with their results. I went for my routine visit to the gynecologist for an annual smear, etc. The doctor managed to fit me in two days later for a small operation called the Letts biopsy. This is where they cut away an even surface of the cervix. I went in for that and hoped that would be it. We have a trace of cervical cancer in the family;

my sister was diagnosed after having her first child. She had the biopsy, the nitrate stick and laser treatment, but none of them 'cured' it. As an almost last resort she opted for ozone treatment. After six treatments she went back for a check up, only to find it normal. This definitely makes you look at ozone as the way to go. I decided that I would follow the ozone route. Whether as a precaution or cure, it would be a good thing. That is when I contacted you. From my first introduction to you, which was a phone call, I knew I had contacted the right person. I only managed to get three treatments. But still the three, which included the slow release and steam bath, made a great difference. I had to have a follow up operation and I am pleased to say the results came back normal. I am totally convinced the ozone is responsible. It worked in my family and now me."

Clostridium, Strongyloids, Trichuria (Parasites)

Dr. J. P. Dummett, Australia—A patient was recently referred to our busy Bio Toxic Reduction Research Center for severe abdominal cramping. On further investigation via Cryptostain blood urine and saliva analysis, we discovered a mass infection of Trichuria and Strongyloids. I had asked this patient to fill out some official forms prior to treatment but due to his discomfort and extreme belching I immediately seized the opportunity to try out my newly improvised O_3 steam sauna unit and O_3 (ozone) hyperthermial whirlpool spa. Upon completion of the treatment my new patient was absolutely sold over to the concept of O_3.

Ozonated Oil

Carcinoma

Dave, Courtesy, Dr. S Pressman, Canada—"Wanted to let you know, again, that the ozonated olive oil appears to be working wonders with the basal cell carcinoma. I have somewhat extensive sun damage on the temples, forehead, and nose and around the ears. It's tough to put a number on these things, but after 10 days, 3 applications per day, I estimate there has been a 70-to-80 percent improvement in the facial skin. Also have a nickel sized, deeper red patch (that tends to bleed) on my back, which is 50 percent improved. In addition, I have been applying some on my chest (upper rib cage, just below left breast) where I have had a stubborn case of Grover's Disease for seven years now. The rash has improved 50 percent. I am becoming quite a believer in the power of ozone! "Thanks again for letting me know about this stuff. I am truly grateful."

Cuts and Scrapes

M. McDonald, U.S.—"I fell yesterday, cut and scraped my legs and arms. I have been applying the ozonated olive oil and these seem to be healing rather quickly."

'Saving Face' Interview

I interviewed Norman on the left coast when he called me up to tell me about his product 'Saving Face' 1 and 2, ozonated olive oil.—"Bentonite clay cleans up the terrain of the skin. Hydrogenated fats, blackheads and transfats all stay in skin because the body doesn't know what to do with them. Saving face contains peppermint, eucalyptus, and other aromatics that tend to be 'openers of the way' so that ozonated olive oil travels in there deeper and yanks the hydrogen bonds and collapses the toxic non-food structures into liquids and gas. Just like removing little sticks of margarine in the skin and puts them back into circulation or forces them out through the upper layers of skin. "I've had people's complexions turn around overnight." he reported.

"I gave some to a film star's massage therapist. She was amazed. She gave some to the film star. He started using it and went down to do a film. The director said, "Man what did you do to your face it's great?"

John Newton got a job on *Melrose Place* and gave it to their makeup artist. She started using it on everybody there, and later Calista Flockhart, (Ally McBeal) and her people started on it. It's making a Hollywood buzz.

"I gave some to my dad. He was having a problem with his knees and arthritis. We kept putting Saving Face 2 on his knees and five days later his pain was gone."

Ozonated Water

Bladder Cancer

Dr. Saul Pressman, Canada—"Dear Ed, The guy who drank the ozonated water for bladder cancer and cured it was a Hell's Angel in Wolverhampton, England. He was a very nice chap, very large and very grateful. Whacked me on the back hard enough to crack ribs."

Stomach Cancer

A man called me up. He said, "Ed, I'm so glad you wrote that book."
"Why?" I asked.
"Well, I went out and I started ozonating my drinking water. I drank about 20 glasses of fresh ozonated water a day and my stomach cancer disappeared."

Violet/Neon Ray

Rev. Mary Seid is a certified Vibrational Therapist in Melbourne, Florida specializing in the application of the Violet and Neon Ray corona glow high frequency transdermal application device. (She does not treat or give sessions, only operational demonstrations prior to the sale of this and 25 other types of devices.

Here are high frequency photos of her energy aura before and after use of the oxygenating Violet Ray. It showed up better in the original color photos as a brilliant violet, but the increase in aura strength was remarkable after she applied the Violet Ray device to herself. Perhaps this energy increase of some kind (Chi, Prana?) is part of the reason for the good results obtained according to the testimonials below narrated by Reverend Seid. She states Edgar Cayce said the Rays must be used a minimum of 5-to-10 minutes a day for 30 days, but many good results have been obtained in one application. She makes no medical claims. She simply recounts what she personally has seen and experienced.

Reverend Seid has observed the following, and reminds you that these testimonials do not mean that you will get the same results:

Allergies

"My own testimony is that I use it when I get an allergy attack. If I put the Ray, either the neon orange, or the violet one on my sinuses after sneezing a lot, they kill the bacteria that have collected in my sinuses and immediately dry them up. Every time I get an attack, this ozonation stops my sneezing or runny nose immediately. Once I had to apply it again an hour later in order to stop my sneezing.

Bone Spur

"At a small party, a woman in her '60s had a bone spur and the bottom of her right foot was purplish. After she rubbed the Neon Ray on the bottom of her foot for about 10 minutes, we compared the color to her other foot and her normal color came back.

Breast Cancer

"I have been using the violet ray for approximately one week now, and have noticed the following changes. The tumor in the right breast has reduced in size by what I believe to be half, and the tissue area around this mass has also changed and softened. I use my   Ray three times per day. Thank you for introducing me to this wonderful energy enhancing device.

Carpal Tunnel Syndrome

"On 3/5/2002 a women called to say that after a week of using the Ray she no longer has this wrist pain.

Crippled Hand

"Maryjo called me from Indiana to say she placed it on a man whose hand was crippled. After 15 minutes of treatment he could move his fingers and with less pain. Many people observed this.

Crohn's Disease

"A woman with Crohn's Disease had pain on her right big toe. Her right foot was swollen. She applied the Neon Ray to the toe and swelling. Within minutes her toe pain was much relieved and about five minutes later her swelling was down. She was amazed. She could not believe it. I told her the Ray drew the circulation to the foot and toe.

Dental

"Twice I had a tooth abscess with lots of pain and by rubbing the Ray on my jaw over the area, it helped the pain greatly by ozonating the infectious bacteria.

Diabetes

"We were demonstrating the Ray today and taking aura photos, showing people how well it works to reduce wrinkles and complexion problems and how it promotes hair growth. A man came over and wanted to try it on his hands. Warren is his name. He may be in his '60s and has diabetes. His hands have painful arthritis and the top of the hands had big spots of black and blue bruising covering the whole top of each hand from lack of circulation. He applied the Violet Ray I think about 10 minutes on each hand. His family was with him, watching. Soon more people gathered around as everyone watched the bruising disappear, little by little. It was awesome! He could move his fingers and wrists without much pain and with more flexibility.

"His flesh color came back to his hands except for one spot. They walked around, came back an hour later and that spot was less noticeable. The Ray drew the circulation to the hands, ozonating the blood which means it gave his oxygen-starved hands oxygen. This also moved the stagnated blood.

"Steve in Orlando gave his Violet Ray to his 83 year old mom. After putting it on her diabetic wounds once a day for a week, they healed up, probably because the Ray draws the circulation to the wound. The Ray ozonated the wounds, which meant the ozone killed the bacteria in the wound. This enabled her immune system to take over [so she was able] to heal herself. The ray just gave her the extra energy she needed.

Earache

"Once I had the beginning of an earache. It bothered me for two hours, getting worse as the time passed. I put the Violet Ray on the outside of my ear for a few minutes. About an hour later the earache was completely gone.

Elderly Shaking

"Maryjo put the Violet Ray on her dad who is in a nursing home. He shook so much he could not feed himself. She put the Ray for a few minutes on each of his hands. That night he was feeding himself.

Fight injury

"Alan, my husband, was in a fight. Three men tried to mug him. Alan got punched above his eye. He used the Violet Ray as soon as he got home to help quickly reduce the pain. His knuckles were scratched and swollen from the fight. Minutes after putting it on his knuckles, the swelling had diminished greatly.

Foot and Hip Pain

"My brother Fred had a pain on the bottom of his foot for weeks. He did not know the cause. After a few minutes of the Violet Ray, the pain left and never came back. The point on that part of the foot was related to the Kidney acupuncture point. Andrea says she puts the Violet Rayon her hip that had been bothering her for years and that she gets pain relief.

Fur Loss

"Melanie's cat had no hair on her back, from allergies. The conventional vet and the holistic vet could not help her. She called to tell me that after a few days of rubbing the Violet Ray comb electrode on her cat, the hair is growing back and the cat has more energy like when it was a kitten. She put her testimony on video.

Gout

"Wanda placed the Violet Ray on her husband's feet because he has gout. He was very skeptical, until after his treatment his pain was reduced and he could move his toes easier. The next morning he woke up with much less pain in his feet than normal. He was impressed.

Hamstring, Reproductive Organs, Back Pains

"Mark M. used the Violet Ray to stop the pain of his hamstring; it took a couple days of using it twice a day for it to stop the pain. Also, he says his sperm looks healthier, like when he was younger. He also uses it for his lower back.

Pain and Energy

"A woman with shoulder pain became pain free in minutes. Another woman wanted to try it for energy and was successful]. Mostly she put it on her face. A woman who owns a health food store in Florida was tired and she placed the Neon Ray on her third eye, or brow, for about a minute. Right away she felt an increase in energy; now she wants to sell them.

Polio

"On 02/19/02 a women name Paula called me to ask if this was the same Violet Ray doctors used many years ago. I said, 'Yes, it is. How did you know that?' She said she realized this because when she was a little girl she had polio. A doctor would come by her house, massage her and use the Violet Ray and the Ray cured her, she said. WOW! I said. I'm looking forward to meeting her. She wants a Ray now.

Spider Bite

"My ex-husband was bitten by a brown recluse spider which is poisonous. It swelled very fast and became painful. I used my ray on his leg from his knee to his foot five times the first day and a few times the next day. The swelling and pain was gone. Those spider bites have caused many people to lose a limb.

Yeast

"Rob had some form of yeast on his mouth for a few weeks. He called it Candida, and said he had tried creams and prescription from the doctor with no success. His friend had him place the Violet Ray around his mouth for a few minutes. By that night, the problem was almost gone and the next day it was gone completely."

Low Level Lasers

All low level laser products used in the testimonials below were sold for industrial/veterinary use—any decision to use low level laser products on humans was made solely by the purchaser.

Arthritis

B. D., California.—"I have suffered from Psoriatic Rheumatoid Arthritis for eight years. I received chemotherapy, many anti-inflammatory drugs and cortisone and none of them worked. I met Dr. Larry and he demonstrated the laser on my hand that I could not open. After the demonstration, I could open my hand and the pain was gone!"

Arthritis

S. S., Florida.—"I had severe crippling arthritis with considerable pain. After just two days with the Resonator Laser and Laser Assist, the pain was 80 percent gone! When I was without my laser for a few days, all of the pain returned. When I started using the laser again, I was nearly pain free for the first time in years!"

Bicycle Injury

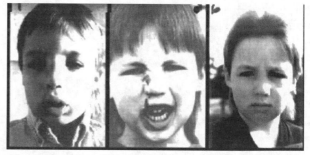

K. L., Wyoming.—"A bicycle accident resulted in the baby teeth being knocked back into the roof of the mouth. A surgeon repositioned the teeth and sutured the gums. The Resonator laser was used for 10 minutes five hours after the accident and daily thereafter. By day two the swelling and pain were gone. The scab was gone by day three and by day six the injury was completely healed. The baby teeth did not turn black nor were the permanent teeth damaged!"

Carcinoma

Dr. W. Watson, Florida.—"My 29 year old nephew had a large cell carcinoma (6cm X 2.3cm) on his tongue. He rejected radiation, chemotherapy and surgery. He used the Rotary Multiplex with nutrition and chelation. Six months later the cancer had nearly disappeared. The remaining cancer was scraped off surgically. One and a half years later there has been no reoccurrence of any cancer!"

Chest/Arm Pain in Elderly Patient

Dr. Luis Joy, Puerto Rico.—"An 85 year-old female fell and severely bruised her chest and rib cage. Tests reveled that nothing was broke, but after two weeks she failed to heal. She could not raise her arm or breathe without pain. The Rotary Multiplex was used one time for a total of six minutes. Fifteen minutes later she could raise her arm and breathe without pain!"

Deafness

T. M., Oregon.—"I had nearly lost my hearing and after tests, ordered $4,000 worth of hearing aids. After just one demonstration of a combination of the Rotary and Resonator lasers on my ears, my hearing returned. I went back to the doctor and he rechecked my ears and said that I no longer needed the hearing aids and [the doctor] refunded my money."

Inflammation

S. L., Oregon.—"Following heart surgery, my husband began experiencing severe headaches which the doctor said was caused by a condition called arteritis (inflammation of the arteries in the head). The prescribed medication took about one hour to stop the terrible pain. The last time he had one of these headaches, we used the laser on his head. The pain was GONE in less than two minutes! This machine is wonderful! We are so relieved to have this machine—if for nothing else but to relieve his headaches. However, it also eliminated my arthritis pain and his sciatic nerve problem as well."

Lower Back Pain/Migraine Headaches

Dr. P. K., Alabama.—"I bought a laser only weeks ago and have already seen wonderful results in lower back pain, migraine headaches, and carpal tunnel. I bought a Pro Pack only weeks ago and have already seen wonderful results in lower back pain, migraine headaches and carpal tunnel."

Shattered Jaw Pain

Dr. S. Dardin, Illinois.—"My friend had his jaw shattered in the uprising In Panama 15 years ago. Ten surgeries later he still had constant pain. I demonstrated the Rotary Multiplex laser for 10 minutes on his jaw; the pain was gone and has remained gone for over two months."

D K., U.S.—"I broke my right foot. It was a small fracture, but when any pressure was applied, I experienced considerable pain. Dr. Lytle demonstrated the laser and afterwards I was able to tolerate much harder pressure. I returned a day later and lasered my LEFT foot, which has given me extreme pain for 15 years, especially in the mornings. After two demonstrations, I could stand on the concrete floor without pain!"

Car Injury

Dr. D. H., South Dakota.—"I was stopped and hit from the rear at 35 mph, causing a serious whiplash injury. After a week of chiropractic adjustments, acupuncture and analgesics, the pain was still severe. After one demonstration of the Resonator and Rotary lasers combined, the pain was gone and has not returned after one week. A side benefit was that my energy levels really improved as well."

Intense Neck/Shoulder Pain

P.C., California.—"I had been experiencing intense neck and shoulder pain which would radiate down my arm, causing my hand to become numb. I tried three different types of prescribed pain pills and none gave me any relief. At the Atlanta Market, I was given a demonstration of your laser. Within the hour, the pain was almost gone. It was the first relief I'd had in four weeks. I had a second demonstration and the pain was completely gone within one hour."

Urination Frequency

Dr. J. L., Wisconsin.—"I have seen changes in my night urination frequency from three times per night to zero times per night after using the Rotary laser just four times on my lower abdominal area."

Wire Cut

A. A., South Dakota. —"This horse received a deep wire cut. Due to the severity of the wound, the vet recommended putting the horse down. With continued laser treatment of the 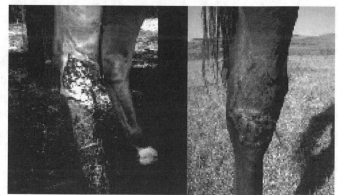 infected granulation tissue, the wound closed rapidly. With just three more laser treatments, <u>hair grew in the scar tissue</u>!"

Chapter 23

Medical Reports

News: CNN "Extra Oxygen Cuts Post-Surgical Infection Risk"

Extra oxygen cuts post-surgical infection risk, study says, January 20, 2000. Giving patients an extra boost of *oxygen* during and after surgery can cut the rate of post-surgical infection by half, says a study experts are calling dramatic and surprising.

(I guess they didn't get their copies of my first book yet!)

The research, published in this week's New England Journal of Medicine, suggests hospitals could potentially reduce the number of surgical site infections for just pennies per patient.

In the trial, 500 colorectal surgery patients from three European hospitals received either 30 percent or 80 percent oxygen both during the operation and during the first two hours of recovery.

Oxygen makes up 21 percent of the air people breathe in everyday life; anesthesiologists routinely give patients 30 percent oxygen during surgery to compensate for reduced blood circulation.

In addition to monitoring the level of oxygen in the blood, researchers also tested the oxygen concentration in muscle tissue for two sub-groups of patients.

Of the patients who received 30 percent oxygen, 28 developed infections within 15 days of the operation. But among those who got 80 percent oxygen, *only 13* became infected.

"This is a dramatic finding," said Robert Haley, MD, associate professor of internal medicine at the University of Texas Southwestern Medical Center, Dallas, and an expert on surgical infections. "While scientists have long known about oxygen's role in fighting established infections, this study is *the first to reveal oxygen's ability to prevent infection.*" He said.

Daniel Sessler, MD, senior author and an anesthesiologist at the University of California, San Francisco, said—"A radical discovery that ability can be attributed to the immune system's use of free radicals, which are chemical byproducts of oxygen.

"Infection-fighting cells take the oxygen, convert it into a type of free radical and inject those substances into bacteria to destroy them. These radicals are different from those associated with cell damage and aging." he said.

(Go ahead, doc. Say 'active oxygen sub-species' and 'peroxide.')

Finn Gottrup, MD, professor of surgery and an expert in wound healing at Bispebjerg University Hospital in Denmark.—"What surprised me with this study is it's only for two hours that they're giving (oxygen), making it a virtually effortless way to prevent infection." He said.

Gottrup, who wrote an editorial accompanying the study, said in an interview that "High concentrations of oxygen given continuously for several days could scar lung tissue in some adults and eye tissue in premature babies." But he agreed with other experts that inhaling oxygen for the relatively brief duration tested in the study poses little or no danger.

(They are talking about unhumidified oxygen which is drying and it can dry out tissue. We never advocate going against the protocols.)

Gottrup regards the research as a new approach in the fight against bacteria which have become increasingly resistant to antibiotics. "With oxygen, there's very little risk and very little extra problems. You only have to put on a mask in the recovery room." He said.

Other factors involved —According to Sessler, who is also director of the Outcomes Research Group, an international collaboration of scientists which conducted the trial, further studies may show that giving oxygen during surgery alone is enough to stop infections before they set in. A few years ago, the same group published a study that showed a two-thirds reduction in infection risk by keeping the body at normal temperature during surgery.

Lennox Archibald, MD, medical epidemiologist at the Hospital Infections Program of the Centers for Disease Control and Prevention in Atlanta— said "Poorer nations may be the biggest beneficiaries of such low-cost prevention techniques.

"In sub-Saharan Africa and Southeast Asia, one of the biggest causes of mortality and morbidity is surgical site infections." He said.

In the United States, such infections are listed as the third most frequently reported infection acquired in hospitals, according to CDC reports.

Surgeons may start using increased oxygen for this purpose at any time. But Archibald said more studies are needed before the CDC can officially recommend the practice in guidelines for preventing infections."

Will somebody *please* send all these people and people like them this book? Are they starting to leave the dark ages, at last? Will one of them keep following through? Will they actually come right out and take the next step to acknowledging the beautiful power and simplicity and safety of healing active oxygen forms?

Water-Dehydration Reports

Asthma —Frank J. Cerny, Ph.D., and colleagues from the University of Buffalo, presenting at the June 2000 American College of Sports Medicine meeting, reported that—exercise-induced asthma (which is very common in asthmatics) is caused by water loss from the airways—and can be prevented by pre-exercise hydration.

Cancer —A 1999 study published in the *New England Journal of Medicine*—examined the association between risk of bladder cancer and fluid intake in almost 48,000 men over 10 years. The men who drank the most had the lowest risk of developing cancer.

Constipation —An Italian study found that in—patients eating a high-fiber diet (25 grams per day) as a therapy for constipation, increasing fluid intake to 1.5-to-2 liters per day (6-to-8 cups) significantly improved the condition, compared to fiber alone.

Heart Disease and Stroke —A study of 34,000 Seventh Day Adventists revealed that—individuals who drank a minimum of five glasses of water a day had half the risk of heart attack and stroke as those who drank only two glasses daily. The researchers reported that adequate hydration decreased the viscosity, or stickiness, of the blood and improved blood flow.

Kidney Stone —Gary C. Curhan, M.D., of Brigham and Women's Hospital in Boston—has found that 8-to-10 glasses of fluids a day decreases the likelihood of developing kidney stones. Water dilutes the urine, minimizing the risk of the formation of crystals that can cause kidney stones.

Bottled Oxygen Water Report

Oxy-Water

Dr. Theresa Dale, Ph.D., N.D.

The Wellness Center of Research & Education, Inc.

"Apparently some questions have arisen concerning the ability of the human body to absorb oxygen in water. The oxygen in the OXY-WATER is absorbed instantly through the mucosa of the mouth just as homeopathic remedies and nutrients are. Actually, the moment the water touches the inside of the mouth (especially under the tongue), the transference of information present in any substance begins to absorb. It has been noted that some people have allergic reactions to foods when a minimal amount of the substance is merely placed in the mouth.

"I recommend anyone who wants better health to drink OXY-WATER."

Atomic Absorption Laboratory Services, Inc.

Re: Analytical results of sample I.D. Oxy-Water 2-21-98

AA I.D. #0T1708

We have completed our analysis on the above referenced sample. The results are as follows:

PARAMETER: Dissolved Oxygen

RESULTS:34 mg/l

METHOD: 360.1

All results are expressed in mg/l.

"The above results have been checked and are certified as true results based on the designated samples.

Lu Allen, Lab Manager and Charles Allen, QA/QC Manager

"Bob, Normally a typical water sample at this temperature would contain approximately 8-to-9 mg/l of dissolved oxygen. I personally have never seen dissolved oxygen level this high–Lu."

Oxy-Water Awarded Honors

In a judging of national brands of oxygenated waters, Oxy-Water was awarded the distinguished 2002 *Gold Taste Award* by the American Tasting Institute. ATI's judging panel is comprised only of accomplished chefs and all products are judged double blind. The oxygenated waters were judged on appearance, flavor, taste and texture.

OxyMune Clinic Test

Increased Oxygenation

Dr. M. Dayton, U.S.—"I have had a few people with relatively low oximetry readings who have increased oxygenation per pulse oximetry within 20 or so minutes. We have a few anecdotal findings using the pulse oximetry system. I have given the OxyMune to people with fatigue and fibromyalgia and am waiting results. I took the stuff myself and was able to exercise better.

"Perhaps, the difference on a longer-term basis is best noted when first starting and after stopping. The taste to acquire tends to easily disappear in juice. No unpleasant untoward effects, except for the temporary challenge to the palate when taken alone or with water. I have enough confidence in this product to get involved."

Hyperbaric References

Contact hyperbaric practitioners in Resources section.

Hydrogen Peroxide References

Seven-Thousand Available!

Contact I.B.O.M. in Resources section.

Chapter 24

Ozone Citations
Seven Good Clinical Ozone References

The government website Medline has 163 ozone references, and 30 include a control group! I have separated out seven of the best ozone medical references. I have done this so you can copy them quickly if needed to give to others. There are over 500 Oxygen Therapy references on my website: www.oxygenhealth.com

There are more than **3,000 medical references** in the German literature over the past 50 years showing oxygen/ozone's safe successful use on hundreds of thousands of humans who were given *millions* of dosages!

1. The German Medical Society published "Adverse Effects and Typical Complications in Ozone Therapy" by Marie Theresa Jacobs in January of 1980. One-thousand therapists treated 384,775 patients with ozone with a minimum of 5,579,238 applications and that the side effect rate observed was only .000005 per application! Ninety percent of the therapists reported the treatment was effective. This is one of the lowest side effect rates in existence. The 1980 report also stated, "The majority of adverse effects were caused by ignorance about Ozone Therapy (operator error)." The University of Innsbruck's Forensic Institute published Dr. Jacob's dissertation quoting this in The Empirical Medical Acts of Germany.

2. The International Ozone Association and the ozone machine manufacturers report **more than 7,000 doctors** in Europe using medical ozone safely and effectively, some for more than 40 years.

3. 1980, Aug 22nd—Sweet F, Kao M S, Lee S-CD (Dept. of obstetrics and Gynecology, Washington University School of Medicine, St. Louis, Mo.) and W. Hagar (St. Louis Air Pollution Control) published in Science Vol. 209:931-933, a U.S. peer reviewed scientific journal—"*Ozone Selectively Inhibits Human Cancer Cell Growth.*" They announced, "Evidently the mechanisms for defense against ozone damage are impaired in human cancer cells. . . . All of the cancer cells (lung, breast, uterine and dometrial) showed marked dose-dependent growth inhibition in ozone at .3 and .5 ppm while the normal cells were not affected. "Evidently cancer cells are less able to compensate for the oxidative burden of ozone than normal cells." They also stated that "Ozone inhibits cancer 40-to-60 percent, and up to 90 percent in a dose dependent manner, yet there is no response from mainstream medicine."

4. 1983, May 24th-25th—Proceedings, Sixth World Ozone Conference Washington, D.C., 412 page book, *Medical Applications of Ozone—* INTERNATIONAL MDs LIST 33 MAJOR DISEASES SUCCESSFULLY TREATED WITH OZONE.

"OZONE Removes viruses and bacteria from blood, human and stored.... Successfully used on AIDS, herpes, hepatitis, mononucleosis, cirrhosis of the liver, gangrene, cardiovascular disease, arteriosclerosis, high cholesterol, cancerous tumors, lymphomas, leukemia. Highly effective on rheumatoid and other arthritis, allergies of all types... Improves multiple sclerosis, ameliorates Alzheimer's disease, senility, and Parkinson's.... Effective on proctitis, colitis, prostrate, candidiasis, trichomoniasis, cystitis.... Externally, ozone is effective in treating acne, burns, leg ulcers, open sores and wounds, eczema, and fungus."

5. 1988, Oct 26—Associated Press—"Ozone may limit AIDS Symptoms" Bethesda Naval Hospital's Dr. Kenneth Wagner reports—**ozone stopped the HIV virus from multiplying, and left cells undamaged.** Dr. Steven Kleinman from the American Red Cross says it should be experimented with further. AP then reports on Dr. Carpendale, Chief of Rehabilitative Services at the San Francisco VA Hospital's ozone study controlling diarrhea, and possibly hindering the AIDS virus.

6. 1991, Oct 1—The peer-reviewed Journal of the American Society of Hematology published the ozone\HIV work of MDs Wells, Latino, Galvachin and Poiesz. —Their article *Inactivation Of HIV Type 1 by Ozone In Vitro,* appears in Blood Journal, Volume 78 Number 7, Oct 11, 1991, pg. 1882 describing the research coordinated by Dr. Bernard Poiesz State University of New York at Syracuse Research Hospital.—They performed 15 replications of an ozone study that interfaced ozone with HIV infected factor 8 blood. **The ozone completely removed the HIV virus from the blood 97-to-100 percent of the time, yet was non-toxic to normal healthy blood components**. Ed McCabe announced this study back in 1988, in his *"Oxygen Therapies"* book.

7. 1993, Aug-Sept—International Ozone Association. Eleventh Ozone World Congress & Exhibition, August 29 to September 3rd, 1993, San Francisco.—Tours were given at nine San Francisco water treatment plants using ozone before the formal meeting. Dr. Michael Carpendale, from San Francisco, California's Veterans Administration Hospital, was honored for his research work and for helping to put on this year's meeting. From his talk, *Ozone, HIV and AIDS*. "Ozone inactivates HIV at concentrations non toxic to blood cells. In saline .4 ug/ml. In serum 4.0 ug/ml. In blood 50 ug/ml. Ozone is an incredibly safe drug, increases oxygen between cells, increases vaso dilation and perfusion, increases cellular membrane permeability. T4 cell is the best marker to use." And, "Veterans Administration Hospital in Florida treated a referred patient with rectal ozone; his mother states: 'It cured his cryptosporidium diarrhea.' Doctor Frank Shallenberger reported using IV Ozone Therapy via infusion pumps set at 100cc per hour going into a pickline. He suggests freshly refilled multiple 10cc syringes instead of one large syringe so the ozone doesn't degrade in the syringe over time. He noted that the body's antioxidant reaction sets in within five to six days. Dr. Bernie Kershbaum, Philadelphia, PA, was elected head of the IOA Pan American Medical Group. The two-day medical ozone meeting was well attended by participating doctors from all over the world.

Comment: O3OHATOP—that's ozone in Russian and the meeting plainly demonstrated to us that in 1993 even the Russians had more health care freedom of choice than we did.

Dr. Claudia Koscherkova, head of research and development at the Nisnerokinov Medical Institute treated 39 arteriosclerosis patients with ozone. He and C.N. Kontorschikova M.D., Head of Central Research, and especially Sergi P. Peretyagin, M.D., Manager of the research center, are Russian oxy-pioneers. Sergi P. Peretyagin is the father of modern Russian ozone, and he and others are using medical ozone on babies, animals, adults, and even gunshot wounds! Their video detailed simple quick and inexpensive techniques like an ER technician sticking the ozone output tube down into a bullet hole to instantly sterilize and cauterize the wound. They have four major medical ozone treatment centers in their largest cities; Nezhni Novgorod, Ivanovo, Kirov, and Smolenskj, plus others in the Baltic Republics.

They are also using Ozone Inhalation Therapy—a 1-to-4 mcg/ml ozone concentration is humidified by ultrasound and inhaled. They are also pioneering a recirculatory extracorporeal ozonation of human blood for up to 1½ hours at a low concentration of 1½-to-2 micrograms per milliliter.

Hundreds of Russians have, during the past 15 years, treated thousands of people with ozone, and along with the other countries using it they are showing up America's so-called 'superior' medicine.

Chapter 25

North American Ozone Medical Citations
Prove Oxygen Therapies Safe and Effective

1929, *"Ozone and Its Therapeutic Action"*—A 40 page book with a different medical paper from a different expert on each page. The book names 114 diseases and applications of ozone and prints the research from the centers and doctors centers using ozone. Armor Research Foundation Institute of Technology, (IL), American College of Physical Therapy, Berlin University, Behren Memorial Hospital (Glendale, CA), Board of Education, (St. Louis), Bouvicant First Hospital (Paris), British Army Medical Service, (London), Chicago College of Medicine and Surgery, (Chicago), Harvard University, (Cambridge, MA), Polytechnic Institute, (Brooklyn, NY), Physical Chemistry University, (IL), Post Graduate Medical School, (NY), S. California University of Los Angeles, Salaberry Hospital, (Buenos Aires), Spaulding General Hospital, (Portland, OR), Western Reserve University, (Cleveland, OH), Washington University, (Seattle, WA).

1961, *Encyclopedia of Chemical Technology*, **Volume 16, Third Edition. John Wiley & Sons**—"The symptoms of breathing high concentrations of ozone are acute; there appear to be no chronic affects among normally healthy people because the body has the ability to repair such damages. No free radical reactions which directly involve ozone have been observed. During the 80 year history of the large scale usage of ozone, there has never been a human death attributed to it."

1987, March 8-12—K.S. Zanker presents, *Ozone has benefit in cancer treatment,* paper at the 2nd International Conference on Anticarcinogenesis and Radiation Protection Gaithersburg Maryland. — "The selectively inhibitive effect of medical ozone prepared from pure oxygen on tumorous human cell cultures has been known for a long time, and has recently been confirmed once more." Washuttl, Viebahn, and Steiner, reviewing this study.

1988,—Dr. Gerard Sunnen publishes, *Ozone in Medicine: Overview and Future Directions* **in The Journal of Advancement in Medicine**— Dr. Sunnen, at the Bellevue Medical Center, New York City, lists medical ozone as commonly used worldwide on "Herpes, AIDS, and Flu. Also, used on wounds, burns, staph infections, fungal, and radiation injuries, and gangrene, as well as on colitis, fistulae, hemorrhoids and anal infections. It promotes healing. Blood ozone treatments have been used to treat virus infections, including AIDS, hepatitis, flu, some cancers, diabetes, and arteriosclerosis. Used in dental surgery, periodontal disease, mixed in water and swallowed for use on gastric cancer, and applied as a wash in intestinal or bladder inflammation. Mixed with olive oil, it is used on fungal growths and skin ulcers. Ozone baths are used to irrigate the skin, to disinfect and treat eczema and skin ulcers. All of the world's blood supplies may be made bacteria and virus free (AIDS, etc.) by passing 40-to-50 mcg/ml of ozone through them."

1988, Nov 28 Insight, Vol. 4, No. 48, P. 56, by Dina Van Pelt—Ozone may be able to kill HIV without harming infected blood cells. The Bethesda Naval Hospital in Maryland conducted tests on HIV infected blood in which ozone killed the virus without damaging the cells that contained it. Dr. Kenneth F. Wagner, senior research physician for HIV research at the Henry M. Jackson Foundation for the Advancement of Military Medicine, Rockville, MD., says "European physicians have used ozone to safely treat viruses for years. In a study of five patients at San Francisco's VA Hospital, all five showed significant improvement without signs of toxicity."

1989, Aug Poult Sci 68(8): 1068-73, *Biocidal activity of ozone versus formaldehyde against poultry pathogens inoculated in a prototype setter.* Whistler PE, Sheldon BW Department of Food Science, North Carolina State University, Raleigh 27695-7624—Ozone was evaluated as an alternative hatchery disinfectant to replace formaldehyde in the event that the Environmental Protection Agency regulates the use of formaldehyde under the Toxic Substances Control Act. Cultures of Staphylococcus, Streptococcus, and Bacillus species previously isolated from poultry hatcheries and selected culture collections of Escherichia coli, Pseudomonas fluorescences, Salmonella typhimurium, Proteus species, and Aspergillus fumigatus were spread-plated on open petri plates and independently fumigated with ozone or formaldehyde in a prototype laboratory poultry setter. Ozone (1.52-to-1.65 percent by weight) resulted in bacterial reductions of greater than 4-to-7 log10 and fungal reductions of greater than 4 log10, whereas formaldehyde achieved reductions of greater than 7 log10 and greater than 5 log10, respectively, after eight minutes of exposure to either disinfectant. Potential mutagenic effects were observed in ozonated E. coli colonies resulting in decreased superoxide dismutase activity and increased catalase activity when compared with non-ozonated control colonies. In this study ozone reduced microorganism counts but not as much as formaldehyde. Ozone may be used as a disinfectant against selected microorganisms, although further testing under actual hatchery conditions is needed before making recommendations to the industry. NATIONAL LIBRARY OF MEDICINE PUB MED ID#: 2506541

1991, Oct—Dr. Michael T. Carpendale, MD. Veterans Administration Hospital, San Francisco, and Joel Freeberg, M.D., UC Medical School San Francisco, Bay Medical Research Foundation, San Francisco, published in The Journal of Antiviral Research 1991, Volume 16 Number 3: 281-292 the following medical paper—*Ozone Inactivates HIV at Noncytotoxic Concentrations* "HIV (p24) was reduced in all ozone treated cultures compared to controls." Dr. Carpendale also privately published, *Ozone Treated HIV+ patient becomes PCR Negative*, the stories of two ozone rectal insufflations using AIDS patients—one who became PCR negative.

1993, Jan—Rev Fr Transfus Hemobiol 36(1): 83-91, *Viral inactivation and reduction in cellular blood products*. Friedman LI, Stromberg RR, American Red Cross, Holland Laboratory, Rockville, MD.—Even though the risks associated with the transfusion of blood products are lower than ever before, considerable efforts are being employed to improve the safety of the blood supply. Based upon available data, a six log (99.9999 percent) reduction in virus level from screened and tested blood components should significantly reduce or eliminate the risk of post-transfusion infection. The objective has been to identify 'generic' methods, that is, one that would be applicable to all viruses. For red cells, physical and chemical approaches have been studied; for platelets, the approaches have been limited to chemical. The physical methods include depletion of leukocytes by filtration, removal of plasma by washing, and viral inactivation by heat. Among the chemicals investigated to inactivate or help displace virus are **ozone**, detergents and hypochlorous acid. Several photochemicals have also received intensive investigation: merocyanine 540, a benzoporphyrin derivative, aluminum phthalocyanine, and methylene blue. For platelets, photochemical inactivation methods using merocyanine 540, and two psoralen derivatives, 8-methoxypsoralen(8-MOP) and aminomethyl trimethyl psoralen (AMT), have also been studied. Approaches which include washing are not suitable. For the most part, either viral removal or inactivation has been insufficient or red cell or platelet damage unacceptable. However, there are a few indications that at least inactivation of a specific virus, such as HIV, may be possible without major cell damage. These studies are in their early stages and significant work remains. If feasibility is clearly shown in vitro, it is likely that in vivo primate studies to demonstrate safety and efficacy will be required. Publication Types: Review, tutorial NATIONAL LIBRARY OF MEDICINE PUB MED ID#: 8476492

1993, May—Am Surg: 59(5): 297-303, *Irrigation of the abdominal cavity in the treatment of experimentally induced microbial* **peritonitis: efficacy** *of ozonated saline,* Ozmen V, Thomas WO, Healy JT, Fish JM, Chambers R, Tacchi E, Nichols RL, Flint LM, Ferrara JJ, Department of Surgery, Tulane University School of Medicine, New Orleans, LA 70012.—Ozone is an oxidizing agent possessing potent in vitro microbicidal capacity. This study was designed to address the extent to which irrigation of the contaminated abdominal cavity (using a saline solution primed with ozone) is effective in reducing morbidity and mortality. Gelatin capsules containing different quantities of a premixed slurry of filtered human fecal material were implanted in the peritoneal cavities of a preliminary series of rats. Three inocula concentrations were selected for later experiments, based upon their ability to produce morbid consequences: (1) high (100 percent 1-day mortality), (2) medium (70 percent 3-day mortality, 100 percent abscess rate in survivors) and (3) low (100 percent 10-day survival, 100 percent abscess rate). Fecal and abscess bacteriology were similar in all rats. The peritoneal cavities of 240 rats then underwent fecal-capsule implantation (three groups of 80 rats/inoculum concentration). At celiotomy four hours later, equal numbers of rats from each group were randomly assigned to one of four protocols: (1) no irrigation, (2) normal saline irrigation, (3) saline-cephalothin irrigation and (4) ozonated saline irrigation. Each treatment lasted five minutes, using 100 ml of irrigation fluid. Mortality was significantly reduced when, in lieu of no irrigation, any of the irrigation solutions were used. Additionally, ozonated saline statistically proved the most effective irrigating solution for reducing abscess formation in survivors. NATIONAL LIBRARY OF MEDICINE PUB MED ID#: 8489098

1993, Sep 17—J Clin Gastroenterol (2): 142-5, *Does ozone alleviate AIDS diarrhea?* Carpendale MT, Freeberg J, Griffiss JM Rehabilitation Medicine Service, San Francisco Veterans Administration Medical Center (SFVAMC) 94121—Five patients with acquired immune deficiency syndrome (AIDS) or AIDS-related complex (ARC) and intractable diarrhea were treated with daily colonic insufflations of medical ozone (oxygen/ozone mixture) for 21-to-28 days. The daily dose of ozone (O3) ranged from 2.7-to-30 mg. Three of the four patients whose diarrhea was of unknown etiology experienced complete resolution and one patient had marked improvement. The fifth patient, whose diarrhea was due to

Cryptosporidium, experienced no change. No consistent change in the absolute number of helper (CD4) or suppressor (CD8) lymphocytes was detected and no obvious changes were seen in the PO2 or the results of routine hematologic and blood chemistry studies. Patients had mild to moderate local discomfort during ozone administration early in the course of treatment, but no adverse systemic effects were observed. **The results of this series suggest that medical ozone administered by rectal insufflation is simple, safe and effective**. Should this simple treatment be used routinely to treat chronic intractable ARC/AIDS diarrhea? NATIONAL LIBRARY OF MEDICINE PUB MED ID#: 8409316

1994, March 25—Dallas, Texas. IBOM International Bio-Oxidative Medicine Foundation's 5th annual meeting—MDs from all over the world highlighted their own work successfully using ozone and/or hydrogen peroxide and other oxidative compounds in medicine and attending special educational workshops. Among the papers presented were: *Spontaneity of Oxidation in Nature*, Majid Ali, MDv; *Ozone in Medicine*, Frank Shallenberger, MDv; *Hydrogen Peroxide and Free Radicals*, Charles H. Farr, MD Ph.D.; *Complex Oxidative Compounds*, George Freibott ND; *Experiences in the further Treatment of AIDS, Cancer and Chemical Toxicity/Hypersensitivity Using Bio-Oxidative and Nutritional Therapies,* Robert Allen MBBS (Australia); *The Cause of All Disease from a Holistic Perspective*, Ed McCabe; *Oxidative Therapy and the Answer to AIDS*, Robert Willner, MD, Ph.D; *AIDS, Immunology and Ozone*, Frank Shallenberger, MD; *Experiences With Medical Ozone*, Stanley W. Beyrle, N.M.D.; *Ozone May Inactivate HIV by Reducing p120-CD4 Binding Affinity, Lysing the HIV Lipid Envelope and Oxidizing the HIV core*, Oscar K. M. Hsu (Harvard).

1994, Sep-Oct—J Eukaryot Microbiol; 41(5): 56S-57S *Pilot-scale ozone inactivation of Cryptosporidium*—J.H. Owens, RJ Miltner, F.W. Schaefer and E.W. Rice., Risk Reduction Engineering Laboratory, U.S. Environmental Protection Agency, Cincinnati, Ohio 45268. NATIONAL LIBRARY OF MEDICINE PUB MED ID#: 7804257

1998, Jan—Med Hypotheses; 50(1): 67-80, *Selective compartmental dominance: an explanation for a noninfectious, multifactorial etiology for acquired immune deficiency syndrome (AIDS and a rationale for Ozone Therapy and other immune modulating therapies.* **Shallenberger, F**—the most widely accepted etiological explanation for acquired immune deficiency syndrome (AIDS) currently invokes an infectious model involving the human immunodeficiency virus (HIV). Because this infectious model has failed to meet any conventional criteria for establishing microbial causation, this theory still relies on the high, though not perfect, statistical correlation linking presence of HIV antibodies with patients diagnosed with, and at risk for the syndrome. Many scientists and clinicians now doubt the HIV theory and propose instead a multifactoral causation similar to that seen in cancer and heart disease. In order to discard the HIV model, however, it is necessary to explain the high statistical correlation mentioned above. Recent studies involving cellular mediated immunity and cytokine modulation may explain this statistical relationship without the need to invoke infectious causation, by suggesting certain functional characteristics and feedback loops in the immune system, which the author calls selective compartmental dominance (SCD). SCD provides a model in which chronic dominance of the humoral immune compartment secondary to chronic high-dose antigenic challenge results in chronic suppression of the cellular immune compartment. This model predicts that even HIV-negative members of the risk groups are susceptible to AIDS, assigns no special causal role for HIV in AIDS, and suggests a rational course of nontoxic therapy that can potentially reverse cases in the earlier stages. Publication Types: Review, academic NATIONAL LIBRARY OF MEDICINE PUB MED ID#: 9488185

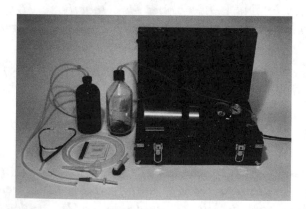

Chapter 26

Oxy-Experiences in Other Countries
Germany

Thousands of doctors, over 50 plus years, have safely and successfully delivered millions of ozone treatments for so many different diseases and conditions that I can't list them all. Dr. Gerd Wasser told me there are five to seven thousand doctors and 2,000 or so paramedics using it there, as reported to him by the top five ozone generator manufacturers.

I must of necessity advise you to research all the English and German language ozone literature from Germany starting with Viebahn-Haensler's classic, *The Use of Ozone in Medicine* and to also call all of the German ozone oxygen societies and manufacturers on your own.

Remember, in Germany they have a great deal of experience, but mostly still use the old slow autohemotherapy method. This delivery system has merit but has already been eclipsed by direct IV and Recirculatory Hemoperfusion ozone delivery systems.

Cuba

In addition to the other Cuban stories in this book, including the long list of references in the back, at the Center of Medical and Surgical Research in Havana, 15 adult patients suffering from herpes zoster were treated with injections of ozone/oxygen for a period of 15 days. All patients were completely symptom free after the treatment, AND follow-up examinations a year later showed not one relapse.

Italy

Professor Velio Bocci

Professor Velio Bocci is well respected in the international medical community for his publications and research investigating the pharmacology and physiological significance of interferons, and more recently, the induction of endogenous cytokines using ozone. His CV is impressive, and he has written several papers documenting parts of what we all know to be true about ozone by creating a bridge between historic ozone usage and modern breakthroughs in immunology. He has focused his ozone research on mapping the metabolic pathways of the immune system that are activated by Ozone Therapy. Here are the particulars.

Velio Bocci has been Professor of General Physiology since 1971 and since 1978 Director of the Institute at the Faculty of Pharmacy at the University of Siena, Italy. He was awarded the Medical Doctor degree at the University of Siena in 1954, and specialized in Respiratory Diseases and Clinical Hematology at the University of Rome (1958) and at the University of Florence (1975), respectively.

After a brief training in Surgery, in 1956 he went back to the Institute of General Physiology at the University of Siena where, with some intermissions, he has worked since then. He received further training in Biophysics at the National Institute for Medical Research in London (1959-1961), in Biochemistry at the Maudsley Institute in London (1964-1965) with a Nato fellowship, and in the Department of Biochemistry at the University of Buffalo, US (1967-1969) with a Buswell Fellowship. He has spent also brief periods in Biochemistry at the University of Uppsala (1972) and in Neurochemistry at the University of Goettingen in 1976.

He is the author of about 305 publications and monographs mostly published in international, peer-reviewed journals, including a book on Interferons (1993) and a book on Oxygen-Ozone Therapy (2000). His fields of research include important subjects such as plasma protein separation, labeling and metabolism, neurochemistry of isolated neuronal cells, erythrocyte metabolism, pharmacology and physiological significance of interferons and, more recently, the induction of endogenous cytokines using ozone. Since 1991, he has contributed crucial research papers regarding the biologic effects of ozone aiming at making Ozone Therapy a truly scientific endeavor.

He has given seminars on problems related to interferons in several Italian, English, American and German Universities and has participated during numerous international and national Congresses as session chairman and speaker.

He is on the Editorial Board of the Journal of Biological Regulators and Homeostatic Agents, Mediators of Inflammation and has served as Coeditor of Kidney, Proteins and Diseases (Karger, Basel) from 1987 to 1994.

On 1995 he was awarded the Hans Wolff prize for innovative researches in the field of Ozone Therapy.

By considering him the top expert of Ozone Therapy the International Ozone Association (Zurich) invited him to hold the lecture on "Ozone in Biology and Medicine" at the International Ozone Symposium (Basel, October 21-22 1999) during the 200th Anniversary of the birth of the ozone discover C.F. Schönbein.

On June 1999 he was elected president of the International Medical Ozone Society (IMOS- Italia). He directs and teaches a theoretical and practical course on Ozone Therapy during September-October of every academic year.

1. Bocci, V., C. Aldinucci, F. Corradeschi, G. Valacchi, and G. Fanetti. Ozone Therapy as a complementary medical approach. Where do we stand today? Where are we going? Atti Congressuali 1998.(In Press)

2. Bocci, V., G. Valacchi, F. Corradeschi, C. Aldinucci, S. Silvestri, E. Paccagnini, and R. Gerli. Studies on the biological effects of ozone: Generation of reactive oxygen species (ROS) after exposure of human blood to ozone. J.Biol.Regulat.Homeost.Agent. 12: 67-75, 1998

3. Bocci, V., C. Aldinucci, F. Corradeschi, L. Paulesu, and G.P. Pessina. Il fumo di tabacco induce citochine proinfiammatorie a livello polmonare: cinetica di produzione, passaggio transalveolare ed effetti sui macrofagi. CNR, Progetto Finalizzato "Prevenzione e controllo dei fattori di malattia (SP2-ambiente e salute)", Follonica (GR), 27-28-29 marzo 1996. Atti Congressuali 18-19, 1996.(Abstract)

4. Bocci, V., S. Silvestri, F. Corradeschi, E. Luzzi, and C. Aldinucci. Verso una razionalizzazione della ozonoterapia. Atti del VI Congresso della Societa' Italiana di Ossigeno-Ozono Terapia, Roma, 29 settembre/10 ottobre 1994. Acta Toxicol.Ther. XVII(2-3): 185-199, 1996.

5. Bocci, V., E. Luzzi, F. Corradeschi, L. Paulesu, A. Di Stefano, S. Silvestri, and C. Aldinucci. Development of a biological response modifiers system based upon stimulation of blood with ozone ex vivo and reinfusion. 3rd Symposium on Biological Therapy of Cancer, Munich. April 19-22, 1995. 1996, (UnPub)

6. Bocci, V., F. Corradeschi, E. Luzzi, C. Aldinucci, L. Paulesu, and S.D. Silvestri. Ozonoterapia: ieri, oggi e domani ? Ossigeno Ozono/ Fitness & News VI: 1-2, 1994

7. Bocci V. Ozone Therapy today. Proceedings 12th World Congress of the International Ozone Association. Ozone in Medicine. Lille, France 1995: 13-27

1. Bocci V. Ozonization of blood for the therapy of viral diseases and immunodeficiencies. A hypothesis. Medical Hypotheses 1992;39: 30-34.

2. Bocci V. Autohemotherapy after treatment of blood with ozone. A reappraisal. The J of Intern Med Res 1994; 22:131-144.

3. Bocci V. A reasonable approach for the treatment of HIV infection in the early phase with Ozone Therapy (autohemotherapy). How 'inflammatory' cytokines may have a therapeutic role. Mediators of inflammation 1994;3: 315-321.

Also from Italy, Dr. R. Mattassi of the Division of Vascular Surgery at the Santa Corona Hospital in Milan, Italy treated 27 herpes patients with intravenous injections of oxygen and ozone. All patients healed completely after a minimum of one and a maximum of five injections. Five years later 24 of the 27 were still outbreak free. Re-infection was suspected in the other three.

Russia

And as if life isn't confusing enough at times, here in the U.S. we have the **Medizone** International ozone company:

And in Russia, in the Moscow area, they have **Medozon(e)**:

How Russia Is Beating Nuclear Swords Into Ozone Plowshares

In addition to the other references in this book about how the medical community in Russia has discovered and wholeheartedly embraced Ozone Therapy, I found this interesting item on a U.S. government 'trade' website during a web search!

RUSSIA: (ISA) HOME HEALTH CARE PRODUCTS AND EQUIPMENT MARKET

August 2000, Prepared by Ludmila Maksimova, Commercial Specialist
The Commercial Service Moscow 23/38 Bolshaya Molchanovka Ul., Moscow 121069, Russia

"There are numerous examples of successful production of home health care devices at the **Russian former defense enterprises**. One such example is the Medozones company from Nizhny Novgorod, which specializes in production of medical oxygen generators. The company was created by a group of Russian medical scientists in cooperation with the nuclear scientists from the Federal Atomic Center. **In response to the growing popularity of Ozone Therapy** the above groups of scientists formed the Russian Association for Ozone Therapy in 1995."

The Russian World of Ozone Today

The cold war may be over, but America is way behind in the medical ozone race so far. The Russians have a huge lead. Firstly, there are about 1,300 medical ozone machines used in Russia, 600 of which are Russian Medozons series medical ozone generators. There are about 1,950 MD doctors and about 1,300 nurses practicing Ozone Therapy there.

There are about 1,100 medical organizations where Ozone Therapy has been used as a monotherapy or in addition to other therapeutic methods, 660 of which are big state regional or city hospitals. Some of these big state hospitals have departments for Ozone Therapy, and the remaining 440 are private clinics or surgeries or medical centers specialized only in Ozone Therapy or providing ozone treatments along with other methods.

Also, there are about 40 state medical institutes, universities and academies teaching Ozone Therapy to doctors, for example four in Moscow, and two in Nizhny Novgorod. There are about 200 doctoral dissertations on Ozone Therapy. During the last few years, six International and three Russian congresses on Ozone Therapy took place: in New York, 1989; Gavana, 1989, 1997; Monaco, 1991; San Francisco, 1993; Lille, 1995; Tokyo, 1997; N-Novgorod, 1992, 1995, and 1998. The Association of Ozone Therapeutists was founded, and the practical manuals were issued. Several companies both in Russia and abroad have begun serial manufacturing of devices for Ozone Therapy.

Ozone Therapy began in Nizhny Novgorod, which remains the homebase of Russian Ozone Therapy. Russia formally started with fundamental scientific experimental and clinical investigations into ozone about 25 years ago. They told me that in 1977 they developed the methods of intravenous ozonated saline infusions and the original method of ozonated extracorporeal blood circulation (RHP).

It was only after a long series of investigations were undertaken that confirmed ozone's safety and efficiency that Ozone Therapy was successfully introduced into the entire Russian health care system. Nowadays Ozone Therapy is being routinely carried out on a daily basis, on thousands of patients and yielding remarkable clinical results. During the last 10 years Ozone Therapy has come of age in Russia. And its progress is intensifying each year, so much so

Dr. Sergey P. Peretyagin.

that the Russian medical experts expect that the above mentioned data will increase by approx. 30 percent in the next year.

When I corresponded with the Russian Medozone Company, I found that my previous 14 years of work spreading the good word about Oxygen-Ozone Therapies all over the world had even reached into mother Russia and brought me personal notice there. I received the following correspondence from Natalia Berdnikova, Manager of Foreign Business at Medozons Ltd. on 9/23/02:

"Dear Mr. McCabe, thank you very much for visiting our web site and expressing an interest in our company and our ozone products. Your name is well-known in Russia particularly among the specialists in the field of Ozone Therapy. And you do us proud to focus your attention on our company.

"We are really interested in expanding our commercial activities into North America. We actually have much experience working in cooperation with the Russian Association of Ozone Therapy which has developed its own school of Ozone Therapy existing for about 25 years.

"There are a great number of experimental and practical clinical investigations of Ozone Therapy being conducted on the hospital grounds of Nizhniy Novgorod Medical Academy in each field of medicine. There are a great number of doctoral dissertations being published on Ozone Therapy. Our doctors participate in all the most important World Ozone Congresses where they introduce their papers on the use of medical ozone.

*"The priority of the Russian school of Ozone Therapy is the method of **intravenous infusion of ozonated physiological 0,9% sodium chloride solution** developed in 1977 by Prof. Peretyagin and co-authors. It incorporates possibly low, absolutely safe and therapeutically active ozone concentrations up to 10 000 mcg/L. This approach employs a long-lasting contact of medical ozone with the sick human body that allows the achieving of positive therapeutical effects by using considerably lower ozone concentrations in comparison with the Western school. The safety and high efficiency of this ozone concentration range has been confirmed by a great number of experimental and clinical investigations. Along with this approach we successfully perform all known traditional methods of Ozone Therapy as a monotherapy or in treatment complex.*

"Medical ozone generators of Medozons series have been designed and manufactured specially for the Russian school of Ozone Therapy. Our devices have a completely automated operating system: before treatment you need only to set treatment time, ozone concentration and flow rate which will be automatically maintained with attested high accuracy (+/- 5%) during the treatment, that means, you don't need a photometer to perform a check-up every 100 hours of operation, because traditional ozone machines show a decrease in ozone generation with time (as Prof. V. Bocci wrote in his paper 'Has Ozone Therapy any future in the West?')

"What primarily makes our machines different from the traditional Western ones is the new conception of ozone production. We have used in our ozone generators not traditional ozonization tubes, but an ozonization chamber with attested durability of more than 8000 continuous operation hours before scheduled servicing. Unlike the traditional generators where the regulation of ozone concentration delivered depends on the tension and oxygen flow rate used, our machines at a constant tension and constant gas flow rate (three possible variants: 0,25; 0,5 and 1 L/min) allow the creation of 10 000 values of ozone concentration (at 1 step per mcg/L) within the range 0 to. 10 000 mcg/L through the regulation of electric discharge that occurs automatically after setting the required ozone concentration and flow rate on the keyboard of the unit, and this is thanks to the revolutionary new construction of ozonization chamber employed...

Thanks.
Awaiting your soonest reply,
Natalia Berdnikova, Manager Foreign Business."

From the Russian Medozone sites:

Фирма "Медозон"

MEDOZONS Ltd. is a medical company engaged in the development of innovative methods of Ozone/Oxygen Therapy and the manufacture of medical ozone devices as well as accessories for Ozone Therapy successfully used in various pathological situations. The company was formed in 1995 as a result of the successful cooperation between the Russian Association of Ozone Therapy provided a scientific-medical support and the Russian Federal Nuclear Centre ARZAMAS-16 provided a scientific-technical support.

The head office is in Nizhny Novgorod (approx. 450 km easterly from Moscow, with approx. two million inhabitants, the capital of Povolzhsky Federal Region of Russian Federation). The production facilities are located in Arzamas (Nizhny Novgorod Region), the factory was certified on April 19th, 2000 by the Notified Body TUEV Thueringen E.V. according to EN ISO 9001 (1994-08). The company has an official representation in Ukraine.

'Medozone' is approved by the <u>Ministry of Health of Russia</u> and recommended for production and use in medical practice.

Only recently, our main activities have been concentrated within Russian Federation and CIS, but now the situation is changing and the company is striving for the further expansion of its sales and distribution structures worldwide.

Russian Ozone Products

(From the Medozons Website)

Medical Ozone Generators of the Medozons Series have been developed in close cooperation with the Russian Association of Ozone Therapy especially for the Russian school of Ozone Therapy. The new Medozons-Universal medical ozone generators have a name that reflect their features; the ability to realize both the Russian and Western schools of Ozone Therapy.

The medical ozone generator model 'Medozons-Universal' has been manufactured since 2001. Reflecting its name, this model allows medical work within the range of ozone concentrations accepted both by the Russian and Western school of Ozone Therapy as well. This model is manufactured only on individual request. Like other medical ozone generators of Medozons series it ensures an absolute purity of ozone/oxygen gas mixture generated, and high accuracy in maintaining the given ozone concentration. It is reliable and safe to use. The range of delivered ozone concentrations is 0,1-to-80 mg/L. Accuracy in maintaining the given ozone concentration is +/- 5 percent.

Ozone generator Ozon-250 is used for disinfection, decontamination and deodorization of operating and other treatment rooms through air ozonization. The air unit is ensures 100 percent disinfection of air in the room to be treated and effective suppression of micro flora on surfaces and deodorization of air. Ozon 250 is approved for use in the practice of medical disinfection. It can completely replace the use of chemical disinfectants and ultraviolet radiation.

'Fit' apparatus is for delivering fine dispersions of ozonized water and was

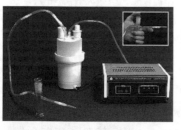

designed on the basis of an ultrasound inhalator using the SAWRS-effect (resonance structural effect on the physical state of material surface). The ultrasonic ozonized water disperser can considerably improve the therapeutic properties of medical ozone applied by local, intracavitary and inhalative routes. Fit comes with one set of accessories for external and intracavitary irrigation as well as for **inhalation methods**. Fit is used in gynecology, dermatology, cosmetology, stomatology, combustiology, and surgery. Aerosol particles delivered in this way are of submicroscopic size (less than 1 mcm) with stimulated surfaces providing a long life-time of the induced drops due to 'self-coordinated' linked stimulations 'volume – surface.'

The Fit apparatus allows the delivery of aerosol particles of ozonized water having advantageous properties:

- Owing to sizes of less than 1 mcm, drops are able to penetrate into wrinkles of mucous membrane, pores and even through skin, factually it's a new way of introduction of medicine into the organisms.
- Therapeutical effect of ozonized water can be considerably improved thanks to ultrasound induction of production of particles by powerful energy impulse.

Inter-action between stimulated aerosol surface and bacteria/virus membrane lipids leads first to disbalance of aerosol drops resulting in so called 'drop explosion' and finally, to efficient destruction of bacteria/virus membranes by the 'cumulative stream' produced, thus improving the biological effects of the medicine used.

Medozons, in cooperation with the Russian Association of Ozone Therapy, has worked out a Physician's Manual for Ozone Therapy containing experimentally verified and clinically proved treatment programs. The manual is supplied free of charge with the purchase of each medical ozone generator.

Medozons, again in cooperation with the Russian Association of Ozone Therapy, also provides educational courses to medical practitioners under the supervision of Dr. Sergey P. Peretyagin, President of the Russian Association of Ozone Therapy, an ozone pioneer with about 25 years of experience in Russian Ozone Therapy.

Innovative Russian Ozone Application Accessories

Ozone can be applied externally to specific diseased areas via 'cupping.'

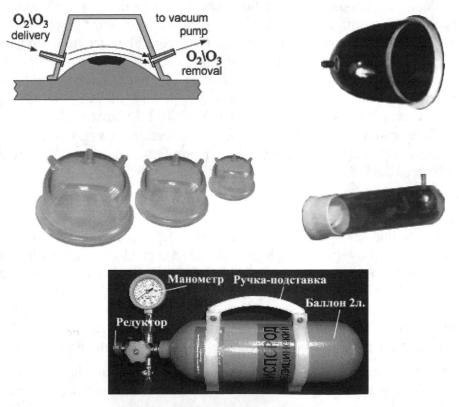

England

London Therapist's Experience with Ozone Steam Saunas

Mark G. Lester, The Finchley Clinic, England—"I am a holistic therapist in London who has been working with Ozone Therapy for the last three years. I first took an interest in Ozone Therapy after my mother's death from cancer in 1997. Inspired by the work of Ed McCabe and a number of patients I spoke to; I decided to introduce the therapy to my clinic. The decision to buy an ozone generator and steam bath was sealed after I spoke to a man in Texas who had cured himself of cancer (lymphoma) using this method.

"My results were spectacular from day one. My first patient was a lady with shingles whose sores dramatically improved after one session (and subsequently disappeared altogether after some more sessions).

"Very soon after that I started getting a lot of inquiries about chronic fatigue syndrome and M.E. and this is where the ozone steam sauna method really shines. It is also very effective for those not suffering with M.E. but who are run down and lacking vitality generally. I found that the vast majority of people who give the therapy enough time would improve tremendously, reporting increased energy levels and overall well-being with 4-to-8 weeks of weekly treatment. Many were in effect cured. Others were not 'cured' but did improve beyond recognition. However, some need to come once a month to avoid slipping back. I see that as natural and normal, in the same way that someone will be 'cured' of scurvy by taking vitamin C and the disease will return if vitamin C is absent in the diet; people are 'cured' by Ozone Therapy, but, if treatment stops altogether, the oxygen deficiency which caused their problem in the first place will start to creep back.

"Although Ozone Therapy does help with many diseases such as chronic fatigue mentioned above, I prefer to see it as a method of detoxifying the body rather than as a medicine for any specific disease. Therefore any problems caused by toxicity will respond. One very obvious area where I have found the ozone steam sauna highly beneficial is silicon poisoning from artificial breast implants. Ladies who make the mistake of having these dreadful things usually end up regretting it later in life and are beset by illness and pain. However I have found the ozone steam sauna to be very effective and getting such people feeling better within a few sessions.

"I find that virtually every practitioner insists that their method of using ozone is the best way, and in some cases the claim is made that their method is the *only* way. Personally I am not prepared to subscribe to such nonsense. The fact is that despite the many studies carried out on the effectiveness of ozone generally, no study has ever been done comparing the different methods, and personally I believe in them all.

"One of the few organizations to be honest about the fact that all methods work is the "CENTRO INTERNACIONAL DE RETINOSIS PIGMENTARIA" in Cuba who state, "For Ozone Therapy we use low concentration ozone given in intravenous and rectal routes. Patients may choose between these two routes of administration since there is no significant difference between them in terms of efficacy." [In one of their retinitis pigmentosa studies 89 percent of 175 patients studied showed marked improvement for two years after treatment –Rev CENIC Cien Biol, 1989; 20: 84-90.]

"My own belief is that the same could probably be said of the steam sauna method. I believe the main thing is to get the ozone into the body by any means. However, I am willing to say that I have found the steam sauna method very effective and useful to my patients.

"Insufflations are also used (vaginal, ear, and rectal), where necessary in my clinic. However the steam sauna method has the advantage of hyperthermia at the same time. To quote from my website, 'During a fever, the functioning of the immune system is stimulated, while the growth of bacteria and virus is forced to slow down.'

"In a 1959 review of studies on the effects of heat treatments, Mayo Clinic researcher Dr. Wakim and colleagues cite findings indicating that the number of white blood cells in the blood increased by an average of 58 percent during artificially induced fever. The generation of antibodies speeds up, as does the production of interferon, an anti viral protein that also has powerful cancer-fighting properties. However with the addition of ozone the effects are vastly more powerful. People sometimes ask how transdermal ozone compares with Turkish steam baths, but the truth is that there is no comparison. My experience is that the combination of hyperthermia and ozone *at the same time* works very well indeed, and is a lot cheaper for patients than autohemotherapy or directs IV as the majority of therapists are not medical doctors and do not demand such high fees.

"Skin diseases are another area where the steam sauna method really shines. Because the ozone can work directly on the skin to heal the lesions and detoxifies the body at the same time, my experience is that conditions such as eczema and psoriasis improve very fast. People usually notice the severity reducing within about three sessions. In some cases people have been 'cured' within three sessions, though as previously mentioned in many cases monthly maintenance treatment is required to continue the effects.

"Arthritis is another ailment that I have seen the ozone steam sauna help enormously. I have seen people who struggle to walk up my stairs bounding up, pain free, within a few sessions.

"MS is a disease where heat is supposed to be potentially harmful. I am sure this is true, but I have found many times that the benefits of steamed ozone clearly outweigh the harm of high heat levels. I have used the steam sauna a number of times with MS patients and always found that patients report greater energy and mobility with no untoward effects. (In cases with this disease, I do start cautiously with the heat in the early sessions however).

"Finally, many people comment on the emotional benefits. A lot of people have reported to me that depression has been alleviated (without the need to spend thousands of pounds talking about their relationship with their mother!) I have two theories about this. I believe either it is because once the body is detoxed the brain functions better. The second is that once the body (and the brain, especially,) finally have that which has been lacking for so long –sufficient oxygen—there is a mental lifting and depression improves. It is also a great stress reliever. No doubt this is partly due to the oxidizing effects of ozone upon adrenaline. However it is also due the relaxing effect that heat has generally, one of the reasons why Turkish steam baths are so enjoyable."

Australia

Bob Graham reported to me that Dr. Noel Campbell and Dr. Ralph Ballard at Swinburne University in Melbourne, Victoria, Australia gave six bone cancer patients one hour ozonated steam sauna treatments over five days and all six patients were able to stop using morphine for pain.

VALENTINE SERIES No. 2588   OZONE THEATRE, MILDURA.

One patient had breast cancer and unfortunately a sloppy biopsy had spread it all over her exterior skin as they drew the needle out (without flooding the area with ozone) and she left and was in a lot of pain. She had her chest cupped (bagged) and ozonated. In two weeks the cancer had turned jet black, she came off morphine, and they continued ozonating as the cancer cracked and fell off. New tissue was growing in at 12 weeks. Lucky for her that the university had already seen chronic fatigue removed by ozone so they were open to it. After 9 years, the Australian Therapeutic Goods Administration (TGA, like our FDA) has approved ozone research at Swinburne University.

Czechoslovakia

Dr. Capek's Clinical IV Ozone Results

"The Injections of ozonated blood are indicated especially for sleeping disorders, regulatory (vegetative) disorders and for a support of general immunity. Can be used effectively in chronic HSV/Herpes, Zoster affections, in chronic tonsillitis, allopetia areata, and a male infertility.

"Rectal insufflation I find especially suitable for children, where we can address chronic intestinal dysmicrobia with consequent immune disorders, again HSV/HZV eruptions, eczema, chronic colitis, colitis ulcerosa and even Mb. Crohn.

"All three above mentioned applications can be used in different types of cancer.

"Direct injections of gas oxygen/ozone mixture into femoral artery bring instant relief and improvement in cases of arterial obstruction (obliteration) in the lower extremities. Once to twice a week is sufficient frequency, if we can give a complementary homeopathic/isopathic support. This application can save a leg affected by diabetic gangrene, a diabetic angiopathy. Also, it may be the only access, useful in patients who are/were using IV drugs (here I mean both chemotherapy and hard-drugs as well). Some of these patients responded very well and demanded this therapy in order to overcome the physical hangover-stress. I remember one case of a middle-aged woman, who suffered for years from edemas in lower abdominal/femoral area. She tried everything in the world and nothing helped. After single Hyperbaric Ozone Therapy (and a dietary advice) most of her edemas disappeared. I have treated her then several times until she was completely symptom-free! (She worked at the post-office and brought me thereafter the whole department for examination and therapies.)"—Oldrich Capek

Canada

1991—Medizone ozone company negotiated with the Canadian Armed Forces—to commence 2½ million dollar ozone research study to begin in 1992, comprised of animal studies, blood sterilization studies, human trials, the works.

1992—Jan Captain Michael Shannon, Deputy Surgeon General for the Canadian Naval Forces—has 350 scientists under him. He wants to commence a 2½ million-dollar ozone medical research study that will include blood sterilization, animal trials and more. The Canadian Government realized that if a major event occurs, their army's blood supply would be inadequate in its present condition. He states, "With thousands of people going to Europe yearly at a considerable cost to get this treatment, there must be something to it."

1992—Bolton, A. Report on *Scientific Studies to Elucidate"Ozon-O-Med" Treatment of Peripheral Vascular Disease*—Intermune Life Sciences, Etobicoke, Ontario, 1992.

1992, Sept—Research & Development Bulletin, No. 234. Science and Technology section of the Canadian Government's Supply and Services Dept., publishes—*Better Blood Sterilization with Ozone*—"Under a $303,943 contract with the Surgeon General's Branch of Department of National Defense [DND] Headquarters, researchers from the National Reference Laboratory at the CRCS are investigating two ozone sterilization technologies to confirm their reported efficacy in deactivating a variety of potential viral contaminants in blood, including HIV-1 and hepatitis.

1993, April—Journal of the Canadian Medical Association: Medical Science News (Can Med Assoc J 1993; 148(7) pg 1155)—*Are Worry Free Transfusions Just A Whiff of Ozone Away?* by Albert C Baggs, B.Sc. — "Scientists in the U.S. and Canada are investigating the use of ozone to destroy the HIV virus, the hepatitis and herpes viruses and other infectious agents in the blood used for transfusion. The studies were endorsed by medical circles of the North Atlantic Treaty Organization (NATO) because of a concern that viral pandemics have compromised the ability of world blood banks to meet urgent and heavy military demands." In a brief to the NATO Blood Committee, the Surgeon General of the Canadian Armed Forces reported upon Canadian findings that... "A three minute ozonation of serum spiked with one million HIV-1 particles per milliliter would achieve virtually 100 percent viral inactivation." It was also found that "the procedure would destroy several other lipid-encapsulated viruses, including simian immunodeficiency virus and various strains of interest to veterinarians." The journal report described the work of Mueller Medical and Medizone International.

Comment: The Surgeon General of Canada is so impressed with ozone he is telling NATO about it! How much more *Establishment* can we get?

1993, June 2—Headline: Medizone's Blood Decontamination Technology Proven Successful in Canadian Monkey Trial—Still trying to inch their way through the system, Medizone announces its successful trials on monkeys. This was one of the requirements imposed last August when McCabe/Bedell/Latino, et. al. met with Dr. Fauci at the NIH. Dr. Fauci said, "Why can't you do a simple monkey trial?" So Medizone did and they announced the successful completion of the first two phases of a Canadian research project overseen by scientists representing the Canadian Red Cross, Canadian Departments of Defense and Agriculture, Cornell University Veterinarian Medical College and Medizone Canada Ltd. Two groups of monkeys were infused with plasma infected with highly virulent strains of Simian Immunodeficiency Virus (monkey equivalent of HIV). The first group died within 12 days. The second group's infected plasma was first infused with ozone through Medizone's process. None of the second group showed any sign of infection. Dr. Latino, Medizone's President stated: "These preliminary research results indicate the capability of Medizone's patented scientific and technological process to inactivate blood and blood products of certain viral contaminants, including the AIDS virus. Including the above monkey trial, Medizone has so far submitted to the FDA all the following data:

1. Long Island College of Pharmacy rabbit study showing no toxicity at concentrations up to 10 times the dose proposed in man.

2. Cornell University Veterinary College feline study, showing no detectable toxic effects.

3. The Hematology Journal HIV inactivation study showing 100 percent inactivation.

4. Mount Sinai School of Medicine, New York revealing hemolysis and coagulation changes well within the standard for re-infusion of packed human blood.

1993, Sept 2—The world premier of the Canadian video, *Ozone and the Politics of Medicine*—was shown at the International Ozone Association meeting. The video was subsequently shown in Los Angeles, New York, Salt Lake, other health shows, and won The Atlanta Film

Festival's 'Best Documentary Award'. Although this documentary has won four international film awards, none of the networks, including PBS, would air it.

1996, December 12, Life Tech's Corporation—announced today that Dr. Fred Quimby, the leading researcher at Cornell University—has confirmed that ozone-induced oxidative stress inactivates Simian immune deficiency virus, SIV, in human blood. Dr. Quimby's research provides independent proof of concept that Life Tech's technology, which is an ozone technology, sterilizes human blood of viruses that cause illness and death in humans. Walter Dermouth, Life Tech's President and CEO, summarized the significance of this scientific validation. Dr. Quimby's data has verified the unique ability of ozone-induced oxidative stress to kill viruses in human blood and verified the unique ability of ozone to kill viruses in human blood efficiently, effectively, and without damaging the blood. It is particularly notable that Dr. Quimby inactivated SIV, which is virtually identical to the Human Immune Deficiency virus, HIV. We have now taken an important step toward eliminating the risk of getting HIV and other deadly diseases from the transfusion of blood and blood products. Life Tech Corporation is a Canadian company dedicated to developing, commercializing innovative technologies to sterilize blood and other biological fluids. So, here's another company that's taken another form of an ozone delivery system and is very slowly trying to make its way through the medical establishment step by step by step. Proving what millions of people on this planet already know, though somehow the American Medical Establishment can't seem to listen very well . . . that when you flood the body with oxygen and ozone, the virus or bacteria is anaerobic, so it can't live in oxygen."

Mysterious Canadian Study

In Canada, Commander Michael Shannon, former Deputy Surgeon General for the Canadian Naval Forces was and is successfully researching ozone. He found that in Europe an estimated 350,000 people were treated with ozone between 1980 and 1985. The University of Bonn reviewed these cases and reported virtually no side effects of Ozone Therapy when properly administered. In 1990 he said, *"The products of this research have worldwide applications. In the right concentration, ozone sounds almost too good to be true. We're trying not to be overly enthusiastic, but the data so far is very compelling."*

My 1994 Article
Debunking the Flawed Canadian Study

Canadian Medical Research Mystery
Department of Defense Positive Ozone - HIV Results Ignored

The story of a mystery within the modern Canadian medical establishment. A proven safe and effective treatment for alleviating AIDS symptoms is being hidden from view.

Summary: Various Canadian government and military and commercial representatives got together in 1989-1990 to allegedly test the safety and effectiveness of 'Ozone Therapy' in the treatment of AIDS. Two small pilot studies were undertaken. This resulted in the 1991 Publication, *The use of ozone-treated blood in the therapy of HIV infection and immune disease—a small pilot study* (phase one was 10 patients, phase two was 14 patients) *of medical ozone's safety and efficacy. AIDS* 5:981-984. G. Garber, D Cameron, N Hawley-Foss, D Greenway and M. Shannon.

The methods used were antiquated and the device quality control was non-existent by any objective standard, and the actual published conclusions were exactly opposite to the conclusions written up by one of the chief investigators!

This is a study the rare detractors of medical ozone like to quote, to falsely try and promote a viewpoint that it doesn't work. I say 'falsely,' because we know for a fact that such detractors parade the false published version of this study out instantly, while hundreds of positive medical ozone references held in their possession remain strangely ignored. Hundreds of references were sent to one of the detractors, the U.S. FDA, and they have admitted having them in writing, yet they never mention them, and only send this incorrect one to government officials that query them.

Modern medical Ozone Therapy consists of taking three combined very active atoms of pure oxygen (ozone) and surrounding all the anaerobic (can't live in oxygen) primitive cell bacteria, viruses, funguses, and parasites with it. This active form of pure oxygen makes it impossible for the quickly oxidized microbes to live and ozone is also harmless to normal healthy human cells when used correctly. All the secondary infections and possibly even the prime infection, either go away or enter a level of remission concurrent with the procedures and methods applied, including the pre-treatment health of the patient. When applied properly—meaning using correct procedures, delivery methods, concentrations, volumes, and durations—ozone is extremely safe and effective, according to published animal and human studies over the past 100 years.

In an early (early for North American research) Canadian establishment attempt to see if there is any validity to the overwhelmingly positive reports on medical ozone coming out of Europe, and to see if it was safe, an outdated and poorly executed ozone protocol was used in a 'small pilot' study by researchers inexperienced in the use of Ozone Therapies. Outdated because the method chosen was long ago deemed painful and ineffective in 1938 and inexperienced, because they had never used ozone before.

Modern medical ozone application has a few significant procedures that classically trained scientists outside the field know nothing about, since the many Ozone Therapy procedures are not taught in North American and Canadian medical schools. These investigators incorrectly chose the delivery method of Minor Autohemotherapy as their starting point. This was a backward decision when we consider that ozone has been in use by thousands of European physicians for over 50 years and far better methods are currently in use worldwide. The Minor Autohemotherapy method (min AH) they chose involved withdrawing a small amount of blood, mixing ozone into it—to kill the viruses, and then re-injecting the dead viruses—this is the immunization theory, not Ozone Therapy.

Minor Autohemotherapy is a poor cousin to Major Autohemotherapy, which is now giving ground to even more successful modern delivery methods. The most modern of the successful Ozone Therapies skip this unneeded dead virus inoculation step and directly flood the blood, lymph and cells with virus destroying pure medical ozone/oxygen gas. This is done in a variety of ways, IV, dialysis type recirculatory systems, ear, vaginal, penile and rectal insufflation, sauna bags, and devices, breathing ozonated air and drinking ozonated water. Our bodies soak these higher oxygen forms right up harmlessly because we have evolved in an oxygen environment.

Starting in the late 1800s and up to the present, hundreds of thousands, perhaps millions use some of these methods daily. The small pilot trial we are discussing was sponsored by the Ottawa General Hospital Infectious Disease Division, the Canadian Department of Health and Welfare, the Canadian Federal Center for AIDS, the Canadian Department of National Defense, and The Mueller Medical Company of Canada, now Vas-O-Gen.

My analysis: If you compare the protocols used in this study with the known to be more effective modern ozone methods and then also compare the internal letters of the investigators reporting their documented findings against the strange pronouncements of the final published version of the study, it immediately becomes apparent that the published study was, if not deliberately fraudulent, then close to it via negligence.

1990 Antiquated Protocol of Minor Autohemotherapy used by Canadian DOD	Modern Methods of Medical Ozone Therapy Ozone Direct IV or Recirculatory Ozone
Too little ozone, too weak a concentration	Perfect ozone amount & concentration
Treated only a small amount of blood	Flood all the body fluids and cells with ozone injected or recirculated in the veins
Withdrew and treated only 10 cc of blood	Perfuse the whole blood supply, or inject, or re-infuse at least 500 cc of treated blood (1/2 quart)
Treated the blood with a less effective and too-low concentration of ozone, only 3 mcg/ml^3	Treat with proper concentrations – always use a minimum of 27 to 42 mcg/ml^3
The ozone was heated first	Ozone is never heated, heat destroys ozone.
The tiny amounts injected into the wrong place (huge gluteus muscle)	Ozone is always infused only into veins or arteries and rarely or never into the muscles
A scant total of only three tiny injections given	Full amounts of ozone are best applied twice daily or at least every other day until healed
Study used a broken machine which intermittently produced no ozone	Use a working machine

How can I make such an accusation? Let's look closely at the facts. Of prime importance is the fact that the very design of the study was so out of touch with current known worldwide private ozone medical practices that it should never have been labeled 'Ozone Therapy'. It is a travesty to call this anything other than a poor distant cousin to modern Ozone Therapy.

However, on the plus side, one fact stands out on the very first page (981) of this study published in 'AIDS.'

The authors plainly state: *"Preliminary work has suggested that ozone does inactivate HIV in vitro."*

Then they also state that they proved ozone does indeed kill HIV-1. They withdrew 10cc of blood, and interfaced 3 mcg/ml^3 of ozone with it and the ozone destroyed all the HIV-1 viruses, and didn't hurt the blood. *"The resulting inoculation presents a killed virus antigen preparation."*

These statements alone prove ozone deserves further study! Let's examine the materials and methods used by doctors who assume, without any training, that they know everything about ozone. Look at the above table comparing the outdated protocols employed in the DOD study and the normal modern medical ozone delivery methods.

As I have shown, when you examined this sort of 'test,' you knew it would be doomed to failure from the start since it is too little, too weak, too few. I knew for sure that when the 'study' was over, all the patient blood was still black and diseased.

Results Ignored

The published document ignored the results of Phase 1A wherein three of the few patients who had any immune system left—each with CD4 T-cells above 200—had their counts go from 220-to-230 up to 500. The patients gained weight and reported feeling great. Instead, the published document stated: "No difference was seen between placebo and ozone treated patients." The reason for the seeming mystery: I learned from a personal interview with the ozone generator manufacturer's technician that in Phase 1B, the second half of the study, either the ozone generator mysteriously 'broke,' or someone deliberately sabotaged it, because the ozone generator was producing very little or no ozone! And when the Mueller medical technician dutifully reported this to the investigators, he and this fact were ignored and the study was written up without reflecting the facts!

Incorrect Dosage Schedules and Volumes

Phase 1A and Phase 1B treated only 10cc of patient blood during only three times a week treatment days. Wrong procedure. This is fine to make an inoculation, but inoculations only work on people whose immune system are fully functional; certainly inoculations are not applicable to a study of AIDS patients and their compromised immune systems demonstrating only 50-to-500 T cell count ranges.

Ozone is used to sterilize municipal drinking water all over the world. How are you going to clean up the microbe infested waters that these human patients were made up of, by putting only 10cc (less than a teaspoon) of barely-touched-with-ozone blood into a muscle—and only three times a week? Compare this choice of Minor Autohemotherapy (MinAH), with its only thrice weekly injections against the obvious objective of getting rid of this disease by cleaning all the viruses, bacteria, funguses and parasites out of the 100+ POUNDS of water that the human body consists of. The injected (MinAH) small amount of oxygen/ozone is used up immediately when the oxygen tries to oxidize the existing and incoming pollution and microbes.

Also, the total cleansing objective is challenged daily by the added burden of leaving 2⅓ days of normal daily living between treatments. These 'skipping treatment' days allows the environmental and dietary toxic intake load to continually tend to undo this minuscule attempt at the cleaning process. There is no way you could ever hope to 'keep up' with this method by cleaning faster than the body absorbs new toxic burdens, especially under the stress of a disease like AIDS and its constant bacterial and viral replications! 10cc of Minor Autohemotherapy is to be considered only a drop in the pond of the diseased body waters.

Incorrect Concentrations

The tiny 10cc of withdrawn patient blood was treated with an equally weak and tiny 3 microgram per cubic millimeter by weight, ozone concentration (assuming a functioning ozone generator!) Private clinics using ozone know that a minimum of 27-to-42 mcg/ml^3 concentration is necessary for maximum viral kill with a minimum of hemolysis (standard acceptable levels of normal cell damage). This study used only a drop in the pond of acceptable concentrations.

Wrong Delivery Method

The tiny amount of blood with the (possibly intermittent or non-existent) tiny concentration of ozone was introduced into the body by injecting it into a large muscle. No one who really knows how to use ozone has employed this method since 1938, when Dr. Paul Aubourg used it in his study in two Paris hospitals. He proved that although other methods of the application of ozone, like rectal insufflation, gave excellent effectiveness, the intramuscular injection method was 'painful and ineffective.'

The actual data does not match the published conclusions. Even more suspect than the above errors is a detailed comparison between the actual investigators' internal inter-office correspondence and the final published document. Let's look at excerpts from a letter by Captain Michael Shannon, now Commodore Shannon of the Canadian Department of National Defense (the Canadian military forces) written to Dr. D.W. Boucher on January 24, 1990. Note: A copy of the letter to Dr. Boucher and the accompanying data was stamped CONFIDENTIAL and handed/ leaked to me at a health show in a plain brown envelope by an interested party who said, "You don't know where you got this." The party was outraged at the following duplicity, but too self-protective to directly challenge 'the system.'

Actually, there was no need for all the cloak and dagger stuff, because a copy of the exact same letter—not marked confidential—had previously been published by Barry Bruder in his work *Ozone Therapeutics, a Current Compendium* in August of 1993. M.E. Shannon CD, MA, MSC, MD, one of the principal investigators, wrote the following in his final report and recommendations to the superior official representing the government funding, Dr. D.W. Boucher, of the Bureau of Biologics, Health Protection Branch, Health and Welfare, Tunney's Pasture, Ottawa, Canada, on January 24th, 1990: *"Ozone Therapy in AIDS/Project #231 Summary of Findings."* Dr. Shannon: This trial yielded...

"Encouraging results."

"There has been no clinical, biochemical or immunological evidence of adverse/toxicological effects....

"An improved sense of well being characterized the clinical responses of all patients....

".... Several patients reported a return of appetite and concomitant weight gain....

"Four patients suffering from arthralgic pain reported a significant amelioration of symptoms...." Three out of Four reporting complete relief of what was well documented to have been a chronic condition....

"The lack of bruising at the site of injection was somewhat surprising....

"Three patients showed a significant positive response...in their CD4 measures....

"There were no detrimental effects on absolute CD4 counts for any of the patients....

"One patient showed a 52 percent reduction in the initial P24 antigen levels with a corresponding increase in absolute CD4 count."

The earlier September to October 1989 series of investigations by Dr. M.O. Shaughnessy, Virology Division of the Bureau of Laboratories and Research Services, *"clearly support the contention that the technology has potent virucidal (virus inactivating) effects. . . . It would appear that this form of therapy constitutes a potent means of inactivating HIV-1 in contaminated blood supplies, and may also provide a means for patient specific 'auto vaccination' in selected cases."* ('Selected cases' meaning those with enough of an immune system left so that an inoculation will make the immune system respond.)

"These results are considered well beyond the error limits for the particular assays and indicative of potential therapeutic benefits which should be further investigated.... As reported in earlier correspondence, (1988/89 Ottawa General/NDMC) several cases of long standing sciatica and one case of severe facial pain secondary to an invasive naso-pharyngeal carcinoma responded dramatically to this form of therapy....

"As the understanding of ozone biochemistry increases and potential toxicological concerns dissipate, analgesic applications of this therapy should be pursued.... Since a subgroup responded, consideration should be given to the need for extended follow-up, and administration of a 'booster cycle' to commence as soon as possible."

Dr. Shannon's Recommendations

After compiling the trial data: *"The results of this Phase I clinical trial [are] sufficiently encouraging that the research team at the Ottawa General Hospital would like to pursue an extension to the subject trial as outlined... The potential benefits of this inexpensive, safe and possibly efficacious treatment for the rapidly growing HIV-1 pandemic warrants further attention. Your assistance in this regard is respectfully solicited."*

The Actual Written and Published Paper

Remember, these two following statements are being quoted and passed around by some in government agencies as 'proof' that ozone 'doesn't work.'

*"In summary, these small pilot studies have shown that the Mueller Ozon-O-Med Ozone Therapy protocol appeared to have **no detectable beneficial effect**."*

"Our work does not support the continued use of this technique in patients with HIV associated immune disease." (Emphasis mine.)

Even with all the problems this study had, they never said, 'ozone doesn't work,' only that the delivery method didn't work! So, when someone or some agency (again, as we have seen happen) tries to use this study as proof that ozone doesn't work, they are blatantly guilty of deception.

Although five investigators were listed as principal authors on the published paper, exactly who were the actual final paper-writing authors? From Dr. Shannon's communication to The Bureau of Biologics: *"Be advised that Doctors Garber and Cameron (Ottawa General Hospital) have formally submitted an abstract related to this trial to the International Conference on AIDS presentation this June."* It is also extremely interesting to further note that Dr. Shannon was never given a review copy to sign off on before the paper was published. In other words, although his name appears upon the published version, he was denied any input into the final version of what was said. What mysterious forces would promulgate this obvious perversion of truth?

One source I interviewed, an enthusiast for Canadian ozone research, stated that he understood 'the word on the street' to be that Dr. Garber was looking forward to proudly announcing the positive results at the upcoming big AIDS conference, but when the second phase didn't produce as good results as the first phase, he became crestfallen and couldn't make the announcement. He turned his back on the project.

Of course, the non-existent daily quality checking of the ozone generator was his responsibility. So then he either had to go on record admitting that the trial he was responsible for was flawed, or alternatively, make the claim that ozone is worthless. History shows the decision that was made.

Were mysterious forces at work when Dr. Shannon was seeking labs to possibly continue the research, yet was *told? "All the labs are booked up for years on other work."* Who would have enough money to tie up all the labs and lock ozone out? Why wouldn't Captain—now Commodore—Shannon come forward and publicly withdraw his support of the study? Who can blame him? We all know what happens to military 'whistle blowers.' Even though he hasn't directly spoken out, he remains absolutely pro ozone to this day.

Why would the investigators, and all the connected agencies, ignore and continue today to ignore, the notification of the broken machine? Perhaps to have spent or taken the money to do such an expensive study, and being known as a 'respected department' or 'respected investigator' with a reputation to protect and above all a need to continue the funding, maybe it is just too hard to admit your people, or you, personally, didn't do an expensive study correctly by completing such a basic daily task as quality checking the ozone producing machine. Let us hope that more nefarious forces were not in play.

And finally, why would the published data suddenly and mysteriously change its obviously positive data into a negative published summation? Unfortunately, the stench from this rotting carcass now infests my country as well. The U.S. FDA, uses this same study as a reference and seriously oversteps the truth and their boundaries to make the unjustified and way too broad pronouncement that, based upon this Canadian report, "Ozone Therapy does not enhance parameters of immune activation nor does it diminish measurable p24 antigen in HIV-infected individuals." His brazen pronouncement was quoted from an actual FDA letter to one of our elected U.S. representatives, Congressman Sherwood Boehlert.

Remember, the false published paper clearly never says that ozone doesn't work, but that only the particular delivery system of Minor Autohemotherapy, as used, incorrectly, and with its broken generator, doesn't work. Are we to be surprised? I agree wholeheartedly that the protocols, as used here, are terribly ineffectual when compared to normal medical Ozone Therapies as practiced daily worldwide by thousands; especially if you try it with a faulty generator. Even with this handicap, the plain facts remain that the investigators used a comparatively ineffectual ozone delivery method, an antiquated protocol and a possibly broken generator to create only a killed virus antigen preparation, and then injected too little of this mixture in the wrong place. Even these inadequate methods yielded surprisingly positive results (when given time to do their inoculation work) in a few of the patients whose immune systems were still functioning.

So, in summary, a protocol that any serious practitioner would laugh at was used. Their published conclusions were based upon false data—if the machine was broken. And their fraudulent conclusions were written in total disregard for human suffering, not caring how far back ozone research would be set, or how many lives were at stake.

The tragedy of allowing such aberrant summations to be published, and to allow such pronouncements to be made based upon the mysterious summations, is that this errant information is repeated by supposedly impartial agency officials to our elected representatives and to news reporters, while these very agencies ignore the thousands of studies that show ozone does work.

This downright intellectual dishonesty is used to politically justify barring further real research into ozone in North America that would immediately prove we have something ready right now to eliminate suffering and save lives. We know this is true, because ozone has been in use for 100 years by thousands of physicians.

Where are the positive-thinking Canadians attempting to go from here? Quoting Commodore Shannon: *"Allister Clayton, the Director of the Federal Center for AIDS went to bat for our funding. . . . The book is not closed on the efficacy of Ozone Therapy for the treatment of AIDS. . . . We need more funding…. There is a role for ozone in medicine."*

References:

The Use of Ozone-Treated Blood in the Therapy of HIV Infection and Immune Disease: A Pilot Study of Safety and Efficacy. AIDS 1991, 5:981—9842.

Safety: January 1980—The German Medical Society for Ozone Therapy commissioned Marie Theresa Jacobs and Prof. Dr. Hergetbegan from the University Kilnikum Giessen and the Institute for Medical Statistics and

Documentation of Giessen University to begin an inquiry entitled—*Adverse Effects and Typical Complications in Ozone Therapy*. 2,815 questionnaires were sent out to all Western German ozone therapists known by the Medical Society for Ozone Therapy. (AGO, Arztliche Gesellschaft fur Ozontherapie). 884 went to physicians and 1,931 to therapists. They collected 1,044 replies, or 37 percent of the total. The replies that were returned stated that 384,775 patients were treated with ozone with a minimum of 5,579,238 applications and that the side effect rate observed was only .000005 per application! The report also stated, *"The majority of adverse effects were caused by ignorance about Ozone Therapy* (operator error)." The University of Innsbruck's Forensic Institute published Dr. Jacob's dissertation quoting this in *The Empirical Medical Acts of Germany*.

Ozone vs. AIDS, the History and Suppression of Ozone Therapy in the United States as of May 1994. Energy Publications

Medical Ozone: Production, Dosage, and Methods of Clinical Application. Parisian Medical Bulletin—'Bulletin of Medicine, Paris' 52 or 42:745-749.

1993, Sept 2—World premier of Canadian video, *Ozone and The Politics of Medicine*—by Geoff Rogers and Riener Diedrau at the International Ozone Association meeting, San Francisco California. Dr. Horst Kief, of Germany, states there are more 8,000 doctors using ozone in Germany and Austria alone today.

1994—Cmdr. Shannon. Personal interview with Ed McCabe©1967, 12/19/94.

Commander Shannon, Canadian Armed Forces Deputy Surgeon General, comments again in 1995—"Notwithstanding the negative findings of Dr. Garber's 1991 clinical trial, I firmly believe that Ozone Therapy has potential to play a valuable role in the medical management of AIDS. From a regulatory point of view, it is clear that not all forms of Ozone Therapy will be considered sufficiently safe and/or efficacious in this regard; however, there is no doubt in my mind that a protocol will eventually emerge with proven benefit.

"Looking back at my past experience with Minor Autohemotherapy in the treatment of AIDS, there still remains a discrepancy between the Phase 1A and 1B trial results, which may, in part, relate to the lack of sophisticated technology to control for O_3 concentrations in both trials. Given the lack of any significant therapeutic breakthroughs in the treatment of AIDS since that ill-fated trial and the growing testimonial support for its efficacy, the need for further clinical research with ozone is certainly indicated. It is indeed unfortunate that the North American medical community and its funding agencies could not take a more neutral stance on this subject; tragically, professional opinion has been somewhat polarized on this issue.

"I believe it is time to take the emotion out of the arguments, both pro and con and commence a systematic examination of the evidence currently available on the merits of this therapy. Where information gaps exist (particularly in peer-reviewed scientific studies) which might preclude any regulatory decision on the validity of certain claims, properly designed research initiatives should be encouraged with the same kind of public support normally afforded any other scientific endeavor of this import."
—M.E. Shannon

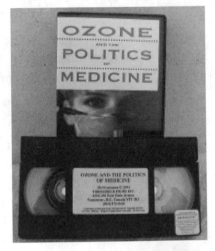

Geoff Rogers' first award-winning Canadian Film Board video that the U.S. TV networks refused to show. Geoff has not given up either.

NEW!
Ozone, A Medical Breakthrough?
(Film and Video)
The sequel to:

Ozone & The Politics of Medicine
(Film and Video)
by
Geoff Rogers
Threshold Films
In association with The Canadian Film Board.

Ozone in Veterinary Medicine Worldwide

Oxygen Therapies (especially Ozone Therapy) in Veterinary Medicine is not new. Animals have a lot of pathology which can be superimposed to human being diseases; moreover, there are no placebo effects in Veterinary Medicine.

In 1997, Prof. Paolo Scrollavezza and his colleagues presented the paper—*Ozone Treatment and Blood Lactate Variation after Thoroughbred Racehorses.* This paper was presented at the World Equine Veterinary Association Mondial Congress.

In 1997, Havana, Cuba, Prof. Paolo Scrollavezza and his colleagues presented their research paper—*Ozone Treatment in Mastitis, Metritis and Retention of Fetal Membranes in the Dairy Cow.*

In July, 2000, Japanese researchers successfully used Ozone Therapy and documented their findings in the paper—*Intramammary application of Ozone Therapy to acute clinical mastitis in dairy cows.*

Parvo

Parvo strangely appeared in 1978. It smells terribly and puppies and dogs that get it show vomiting first, then fever, lethargy, and lack of appetite and then usually die from septicemia and dehydration. The virus that causes it is very similar to feline distemper. Parvo grows in rapidly dividing cells in the intestine and immune system and can cause bloody (or not) watery diarrhea and sudden death "There is no cure." "There is no treatment to kill the virus once it infects the dog."

OK… so somebody explain this. A man's puppy gets Parvo, confirmed by the Veterinarian. The man takes his puppy to the Vet on Monday; the pup gets injected with ozone, and also receives ozone rectal insufflation. The dog starts to improve. On Wednesday the pup got the same treatment. On Friday the vet called the owner and said I took your pup home with me last night and the kids played with him. Come and get your dog. True story. It's also true the Vet later said that he'd lose his license if the owner ever told anyone about him doing this. Names withheld to protect the innocent.

Part Four

The
Politics

Chapter 27

The Struggle

First they ignore you, then they laugh at you,

then they fight you, then you win.—Mahatma Ghandi

Shouldn't You Be Rich and Famous?

Numerous people say to me, "Ed, if you *really* had a cure for AIDS, or cancer, or all these other diseases, you'd be rich and famous overnight, wouldn't you? We'd all know about it. It would be all over the press." I have often had to try and answer this most popular and seemingly very logical question. As children, in our young minds we all want good to be immediately rewarded. It always worked this way in the movies, so it must be true. Wouldn't it be great to live in a world where good is immediately rewarded, and that way more and more of us would always be encouraged to do more and more good things? This well-meaning question is asked sincerely, but its underlying assumption precludes the subtle and overt political and societal realities of today. For example, a Tufts University study estimates it costs 800 million dollars to get FDA approval and bring a new drug to market. I don't have 800 million to donate to the cause, do you? Well, who does have this kind of clout? The drug companies and they're not Boy Scouts. They make 37 billion dollars in profit yearly. Are they going to promote something that negates a lot of the need for what they sell?

It is true that I have seen people reverse just about every disease that afflicts mankind and I have told many people about these things. It is also true that unlike the corporations I do not at the moment of this writing have unlimited money to hire public relations flacks to secretly create and pay phony front groups of 'concerned citizens' to pretend to advocate the latest corporate viewpoint. This is where some major 'public opinions' displayed on the news that everybody thinks everybody else is thinking come from. It's not my opinion, it's all documented. There are books on it.

The control of the media is the control of many human archetypes. Why do you think Americans spend 532 million a year on over the counter drugs? And why do you think Americans spent 90 billion more on drugs last year than the 64 billion that was spent six years ago? Not because we're getting healthier, but because more and more powerfully convincing TV ads tell us to 'ask our doctors about' the latest, more expensive packaging of an old drug. The drug companies spend three billion marketing their products to us. More importantly to them they spend five billion to market to the doctors. Free trips and other inducements are said to be common, as everybody feeds at the drug business trough.

First we are instructed what to believe and how to think by 'news' stories, then we are pummeled with TV doctor shows, and then we are polled to prove the message got through—and the results are announced that the majority believes it, in order to convince the holdouts. I do not control the media, so I'm not rich and famous. Those who do control the media work for somebody who is selling something. We don't have room in the book to list all the tentacles of the greed that opposes good. Of course, not everyone in the media is corrupt but this is the standard operating procedure for the aggregate. Perhaps it all boils down to our own human failings wherein good stories don't sell as well as bad stories.

Still think the news is honest? Let's trace exactly how one country's media were ordered by somebody, somewhere, to act in unison to steer public opinion negatively, destroying any chance that our safe, effective ozone therapy would ever undergo a fair test or see the light of day in Australia until 2003.

Australian Patients Denied Ozone by Media Campaigns

How the Australians were cheated out of Ozone Therapy by the Media.

1993, Feb 25th—On the strength of my numerous public lectures all over Australia and New Zealand, 20 AIDS patients and three cancer patients traveled from Australia to Cebu City in the Philippines to start treatment on the Polyatomic Apheresis ozone equipment. The clinic's organizer, Mr. Bob Graham, had attended almost every lecture I held while traveling all over southeastern Australia. Mr. Graham was such a humanitarian and so impressed with the data (he took copious notes) which he independently verified, that he took it upon himself to mortgage his own house to buy the equipment and open the clinic. He chose the Philippines as a site, to try and get away from the power of the pharmaceutical industry's—recently passed in secret, without any significant debate, during the holidays—highly restrictive status-quo-favoring 'Therapeutic Goods Administration' laws in Australia. He wanted to research the medical use of ozone himself, try and help some volunteer AIDS sufferers, and produce solid data showing the results.

Mr. Graham invited the Australian *Current Affair* TV program to come along and film the trials so that the public could see the dramatic benefits from Ozone Therapy. On February 19th, Peter Wilkinson, from Nine Network Australia Limited, one of the producers of the show, wrote to Mr. Graham: "You have invited us to come to Cebu to view the trial for ourselves. You have agreed to give us full access to anything we will require in order to independently assess your trial. In turn, we have agreed to guarantee anonymity to any of your patients who desire it. We will pay all our costs. I understand that some of the people traveling are concerned that pressure will be applied to drop the story. Such pressure has never been applied to this program. The fact that we are making the heavy financial commitment to go with you is evidence enough that we intend to finish the job. If any pressure was applied, in itself that would be another (TV) story."

Soon after this correspondence, unknown to Mr. Graham the clinic owner, four of his potential patient participants had gotten wind of the upcoming TV story. Fueled by their ego-driven-greed, they hired a lawyer and demanded a quarter-of-a-million dollars each from the TV people to permit their being filmed undergoing treatment. The TV lawyer informed the top brass at the network, and now Current Affair had a BIG problem. After all the preparatory investment and man hours already in this story, they couldn't afford to drop it; yet, to go ahead was to be involved with expensive litigation against publicly visible poor suffering AIDS patients.

A terribly evil decision was made. The call comes down from the top: Tell the producers the only way we can recover our investment is to launch a full bore attack and discredit the people, the clinic, the therapies and everything connected to it. That way we get a great story, don't have to pay anyone, make ourselves look like heroes to the sheep-people while posing as their 'guardians against quackery' and their health and suffering be damned!

This was the actual scenario which was later confirmed to Duncan Roads, publisher of *Nexus Magazine*, when he was approached by one of the staff researchers on the TV program that repeatedly delivered the deadly punches. What goes around comes around, and soon after the smear he helped create he had to call Duncan for personal help, since Nexus published my ozone articles and his brother needed treatment. When pressed as to why no reporters had called me for any of the volumes of positive medical reference documentation, the researcher at first claimed he 'couldn't find my phone number.' Duncan immediately sensed falsity, since my number was plastered everywhere in print, so he kept hammering away at the researcher's conscience. The researcher finally admitted he was anguishing over being forced to smear ozone by his TV bosses when his own brother had AIDS and that the producers said they couldn't care less about ozone or the truth, and that during the staff briefing he was instructed that their mission was to only (actual quote) "Hang Basil Wainwright out to dry." No looking for safety or efficacy. Just the dirt on Basil. Basil, the manufacturer of the Polyatomic Apheresis ozone machine was hounded by the British Government for previous ideological political clashes with them when he was their political prisoner. They forced him to invent for them while in their jail, and ever since, he has been hounded by the power of the press. Hounded out of the U.S. by Interpol, Basil recently resided in Africa [now he's in Asia] and claims over 600 documented AIDS cures with ozone.

I pity the finely dressed, polite, suntanned, cultured person riding in the chauffeured car to his plush office on top of some corporate structure; someone who never raises his voice, who single-handedly put money and pleasing his masters ahead of the suffering and dying of the Australian people; and in one decision ordered his TV producers to execute the TV smear on ozone. This person poisoned ozone's public image in medicine and rejected the suffering Australians who desperately needed a safe therapy. These persons will witness the untold suffering that they created and that could have been prevented had they chosen wisely and compassionately.

Do you think *their* hidden criminal misdeeds will ever see the exposing light of a TV mini-movie?

After it was over, this large commercial station, Channel Nine, was caught red handed by a small government channel TV show that acts as a media watchdog. The government public TV media watchdogs featured one of the Cebu City ozone patients in the original news footage which was shot by Channel Nine. The patient was gushing about how great he felt—he felt fantastic and was really positive on the Ozone Therapy he received for only a short while. The watchdog show plainly demonstrated how Channel Nine's 'Current Affair' had re-edited and re-sequenced the footage to make the patient appear to say exactly the opposite of the good things he said, and aired the phony footage all over Australia to discredit ozone! The show was revealing, but the small government station with fewer viewers was no match for big money and big media, so truth went no further that day.

Unknowledgeable about all this evil about to descend upon his project, and ready to go to the Philippines, clinic owner Graham held a meeting just before leaving under the auspices of the Victorian AIDS Council, Inc. The Victorian AIDS Council, Inc., Gay Men's Health Center, Inc., serves as the self appointed guardians of the gay community's health. Their newsletter announced events such as a Gala gay Easter party at the Princess Hotel, and 'advises' the Gay community which treatments to get for AIDS, as a 'public service.'

At first, The VAC self appointed 'Treatments Officer' Tony Maynard was all for the ozone research and asked for documentation and got it. He was so impressed he advertised for people to go to the trials in the Philippines. He called my lecture on ozone 'great' in a letter to Mr. Graham on November 30th, 1992.

Later on, Mr. Maynard turned 180 degrees against ozone. He announced by writing in the local *AIDS Council News* gay newspaper that after the first meetings he was soon approached and suspiciously influenced by the TV people and placed an undercover TV 'agent' with a hidden camera in the group meeting of those going to Cebu. Mr. Maynard apparently suddenly saw an opportunity to be 'the big shot expert' in a much bigger way than his little newsletter could ever offer. Envisioning TV grandeur, he helped the Australian Current Affair TV program destroy the legitimate attempt at doing much needed research, through personally consciously generating negative public opinion.

He soon went against the reputable medical study documentation provided to him by me and joined the media feeding frenzy. Now he was just like the big dogs and quickly put out his own newsletter article and used such fraudulent statements as "Ozone treatment has been discredited by every major medical organization in the world." This is a blatant lie, but he needed to justify his selling out.

Later, it also came out that the very same Victorian AIDS Council, a private, supposedly 'grass roots organization' in reality is a corporation. A corporation which strongly and constantly pushes the competitive-to-ozone and expensive and highly toxic drug AZT and gets support directly by *donations* from major pharmaceutical interests who hate ozone's successes!

Returning to Mr. Graham. Still unaware of the plots against his efforts to prove ozone's efficacy, one day he went out to his mailbox and found a TV camera crew racing down the street toward him. They were trying to make it look like he was on the run, all the while shouting at him, "What about the scam ozone clinic you are running to cheat AIDS patients!" He stood, not ran, and calmly said, "Gentlemen, you have been seriously misinformed. Would you care to come inside and discuss this?" He bought them inside, sat them on the couch and served tea. He then showed them ozone documentary videos which provided them with legitimate positive medical ozone reference literature. They left supposedly on good terms.

Next thing he knows—BEFORE the clinical trials even started—Mr. Graham is the lead story on Australia's version of the Current Affair TV program, and he is being accused of running some sort of a scam! This was the night before they had even left the country. He decided that to really find out for himself if ozone works to end suffering is far more important than protecting himself or hiding. He left for the Philippines, and the trials started. The scant money raised on a prayer to support the clinic runs out halfway through.

On to the scene arrives 'Johnny Young.' Apparently, he is the Aussie equivalent of aging U.S. TV's Dick Clark teen idol. He came to help comfort one of his AIDS stricken buddies (saying, he's 'just a friend') who was at the clinic. Seeing the amazing positive results from ozone for himself and being in his own personal acute financial distress, he recognized a golden opportunity. As soon as the clinic got into financial difficulty, he schemed to buy it all and take over, thinking he could get rich owning what he is convinced is a cure for AIDS, after seeing the patients' miraculous and dramatic improvement.

Johnny Young, in Cebu, called Mr. Graham in Australia at 3:00 p.m. telling Mr. Graham that he and his partners had enough finances to complete the trial IF the following conditions were met.

1) All clinic directors must resign.

2) The present shareholders would keep 25 percent and Johnny's group would take over 75 percent of the company, IF Mr. Graham would turn over the cash flow figures within two hours, showing what it would cost to complete the trial.

3) Johnny wanted the sole rights for Polyatomic Apheresis, Inc., for Australia and all of Southeast Asia.

People who can confirm this all as true: Sue Anne Taylor, Linda and Lou Lester, Mary Baker, Basil Wainwright (PAI), Richard Bernard (PAI) and Mr. Graham and his accountants in Australia. I also heard all of this by phone as it unfolded and as he kept in touch with the clinic's progress.

To keep the trial going and to not have to abandon the poor suffering and dying patients who were slowly starting to turn around, over the following week all directors sadly resigned as requested and the feasibility study was prepared by the accountants. One of the now ex-directors flew to the Philippines with all the documentation for Johnny Young. He arrived in Cebu on Monday morning. Johnny Young, in his enthusiasm to take over the clinic, gathered the staff together and informed them that he was taking over the clinic and Mary Baker would be the sister-in-charge. All doctors and nursing staff were told they were now answerable to him. Australia's first test of this legitimate Ozone Therapy Recirculatory Hemoperfusion delivery method was stopped in its tracks by ego.

Someday we will have a fully open and accountable major U.S. hospital testing of this therapy that already works all over the world. Want to help?

Our National Dysfunction

Curiously, it's a very interesting political situation in the U.S. Oxygen Therapies have personally been widely used by many very high profile persons from within government and the big illness organizations. The German doctors that treat them with ozone tell us so. These bigwigs are the same people who publicly advise us that the Oxygen Therapies are still 'unproven.' I wonder why?

Historian Eustace Mullins' exceedingly detailed history books chronicle the long-standing incestuous relationships between politicians, big drug, and health societies. And now the far ranging corruption accepted as a normal and necessary way of life inside the medical establishment which pontificates against other competing therapies is so bad that their own medical journals are doing exposé's on it. *The Lancet* is said to be *the most prestigious medical journal in the world.*

Who is Rich and Famous?

THE LANCET • Vol. 359 • April 6, 2002

"Chief executives of multinational pharmaceutical companies have much to celebrate this week. They saw spending on prescription drugs in the USA soar by a remarkable *17 percent* in 2001, according to figures recently released by the National Institute for Health Care Management Foundation. As bonuses for corporate leaders ratchet upwards, so does the unalleviated financial pressure on the elderly, the largest users of these drugs. Direct-to-consumer advertising campaigns for cholesterol-lowering agents, anti-ulcer medications; anti-arthritics and antidepressants have been strikingly successful. Total retail spending on prescription drugs was US $155 *billion* in 2001, almost *double* what it was in 1997."

The drug bosses and heavily funded politicians drain more and more out of the pockets of the elderly through lavish advertising we can't escape. Advertising so all pervasive and effective that a huge portion of our mainstream society now lives their lives as if being drugged is 'normal' and if you aren't on drugs, then you must be missing out on something wonderful. You are constantly conditioned this way, starting with mom, TV, and your school officials—all the way through to the grave.

What's wrong that YOU, all by yourself—just the way God made you— are not enough? What lie about your own inadequacy did you swallow long ago? And who sold it to you? And who gets rich off our believing errors and our resulting misery? Follow the money.

The Lancet continues: "A study of the interactions between authors of clinical practice guidelines and the pharmaceutical industry, published in *JAMA* (Journal of The American Medical Association) in February, found serious omissions in declarations of conflicts of interest. Almost *90 percent* of authors received research funding from or acted as consultants for a drug company."

So the people with the responsibility to always tell us the whole truth about health and the drugs are 90 percent influenced. How can you blindly and automatically trust this industry and these people with your life? How can you trust them to tell you which ones are the good therapies? How many times have I heard "My doctor says..." after he listened to the industry pundits.

I understand the researchers first have to earn their livings and feed their families before they can do anything good, but the real question is "Are they always being honest?" Honesty means full disclosure of all side effects and harm and worth.

I personally understand how this having to eat first thing works, since here I am telling you I prefer Oxygen Therapies, and at the same time consulting to the industry. The difference is I'm honest and disclose all. It takes a brave person to walk that line with integrity. As *The Lancet* says, "A more important question arises: do those doctors who support this culture for the best of intentions—e.g., to undertake important research that would otherwise remain unfunded—have the courage to oppose practices that bring the whole of medicine into disrepute?"

The Lancet goes on to give just one example, The American Heart Association accepts U.S. $11 million in donations from Genentech while at the same time recommending their product for stroke management. The Lancet then asks and answers, "How tainted by commercial conflicts has medicine become? *Heavily, and damagingly so*, is the answer."

Yet when some health crisis scares you, these are the first people you go running to. Designed conditioning. And the politicians are paid very well to keep it all going just the way it is.

"Ozone is a toxic gas with no medical applications."

—Statement published in the Federal Register and repeated in front of a Congressional investigating committee while Ozone Therapy was being used daily by thousands of physicians in Europe and Russia for over 50 years.

Government Response to Natural Food and Health Explosion

Exploding public interest in finding real solutions for the chronic suffering and disease plagues has created more and more healthy buying decisions that are in turn creating enormous pressure on established markets. Money makes things happen. The loss of market share has forced many changes in the food and health areas.

The tremendous demand for alternatives to business as usual has forced the government's National Institute of Health to open an Office of Alternative Medicine. I was skeptical that they were only window dressing for political correctness and not really seeking solutions, but former Congressman Bedell was very persuasive and convinced me to accept an invitation to come and testify at their initial formative meetings. I told them in no uncertain terms about how good the Oxygen Therapies are. I brought documents and doctors as proof of the effectiveness of Ozone Therapies. On the next page I'll show you how the Washington Post wrote up my testimony.

HEALTH NEWS

A New Look at Alternative Therapies

NIH Probes Treatments Outside Mainstream Medicine From Massage to Acupuncture

By David Brown
Washington Post Staff Writer

Two weeks ago, John C. Pittman, a 33-year-old physician in Raleigh, N.C., stopped giving ozone gas to patients with the AIDS virus after the North Carolina Board of Medical Examiners, acting on the complaint of another doctor, told him they were looking into his controversial practice.

Yet an advisory board at the National Institutes of Health last week expressed keen interest in Pittman's work and asked him to provide more information on his claim that of 25 patients with the virus, three had overcome it after having the highly reactive gas infused into their blood.

Pittman's role as both pariah and pioneer of new treatments represents just one of many ironies—and contradictions—the medical establishment is likely to encounter as it begins its first organized look at "unconventional" medicine.

Over the next few months, a special NIH panel will attempt to glean from nutritional therapy, touch therapy, acupuncture, naturopathy, homeopathy, herbal medicine and other alternative modes of healing a list of treatments that may be promising and can be tested in conventional experiments.

The inquiry is headed by the Office for the Study of Unconventional Medical Practices, which was established after the 1992 federal budget requested that NIH spend at least $2 million on such an effort. NIH's total budget is $8.9 billion.

"Throughout the history of medicine, many great discoveries were made based on theories that were ridiculed early in their use because they were viewed as radical for the conventional thinking of the day," Jay Moskowitz, an NIH associate director, said last week at the start of a two-day hearing in which more than 40 practitioners spoke briefly about their therapies.

The agency's attitude is one of rapprochement with practitioners outside the medical mainstream as, in the past, the NIH has occasionally investigated unorthodox treatments. The National Cancer Institute is now designing at least one, and perhaps three, trials of substances called neoplastins, which a Houston physician has used for years in treating brain tumors. A delegation from the cancer institute reviewed seven of the doctor's cases and found that in each patient the tumor shrunk with treatment.

At the same time, several people at the hearing said that the longtime suspicion—and often contempt—for alternative medicine will be hard to overcome.

"There are a lot of people who don't even want to show up here. They're afraid of finding their names written down on a blacklist," said Michael L. Culbert of the Committee for Freedom of Choice in Medicine, who works at a clinic in Tijuana, Mexico. He called the Food and Drug Administration, which approves drugs for use in the United States, "little more than a policing agency used to clear the marketplace of non-toxic, nutritional, herbal, chiropractic and other competitors whenever possible, hence securing obscene profits for the drug monopoly."

Ed McCabe, author of a book on unconventional uses of oxygen, including ozone, said he asked six doctors to come and present their findings, but all refused because they feared losing their licenses if their unconventional practices were made public. "I came because I'm not an MD, so nobody can take my license away," he told the panel before going on to talk about how ozone treatment had markedly improved the conditions of 300 patients who tested positive for the human immunodeficiency virus (HIV), which causes AIDS.

Ozone, a molecule made of three atoms of oxygen, decomposes to form peroxides and superoxides, which are highly reactive substances capable of destroying bacteria and viruses. Certain white blood cells also use peroxides and other oxygen derivatives to fight infection, so ozone's use in HIV treatment is not entirely far-fetched.

Although ozone has not been formally tested in humans, a report published last year in the journal Blood described an experiment in which the gas inactivated the AIDS virus in cell culture.

Practitioners of ozone therapy either treat a blood sample with the gas and return it to the patient, or infuse a small volume of gas directly into a vein. For HIV infection, the treatment should be done twice daily for at least three weeks, according to its proponents.

Other treatments described at the hearing range from acupuncture and massage, occasionally used in conventional medical regimens, to some virtually unknown outside the universe of alternative medicine. Among the latter is the use of the "somato-scope"—an altered dark-field microscope—to observe small organisms in the blood, called "somatides," that go through a visible 16-stage cycle in response to cancer and other toxic states. This theory's main backer is Gaston Naessens, of Sherbrooke, Quebec, who developed a cancer drug called 714X, derived from camphor. Naessens did not attend the hearing, but David W. Ganong, a 47-year-old dentist from outside Boston, described how his father's lung cancer had been arrested by 714X and his recovery monitored by the somatoscope.

Several speakers noted that mainstream medicine may find it hard to pluck a few useful treatments from these alternative healing approaches without having some of its own basic tenets severely shaken.

Homeopathic physicians, for example, often prescribe chemicals so diluted that they are undetectable by routine measurement. Homeopathic drugs have been used by thousands of people for hundreds of years, and it seems unlikely the placebo effect explains their success entirely. A review published last year in the British Medical Journal looked at "methodologically sound" trials of homeopathic treatment and found that in more than 80 percent of them the treatment showed clinical benefit.

Such results question the reigning theory that there is a minimum dose for chemical substances, below which the compound could not possibly have an effect. Homeopathy argues, instead, that a chemical's impact depends not on dose but on an individual's sensitivity to it. "This complicates all the assumptions of the FDA," said Edward H. Chapman, of the American Institute of Homeopathy, in Massachusetts.

Information about unconventional treatment is not easily available to mainstream physicians, many speakers said, noting that journals of alternative therapy are rarely included in computerized databases of medical literature. Other speakers, who practice unorthodox therapies, said in turn, that they were similarly unfamiliar with some of the norms of contemporary medicine, in particular the methods of clinical research.

"We need to be involved in research training programs. We need research facilities. We need guidance in writing grants. We need our work to be seen and evaluated," said Tori Hudson, associate dean of the National College of Naturopathic Medicine in Portland, Ore.

The NIH office will spend at least part of its money helping alternative practitioners "develop methodologies so that their results are interpretable and analyzable," Moskowitz said. Several speakers observed, however, that many unconventional treatments will be difficult to evaluate with methods based on the idea of isolating a single treatment variable for study.

"We are incarcerated in the double-blind, crossover model," said Majid Ali, a New Jersey pathologist whose clinical practice stresses nutrition, fitness and environmental therapy. "It is not appropriate for holistic therapy, in which there are many variables and neither the practitioner nor the patient can be blinded to the treatment." ∎

Statue of the father of homeopathic medicine Samuel Hahnemann. Homeopathic treatments are one of the therapies being studied by NIH.

BY LARRY MORRIS—THE WASHINGTON POST

WASHINGTON POST HEALTH/JUNE 23, 1992

8

(* Section enlargement next page)

On June, 23rd 1992 the *Washington Post* printed the high point of my testimony at the National Institute of Health's Office of Alternative Medicine's formative meeting. I reported that I had interviewed six doctors quietly using Ozone Therapy successfully on more than three-hundred people with AIDS. Nothing happened.

Enlarged *Washington Post* article, June 23rd, 1992

ropractic and other competitors whenever possible, hence securing obscene profits for the drug monopoly."

➤Ed McCabe, author of a book on unconventional uses of oxygen, including ozone, said he asked six doctors to come and present their findings, but all refused because they feared losing their licenses if their unconventional practices were made public. "I came because I'm not an MD, so nobody can take my license away," he told the panel before going on to talk about how ozone treatment had markedly improved the conditions of 300 patients who tested positive for the human immunodeficiency virus (HIV), which causes AIDS.

Ozone, a molecule made of three atoms of oxygen, decomposes to form peroxides and superoxides, which are highly reactive substances capable of destroying bacteria and viruses. Certain white blood cells also use peroxides and other oxygen derivatives to fight infection, so ozone's use in HIV treatment is not entirely far-fetched.

Although ozone has not been formally tested in humans, a report published last year in the journal Blood described an experiment in which the gas inactivated the AIDS virus in cell culture.

Practitioners of ozone therapy either treat a blood sample with the gas and return it to the patient, or infuse a small volume of gas directly into a vein. For HIV infection, the treatment should be done twice daily for at least three weeks, according to its proponents.

Other treatments described at the hearing range from acupuncture and massage, occasionally used in conventional medical regimens, to some virtually unknown outside the universe of alternative medicine.

Among the latter is the use of the "somato-

I had to go all the way to DC at my own expense and try, but, as I suspected, they were not looking for real answers. No one cared, no one followed up, no one investigated, and when I officially applied to be on the newly forming NIH Alternative advisory panel as a lay advisor, no one would respond to any of my letters. Several of them glared at me whenever I went there and spoke my truth. The fix was in. How many died since then because of rejecting our information?

The only people being promoted are academics who match the hidden objectives of appearing alternative, but who do not really threaten anything with actual solutions. The bureaucrats do not really know how hard it is in the daily world of suffering experienced by those affected from their policies.

The next article (a reprint of my NIH testimony) strangely disappeared from the CDC public access database after I saw it there. I do not like to get after those in error, but too much greed and suffering has held sway for too long. Thank goodness many honest people work long and hard, and selflessly, and lovingly in the medical industry and in the local health societies like the cancer, heart, MS, and AIDS groups. It's probably a good thing they don't know what's above them, for they might become disheartened if they did.

Point of View

Ozone Therapies for AIDS

Vol. 6, No. 6   December, 1992

AIDS PATIENT CARE

A JOURNAL FOR HEALTH CARE PROFESSIONALS

By Ed McCabe

Oxygen therapies have been in use medically for over 100 years. The basis of the therapies is quite simple. Ultimately, toxicity collects within the body over time. This buildup of toxicity can have physical, emotional, and behavioral cofactors which must also be addressed. The buildup of toxicity within the body invites the proliferation of anaerobic disease-causing microorganisms. Oxygen therapies are used to flood the body with active forms of oxygen. When the body becomes safely saturated with these special forms of oxygen, it reaches a state of purity wherein the disease microorganisms—such as HIV—are killed, and the underlying toxicity is oxidized and eliminated.

Healthy human cells love oxygen, but since the harmful viruses are anaerobic, once these viruses are surrounded with active forms of oxygen they must die. Active forms of oxygen also eliminate any cells with the virus hiding *inside* them since, unlike healthy cells, they no longer are able to protect themselves from oxidation with the body's natural antioxidants.

Originally presented as testimony to NIH sub-committee on overlooked but effective therapies on June 18, 1992, Bethesda, MD.

254

Once the viruses and bacteria are all killed, the patient's immune system is then rebuilt with vitamin, mineral, and, if needed, behavioral modification and other therapies. In this way diseases such as AIDS, cancer, and multiple sclerosis are being eliminated or at least controlled on the international scene, and within the U.S. medical underground.

Ozone therapy is one of the oxygen therapies. Over 5000 medical ozone generators have been sold in Europe and are presently in use by as many doctors worldwide. Ozone is not smog. It is a clear gas (O_3) made from pure medical oxygen (O_2). By bypassing the lungs, specific concentrations and volumes of ozone gas made from pure oxygen are readily applied to the body fluids and tissues through IV infusions without any harm, usually during a minimum treatment schedule consisting of application every day for 21 days. The medical ozone breaks down into various oxygen subspecies which destroy (oxidize) anaerobic viruses, microbes, or diseased cells and yet leave the normal healthy cells alone. The standard 21-day minimum treatment requirement is being challenged by recent developments, notably in the field of ozone generation equipment and other protocol refinements.

Ozone therapy was used medically in the U.S. by Nikola Tesla in 1900. It has been in continual use since then in our country, but due to legal pressures from

our FDA, ozone is presently only widely used in Europe. Ozone therapies are among the safest therapies ever used. One European study of over 5.5 million treatments showed a side effect rate of .0007 percent, probably among the lowest of any therapy known. Side effects such as fever and weakness are minor and temporary. Physicians characterize ozone as "blatantly nontoxic." When the proper protocols are followed, ozone has been proven highly effective in the treatment—and possibly the elimination—of over 40 common diseases. There are about 300 doctors in the U.S. using some form of oxygen therapies daily.

AIDS Experience

I have kept in continual contact with six U.S. M.D.'s who, independent of each other, have collectively reported bringing over 300 AIDS patients to HIV negative status (using Western blot, ELISA, and PCR), including complete eradication of any secondary disease factors such as energy loss, weight loss, diarrhea, etc. I have also interviewed many AIDS patients who have been undergoing ozone therapy on their own, and all those who followed good protocols reported an immediate increase in energy, weight gain, and T cell stabilization or increase in a few weeks.

The most notable of the ozone physicians to date is Dr. James Boyce, who

Continued-

*"*Ozone therapies are among the safest therapies ever used.*"*

brought 237 AIDS patients to health, each within 60 days. His results were verified by major independent lab testings before and after treatment. Strangely, like most doctors in our country who have bravely tried to help their patients with ozone therapy, Dr. Boyce has subsequently been attacked by several government agencies trying to put him out of business. While our country reels from the impact of AIDS, this physician, who has the best documented track record in the country of treating and perhaps even eliminating AIDS, now spends his days farming instead of practicing the medicine he was so proficient at.

Another typical modern case is that of Dr. John Pittman from Raleigh, NC. He started using ozone in 1991 with excellent results, but had his research project cut short when local physicians who never heard of ozone—and did not bother to find out anything about it—turned him in to the state medical board which subsequently asked him to stop using ozone, since it is not FDA approved. A local self-proclaimed AIDS activist also made the same mistake of turning in Dr. Pittman for the same reasons. Dr. Pittman is hopeful of finding a way to continue his research.

The Medizone company of New York, which holds a patent on blood ozonation, has continually been asking the FDA for permission to begin ozone trials on AIDS patients since 1986. For six years, the FDA has stonewalled, telling them to "go back

and do even more (flawless) ozone animal studies." This is in spite of the fact that ozone therapy has been used safely on humans on a daily basis for over 50 years in Europe.

Safety of application or effectiveness is not questionable among the physicians who have personally used ozone daily for 30 years on humans. Medizone and others are ready to begin strictly designed and monitored human trials. And, like many U.S. physicians I have interviewed, they would have commenced clinical trials six years ago—if not for the FDA actively preventing it.

Physician Fears

The biggest problem is that in 1976 the FDA declared ozone a "toxic gas with no medical uses" via publication in the Federal Register. I have written that: "Printing this statement in a publication paid for with our taxes is either a blatant attempt at suppression of truth from the highest levels, or one of the poorest research jobs ever done. It obviously favors competitive drug therapies, and ignores well over 50 years of safe and effective medical use on hundreds of thousands of humans—backed up with medical references and clinical studies in Switzerland, Italy, France, Germany, Australia, New Zealand, Mexico, and the U.S."

Although legally, ozone is believed by some to be "grandfathered" since its use in U.S. medicine predates the existence of

the FDA itself, no one has tested this position in court. So the FDA, and the state medical agencies and boards that take their lead from the FDA, are actively persecuting any physician using ozone. Under these prevailing "siege" type conditions, I could get none of the U.S. doctors who have achieved great successes with ozone therapy to come here today to the NIH. When asked to appear, the doctors all reacted in fear of the government, fear of the loss of their licenses, and fear of being raided and cleaned out by SWAT teams in the same way that Dr. Jonathan Wright and other M.D.'s have been treated most recently. Upon inquiry, I commonly heard: "I am not ready to appear." "Do not mention my name!" "The doctor already fled the country." "I'm scared to death they will take away my license."

On the bright side, if immunity from state, federal, agency, and medical board prosecution could be assured, these humanitarian doctors would gladly make available their knowledge for the public good. I stand ready to do the same. | |

Mr. McCabe is an investigative journalist and leading lay expert on oxygen therapies.

Reprints of this article are available in bulk quantity. For information and prices, write to or call: Karen Ballen, *AIDS PATIENT CARE*, 1651 Third Avenue, New York, NY 10128, (212) 289-2300.

The Business of Medicine Pretends to Change While Fighting and Ignoring the Real Solutions

New York's Memorial Sloan-Kettering Cancer Center and other major city hospitals like Miami also had to open 'integrative medicine' departments. I talked to a Miami doctor and nurse who said they used Oxygen Therapies in the new 'Alternative' wing, and the Oxygen Therapies were working, but the administrators removed them from the new program. Apparently the purpose of opening these new departments is not to institute any real change, but to diffuse and silence the critics. The public has anguished over too many family and friends that suffer and die as hospital failures, despite getting the 'best of care.' They remember them all very well.

The old ways make profits, but not health, and the big boys had to do something to stem and spin the tide of public awakening. They make a big deal in their brochures and expensive ads about changing their ways and they promote how enlightened they are because they mention massage and acupuncture and how they can make you feel better after being poisoned with chemo. At the top the big decision makers know there is no real danger of the minor alternatives they advocate *actually* curing your disease—and thereby upsetting the lucrative revenue stream coming from the entrenched business as usual.

Certain Ph.Ds and MDs are heavily promoted and brought forward as so-called national authorities on—or champions of—alternative and complimentary and integrative medicine in bookstores and newspaper articles. But the only ones promoted heavily nationally are the ones who disparage the real solutions like Oxygen Therapies.

"I wouldn't want them in my body [referring to ozone and peroxide]." So proclaimed the establishment's darling 'alternative' smiling bearded doctor, Andrew Weil, on 09/24/1996. After reading what he wrote, I went to great lengths to write to him on a number of occasions and sent him documentation proving the value of the Oxygen Therapies and offered to help him learn this information in any way possible. This so-called 'man of the people' did not respond to any of my correspondence sent to several addresses. Maybe he secretly read it, however, because he seems to at least have one foot on the oxy-bandwagon by selling tapes about breathing and self healing. He says Harvard Medical School taught him nothing about breathing.

Another author, a Ph.D., was set up to appear as an expert by her writing a book about alternatives and making fun of half of them. This made her another darling of the establishment—as the cancer industry would love to recoup their increasingly lost-to-alternatives revenue. Their PR agencies desperately needed someone to give them the window dressing appearance of being 'friendly' to other treatments.

The Los Angeles Times recently ran a column with the following very poorly researched, absolutely incorrect, scientifically substandard and totally outdated statements written by this same Ph.D. author. She postures as an expert, but from what she says, we know she has no idea what Oxygen Therapies are, let alone how they work, or how they have been used successfully worldwide for over 50 years. You can tell that she hasn't interviewed anyone using them, hasn't been to any oxy-clinics and yet she assumes the authority to tell you they are no good and dangerous.

And is it any surprise she makes her living as chief of integrative medicine at Memorial Sloan-Kettering Cancer Center in New York, and is a member of the complementary and alternative medicine committee of the American Cancer Society? Surely we can trust her, surely she has no agenda when she says:

"The assumption that all disease, even disease as serious and complex as cancer, is caused by a lack of oxygen in body tissue is *unfounded*.

"The use of Oxygen Therapies are *groundless and a waste of money*, some of them are *potentially harmful*.

"Such treatments make *no contribution* to the body's oxygen needs.

"If we were like fish and had gills, maybe we could separate oxygen from liquids. We can't... we *can't bypass the lungs*. We can get oxygen into our bodies by one route: the lungs... Remember, too, the body does *not absorb oxygen into the bloodstream through the intestines*." Oxygen is absorbed into our bodies through our lungs, not through the stomach or intestines.

"But ozone is *toxic*... Ozone is fine in the upper atmosphere, but at ground level it's a harmful *pollutant*. Ozone is a powerful oxidizing agent because it generates 'free radicals' when it decomposes. Free radicals *destroy natural biological substances*.

"There is *little, if any reliable evidence* that the claims made for Ozone Therapy as an AIDS treatment are valid."

I italicized the old hackneyed phrases so you can easily recognize them when they pop up. If one was too deceptive to update themselves on the current international ozone research, you might make such harmful statements out of ignorance, but the Los Angeles Times really hurt its own credibility by printing her. I use the word harmful, because such deliberately un-researched statements might scare suffering people away from ozone solutions we know to exist—but unfortunately do not have billions of advertising dollars behind them. Oh, did I mention this author I'm quoting was also a founding member of the Advisory Council to the National Institutes of Health Office of Alternative Medicine, you know, the one that refused to respond to queries and testimonies and ignored investigating a proven and successful ozone AIDS treatment?

The establishment pundits publicly and desperately positioned as the trusted voices of the alternative therapies scene magically find it very easy to get published quickly and distributed everywhere. With big establishment money behind them they are shamelessly cashing in on the public's disenchantment with the status quo, and like wolves in sheep's clothing they appear on the bookshelves of health food stores in glossy color—smiling away—and leading you to business as usual. Their common purpose whether they are aware of it or not, is to contain the growth of the competition, by being spin doctors diverting the public's attention away from the *real* solutions and issues and into minor alternatives.

They were nowhere to be seen while so many of us fought so hard for existence and for you in the trenches of shows and meetings and publications for so many years. They have no experience in Oxygen Therapies, have never used them, don't know anyone who does, don't know the difference between active and regular oxygen forms or air, and yet suddenly appear out of nowhere, reach into our industry and haughtily pronounce themselves our leaders and your experts while calling our therapies worthless and dangerous.

As another example, this aforementioned chief of integrative medicine also wrote: "So-called 'natural' remedies for illness are becoming increasingly popular with consumers, as well as with some managed care organizations. Although some of these methods, such as meditation and massage, may enhance quality of life, many are useless and some are dangerous." Obviously, 'meditation and massage' do not pose any real threat to her favorite billion-dollar industry.

Other cancer treatment sales agencies are disguised as authoritative free public service information websites. They declare; "The idea that ozone is effective in treating cancer has never been tested scientifically. Ozone is toxic. The safety of injected ozone and its long-term toxic effects are unknown. There are no credible scientific reports showing benefit of ozone."

Ah, the ol' 'never tested' and 'no publication game' gambit. It is quite manipulative to suggest that over 50 years of safe and successful Ozone Therapy on millions does not exist if it isn't published in English—and in American journals. I guess if I purposely never wanted to look, I wouldn't have found all the evidence in this book either. And if I never turned on my computer, I also would not find the hundreds of international Ozone Therapy references available online, including the Federal government's own over 163 positive to ozone Medline references. Somebody is hiding something and trying to protect a lucrative industry.

Why the scarcity of American publications? The American authorities and press barons constantly denigrated Oxygen and Ozone Therapies solely to protect their competitive industries and purposely scared away American research into it. With researchers in the past being afraid of persecution for trying ozone, it is easy to understand why only some references to it ever ended up being published in our country. Normally it's 'just business,' but add all the suffering and, whatever the causes, it rises to the level of evil.

Claiming that a lack of local American publication containing clinical studies on Ozone Therapy therefore equates to ozone being 'worthless and harmful' is to cleverly and blatantly deceive you and instill you with fear. Such disparaging statements obviously purposely ignore the real world results of millions of ozone dosages safely and successfully given for more than 50 years by thousands of physicians all over the world.

If you really want to prove to your own satisfaction that these therapies have benefit, call all the medical ozone generator manufacturers in Europe, Russia, and America. Call all the ozone medical societies and the thousands of ozone using physicians. Call all the Russian hospitals and clinics, and speak with their 1,950 MDs and 1,300 ozone nurses. Ask them why they have dared to have been fraudulently making, selling, buying, and using so many ozone generators for so many years. Do you really think all these doctors and nurses are nothing but liars and frauds?

The trick, from the patients perspective, is to recognize the pitfalls associated with establishment products and practitioners; to select useful modalities; and to be wary of unsubstantiated claims and long lists of very real and very dangerous side effects and 'incidents' (like suffering and death) occurring after enough drug usage.

The FDA Point of Perspective

Our problem is that Ozone Therapy, when applied correctly as a harmless-to-humans natural sterilizer is so good, and so effective, and so safe, on so many things that at first it seems *too* good to be true. Dr. Carpendale at the Veteran's Administration Hospital in San Francisco and Commander Shannon representing the Canadian Military Forces both used to tell me how frustrating this fact was to them all the time while trying to tell their colleagues about ozone.

The FDA has issued guidelines to help consumers choose:

FDA Rule #1 is no problem, namely: "*Have the active oversight of a licensed physician.*" Stop harassing any healers for using ozone and they will oversee all day long.

FDA Rule #2 is "*…promotion and commercialization of <u>unapproved</u> products are characteristics of the kinds of deception and health fraud that expose people to ineffective and dangerous products.*" Sounds ultimately kind, fair, and protective, doesn't it. But what if the product under scrutiny has flawless animal studies and 50+ years of millions of safe and effective applications on real live humans in thousands of clinics and they still refuse to 'approve' it due to politics? Does that change the real world scientific truth of how good it is? Does that make it deceptive and fraudulent, or because we advocate it, does the advocating somehow make it become ineffective and dangerous? I think not. Truth is truth, is truth.

FDA Rule #3 *"The products can't come from unknown sources."* No problem here either, respectable international manufacturers have been supplying quality ozone generators to physicians and naturopaths for many, many years. End of the day it comes down to opinions only, not real world facts or good science. And if the opinions are honest after sincerely researching *all* the evidence, then no problem.

The point of taking this trip through the negativity was to bring out into the light of day for all to see and thereby expose everything in hopes of change. All I advocate is honesty and openness in all dealings by everybody.

Chapter 28

O₃ VS. AIDS

O₃ vs. AIDS was a small publication I put out relating the history of the use and suppression of Ozone Therapy. It had hundreds of references. I may again bring it back as a separate volume if needed. Parts of it follow and there are a few repeats of what I already showed you, but I wanted it to stay pretty much as I wrote it then.

In previous years I've concentrated a lot of energy on the use of ozone in the treatment of AIDS. AIDS is much more widespread than people think it is and it's spreading far more rapidly among many more segments of the population than you know about. I printed this up in *O₃ vs. AIDS* the abbreviated history and suppression of Ozone Therapy in the United States, a proven documented and successful treatment for AIDS and other diseases and how it meets the institutional wall of politics.

I wrote an article for *Explore Magazine, Explore New Dimensions*. The magazine tells you all about my testimony before the National Institute of Health when I went to the start-up-meeting of the Office of Unconventional Medicine. I testified that at the time, there were six doctors who had taken more than three-hundred people to HIV negative. "Every single one of the doctors is scared to death to come here today. So I'm here to represent them. I am now testifying on their behalf," I explained.

So, the Washington Post picked it up and wrote the big article I inserted in this book a few pages ago—Washington Post, 6/23/92, Health Section:— "Ed McCabe, author of a book on unconventional uses of oxygen, said he asked six doctors to come but all refused because they feared losing their licenses if their unconventional practices were made public. I came because I'm not an MD, so nobody can take away my license."

I then told this reporter, "You know we've got three-hundred people HIV negative. They all had night sweats, diarrhea, Coxsackie virus, thrush, candidiasis, all this stuff. It's *all gone*. They're *testing negative*. If that's not the elimination of AIDS, I don't know what is." He wrote: Ozone treatment had 'markedly improved' the conditions of three-hundred patients. I don't really blame him. Unless I had the lab tests in front of me, I wouldn't get behind it either. But nobody ever pays any oxygen people to do the expensive testing everyone requires; despite the fact the NIH has 23 billion a year to spend on research any way it wants to.

NIH Presented with Successful AIDS Treatment in 1992

In July 1992 a trip was made to Washington after setting up private meetings with two U.S. Congressmen to coincide with the meeting that former Iowa Congressman Berkley Bedell (www.nfam.org) and I had set up with U.S. Senator Tom Harkin, the former Presidential Candidate.

We decided to invite two doctors who had each brought a patient from HIV+ to HIV-. We also invited Jim Caplan, the man responsible for convincing the Cubans to approve medical Ozone Therapy for general use, and Dr. John Pittman, ozone using doctor and one of his recently denied treatment AIDS patients. Dr. Pittman's office was closed down by the North Carolina state medical board in the middle of successful clinical ozone trials due to 'ozone not being FDA approved.'

We visited the Congressmen, were warmly received at each meeting and ended up in turn at Senator Harkin's office for our meeting with him, where we were joined at this point by Dr. Michael Carpendale and his boss from the San Francisco Veterans Administration Hospital.

Senator Harkin scheduled us for only a 30 minute meeting, but he was so intrigued by our proof that we discussed ozone successfully treating AIDS with him for 1½ hours. He immediately decided to set up a meeting between us and the NIH's (National Institute of Health) Institute of Allergy and Infectious Diseases Director, Dr. Anthony Fauci. The AIDS 'problem comes under the jurisdiction of this institute, and Dr. Fauci has been referred to as the U.S. Government's 'AIDS Czar.' Senator Harkin is on the NIH appropriations committee, so he has their ear.

On August 20th, 1992, we met in NIH's building 31, wing 7A, and room 24 with Dr. Fauci and his boss, Deputy NIH Director, Dr. Moskowitz. Also present were Dr. Hill, Dr. Killan and other legislative and legal aides. Mike Hall and Marina Metallios were there to observe for Senator Harkin's Office. About 30 people attended.

We presented our two ozone treated patients who no longer were HIV+, no longer had fevers, swollen lymph nodes, diarrhea, pain, night sweats, weight loss, or any other manifestation of the AIDS/ARC disease. We handed Dr. Fauci and the others copies of their medical documentation, and they listened to Dr. Carpendale and one other doctor and their former AIDS patients. Dr. Latino from Medizone spoke of the flawless ozone animal trials that had already been done by Medizone. Dr. Pittman and one of his patients made emotional pleas for the open medical use of ozone so he could finish his clinical ozone trials. I asked for the same, gave them a brief 50 year history of the effectiveness of medical ozone on hundreds of thousands of people in Europe, and cited ozone's perfect safety record in millions of dosages. We made a sound, experienced, documented, and reasonable case for the immediate investigation of ozone's effectiveness in treating illnesses like AIDS successfully. I also asked if anything could be done to influence the FDA to halt its suppressions of ozone-using doctors. so we could supply the studies they required. We were only told that the NIH had no power over the FDA. No offer of making a call, just nothing.

Picture this: Here's our small but dedicated group gathered at a round table with the U.S. Government's official AIDS policy makers. Around the outside of the table are aides, secretaries, assistants and division chiefs There were no big corporations funding us, as is usually the everyday case at these meetings. Although the NIH people only had to walk down the hall to be there, we all had to take time out from work and pay our own considerable travel, hotel and meal expenses. We came from all over the country simply to help our fellow countrymen dying from AIDS. We were sitting right there at the table with two now perfectly healthy former AIDS patients testing HIV negative—one PCR negative (Polymerase Chain Reaction—a test for any of the seven nucleotides of the HIV virus itself), and one Western Blot/ELISA (HIV antibody presence) negative. We were sitting there with the living examples and their records showing complete eradication of all secondary diseases, their actual doctors, a politically harassed doctor and his patient who can't get the treatment and several thick notebooks of ozone medical references from the U.S. and Europe.

What answer did we get? Dr. Fauci: *"We see no reason to pursue this."* Let me repeat that: *"We see no reason to pursue this."* And, *"They are obviously so healthy that they must not have had the disease!"* Also,

"We won't look at this treatment unless you have the patients PCR tested twice before treatment proving the presence of the HIV virus, stored blood from when they were positive, two PCR tests during treatment, and two PCR tests after ozone treatment proving the absence of the virus."

But what about this? *"Contrary to popular belief, PCR cannot determine what portion, if any, of the genetic material it detects represents infectious virus. In fact more than 99 percent of what PCR measures is noninfectious."* —Dr. Kary Mullis, who won the 1993 Nobel Prize for inventing PCR.

What a reply from an institution funded with billions of dollars of our money and your tax dollars to find a cure for AIDS while people suffer and tragically slowly waste away. Not exactly encouraging, was it? The weight of evidence sitting right there was enough to immediately start investigating ozone, without adding all the new requirements on. The NIH has billions of dollars and they could have done something, anything. After all we went through; you would think they could have at least made some phone calls to other doctors and patients as a goodwill gesture. But in effect, all they said was "Go home." Even though they did seem to give us a clearly defined goal to shoot for, I couldn't help feeling that the goal line was just moved way back as quickly as we approached it.

The Problem with the NIH Reasoning:

1. Although they said they were unknowledgeable about the FDA's history of seizing ozone machines, harassing ozone-using doctors and forcing doctors to falsely claim ozonc as worthlcss, they did hear me tell them of all this and how hard it was to get any doctors to show up at all to testify and present evidence to them. How can anyone conduct open trials on this beneficial treatment if the FDA will close them down as soon as they open the doors?

2. Promoting the controversial 'HIV causes AIDS' scenario, they wanted six PCR HIV tests, each costing around 350 dollars. Total of $2,100+ per patient. First of all, there are several PCR tests around and none of this is covered by insurance, so which PCR test will they believe? And who is going to pay for it? The ozone doctors are financially strapped and the AIDS patients had already spent their savings on hospitals, doctors and drugs like AZT, DDI, etc. There were also no PCR tests commonly available back then when the patients we brought in first tested HIV positive, so having PCRs on them was an impossible requirement! Also, many of the patents on Ozone Therapy are now in the public domain, so no pharmaceutical company will support research.

3. They told us, via their announcement that they will not consider ozone unless we met the new PCR requirements and that the HIV virus is the ONLY thing to look for. Nowhere has it ever been proven that the HIV virus is the only definitive cause of AIDS; it was simply announced one day in the media as a probable cause. Dr. Gallo's work 'HIV causes AIDS' was unproved. People die from AIDS who never test positive for HIV. The virus is probably only a promoter or possibly a co-factor of the disease.

So why judge ozone's effectiveness upon the presence or absence of a possibly non-essential virus? Why ignore the most significant facts proving *complete eradication of all secondary diseases and symptoms*? This secondary disease eradication is a far more compelling test of whether or not to immediately begin research into ozone, if those who suffer can at least have their suffering eliminated whether or not they test PCR negative, why not use it NOW!

4. What about the fantastic improvement in the quality of their life all by itself since doing the ozone?

The way the Center for Disease Control has decided to officially classify matters if someone has AIDS or not tells the story. They look for the presence of several 'hallmark' diseases all occurring at once. Both patients that we brought in had completely eliminated their secondary infections and any clinical symptoms. Therefore by definition, besides testing HIV negative, they no longer had AIDS according to CDC guidelines! The real live people with their medical records and blood tests and doctors were sitting right in front of the NIH employees—yet the NIH employees could not see.

Just so you understand clinical ozone in the proper treatment of AIDS, a few shots of clinical ozone are not going to be magic bullets. Successful ozone AIDS treatment has always been two to five hours a day of a number of oxidative and other therapies for around six weeks in a row, depending upon the particular aggregate methods employed. These methods are always combined with lifestyle changes, proper diet, eliminative organ cleansing, a spiritual or moral balance to help eliminate denial of self and the inclusion of an immune system rebuilding regimen.

Seven U.S. doctors back then (early '90s) reported to me that hundreds of people had turned HIV negative under their Ozone Therapy care. I also have interviewed thousands relating that this therapy is remarkably effective on cancer and even diseases such as multiple sclerosis.

To continue to ignore Oxygen Therapies is criminal.

Chapter 29

Ozone VS. Bio-Terrorism

Everyone has been scared by the realization of how vulnerable we are to bio-attacks. There are oxy-solutions available, and we did our best to help out during the recent post office delivered mail Anthrax crisis. Here's an example of our efforts:

Fax: To, Dr. [M. A.]
CDC Emergency Operations Center
Occupational Environmental Team
(Please forward to Dr. [A] in Cincinnati if he has already left for the meeting.)

Ozone Gas Sterilization Proposal: The Proven Destruction of Bacillus

Broad Executive Summary for Section Chief Dr. [M. A.]/OET

Anthrax is a facultative anaerobe. If surrounded by active forms of oxygen it cannot survive. Ozone gas is a highly aggressive pure oxygen molecule comprised of three or more oxygen atoms and non-toxic when used correctly. Ultimately, it is harmless oxygen. Ozone is commonly employed as a sterilizer. Spores 'soften' and are easier to eliminate in a humid environment. The greater the ozone concentration, the more rapidly ozone eliminates spores. We will employ these variables efficiently by coupling proper relative humidity with ozone concentrations greater than were contemplated in the successful quoted studies. Doing so will result in more rapid and greater spore destruction efficiencies, enabling us to decontaminate the mail without risk to postal employees.

Ozone already proven effective against Bacillus spores

We have in our possession eight published papers already proving ozone destroys Bacillus spores in air and water. Five universities and one institute have already completed and published the necessary studies.

Our Proposal

Orion Industries is prepared to immediately (within weeks) begin sterilization of the mail and the post office facilities beginning in the New York area, and later on expanding throughout the whole country. We have the contacts and organization at the ready to use existing equipment in order to accomplish our mutual objectives as stated.

Orion Industries is meeting with senior staff and officials of the Post Office and the supervisor postal union and specific politicians this week and presenting our proposal to them. We need CDC, EPA, or any credible organization that all agencies will respect to study our literature and give the post office the go ahead to implement installation of our proven remedies for the current clean-up crisis.

Please respond at your earliest convenience, so we may implement the known harmless solutions.

Thank You,

Ed McCabe
Consultant to Orion Industries

[No significant responses were forthcoming after numerous contacts were established via letters, faxes and calls, direct and through assistants. This was the same CDC that years ago had my *Washington Post* and *AIDS Patient Care* published NIH testimony mysteriously removed from their public database after someone called them up inquiring about it.]

Chapter 30

The Lights

There are people who see what is and say we can do better. They take the hero's journey into the unknown territory, setting out to change things by overcoming adversity and bringing back the truth, which lights the way for the rest of us. For the record, a lot of people spent their lives tirelessly researching and further promoting these therapies—and got very little reward for it and often worse. Benedict Lust, Robert Koch, F. M. Eugene Blass come to mind. I can't list them all, but here are a few of the stories of some of what I call the modern oxy-lights.

Walter Grotz
The Foremost Hydrogen Peroxide Proponent in the U.S.

Here's a great guy who taught me many things about peroxide. Retired postmaster Walter Grotz became interested in the hydrogen peroxide story after meeting Father Richard Wilhelm; the 'Peroxide Priest' now deceased. Walter had a number of health problems that disappeared after he personally used dilute food grade hydrogen peroxide and he has been lecturing on the oral use of dilute peroxide for many years.

Walter and Ed

In the early 1990s Walter Grotz was busy publishing his ECHO (Educational Concerns for Hydrogen and Oxygen) newsletter. Walter and I got together and double-teamed the English and Scottish people during our famous British Isles traveling Oxygen Therapies lecture, radio and press interview tour in early 1993. Several English practitioners were already using ozone and peroxide's popularity was slowly growing due to its low cost effectiveness in cases where traditional socialized medicine nostrums were producing no relief.

ECHO UK

Our tour sponsor, Mr. Alwyne Pillsworth, had already formed a British foreign office version of Walter Grotz's U.S. efforts, called ECHO UK. He and his dedicated staff were spreading the oxy-gospel over there. He paid and arranged for Walter and me to tour Scotland and England in a big circle, while granting radio and press interviews during the day and lecturing at night. His wife Christine graciously helped out with the tour, and Stewart and Anne Brown, of Dundee coordinated the Scottish portion.

The British crowds we encountered were enthusiastic, but smaller than I had been experiencing in other countries. This perplexed us for a while, until we figured out that socialized medicine (that no one has to directly pay for) created a sort of societal malaise due to lack of competition. The prevailing attitude of, "Even if it only sort of works O.K., at least it's free", does not encourage searching for more efficient healing methods.

The radio 'talk back' shows were fun and I was interviewed several times by reporters, including the main science reporter for the London Times. I spent days supplying him proofs and despite our best combined efforts, the story never appeared in the paper. The same thing happened in Australia. The reporters got all excited, they had discovered a good answer for disease, and they expected to get the credit for doing something really good. They even brought me into the main office and took formal pictures of me. Mysteriously, someone above stepped in and no story appeared.

After we went back to America, the ECHO U.K. office was cruising along answering questions and having a great time supplying dilute food grade peroxide and printing newsletters for the people of Britain. After a while, the British government's Medicines Control Agency—a part of their Department of Health—suddenly announced they were declaring food grade hydrogen peroxide an unlicensed medical agent. Alwyne was told to immediately stop supplying it, to stop promoting it and to stop publishing his *ECHO Newsletter.*

In a direct affront to freedom of the press, and the right to be fully informed, and the right to chose, the British people were no longer to be allowed any more knowledge on the subject of active oxygen, effectively eliminating free choice. Alwyne was told to "Cease forthwith, or in 14 days be arraigned in the High Court for the criminal offense of promoting an unlicensed medicinal agent and be fined a minimum of £5,000."

Following that, the same mysterious behind the scene goons got the Charity Commission to go after the ECHO office itself. For nine years the non-profit ECHO UK office had filed all accounts without incident and suddenly their accounting was 'wrong.' Any charity as small as theirs never had to present their accounts, but they were required to. To stay alive, the little charity switched to supplying aloe vera, an excellent adjunct to any Oxygen Therapy. The local trading standards office had already approved their aloe labels for two years running and yet the same labels were suddenly declared illegal. This was despite scores of other companies openly selling and labeling their aloe the same way.

Alwyne was advised, "They are out to get you." And he relates how two MI5 (British secret agency) agents had called up begging him to send peroxide to their 'dying relatives' and be entrapped. Under these siege conditions, the tiny charity closed in December of 2001. The British owe Alwyne and Christine a debt of gratitude for trying to help them despite the overwhelming force used against them. Alwyne Pilsworth is at Woodside, Melmerby, Ripon, N. Yorks. HG4 5EZ, England.

The Finchley Clinic of London

There are several oxy practitioners and home supplement suppliers in England, and the contact info are listed in our resources section. While we're on the subject of British politics, I would like to give a favorable mention to oxy-pioneer Mark Lester, a natural health practitioner in London who has been favorably written up in many British publications. When presented with the oxy-facts, he also immediately saw the value of proper oxygenation, and started offering home oxygen and mineral supplements along with ozone steam cabinet sessions to his clients. As in the ECHO situation, the local and national medicine devices and control health authorities threatened him as well. Fortunately in his case, we were able to supply enough documentation to turn the situation around and he's working toward British Register of Complementary Medicine approval. But if he had folded at the first sign of trouble, approval would never be considered. Like Alwyne Pillsworth, he is another British oxy-hero who stood up for his people and deserves a thank you from the British for helping to keep their healthcare choice options open.

Plasmafire

Based in Canada, Dr. Saul Pressman, DCh of Plasmafire International is a great guy. He started Plasmafire International, a company located in Langley, B.C., Canada. They manufacture and sell medical ozone generators, steam saunas and ozonated olive oil. They are unique in that they are the only company that has developed Nikola Tesla's cold plasma ozone generation technique and combined it with the 1880 technique of ozone saunas and hyperthermic treatment invented by Dr. John Harvey Kellogg of Battle Creek, Michigan. I have heard reports that older models slowly lost some strength with age, maybe they needed cleaning.

Saul brought me in during May of 1993 to speak in Vancouver and Victoria on Ozone Therapy. It is incredibly beautiful riding the ferries up there while being surrounded with pristine forest wilderness. After I left, he continued to promote Ozone Therapy, resulting in the adoption of

Ozone Therapy as an accepted modality in British Columbia, Canada, by the Association of Naturopathic Physicians of BC.

To further promote the medical aspect, the company initiated an international symposium to educate British Columbia doctors. Five world class physicians were contacted and engaged as presenters and flown in to provide the benefit of their experience with medical ozone to the assembled physicians. The first symposium was held in Vancouver in June 1994, and was a huge success, attended by more than 160 persons, nearly half of them doctors.

From that impetus, medical Ozone Therapy was eventually adopted by the Association of Naturopathic Physicians of B.C., with more than 50 British Columbia naturopaths currently offering Ozone Therapy.

In December of 1999, the Health Protection Branch of Health Canada raided Saul, trying to shut down the Plasmafire factory. They found nothing and left empty-handed. They followed this up with a series of letters ordering Plasmafire to cease production, advertising, and sales of ozone generators. In compliance with this, Plasmafire was forced to remove the very helpful contents of their Internet website, leaving only a front page.

Ken and Mardy Theifault

The FDA website is proud of putting Ken (and his wife Mardy) Theifault previously from Jupiter, Florida in jail. The SWAT team broke into his home because Ken built and sold ozone generators and put out a video saying what he kept seeing; that ozone could successfully treat disease. Like they always do, the attackers get rich stirring up fear, claiming people like me and the jailed docs make millions off the suffering of others, but everyone I know in this area, including me, has *lost* money. Dr. Boyce lost everything.

The old fights the new at every turn because of the ego empires that depend upon the old ways. As the old guard crumbles, the new becomes the next opportunity to do it right and without ego.

Both Sides of the MD/ND Fence Working Together

Leading the way to showing us how easy medical and naturopathic doctors can work together, Dr. Farr, MD, and Dr. Freibott, ND, were both staunch advocates of the Oxygen Therapies. Dr. Farr, Dr. Freibott and I testified on Behalf of the Oxygen Therapies at Dr. Boyce's persecution-trial for curing AIDS over 118 times.

Is There Any Real Justice Left?

Maybe it was pure coincidence, but four days before I was due to go to Mississippi and testify at the trial, I woke up to a squad of dark military helicopters violently shaking my house by hovering just a few feet above it. They were so close I could see their faces. I went to the trial anyway.

I asked Dr. Farr and Dr. Freibott to both testify for Dr. Boyce as well, and they did. Dr. Charles H. Farr MD, of Oklahoma City is no longer with us, but as the founder of the International Bio-Oxidative Medicine Foundation, IBOM, and avid researcher, he is the one who started the modern ball rolling creating respectability for the Oxygen Therapies among the MDs in America.

Dr. George Freibott ND of Priest River, Idaho is the President of the American Naturopathic Association. He, like Dr. Farr, has been one of those who I consider an oxy-light holder. Before I did all I could to make Oxygen Therapy popular he overcame many obstacles and quietly collected—and held and kept safe—much of the flame of Oxygen Therapy knowledge. He taught me many things and always kept focused on helping the people while keeping the knowledge centered only on truth.

At Dr. Boyce's trial I was called to the witness stand. In an outrageous government abuse of power that should send chills through your spine, the judge put the jury out of the room so they couldn't be witnesses to what was about to take place. They put me in the witness chair, and the prosecuting attorney told me that if I dared to remain and testify on behalf of Dr. Boyce (and ozone) that she would direct the district attorney where I lived to indict me. Imagine somebody doing this to you to prevent you from testifying. Of course, mysteriously, the record of this blatant Constitution crushing illegality never made it into the trial transcript, but there are witnesses. I stayed, I testified, they put Dr. Boyce in jail, and later they used fictions of law and mispersonation and fraud to put me in jail as well.

I was subjected to being in 12 prisons in 18 months followed by three years of probation. That means often I had nothing, no mail, no commissary, no phone access, and no one could write to me or find me. I was in arm and leg chains all day long for a total of 18 days, including being repeatedly chained into airplanes and busses with hundreds of others also chained in. I was put into solitary confinement for a total of two months (that you will never understand), and sent to a prison for the criminally insane on the thin Communistic pretense of claiming I made 'bizarre statements' in court every time I asserted my rights.

My statements were simply things such as "I demand the show cause hearing the law guarantees me," or "I demand production of the underlying foundational contract that you are seeking to enforce against me," or "I reject the court indictment papers because I found out under the Freedom of Information Act that this case you are trying to make me liable for is being brought in the name of a fictional drug dealer living in the Virgin Islands. His name is signified on your papers by an all capital letter spelling, and that's absolutely someone else, not me," or "I demand production of the subject matter (the rem) for all to see in open court as required by law, and without producing it the law says you have no jurisdiction." or "I do not consent to any of this, I do not and have never granted in personam jurisdiction, and I have never knowingly or willingly consented to being a surety for any liability," or repeatedly "I do not consent, You are denying me due process." These are just a few examples of my statements I made as I was being railroaded all alone facing an army of jackbooted armed thugs and their masters. Some of the points I raised may seem odd to you, but you would have to understand Constitutional law. They were all completely proper and lawful. Every proper demand for my rights or production of evidence being used against me was met with silence, ignored, and I was further punished for raising the subject.

The biggest fraud was the absence of a required Grand Jury proceeding; they just claimed there was one. In addition, I rejected in writing any of their lawyers representing or speaking for me as was my right, but that was ignored as well. They didn't have me yet so the stymied magistrate held a secret meeting with one of their phony lawyers in order to plan how to trap me into the federal jurisdiction. Right after the secret meeting their stooge lawyer (who the judge forced me to sit next to) illegally claimed to *accept* the fraudulent indictment *for me* and thereby grant them federal jurisdiction. He had already been fired by me in writing and they did it anyway. This was obvious criminal fraud, and conspiracy, and they all knew it. Uninformed well-meaning people really don't understand the extent of the corruption and still suggest I "get a lawyer." It felt like a true secret dictatorship with great public relations convincing everyone its employees are fair and honest. Impossible to face alone.

International Advocates for Health Freedom - IAHF

Also fighting the good fight against the bought off politicians and multinational corporations while guarding your access to freedom of choice of health supplements is Mr. John Hammell, Legislative Advocate, and Founder of IAHF, International Advocates for Health Freedom.

IAHF is for the people and very effective, and exists on your donations and volunteer efforts. Recently they are working against having U.S. laws allowing vitamin access superseded by draconian international treaties.

They can be contacted at: P.O. Box 625 Floyd, Virginia, 24091 United States of America Tel: 800-333-2553 (N. America) Tel: 540-745-6534 Fax: 540-745-6535 Outside USA, and local: jham@iahf.com and at www.iahf.com

The National Foundation for Alternative Medicine

Former Iowa Congressman Berkley Bedell, was responsible for helping me get to Senator Tom Harkin and to my meetings with the NIH at the top levels. Berk is a great guy whose heart is in the right place after seeing alternatives work. He formed The National Foundation for Alternative Medicine. www.nfam.org 1629 K St. NW, Suite 402 Washington, DC 20006, Tel: 202-463-4900 Check it out, they deserve your support.

Chapter 31

Alternative Medicine Practices Acts

I have been advised that naturopaths who practice safely under the Common Law are therefore not under federal corporate statutes. Please seek guidance from the American Naturopathic Association on this issue.

As of September 2002 the following states and provinces have passed Alternative Medicine Practices Acts completely legitimizing your freedom of access to the therapies of your choice when working with doctors.

Alaska, Colorado, Georgia, Nevada, New Mexico, New York, North Carolina, Ohio, Oklahoma, Texas, Washington, Minnesota, and now California.

In Canada, Ontario and British Columbia have passed parallel laws.

Naturopaths in Canada and America have always used ozone since the late 1800s and are not federally regulated.

Let's look at the new Florida statute, which authorizes provision of and access to complementary or alternative health care treatments by all health care licenses. It also requires patients to be provided with certain information concerning such treatments, and revises Florida's Patient's Bill of Rights and Responsibilities to include the right to access any mode of treatment the patient or the patient's health care practitioner believes is in the patient's best interests.

LEGISLATIVE INTENT

It is the intent of the Legislature that citizens be able to make informed choices for any type of health care they deem to be an effective option for treating human disease, pain, injury, deformity, or other physical or mental condition. It is the intent of the Legislature that citizens be able to choose from all health care options, including the prevailing or conventional treatment methods as well as other treatments designed to complement or substitute for the prevailing or conventional treatment methods. It is the intent of the Legislature that health care practitioners be able to offer complementary or alternative health care treatments with the same requirements, provisions, and liabilities as those associated with the prevailing or conventional treatment methods.

RIGHTS OF PATIENTS

Each health care facility or provider shall observe the following standards: [The following are excerpts from the list of standards.]

...(d) Access to health care.

A patient has the right to impartial access to medical treatment or accommodations, regardless of race, national origin, religion, physical handicap, or source of payment.

A patient has the right to treatment for any emergency medical condition that will deteriorate from failure to provide such treatment.

...3. **A patient has the right to access any mode of treatment that is, in his or her own judgment and the judgment of his or her health care practitioner, in the best interests of the patient, including complementary or alternative health care treatments.**

CITIZENS FOR HEALTH FREEDOM

Florida chapter is an organization of patients and practitioners supporting patients' rights to have access to responsible medical alternatives from licensed health care professionals. They were responsible for this bill's passage. Check the net for your local chapter.

California Passes the Best Laws Yet

"September 24, 2002, is a very important date in Health Freedom History. On that day California Governor Gray Davis signed into law the most important Health Freedom bill in world history - California SB 577. The paragraph below contains PART of the language in the bill. For Californians, section (c), below makes health care "wide open." Naturopaths, Homeopaths, Nutritionists, etc., are now free to offer services, without harassment.

"(c) The Legislature intends, by enactment of this act, to allow access by California residents to complementary and alternative health care practitioners who are not providing services that require medical training and credentials. The Legislature further finds that these non-medical complementary and alternative services do not pose a known risk to the health and safety of California residents, and that restricting access to those services due to technical violations of the Medical Practice Act is not warranted.

"California, the fifth largest stand alone economy in the world, has for some time led the nation in percentage (estimated at 75 percent) of health care dollars spent on 'Alternative Medicine.' Now, that percentage is expected to increase."   —Tim Bolen, Consumer Advocate

Chapter 32

Conclusion
Oxygen is good for you!

And that's not all, why is it some people don't seem to respond to any treatments no matter what they try? Some wait too long before investigating any alternatives. They are so plugged up that their acidic disease state has already destroyed so much tissue that they don't have enough bodily defenses to accomplish the necessary repairs.

Unfortunately as well, some of the alternative clinics have had to refuse treatment to patients that are too far gone and have already sustained extensive radiation, chemotherapy, or other damage. This happens not from lack of compassion, but to keep their doors open. Clinics are wary about being blamed for damage done elsewhere.

Some patients don't respond because they believe they really need to suffer, it's psychological. The patient might subconsciously want punishment to satisfy guilt or karma over a questionable previous action. Maybe the poor sufferer has somehow been conditioned to 'know' deep down they are really 'no good,' or some other similar mental control factor is in place. There also exists the condition of being hurt somehow, but never really letting go of the experience. By keeping dead images alive the mental and emotional energy goes back inside endlessly in sort of a feedback loop instead of coming out. These are a few of the common reasons for continually self-perpetuated illness.

I mentioned control factors. Some of us have been so conditioned into letting experts tell us what to do for so long that we've forgotten who's in charge. How many opinions are formed, and how many lives are lived based on the certainty that the only truth around is what's disseminated by television?

We're the only real experts on our own lives. Only listening to influences outside of us is error. It leads to imbalance and this leads to disease. Having said that, don't go too far on your own without backup. Remember, experts and friends are a wonderful resource. They should be learned from and trusted when appropriate. The point is to strike a balance between the inner and outer influences. A slave is totally outer-directed. A madman is totally subconscious-directed. A balanced person regularly dips into their inner well of creativity to remind themselves how life is connected in a glorious harmony, and acts with wisdom and responsibility as a caretaker of that knowledge.

After reading this book, if you are still unsure of the health benefits that will evolve from oxygenating a deficient body or environment properly, then all I can do is strongly encourage you to hold your breath one more time. Then track down all the experts, companies and historical sources listed in this book yourself. This is where it all starts, by you taking action.

> **Black, brown, dark-purple dirty blood is slow death.**
>
> **Bright cherry-red clean blood is life.**
>
> **When you "Flood Your Body with Oxygen"**
>
> **You are choosing life.**

This book is very important. Tell your friends.

Thank you for reading my book. I hope you were informed and entertained. After all my research there is but one thing I conclude:

"WE STILL HAVE A LOT LEFT TO DISCOVER!"

Happy Oxygen!

Mr. Oxygen, Ed McCabe

You may have noticed that I repeated over and over, 'Individual results will vary' because they will. Due to the current legal climate I cannot claim ozone or any other Oxygen Therapy or supplement will do anything other than be what they are and do what they do. Ozone and other active oxygen delivery systems heal nothing all by themselves. However, the body is a miraculous combination of energy and physicality, and it is programmed to spontaneously heal itself—if and when it is given the proper ingredients and conditions. If it does, praise God, not me.

The
Resources

Resources

These resources are presented as a public service solely for educational purposes. Mention in this book or inclusion in this list is not an endorsement. There is no implied or express warranty for any thing or any person or any entity or therapy or any place mentioned in this book. Ed McCabe and Energy Publications are not and cannot be responsible or liable for your use of this information.

Organizations

Mr. Oxygen™ Home Page
www.oxygenhealth.com
www.edmccabe.org
Oxygen Therapy Central. Articles, Info, Devices Supplements, Publications Legalities, 500+ Oxy/Ozone References.

www.oxytherapy.com
Original Oxy-Therapy site. Mailing List Archives. Oxygen Ozone Therapies Including Ask Mr. Oxygen™ Articles.

www.healthfreedom.com
Citizens For Health is the national grassroots advocacy organization committed to protecting and expanding consumer natural health choices. A nationwide network of community-based chapters and members.

International Assn for Oxygen Therapy
Non profit medical missionaries &
World Missionary Assn ASMM
Dr. George Freibott, Pres.
P. O. Box 1360,
Priest River, ID 83856
Tel: 208-443-4319
Ozone Therapy Training
Dr. Blass and Dr. Koch Info.

The Burney University
Dr. Adele Kadans, Pres.
Dr. George Freibott
324 Wild Plum Lane
Las Vegas, NV 89107
oxytherapies@anahosting.com

American Naturopathic Association
(Oldest Naturopathic Assn.)
World Natural Health Org (WNHO)
P. O. Box 78600
North Central Station
Washington, DC
20013-8600 USA
Tel: (202)-537-0771
Tel: (702)-227-9353

International Ozone Association (IOA)
Pan American Group
31 Strawberry Hill Avenue
Stamford, CT 06902-2608
Tel: 1-203-348-3542
Fax: 1-203-967-4845

International Bio-Oxidative Medicine Foundation (IBOM)
P. O. Box 891954
Oklahoma City
OK 73189-2954
Tel: (405)-634-1310
Fax: (405)-634-7320

International Oxidative Medicine Association (IOMA)
P. O. Box 891954
Oklahoma City
OK 73109
Tel: (405)-634-7855
Fax: (495-634-7320
Practitioner Lists. Training Seminars. For physicians.

Medizone International
P. O. Box 742 Stinson Beach
CA 94970
Tel: 415-868-0300
Fax: 415-868-2344
U.S. Public Stock Medical
Ozone Company

International Oxygen Manufacturers Assn, Inc.
1255 23rd Street, NW, #200
Washington, DC 20037-1174
Tel: + (202)-521-9300
Fax: + (202- 833-3636

Vasogen Inc.
2155 Dunwin Drive, Suite 10, Toronto, Ontario
L5L4M1 Canada
Tel: 905-569-2265
Fax: 905-569-9231
Canadian Public Ozone Co.

The Undersea and Hyperbaric Medical Society
(UHMS)
10531 Metropolitan Ave,
Kensington, MD 20895
Tel: 301-942-2980
Fax: 301-942-7804

Hyperbaric Oxygen Association
Helen Gelly, M.D.
Cobb Hyperbaric Medicine, Inc.
790 Church St.
Marietta, GA 30064
Fax: 407-830-0084

Veterinary

American Holistic Veterinary Medical Association (AHVMA)
2218 Old Emmorton Road
Bel Air, MD 21014
Tel: 1-410-569-0795
Fax: 1-410-569-2346

S. Anne Smith, B.S. V.M.D., O.M.D.
29834 N. Cave Creek Rd.
Suite 118, PMB 118
Cave Creek, AZ 85331
Tel: 480-502-355
Oxygen/Ozone Therapy

Western Nevada Veterinary Specialists
Dr. Dennis T Crowe DVM, DACVS, DACVECC
3389 S. Carson Street
Carson City, NV 89701
Tel: 775-883-8238
Fax: 775-883-8275
Veterinary Hyperbaric Oxygen Therapy

European Organizations

International Ozone Association (IOA)
83, av Foch, F-75116
Paris, France
Tel: (33-1-53-70-13-58
Tel: (33)-1-53-70-13-56

IOA Secretariat:
IOA-EA3G Bat. ESIP
40, av. du Recteur Pineau
86022 Poitiers Ced
France
Tel. 33 (0)5-49-45-44-54
Fax. 33(0)5-49-45-40-60
E-mail: ioa@esip.univ-poitiers.fr

IOA Chapters
Germany
The Medical Society for Ozone Application in Prevention and Therapy
Switzerland
Swiss Medical Society for Ozone and Oxygen Therapies
Austria
Ozone Therapist
Interest Group
(Contact Secretariat above)

Department of Hyperbaric Medicine
David P. Downie F.I.M.T.
C.H.T. Fire H.Q. Douglas
Isle of Man, U.K.
Tel: 00441-624-626-394
Fax: 00441-624-670-289
Oxygen Concentrator
Designer/Consultant

International Medical Ozone Society (IMOS)
Sede Legale:
Via C.Battisti N°267
35121 Padova, Italy
Tel: 049-821-3092
Fax: 049-875-4256

Manufacturers & Distributors of Ozone, Air &Water Systems

N America & Canada
The huge number of dealers and manufacturers are too numerous to list. Refer to the oxygenamerica.com website for all products.

Air Purifying Systems
Carlos Jimenez
3190 S. State Road, 7, Bay 16, Hollywood, FL 33023
Tel: 954-962-0450
Fax: 954-963-0366
Commercial Systems
Condos, Office Buildings

Thomas J. Corbett N.D.
The Healing Center
200 W. 57th St.
NY, NY, 10019,
Tel: 212-581-0101
Tel: 516-984-4900
Ozone Medical Clinic
Manufacturer and Design Consultant throughout the Caribbean.

OxyOz
Philip Seifer
4200 Community Drive
West Palm Beach
FL, 33409
Tel: 561-478-8939
UV ozone generators.

Delta Marine International
Kenneth D. Hughes President
Tel: 954-791-0909
Fax: 954-792-9131
P. O. Box 15458
Fort Lauderdale, FL 33318
Chem-Free Purification
Systems equipment

Oxygen Products

International Oxidation Products (IOP) of Idaho
P. O. Box 502
Nordman, ID 83848
Tel: 208-443-4319
Homozon, Cutozons
Sanozon, Equizone, Teslaire
Ozone Generators

Oxygen America
Rick Kroll
sales@oxygenamerica.com
Tel: 954-492-5151
www.oxygenamerica.com
www.globalhealthtrax.org/19753
All products listed in this book. Hydroxygen
OxyMune, Oxygen
Supplementation and
Concentrators. Bottled
Oxygen Water. Ozone Air
and Water Purifiers. Ozone
Steam Cabinets. Oxygenated
Water Coolers. Chi machine
Hothouse. Tapes

Family Health News
John Taggart
9845 N.E. 2nd Ave
Miami, FL 33138
Tel: 800-284-6263
Tel: 305-759-9500
Fax: 305-759-8689
Oxygen Supplementation
Oxy Dan, Oxy Max
Ed McCabe©1967 Books,
Tapes & Videos,
6% peroxide

Tobin Farms
Cheryl Chamberlain
P. O. Box 70
Alpine, CA 91903
Tel: 619-445-6396
Electric Essence Minerals
Velvet Antler, OxyBoost

Portable Hyperbaric Chambers
The Hyperbaric Therapy
Centre, Suite 1-10
2000 Powers Ferry Road
Marietta, GA 30067
Tel: 706-216-8856
Lance@hypertc.com

Hydrogen Peroxide

Eagle Enterprise
P. O. Box 2144
Whitney, Texas 76692
Tel: 800-833-3256
Food Grade Peroxide

Let's Talk Health
Dr. Kurt Donsbach
1229 Third Ave, Suite C
Chula Vista, CA, 91911
Tel: 800-359-6547
Food Grade Peroxide

Green for Life
RR 3 Box 1333
Street Address: 16-2155
Orchid Drive Pahoa, Big
Island of Hawaii 96778
Tel: 808-982-6133
Food Grade Peroxide

Industrias Mexicanas
Maluc S.A. de C.V.
Ave. Morelos #15ACopl
Sta. Ana Talpaltitlan,
Toluca,
Edo. de Mexico. C. P.50160
Food Grade Peroxide

The Nutri Centre
7 Park Crescent
London W1N 3HE
England
Tel: 0207-436-5122
Hydrogen Peroxide

Low Level Lasers

2035 Inc
(Refer to Page 589)

Skin Care

Aura Research West
324 West Portal
San Francisco, CA 94127
Tel: 800-582-6464
Oxygenating Skin Care

MSM

Planet of Health
Keith Ranch
18030 Brookhurst St. # 564
Fountain Valley, CA 92708
Tel: 714-962-2195
VM: 714-647-6616
High Quality MSM Products
www.msmangelskin.com

Udo's Oil

www.udoerasmus.com
Health Food Stores
1-800-446-2110

UK Oxy-Resources

Resonance
Paul Benson
Unit C
The Scope Complex
Wills Rd., Totnes, Devon
TQ9 5X9, England
www.resonance.uk.com
Tel: 011-441-803 840-008
Tel: +0800-038-0303
Various Oxygen Products

The Finchley Clinic
Mark Lester
26 Wentworth Ave
Finchley, London N3 2DS
Tel: 01144-208-349-4730
Tel: 0208-349-4730 (UK)
Ozone Saunas, OxyMune
Electro Crystal Therapy

Cobella Akqa
5 Kensington High Street
London W8 5NP
England
Tel 0207-938-4800
Oxygen Spa, Oxygen Facials

US Clinics & Hospitals

ARIZONA

Arizona Preventative
Medicine Clinic
Dr. Paul Conyette N.D,
N.M.D., RT (CSMLS) RRP
1933 East University Ave
Mesa, Arizona 85203
Tel: 480-633-3948
IV Ozone Therapy

CALIFORNIA

Europa Institute of
Integrated Medicine
Carolyn Borman, N.D., C.P.
P.O. Box 950
Twin Peaks, CA 92391
Tel: 909-338-3533
Fax: 909-338-7343
Whole Range of Ozone
Therapies

Hospital Santa Monica
Dr. Kurt Donbasch
1227 Third Ave
Chula Vista, CA 91911
Tel: 800-359-6547
Ozone Therapies
Hyperbaric chambers
Hydrogen Peroxide Therapies

COLORADO

Frontier Medical Institute
Grossman M.D.
2801 Youngfield S,
Lakewood, CO 80401
Tel: 303-233-4247
Fax: 303-233-4249
IV Oxidative Therapy

WASHINGTON, DC

Family Practice, Preventive
Medicine & Nutritional
Therapies
Dr. Paul V. Beals, M.D.
2639 Connecticut Avenue
N.W. Suite 100
Washington, D.C. 20008
Tel: 202-332-0370
Peroxide Therapies

Naturopathic Health Care
Dr. B. Feeley, M.A., N.D.
Dr. George Freibott
14816 Physician's Ln
Suite 252
Rockville, MD 20850
Tel: (301)-424-6644
Tel: (775)-227-9353
Oxygen Therapies
Homozon

FLORIDA

Medical Center Sunny Isles
Dr. Martin Dayton, D.O.
18600 Collins Avenue
North Miami Beach
FL 33160
Tel: 305-931-8484
Fax: 305-936-1849
Hydrogen Peroxide Therapies
Ozone Therapies

Naples Institute for
Optimum Health & Healing
2335 9th Street North
Naples, FL 34103
Tel: 941-261-1148
Fax: 941-262-4684
Ozone Steam Cabinet
Ozonated Olive Oil Massage
Ozonated Pools and Spas

Oxygen Day Spa
2333 Coral Way
Miami, FL 33145
Tel: 305-858-0907
Oxygen Facials

Ocean Hyperbaric Center
Richard Neubauer M.D
4001 North Ocean,
Lauderdale by the Sea
FL 33308
Tel: 954-771-4000
Fax: 954-776-0670
Hyperbaric Oxygen
Chambers

O₃ Technologies
Jim Brown (Phlebotomist)
1101 S. Rogers Circle
Suite 16 Boca Raton
FL 33487
Tel: 561-997-5966
Fax: 561-997-5966
Ozone Hydrotherapy
Live Cell Microscopy
Oxygen Supplementation

Physicians Health &
Wellness Centers, Inc
Dr. Carlos M. Garcia M.D
36555 US Highway 19
North Palm Harbor
FL 34684
Tel: 727-771-9669
Fax: 727-771-8071
IV Hydrogen Peroxide

Rev. Mary Seid
2304 St. Andrews Circle
Melbourne, FL 32901
Tel: 321-951-4141
Violet and Orange Ray
Electro-Therapy Devices
www.myholistichealthshop.com

The Lane Spa for Wellness
11382 Prosperity Farms Road
Suite 126, Palm Beach
Gardens, FL 33410
Tel: 561-691-0104
Fax: 561-691-9857
Ozone Steam Treatments
Insufflations
Hydrogen Peroxide Baths
Ozone Water

Wellness Works
Carol L Roberts, M.D.
36555 US Highway 19 North
1209 Lakeside Drive
Brandon, FL 33510
Tel: 813-661-3662
Fax: 813-661-0515
IV Hydrogen Peroxide

GEORGIA

Environmental &
Preventive Health Center
of Atlanta
Dr Stephen B. Edelson M.D.
F.A.A.F.P., F.A.A.E.M.
3833 Roswell Road
Suite 110
Atlanta, GA 30342-4432
Tel: 404-841-0088
Fax: 404-841-6416
Hydrogen Peroxide Therapy

Progressive Medical Group
Marvin Reich M.D.
Carl Shenkman M.D.
4646 North Shallowford Rd
#100, Atlanta, GA 30338
Tel: 770-676-6000
Fax: 770-392-9805
Hydrogen Peroxide

HAWAII

Dr David Miyauchi M.D.
1507 S. King Street, # 407
Honolulu, Hawaii 96826
Tel: 808-949-8711
Intravenous Peroxide

Life Craft Naturopathic
Dr.Wayne McCarthy, N.D.
68-1793 Lina Poepoe
Waikoloa, Hawaii 96738
Tel: 808-883-27351
Fax: 808-883-2735
Minor Autohemotherapy etc.

ILLINOIS

Caring Medical &
Rehabilitation Services
Dr. Ross A. Hauser, M.D.
715 Lake Street, Suite 600
Oak Park, IL 60301
Tel: 708-848-7789
Fax; 708-848-7763
Direct IV Ozone Therapy
Major Autohemotherapy
Intramuscular Ozone
IV Hydrogen Peroxide
Rectal, Vaginal, Bladder
Ozone Insufflations

INDIANA

Center for Innovative
Medicine
Dr. Dale Guyer M.D
1235 Parkway Drive, Suite B
Zionsville, IN 46077
Tel: 317-733-5433
Fax: 317-733-2428
Ozone Therapies
Hydrogen Peroxide

MAINE

Dr. Arthur Weisser, D.O.
184 Silver St
Waterville, ME 04901
Tel: 207-873-7721
IV Hydrogen Peroxide
Photo Oxidative Therapy

MARYLAND

Family Practice, Preventive Medicine & Nutrition
Dr. Paul V Beals M.D.
9101 Cherry Lane, Suite 205
Laurel, MD 20708
Tel: 301-490-9911
IV Hydrogen Peroxide

Lifeforce Inc.
Michelle Reillo RN
1006 Morton Street
Baltimore, MD 21201
Tel: 410-528-0150
Hyperbaric Oxygen Therapy

The Sky Center LLC
6711 Bush Ranger Path
Columbia, MD 21046
Tel: 866-759-2368
Oxygen Spa
Oxytherapy Steam

MASSACHUSETTS

Barry Edelman N.D.
44 Everett Street
Great Barrington, MA 01230
Tel: 413-528-8844
Electronic Oxygenation

MISSOURI

Iridologist Food Consultant
Ron Logan– M.H. ID C.R.R.
12274 Fortuna Rd # K
Yuma, AZ 85367
Tel: 928-210-0107
www.internetherbs.net
Books, Herbs, Bernard
Jensen Products.

MONTANA

Nature's Wisdom
Michael Lang ND
P. O. Box 1473
Ennis, MT 59729
Tel: 406-682-5000
IV Hydrogen Peroxide

NEW JERSEY

The Sun Spa Oxygen Bar
241-243 Lorraine Avenue,
Upper Montclair, NJ 07043
Tel: 973-655-1994
O_2 infusions

NEVADA

Angel Healing Center
1840 E. Sahara, Suite 103
Las Vegas, NV 89119
Tel: 702-474-9998
Ozonated steam
Ozone Rectal, Vaginal, Ear
Insufflations, Limb Bagging
Colonics

Nevada Center of Alt. & Anti Aging Medicine
Dr. Frank Shallenebrger
M.D., H.M.D.
896 W. Nye Lane
Carson City, NV 89703
Tel: 775-884-3990
Ozone Therapies
Stepped Oxygen Therapy

N. Nevada Hyperbaric Ctr
Dr. Johnathon Tay M.D
1698 Meadowood Lane
First Floor Reno, NV 89502
Tel: 775-826-2084
Fax: 775-826-2087
Hyperbaric Oxygen Therapy

The Nevada Clinic
F Fuller Royal M.D.
3663 Pecos McLeod
Las Vegas, NV 89121
Tel: 800-641-6661
Tel: 702-732-1400
Fax: 702-732-9661
Hyperbaric Oxygen Chamber

NEW YORK

America's Alternative Medical Center
Anna Alaris. M.D
221 West 57th Street, 11th
Floor, New York City NY
10019
Tel: 646-456-6966
Major Autohemotherapy

Comprehensive Medical Services, PC
Dr. Christopher Calapai D.O.
1900 Hempstead Turnpike
East Meadow, N.Y. 11554
Tel: 516-794-0404
Fax: 516-794-0332
Hyperbaric Oxygen Chamber

Dr. Howard Robbins
200 W 57th St
Suite 1202,
NY, NY 10019
Tel: 212-581-0101
Direct IV Ozone Therapy

H. Chandler Clark, M.D.
400 Webster Avenue,
New Rochelle, NY 10801
Tel: 914-235-8385
Fax: 914-235-3517
Peroxide Therapies
Major Autohemotherapy
Minor Autohemotherapy
Interarticular Ozone
Injections

Preventive Medicine & Wellness Clinic
Dr. Robert Barnes D.O
3487 East Main Road
(US RT 20)
Fredonia, N.Y. 14716
Tel: 716-679-3510
Fax: 716-679-3511
Autohemotherapy Ozone
IV Hydrogen Peroxide
Ozone Sauna

Schachter Center for Complimentary Medicine
Two Executive Boulevard
Suite 202
Suffern, NY 10901
Tel: 914-368-4700
Fax: 914-368-4727
Intravenous Peroxide
Ultraviolet Blood Irradiation
Direct IV Ozone

The New York Health & Healing Center
Dr. Joseph Carozza M.D
80 Fifth Avenue
Suite 1204
New York, NY 10011
Tel: 917-207-0954
Major Autohemotherapy

Viktor Goncharov ND Ph.D
3049 Brighton 6th St
Suite 105, Brooklyn NY
11235
Tel: 718-368-2755
Fax: 718-368-2755
Koch Therapy

NORTH CAROLINA

Optimal Breathing
Box 1551,
Waynesville, NC 28786
Tel: 828-456-5689
Fax 828-456-5689
Accelerated Breathing

OHIO

Spirit of Health
P. O. Box 3021
Dayton, OH 74136
Tel: 937-256-4923
Fax: 937-256-4933
Major Autohemotherapy

West Chester Family Practice, Inc.
Theodore J Cole M.A., D.O.
N.M.D. 9678 Cincinnati-
Columbus Rd
Cincinnati, Ohio 45241
Tel: 513-779-0300
All Ozone Applications
Ionized Oxygen
Hydrogen Peroxide

OKLAHOMA

Alternative Medicine's New Hope Health Clinic
Dr. Kent Bartell Naturopathic
D.C. 7320, South Yale Ave
Tulsa, OK 74136
Tel: 877-544-4673
Tel: 918-488-8844
Hydrogen Peroxide Therapy
Ozone Steam Therapy

Genesis Medical Research Foundation
Dr. Robert L White ND.
Ph.D., P.A.
5419 South Western
Oklahoma City, OK 73109
Tel: 405-634-7855
Hydrogen Peroxide Therapy
Hyperbaric Oxygen Therapy

Health & Wellness Clinic
Dr. Joel Robbins, D.C. M.D.
6711 South Yale Avenue
Suite 106 Tulsa, OK 74136
Tel: 918-488-0444
Fax: 918-488-0470
Ozone Insufflations/Showers

OREGON

Inner Radiance Health
Nancy Adams Ph.D LMT
P.O. Box 3162
Coos Bay, OR 97420
Tel: 541-888-5111
Peroxide Colon Cleansing

PENNSYLVANIA

Dr. Arthur L. Koch, D.O.
57 West Juniper Street
Hazleton, PA 18201
Tel: 717-455-4747
Oxidative Therapy

Narberth Family Medicine & Acupuncture Centre
Dr. Andrew Lipton
822 Montgomery Avenue
Suite 315, Narberth PA
19072
Tel: 610-667-4601
Fax: 610-667-6416
IV Hydrogen Peroxide

Preventative Medicine of Sharon
Andrew M Baer M.D. MACP
92 West Connelly Boulevard
Sharon, PA 16146
Tel: 412-346-6500
Oxidative Therapy

TENNESSEE

James E. Johnson M.D
P. O. Box 70879
Nashville, TN 37207-0879
Tel: 615-650-0830
Fax: 615-650-0269
Hydrogen Peroxide
Ozone Therapy

TEXAS

Alternative & Traditional Medical Centre
Dr. Antonio Borme, M.D.
Dr. Robert Gilbard, M.D.
Dr. Ray Hammon, D.C.
3809 Main Street
Rowlett, Texas 75088
Tel: 972-463-1744
Fax: 972-463-8243
IV Hydrogen Peroxide

Body Language Therapeutic Clinic
9720 Skillman St,
Dallas, Texas 75243
Tel: 214-340-6072
Fax: 972-495-3784
Ozone Steam Sauna
Ozone Cupping, Bagging
Breathing, Insufflations

Body Works Massage/Oxygen Spa
9205 Skillman #115
Dallas, Texas 75243
Tel: 214-342-5651
Tel: 877-891-5651
Ozone Steam Therapy
Ozone Insufflations
Cupping and Limb Bagging

Breath of Life Ozone Clinic
330 Oaks Trail, Suite 106
Garland, Texas 75043
Tel: 469-371-3432
Fax: 800-260-0801
Ozone Steam Sauna
Ozone Ear Insufflations
Ozone Limb Bagging

Coppell Redox Clinic
Dr. David E Winslow D.O
270 N. Denton Tap #140
Coppell, Texas 75019
Tel: 972-393-4686
Hydrogen Peroxide
IV Ozonated Saline

DFW Pain Treatment Center & Wellness Clinic
Barry L Beaty D.O.
4455 Camp Bowie Blvd
Suite 211
Fort Worth, TX 76107
Tel: 817-737-6464
Fax: 817-737-2858
Hydrogen Peroxide
Ozone

Global Healing Centre
Dr Edward, Group III
10101 Harwin Drive,
Suite 298,
Houston, TX 77036
Tel: 713-484-6550
Tel: 713 484-6551
Fax: 713-490-0319
Rectal and Vaginal Ozone
Auricular Ozone, Oxygen

Holistic Wellness Centre
Michael E Truman, D.O
2401 Canton Dr.
Ft. Worth, TX 76112
Tel: 817-446-5500
Fax: 817-446-5509
H_2O_2 IV Therapy

Integrative Medical Associates of Texas
Andrew K Messamore M.D.
12200 Park Central
#200Dallas, TX 75251
Tel: 972-385-2380
Fax: 972-385-2491
Oz Injections, Insufflations
Ozone Sauna
Intravenous Ozonated Saline

The Oxytherapy Centre
606 W. Wheatland Rd
Duncanville
TX 75116
Tel: 972-296-9456
Fax: 972-296-0933
Steam Sauna, Cup Funnel
Limb Bagging

Stan Wolfe, DVM
8880 Main St.
P. O. Box 400
Frisco, TX 75034
Tel: 972-202-9585
Oxygen Therapies

VIRGINIA

Integrated Medical Center
Manjitt R Bajwa M.D.
6391 Little River Tnpk
Annandale, VA 22312
Tel: 703-941-3606
Fax: 703-658-6415
Hyperbaric Chamber
Colon Hydro Therapy

Mount Rogers Clinic Inc.
Elmer M Cranton M.D.
799 Ripshin Road
P. O. Box 44,
Trout Lake, VA 24378
Tel: 540-677-3631
Fax: 540-677-3843
Hydrogen Peroxide Therapy
Hyperbaric Oxygen Therapy

Wound Care & Hyperbaric Oxygen Therapy Centre
Kenneth McIntyre M.D.
150 Kingsley Lane
Norfolk, VA 23505
Tel: 757-889-2300
Fax: 757-889-5019
Hyperbaric Oxygen Therapy

WASHINGTON

Leo J Bolles Clinic Inc
Betty Sy Go, M.D.
15611 Bel Red Road
Bellevue, WA 98008
Tel: 425-881-2224
Fax: 425-881-2216
Hydrogen Peroxide

Mount Rainier Clinic Inc.
Elmer M Cranton, M.D.
P. O. Box 5100
503 First Street, Suite #1
Yelm, WA 98597
Tel: 360-458-1061
Fax: 360-458-1661
Hydrogen Peroxide Therapy
Hyperbaric Oxygen Therapy

Paracelsus Clinic of Washington
Thomas Dorman, M.D.
2505 S. 320th St #100
Federal Way, WA 98003
Tel: 253-529-3050
Fax: 253-529-3104
Ozone Autohemotherapy
Ozone Joint Injections
Cutaneous Application for
Ulcers, Ozone Colonics
Hydrogen Peroxide

CANADA

ALBERTA

Ayurved Naturopathic Medical Clinic
Dr Raj-Inder Rakhra N.D.
G.A.M.S
121-14 Street N.W.
Calgary, Alberta T2N 1Z6
Canada
Tel: 403-270-7033
Ozone Insufflations
Autohemotherapy
Hydrogen Peroxide Therapy

BRITISH COLUMBIA

Angel Hyperbaric Care Ctr
Dr. Szymanski, N.D.
103, 20560 - 56 Ave
Langley, B.C. V3A 3Y8
Tel: 604-534-2155
Fax: 604-534-2209
Hyperbaric Oxygen Therapy
Ozone Therapies

Dr. Jim Chan. N.D
101-3380 Maquinna Drive
Vancouver, B.C. V5S 4C6
Tel: 604-435-3786
Fax: 604-436-2426
Hyperbaric Oxygen Therapy
IV Hydrogen Peroxide
Ozone Autohemotherapy
Rectal Insufflation

Dr. Phoebe Chow, N.D.
298 Newport Avenue
Vancouver, B.C. V5P 2J2
Tel: 604-327-0021
Ozone Therapies
Hydrogen Peroxide Therapies

Dr. Kathy Graham, N.D.
1283 Main St
P. O. Box 3579
Smithcrs, B.C. V0J 2N0
Tel: 250-847-0144
Fax: 250-847-0144
Ozone Autohemotherapy
Ozone Rectal Insufflations

Helios Natural Health Care
Dr. Peter Bennett
#1 7865 Patterson Road,
Saanich, B.C., V8M 2C7
Tel: 250-544-4331
Ozone Rectal Insufflations
Ozone Autohemotherapy

Dr. Christoph Kind, N.D.
3738 Minto Road
S 652, C-19, RR#6
Courtenay, B.C. V9N 8H9
Tel: 250-336-8349
Fax: 250-336-8748
Autohemotherapy Ozone
Hydrogen Peroxide Therapies

**HOC Centre for
Progressive Medicine**
Dr. Thao Nguyen
111-250 Schoolhouse Street
Coquitlam, B.C., V3K 6V7
Tel: 604-520-3941
Fax: 604-520-9869
Hyperbaric Oxygen
Hydrogen Peroxide, Ozone
Bio-Oxidative Medicine

**Dr. Liliane Holemar M.D.
N.D.,** 20326 72nd Avenue
Langley, B.C. V2Y 1S8
Tel: 604-530-3644
Bio-Oxidative Therapies
Ozone Therapies

Kelowna Naturopathic
Dr. Garrwett G. Swetlikoff
160 - 1855 Kirschner Road
Kelowna, B.C. V1Y 4N7
Tel: 604-868-2205
Fax: 604-868-2099
Ozone Therapies
Hydrogen Peroxide Therapies

Richmond Alternative
Dr. Martin Kwok, ND.
#150-7340 Westminster Hw
Richmond, B.C. V6X 1A1
Tel: 604-207-0167
Fax: 604-207-0167
Ozone Therapy

Vital Path Health Centre
Dr. Neil McKinney, N.D
5300 26th Street
Vernon, B.C. V1T 8E3
Tel: 250-549-1400
Fax: 250-549-1409
Hyperbaric Oxygen Therapy

Dr. David Wang, N.D.
Suite 604, 1200 Burrard St
Vancouver, BC. V6Z 2C7
Tel: 604-687-0119
Ozone Therapies

ONTARIO

Chelation Centres of Ont
Dr. Fred Hui, M.D.
421 Bloor St East, Suite 202
Toronto, Ontario, M4W 3T1
Tel: 416-920-4200
Fax: 416-920-4204

18 Wynford Drive, Suite 313
N. York, Ontario M3C 3S2
Tel: 416-443-0811
Fax: 416-443-0482

730 Essa Road
Barrie, Ontario
Tel: 705-721-1969
Fax: 705-721-1859
Hydrogen Peroxide
Ozone Injections

CKC Hyperbaric Centre
Dr. Uday Chadha. M.D.
2100 Ellesmere Road
Scarborough, Ontario
M1H 3B7
Tel: 416-431-4427
Tel: 416-431-0644
Hyperbaric Oxygen Therapy

**Hyperbaric Medical Centre
of Toronto**
Dr. Max della Zazzera
Dr. Barbara Raymond
4800 Leslie Street, Suite 202
Toronto, Ontario, M2J 2K9
Tel: 416-756-4628
Fax: 416-756-4511
Hyperbaric Oxygen Therapy

Naturomed Inc.
Dr. Michael Prytula, N.D.
296 Welland Ave
St. Catharines, Ontario
L2R 7L9
Tel: 905-684-4934
Fax: 905-684-1849
Autohemotherapy Ozone
Rectal & Vaginal Ozone
Insufflations, Intramuscular
Intraarticular Ozone

**Ottawa Hyperbaric Oxygen
Therapy**
Dr. Barbara Raymond
Dr. Max della Zazzera
Dr John Molot
1935 Bank St
Ottawa, Ontario. K1V 8A3
Tel: 613-521-2391
Fax: 613-521-5443
Hyperbaric Oxygen Therapy

O2 Spa Bar
2044 Yonge Street
Toronto, Ontario. M4S 1Z9
Tel: 416-322-7733
Tel: 888-206-0202
Fax: 416-322-7684
Oxygen Therapy Bath
O2 Lounge/Bar

Toronto Hyperbaric Inc.
101 Queensway
West Suite 100
Mississauga, Ontario
L5B 2P7
Tel: 905-306-0770
Fax: 905-858-4419
Hyperbaric Oxygen Therapy

MANITOBA

Canadian Biologics
Dr. Paul Conyette, N.D
N.M.D., R.T (CSMLS) RRP
Box 21017, WEPO
Brandon, Maniotoba
R7B 3W8
Tel: 204-727-3524
Fax: 204-725-2198
IV Ozone Therapy

QUEBEC

The Preventorium Institute
Juergen Buche, N.D N.H.C.
M.I. Phy.D
175 Rue Meloche
Vaudreuil-Dorion, QC
J7V 8P2 Canada
Tel: 450-424-8777
Oxidative Therapies

OTHER COUNTRIES

AUSTRIA

Ordìnation Dr. Friedrich Humhal
Dr.med univ Friedrich
Humhal
Wienerstrasse 24
Neunkirchen, Austria A-2620
Tel: 02635-62350
Fax: 02635-623504
Ozone Therapies

AUSTRALIA

Operation Hope
Prof. Noel Campbell
Level 5, 167 Collins St.
Melbourne, Vic. 3079
Tel: (61) 3-9639-6090
noelc@smile.org.au
www.smile.org.au
ozone cupping, rectal

Research Institute of Diet Disease & Prevention Inc
Dr Jeffrey P Dummett N.D.
Ph.D., D.Sc. (USA)
P. O. Box 159
83 Mi Mi St
Oatley, Sydney, Australia
Tel: 61-2-9585-1355
Fax: 61-2-959-46909
Ozone Rectal Insufflations
Ozone Hyperthermia Bath
Ozone Colonic Irrigation
Ozone Infra Red Therapy
Ozone Violet Ray

BELGIUM

Dr. Alain Ceulemans
Neerstraat 17
Tessenderlo, Limburg B-3980
Tel: +32+136-67461
Fax: +32+136-72519
Ozone Therapies

BRAZIL

Artz Instituto de Medicina Integral
Eduardo Almeida M.D Ph.D.
Rua Lopes Trov
o 52/803 Icarai
Niteroi, Rio de Janeiro
Brazil
24220-071
Tel: 55-21-612-0698
Fax: 55-21-616-1855
Hyperbaric Autohemo Ozone
Colon Ozone Therapy
Minor Autohemotherapy
Local Ozone Therapy
O_2 Multistep
Hydrogen Peroxide

Clinica Tabacow sc Itda
Dra Marca Maria Tabacow
Gomes, N.D
R Alcides Ricardini Neves N
12 5Andar Conj 514
S Paulo, SP
Brazil, 04614000
Tel: 011-550-52615
Fax: 011-550-52615
Hyperbaric Chambers

CARIBBEAN

Pillar Rock Spa Health & Seawater Centre
Pillar Rock P. O. Box 2639
St. John's, Antigua, WI
Tel: 268-463-0444
Fax: 268-460-8881
Oxygen & Ozone Treated
Seawater Whirlpool
Ozone Bagging, Rectal,
Vaginal, & Ear Insufflations
Oil Inhalation

Wellness Center at the Royal Court Hotel & Health Resort
Dr. H. Anthony Vendryes
13 Sewell Avenue
Montego Bay, Jamaica
Tel: 876-979-3333
Fax 876-979-3555
Ozone Therapies
IV Ozone/Hydrogen Peroxide

Wholistic Rejuvination Ctr. Erwin Dorsch
Light Pole 51, Moonan Dr.
Bypass Road, Arima,
Trinidad
Tel: 868-6674394
Aquacizers, Blood Analysis,
Direct IV Ozone, Iridology

COLOMBIA

Consultorio de Medicina Biologica
Dr, Ramon Unzueta Hoffman
Cra. 42A No. 5A-36
Cali, Valle Del Cauca
Tel: (57)-092-553-3403
Tel: (57)-092-553-6252
Endovenous Oxygen Therapy

COSTA RICA

Vidaloha
Boulevard, Jaco, Costa Rica
Central America
Tel: 506-643-4049
www.vidaloha.com
Ozone Insufflations
H_2O_2 Colonics, Ozone Baths
Oxygen Powder Cleanses
Oxygen Bagging Treatments
Ozone Oil Inhalations
Ozone Sauna

Dr. Fgabio Solano M.D.
Calle
24-28 Avenida Primera San
José, Costa Rica
Tel: 506-221-8120
Oxygen/Ozone Therapies

**International Ozone
Therapy Clinic**
230 Ave. and 15 St., Siboney
Havana City, Cuba
Tel: (537)-21-9264
Tel: (537)-21-2089
Fax: (537)-21-0233
Ozone Therapy
Ozone Oils

Ozone Research Center
Head of Biomedicine Dept.
Dr. Frank Hernandez
Calle 230 y Ave. 15
Siboney, Playa. Ciudad de La
Habana, Cuba
Tel: (537)-271-2324
Fax: (537)-271-0233
Major Autohemotherapy
Minor Autohemotherapy
Insufflations
Intraarticular& Intramuscular
Injections etc.

CZECH REPUBLIC
Center of Reflex Therapy
Vesely Milan M.D
Bludovicka St 396
Prague 9, 19900
Tel: ++420-2-859-1368
Fax: ++420-2-858-6433
Ozone Therapy

DENMARK
Humlegaarden
Ny Strandvej 11,
DK-3050 Humlebaek
Denmark
Tel: +4549-132-465
Fax: +4549-134 -498
Hyperbaric Oxygen Therapy
Hydrogen Peroxide
Ozone Therapies

EGYPT
**Al Far Ozone Therapy
Clinic**
Dr. Mahmoud AL FAR
110A -26 July St. Flt.19
Zamalek, Cairo
Tel: +2-012-341-7465
Fax: +2-012-735-4396
Major & Minor
Autohemotherapy
Rectal Ozone Insufflations
Ozonized Olive Oil, Water

GERMANY
**Germany is full of clinics.
Here are just a few:**

Hans Neiper
Doctor Ledwoch who kept
his staff, but moved to
Langenhagen, Germany,
(Outside of Hanover)
Contact the following:

Dr. Med. Joachim Ledwoch
Paracelsus Klinik Am
Silbersee Oertzeweg 24
30851 Langenhaen
Germany
Tel: 011-49-511-348-080
Fax: 011-49-511-318-417

Dr. Med H.G Eberhardt
Fuerstenstr.1A,
Saarbruecken, Saarland,
D-66111FRG
Tel: +49-681-36303
Fax: +49-681-57259
Major Autohemotherapy
Rectal & I.M. Application

Holler Clinic
Dr. H. Ilsa
Edelfinger Strasse, 26
Bad Mergentheim,
Germany 97980
Florida Contact:
Khara Bromiley
Tel: 904-810-2014.
Ozone Autohemotherapy

Hufeland klinik
Dr. Wolfgang Woeppel M.D
Loeffelstelzer St. 1-3
D-97980 Bad Mergentheim,
(Germany)
Tel: +49-7931-5360
Fax: +49-7931-8185
Autohemotherapy Ozone

Praxis v Falkenhayn
G.v.Falkenhayn Heilpraktike'
Dreiecksplatz 2
Schleswig-Holstein, 24103
Kiel
Germany
Tel: 0049-431-305-1112
Fax: 0049-431-305-1113
Autohemotherapy
Ozone/Oxidative Therapies

Gerd H. Wasser, M.D.
Former Vice President
German Medical Association
for use of ozone in
prevention and therapy.
Consulting Specialist for
Internal Medicine to Forum
E.V., Erfurt Germany.

More than 15 Years of
Experience in Using Rectal
and hyperbaric Ozone
An Hacksteinskuhlen 37
47509 Rheurdt, Germany
Tel: +49-2845-968-66
Fax: +49-2066-999-997
www.ozoneforum.com
gerd.wasser@knuut.de

HONG KONG
Optimum Health Center
2/F Prosperous Commercial
Building
54-58 Jardines's Bazaar
Causeway Bay
Hong Kong
Tel: 2577-3798
Fax: 2890-8469
Ozone Therapy
Oxygen Therapy

INDIA

Aakaash Medical and Research Foundation
Peter Jovan
514 Sathy Road
Coimbatore, Tamil Nadu
India 641 006
Tel: +91-42-252-1001
Fax: +91-42-252-0561
humansafe2001@yahoo.com
www.ozonehospital.org
All Ozone Therapies;
Direct IV, RHP, Steam,
Ozonated Saline.

Sai Nursing Home
Dr. Abhay Arora
A-3/18 Janak Puri
Delhi, New Delhi
India 110058
Tel: 00-91-11-559-1664
Oxygen Inhalation Therapy

INDONESIA

Prima Ozone Clinic
Lukas Hartono M.D.
Jl. Lawu No. 35
Pare, Kediri
East Java 62126
Indonesia
Tel: +62-354-39-1812
Fax: +62-354-395-611
Major Autohemotherapy
Minor Autohemotherapy
Ozone Intestinal Insufflation
Oral H_2O_2

ITALY

Dott. Franco Pegolo
Viale Lacchin, 61
33077 Sacile (PN)
Italy
Tel: 0434-70736
Oxygen/ Ozone Therapies

Gabriele Dr. Maximilian
Via Novara 5 20147
Milano, Italy
Tel: 039-2-4871-3025
Oxygen/Ozone Therapy

Medical Centre
Lamberto Re. M.D.
Via Duca degli Abruzzi, 69
Porto Potenza Picena
Macerata, Italy 62016
Tel: +39-338-836-1813
Fax: +39-71-220-4635
Major Autohemotherapy
Oxygen-Ozone Therapy
Minor Autohemotherapy
Oxygen-Ozone Therapy
Oxygen-Ozone Insufflations
Intraarticular IV, Local and
Paravertebral IV Ozone

P.A.S.S.
Dr. Amato De Monte
Via Duino 1 Udine
Italy 33100
Tel: 0432-229-553
Fax: 0432-513-516
Major Autohemotherapy
Intraarticular Injection
Paravertebral Injection

Studio Pashaj
via Smareglia 26
Milano, MI 20131, Italy
Tel: +39-02-730-787
Paravertebral Injection in
Hernia and Disk Lumbar
Protrusion
Knee-Joint Disorder
Oxygen Treatment
Carpal Tunnel Syndrome
Periarthritis of the Shoulder

MALAYSIA

Blossom Portfolio Sdn.Bhd.
Dr. Ali Abdullah
Centre For Ozone Holistic
#36C-Plaza Pekeliling,
Jalan Tun Razak, 50400
Kuala Lumpur Malaysia
Tel: 603-041-10371
Fax: 603-404-10372
Ozone Therapies

Kampung Baru Medical Centre (KBMC)
Dr. Ishal Mas'ud, MBBSA
(W.Aus) MRCP (Ire
No. 85, Jalan Raja Abdullah
Kuala Lumpur
50300, Malaysia
Tel: 603-269-31007
Fax: 603-269-86711
IV Micro Bubbles
Polyatomic Apheresis

Klinik Medizone
Dr. FH Lew
57, Jalan Kapar 41400 Klang
Selangor Malaysia
Tel: 603-334-87442
Fax: 603-334-87396
Bio-Oxidative Therapy
Blood Ozonation
Vaginal, Rectal & Ear
Insufflations
Hyperthermic O_3 (Sauna)
Ozonated Aromatherapy
Ozonated Water
Body Bag Insufflations
Ozone Cupping, Limb
Bagging, Oral H_2O_2

MEXICO

Dr. Alberto Martinez-Palafox
Pedro S.Varela 3007-10
Ciudad Juarez, 32310
Chihuahua, Mexico.
Tel: 011-52-656-617-1003
Tel: 011-52-656-608-5969

(U.S. Address)
Dr.Alberto Martinez-Palafox
5823 N.Mesa 272,
El Paso. Texas 79912
Recirculatory Hemoperfusion
Major Autohemotherapy
Minor Autohemotherapy
Topical Ozone Therapy
Intramuscular Ozone IV
Ozone Insufflations

American Metabolic Institute.
1-800-388-1083
www.amihealth.com
Ozone Sauna, Direct IV
Autohemotherapy
Rectal insufflation.
Ozonated Drinking Water
Colonic Water, Salve
Swimming Pool

Europa Institute of Integrated Medicine
Carolyn Bormann N.D., C.P
P. O. Box 950
Twin Peaks, CA 92391
Tel: 909-338-3533
Fax: 909-338-7343
Clinic: Allen W. Lloyd Bldg.
406 Ave. Paseo de Tijuana
Suite 201, Int'l. Border Zone
Tijuana, B.C., Mexico
Major Autohemotherapy
Minor Autohemotherapy
Topical Ozone Therapy
Subcutaneous Bleb injections
Intramuscular Ozone IV
Ozone Insufflations

Health Restoration Center
Dr. David A. Steenblock
26381 Crown Valley Pkwy
Suite 130
Mission Viejo, CA 92691
Tel: 714-367-8870
Fax: 714-367-9779
Hyperbaric Oxygen

Hospital Santa Monica
880 Canarios Court Suite 210
Chula Vista, CA 91910
Tel: 619-482-8533
Tel: 800-359-6547

IV Hydrogen Peroxide
Hydro Therapy with
Peroxide & Ozone Therapy

Providence Pacific Hospital
Playas de Tijuana, B.C.
Mexico
Tel: 011-52-66-46-300-700
Fax: 011-52-66-46-300-319
Polyatomic Oxygen
UVBI - UV Blood Irradiation

Regenesis
Dr. Abel Jimenez
Basilio Badillo No. 229-B
Puerto Vallarta, Jal., Mexico
Tel: 322-315-05
Tel: 322-237-72
Ozone & Oxygen Therapies

Sana Fe Clinic
Dr. Oscar Leyva M.D.
Saratoga Avenue &
Internacional
Los Algodones, Baja
Mexico
Tel: 011-52-651-775-64
Ozone Therapy
Autohemotherapy
Insufflations Bagging
H_2O_2

William Hitt Center
Dr. William Hitt
Ave. Paseo Tijuana, 405
Suite 403
Tijuana, Mexico 22310
Tel: 888-671-9849
Fax: 888-671-9849
P.O. Box 434357,
San Diego, CA 92143-4357
Ozone Therapy

**Gerson Institute
Oasis of Hope**
P. O. Box 439045 San Ysidro
California 92143-9045
Paseo de Tijuana #19 Playas
de Tijuana Baja, CA Mexico
22700
1572 Second Avenue
San Diego, CA 92101
Tel: 619-685-353
Rectal Insufflation
Adjunctive Ozone

RUSSIA

Clinic of Ozonotherapy
Alex Novgorodsev July
Lobod, Russkaya 71
Vladivostok, Primorskii Krai
Russia 690014
Tel: 4232-32-64-10
Fax: 4232-26-66-89
Autohemotherapy
Rectal Insufflations
Ultra Violet Blood Therapy
Intramuscular and Intra
Articular Injections
Ozonated Oil

**Ozonetherapy Center
Oxygen 3**
Oleg L. Eppelman Juliya
Kraevskaya Juliya
Korosteleva
Voennoe shosse 20 – A
Vladivostok, Primorsky krai
Russia 690088
Tel: 4232-258-872
Fax: 4232-300-383
Major Autohemotherapy
Minor Autohemotherapy
Rectal Insufflations
Intramuscular Injections
Subcutaneus Injections
Ozonated Oil & Water

**State Medical Academy of
Nizhny Novogorod**
Oleg V. Maslennikov
10/1 Minina s Ozone Sauna
Nizhny Novgorod
Nizhegorodskaya oblast
603005, Russia
Tel: 8312-375-333
Fax: 8312-375-205
All Ozone Therapies

Medozons Ltd.
Postal Address:
ul. B.Panina 9, 603089
Nizhny Novgorod, Russia
+7-8312-167-067, 383-003
Fax: +7-8312-167 000
Home page:
http://www.medozons.ru
Medical Ozone Manufacturer
and Accessories

SOUTH AFRICA

Franz Hermanns H.P.
Samson Klip Farm,
Ingogo, Natal,
South Africa, 0560
Tel: 034-341-1884
Fax: 034-341-1884
All Oxygen Therapies

Naboom Medical Center
Dr. E. Pretorius
20 4th Street
Naboomspruit
South Africa 0560
Tel: +27-14-743-2114
Fax: +27-14-743-1957
Intravenous Oxytherapy

Dr. John Taylor (Hmpth)
Botha str 19, Vrede
South Africa 9835
Tel: 0174-31659
Fax: 0174-31659
Ozone Therapy

SOUTH AMERICA

**Centreo de
Rejuvenecimiento Celular
por las Oxygenoterapia
Hiperbarica**
Artuor Martinez Restrepo
Alle 26 Nro. 7A-36 Barrio
Fatima Palmira, Valle del
Cauca, Colombia
Tel: 57-02-272-45-44
Fax: 57-02-273-32-28
Hyperbaric Oxygen Therapy

**Instituto de Medicina
Integral**
Dr. Luis E. Paolini Pisani
M.D. D.Sc.
Carrera 16 Esquina Calle 13
Edificio San Cayetano P.B
San Critóbal, Táchira 5001
Venezuela
Tel: 58-76-56-70-85
Fax: 58-76-56-74-54
Hyperbaric Oxygen Therapy

Heinz Konrad M.D
Rua Dona Eponina 80
Sao Paulo, SP, 04720-010
Brazil
Tel: 011-247-4918
All Ozone Therapies

SPAIN

**Centro De Medicina
Tradicional China "El
Paraiso**
Dr. Jose Infantes Perez. M.D.
Carretera C·diz, 167. Urb. El
Paraìso, Estepona, Malaga
Spain, 29680
Tel: +34-952-885-095
Fax: +34-952-886-279
Intravenous Ozone Therapy
Intramuscular Ozone Therapy
Intrarectal Ozone Therapy
Intradermal Ozone Therapy

**Instituto De Medicina
Biofisica S.L.**
Dr. Juan Luis Calatayud
Carretero
Avda de la Estacion n 8 – 1
Alicante, Spain 03003
Tel: +34-965-125-500
Fax: 57-022-733-228
Various Ozone Therapies

SWITZERLAND

**Paracelsus Klinik Center
for Holistic Medicine and
Dentistry**
CH-9062, Lustmühle bei St.
Gallen
Lustmühle, Switzerland
Tel: 011-41-71-335-71-71
Fax: 011-41-71-335-71-00
Ozone Therapy
Hydrogen Peroxide
Hyperbaric Oxygen Therapy

THAILAND

**Thai Biologics Medical
Care Ltd**
Dr. Oldrich Capek M.D.
73 Nakornsawan Road
Pomprab
Bangkok 10100 Thailand
Tel: +66-(2)-280-0465
Fax: +66-(2)-356-5586
Email:drcapek@loxinfo.co.th
Major Blood Ozone Therapy
Rectal Ozone Insufflations
Ozone Injections
Oxygen/Ozone Bagging
Hyperbaric Oxygen Chamber

UNITED KINGDOM

**The Chiltern Clinic of
Natural Therapeutics**
Dr. Simi Khanna MD, MBBS
DHOM M.B.P.A.
232 West Wycombe Road
High Wycombe, Bucks
HP12 3AR England
Tel: 01494 472110
Fax: 01494 462264
Oxygen/Ozone Therapy

The Finchley Clinic
Mark Lester Holistic Thpst.
26 Wentworth Avenue,
Finchley, London N3 1YL
England
Tel: +44-0208-349-4730
Tel: 0208-349 4730 (UK)
www.the-finchley-
clinic.co.uk
Ozone Therapy
Ozone Steam Cabinet
Homozon

The Health Consultancy
Frank Meredith MHMA
Spiral Lodge Parkham Lane
Brixham, Devon TQ5 9JR
England
Tel: 44 01-803-859889
Fax: 44 01-803-857335
Ionised Oxygen During
Exercise

Liongate Clinic
Dr. Fritz Schellander
8 Chilston Road
Tunbridge Wells
Kent TN4 9LT
England
Tel: 44 01-892-543535
Fax: 44 01-892-545160
Ozone Therapy
Hydrogen Peroxide Therapy
Multi-Step Oxygen Therapy

IBOM

The International Bio-Oxidative Medicine Foundation (IBOMF) is a 501c, not-for-profit scientific and educational foundation, established in 1987 to further research and education in the field of oxidative medicine. IBOMF provides the IOMA referral list for public distribution, but does not provide information or recommendations about specific medical problems or questions, nor does it recommend any product or equipment. Last updated 1999.

International Bio-Oxidative Medicine Foundation
P.O. Box 891954
Oklahoma City, Ok
73189-2954
Tel: (405)-634-1310
Fax: (405)-634-7320

* Denotes Interim Diplomat Status. Interim Diplomat Status is awarded to those physicians who have completed workshop training and have passed a written exam.

ALASKA
Robert Rowen, M. D.*
615 E. 82nd Street, Ste. 300
Anchorage, AK 99518
Tel: 907-344-7775
Fax: 907-522-3114

ARKANSAS
Melissa Taliaferro, M.D.*
P.O. Box 400
Leslie, AR 72645
Tel: 501-447-2599
Fax: 501-447-2917

ARIZONA
Terry Friedman, M.D.
10565 N. Tatum Blvd.
Suite. B-115
Paradise Valley, AZ 85254
Tel: 602-381-0800

Thomas J. Grade M.D.
6644 E. Baywood
Mesa, AZ 85206
Tel: 602-981-4474
Fax: 602-981-4312

CALIFORNIA
John Beneck, M.D.
22107 Old Paint Way
Canyon Lake, CA 92587
Tel: 909-244-3686
Fax: 909-244-0109

Nolan Higa, M.D.
221 Town Center West, #101
Santa Maria, CA 93454
Tel: 805-347-0067
Fax: 805-929-3032

Bernardo C. Majalca
1568 San Cedro Pl.
Chula Vista, CA 91911
Tel: 11-52-66-30-1232

Robert M. Martin, M.D.
2015 E. Florence Ave.
Los Angeles, CA 90001
Tel: 213-277-9096
Fax: 213-277-9098

Martin Mulders, M.D.*
3301 Alta Arden, #3
Sacramento, CA 95825
Tel: 916-489-4400
Fax: 916-489-1710

Francis V. Pau, M.D.
1465 Loma Sola Court
Upland, CA 91786
Tel: 909-987-4262
Fax: 909-987-9542

David A. Steenblock, DO
26381 Crown Valley Pkwy
Suite. 130
Mission Viejo, CA 92691
Tel: 714-367-8870
Fax: 714-770-9775

Thomas R. Yarema, M.D.
1218 Monroe Ave.
San Diego, CA 92116
Tel: 619-299-8607
Fax: 619-299-8671

COLORADO
Terry Grossman, M.D.
255 Union Street, #400
Lakewood, CO 80228
Tel: 303-986-9455

Thomas R. Lawrence, D.C.
2222 E. 18th Ave.
Denver, CO 80206
Tel: 303-333-3733
Fax: 303-333-1352

FLORIDA
Naima Abdel-Ghany, M.D.
340 W. 23rd Street, Suite. E
Panama City, FL 32405
Tel: 904-763-7689

Gary L. Pynkel, M.D.*
3840 Colonial Blvd., Suite. 1
Ft. Myers, FL 33912
Tel: 941-278-3377
Fax: 941-278-3702

Martin Dayton, D.O.*
18600 Collins Avenue
North Miami Beach, FL
33160
Tel: 305-931-8484
Fax: 305-936-1849

**William Campbell Douglass
III, MD**
101 Timberlachen, Suite 101
Lake Mary, FL 32746
Tel: 407-324-0888
Fax: 407-324-8222

Nelson Kraucek, M.D.
8923 NE 134th Ave., Ste. A
Lady Lake, FL 32159
Tel: 352-750-4333
Fax: 352-750-2023

Eteri Melnikov, M.D.
116 Manatee Ave. E.
Braden River, FL 34208
Tel: 813-748-7943

Carlos A. Unzueta, M.D.
1204 Carlton Ave.
Lake Wales, FL 33853
Tel: 941-676-7569
Fax: 941-676-3896

William N. Watson, M.D.
5536 Stewart Street, NE
Milton, FL 32570
Tel: 904-623-3836
Fax: 904-623-2201

GEORGIA
Martin L. Bremer, D.O.
P.O. Box 131
Cornelia, GA 30531
Tel: 770-538-0910
Fax: 770-538-0910

Milton Fried, M.D.
4426 Tilly Mill Rd.
Atlanta, GA 30360
Tel: 770-451-4857
Fax: 770-451-8492

Oliver Lee Gunter, M.D.
P.O. Box 347
Camilla, GA
Tel: 912-336-343
Fax: 912-336-7400

HAWAII
Wendell Foo, M.D.
2357 S. Beratania Street
A-349
Honolulu, HI 96826
Tel: 808-373-4007

ILLINOIS
Robert Filice, M.D.
1280 Iriquois Dr., #200
Naperville, IL 60563
Tel: 708-369-1220

Ross A.Hauser, M.D.
715 Lake Street, Ste. 600
Oak Park, IL 60301
Tel: 708-386-0078
Fax: 708-848-7789

Thomas L. Hesselink, M.D.
888 S. Edgelawn Dr.
Suite. 1743
Aurora, IL 60506
Tel: 708-844-0011
Fax: 708-844-0500

William J. Mauer, D.O.*
3401 N. Kennicot Avenue,
Suite. 800
Arlington Heights, IL 60005
Tel: 708-255-8988
Tel: 708-255-7700

KANSAS
Jerry E. Block, M.D.
F.A.C.P.
1501 W. 4th St.
Coffeyville, KS 67337
Tel: 316-251-2400
Fax: 316-251-1619

KENTUCKY
Ralph G. Ellis, M.D.
112 Stone House Trail
Bardstown, KY 40004
Tel: 502-349-6313
Fax: 502-349-6313

MASSACHUSETTS
Denise D. Cantin, D.O.
415 Boston Tpke., Rte. 9
Shrewsbury, MA 01545
Tel: 508-842-8118
Fax: 508-842-2148

Michael Janson, M.D.
275 Millway
Barnstable, MA 02630
Tel: 508-362-4343
Fax: 508-362-1525

MAINE
Arthur B. Weisser, D.O.
184 Silver Street
Waterville, ME 04901
Tel: 207-873-7721
Fax: 207-873-7724

MICHIGAN
Vahagn Agbabian, D.O.*
28 Saginaw, Suite. 1105
Pontiac, MI 48058
Tel: 810-334-2424
Fax: 810-258-0488

MISSOURI
Ralph Cooper, D.O.
1608 E. 20th Street
Joplin, MO 64804
Tel: 417-624-4323

Harvey Walker Jr., M.D.
138 N. Meramac Ave.
St. Louis, MO 63105
Tel: 314-721-7227
Fax:314-721-7247

Simon M. Yu, M.D.
138 N. Meramac Ave.
St. Louis, MO 63105
Tel: 314-721-7227
Fax: 314-721-7247

NEBRASKA
Otis Miller, M.D.
1001 S. 14th Street
Ord, NE 68862
Tel: 308-728-3251

Jeffry Passer, M.D.
9300 Underwood Ave. #520
Omaha, NE 68114
Tel: 402-398-1200
Fax: 402-398-9119

NEVADA

Thomas Lodi, M.D.
3663 Calico Cove Ct.
Las Vegas, NV 89117
Tel: 702-228-4139

Robert D. Milne, M.D.
2110 Pinto Lane
Las Vegas, NV 89106
Tel: 702-385-1393
Fax: 702-385-4170

NORTH CAROLINA

John C. Pittman, M.D.*
4505 Fair Meadow Lane,
#111
Raleigh, NC 27622
Tel: 919-571-4391
Fax: 919-571-8968

NEW JERSEY

Stuart Weg, M.D.*
1250 E. Ridgewood Avenue
Ridgewood, NJ 07450
Tel: 201-447-5558
Fax 201-447-9011

NEW YORK

Richard Ash, M.D.
800 5th Ave.
New York, NY 10021
Tel: 212-628-3113
Fax: 212-249-3805

Kenneth A. Bock, M.D.*
108 Montgomery Street
Rhinebeck, NY 12472
Tel: 914-876-7082
Fax: 914-876-4615

Mitchell Kurk, M.D.*
310 Broadway
Lawerence, NY 11555
Tel: 516-239-5540
Fax: 516-371-2919

Joyce H. Marshall, N.D.
23 Madison Street
Hamilton, NY 13346
Tel: 315-824-3007

Bruce D. Oran, D.O.
Two Executive Blvd.
Ste. 202
Suffern, NY 10901
Tel: 914-368-4700
Fax: 914-368-4727

Robert W. Snider, M.D.
284 Andrews Street
Massena, NY 13662
Tel: 315-764-7328
Fax: 315-769-6713

Michael J. Teplitsky, M.D.
415 Oceanview Ave.
Brooklyn, NY 11235
Tel: 718-769-0997
Fax: 718-646-2352

Richard J. Ucci, M.D.*
521 Main Street
Oneonta, NY 13820
Tel: 607-431-9641

Pavel I. Yutsis, M.D.
1309 W. 7th Street
Brooklyn, NY 11204
Tel: 718-259-2122
Fax: 718-259-3933

OHIO

John Baron, D.O.
4807 Rockside Rd., Ste. 100
Cleveland, OH 44131
Tel: 216-642-0082
Fax: 216-642-1415

Bruce Massau, D.O.
E.M.B.A.
1470-B Hawthorne Ave.
Columbus, OH 43203
Tel: 614-252-1500
Fax: 614-252-1685

James C. Roberts, M.D.
4607 Sylvania Ave., Ste. 200
Toledo, OH 43623
Tel: 419-882-9620
Fax: 419-882-9628

Sherri Tenpenny, D.O.
13550 Falling Waters Rd.
Strongsville, OH 44136
Tel: 216-572-1136
Fax: 216-572-2195

OKLAHOMA

Leon Anderson, D.O.*
121 South Second
Jenks, OK 74037
Tel: 918-299-5038
Fax: 918-299-5030

**Charles H. Farr, M.D.
Ph.D.* (clinic)**
5419 So. Western
Oklahoma City, OK 73109
Tel: 405-634-7855
Fax: 405-634-7320

Charles Hathaway, D.C.
1607 S. Muskogee
Tahlaquah, OK 74464
Tel: 918-456-8090
Fax: 918-456-6060

James W. Hogin, D.O. 937
S.W. 89th, Suite. C
Oklahoma City, OK 73139
Tel: 405-631-0524
Fax: 405-631-9465

Gordon P. Laird, D.O.
304 Boulder
Pawnee, OK 74058
Tel: 918-762-3601
Fax: 918-762-2544

Maged H. Maged, M.D.*
5419 So. Western
OKC, OK 73109
Tel: 405-634-7855
Fax: 405-634-7320

Richard Santelli, D.C.*
8216 N.W. 104th
OKC, OK 73162
Tel: 405-789-5114

Charles Taylor, M.D.
3715 N. Classen
Oklahoma City, OK 73118
Tel: 405-525-7751
Fax: 405-747-2717

Michael Taylor, D.C.
3808 E. 51st Street
Tulsa, OK 74119
Tel: 918-749-4657
Fax: 918-749-6263

Robert L. White, Ph.D.N.D.
5419 So Western
Oklahoma City, OK 73209
Tel: 405 634 7855
Fax: 405 634 7320

OREGON

Robert Jamison, M.D.
628 Pacific Terrace
Klamath Falls, OR 97601

J. Stephen Schaub, M.D.
9310 SE Stark Street
Portland, OR 97216
Tel: 503-256-9666

PENNSYLVANIA

Harold Buttram, M.D.*
5724 Clymer Rd.
Quakertown, PA 18951
Tel: 215-536-1890
Fax: 215-529-9034

Arthur L. Koch, D.O.*
57 W. Juniper Street
Hazelton, PA 18201
Tel: 717-455-4747
Fax: 717-455-6312

John M. Sullivan, M.D.
1001 S. Market Street, Ste. B
Mechanicsburg, PA 17055
Tel: 717-697-5050

TEXAS

Antonio Acevedo, M.D.
P. O. Box 707
Bedford, TX 76021
Tel: 817-595-2580
Fax: 817-589-2913

Robert M. Battle, M.D.
9910 Long Point Rd.
Houston, TX 77055
Tel: 713-932-0552
Fax: 713-932-0551

Ronald W. Bowen, D.O.
7121 S. Padre Island Dr.
Suite. 104
Corpus Christi, TX 78412
Tel: 512-985-1115
Fax: 512-985-1467

Patricia Braun, M.D.
1212 Coit Rd., Ste. 110
Plano, Republic of Texas
TP2 (75075)
Tel: 214-612-0399
Fax: 214-985-1207

Elisabeth Ann Cole, M.D.*
1002 Brockman
Sweeney, TX 77480
Tel: 409-548-8610
Fax: 409-549-8614

Ronald M. Davis, M.D.
5002 Toddville
Seabrook, TX 77586
Tel: 713-474-3495

John Galewaler, D.O.
P.O. Box 488
Celina, TX 75009
Tel: 214-382-2345

Charles M. Hawes, D.O.*
6451 Brentwood Stair Rd.
Suite. 115
Ft. Worth, TX 76112
Tel: 817-446-8416
Fax: 817-446-8413

T. Roger Humphrey, M.D.
2400 Rushing
Wichita Falls, TX 76308
Tel: 817-766-4329
Fax: 817-767-3227

George Lofgren, ND
1220 Town East Blvd.,
#250B
Mesquite, TX 75150
Tel: 214-636-2696
Fax: 214-635-9238

James J. Mahoney, D.O.
6451 Brentwood Stair Rd.
Suite. 115
Ft. Worth, TX 76112
Tel: 817-446-8416
Fax: 817-446-8413

Ron Manzanero, MD
3845 FM 2222, #23
Austin, TX 78731
Tel: 512-258-1647
Fax: 512-453-3450

**Frank J. Morales, Jr.
M.D.***
2805 Hackberry Rd.
Brownsville, TX 78521
Tel: 210-504-2330
Fax: 210-548-1227

Carlos Nossa, M.D.
4010 Fairmont Pkwy., #274
Pasadena, TX 77504
Tel: 713-334-1456

Benjamin Thurman, M.D.
102 N Magdalen, #290
San Angelo, TX 76903
Tel: 915-653-3562
Fax: 915-944-1162

Barbara Weeden, C.C.N.
860 Secretary Dr.
Arlington, TX 76015
Tel: 817-265-5261
Fax: 817-274-9971

Stan Wolfe DVM
8880 Main St.
P.O. Box 400
Frisco, Texas, 75034
Tel: 972-202-9585

UTAH

Dennis Harper, D.O.*
5263 S. 300w., #203
Murray, UT 84107
Tel: 801-288-8881

WASHINGTON

**Patrick H. Ranch, M.D.,
D.C.**
9629 N. Indian Trail Rd.
Spokane, WA 99208
Tel: 208-777-8297
Fax: 208-466-6043

VETERINARY

Norman Jason Ward D.V.M.
7030 E. 5th Ave., Suite 3
Scottsdale, AZ 85251
Tel: 602-946-0663
Fax: 602-946-0841

International

CANADA

John Cline, M.D.
5996 Island Hwy. W.
Qualicum Bay, BC V9K 2E1
Tel: 604-757-2388

MALAYSIA

M.S. Balajeygaran MBBS
No. 7 Jalan PJS (Bombay)
2C\28 Kg Medan
Petealing Jaya Selangor
46000
Tel: 011-603-736-9934

Soon Tong Kho, M.D.
183 Batie 17 Jln. Ipoh
48000 Rawang, Selangor
Tel: 001-603-691-8705
Fax: 001-603-617-5617

EAST AFRICA

NAIROBI, KENYA

Geeta Shah, M.D.
P.O. Box 33149
P 11-254-2-742622

PHILLIPINES

Rosario Austria, M.D.
18 Mariposa Street, Cubao
Quezon
Arturo Estuita, M.D.
1986 Taft Ave. Unit 105

Metro Manila
Maria Tablan, N.D.
829 Tangier Street
Las Pinas
Tel: 11-632-827-1011/328
Fax: 11-632-827-1011/227

Garrett G. Swetlikoff, N.D.
160-1855 Kirschner Rd.
Kelowna, BC V1Y 4N7
Tel: 604-868-2205
Fax: 604-868-2099

Robert Ewing, ND
8045 Clegg Street
Mission, BC V2V 3R4
Tel: 604-820-8161
Fax: 604-826-6182

Donald R. Horton, M.D.
2633 Beach Dr.
Victoria, BC V8R-6K3
Tel: 604-384-8421
Fax: 604-592-4961

Alex A. Neil, M.D.
216-3121 Hill Rd.
BC, V4V 1G1

Jim Chan, N.D., DiPi.Ac.*
101 - 3380 Maquinna Dr.
Vancouver, BC V5S 4C6
Tel: 604-435-3786
Fax: 604-436-2426

UNITED KINGDOM

Hugh J.E. Cox, M.D.*
14 Ayleswater, Watermead
Aylesbury, Bucks 4P19 3FB
Tel: 01296 399317
Fax: 01296 399291

Eighteen pages of three columns each filled with clinics and doctors using these Oxygen Therapies successfully day after day worldwide for years and years. And these are just a *few* of the doctors.

"There is no proof" sounds pretty hollow and manipulative by now, doesn't it? At the end of the day, why can't these vaingloriously pontificating experts—with all their unlimited money and staffs and resources—find everything that's in this book? After all, I'm only one person and you can see how much I was able to find. Where's their compassion for all the suffering and heartbreak?

Start by searching over 500 references freely available on my website: www.oxygenhealth.com

B

C

N

O

Late Addition! **Well, well, well...**

Ozone is Produced by Antibodies During Bacterial Killing
The Scripps Research Institute La Jolla, California November 14, 2002:

The Scripps Research Institute (TSRI) is reporting that antibodies can destroy bacteria, playing a hitherto unknown role in immune protection. Furthermore, when antibodies do this, they appear to produce the reactive gas ozone.

The ozone may be part of a previously unrecognized killing mechanism that would enhance the defensive role of antibodies by allowing them to participate directly in the killing. Previously, antibodies were believed only to signal an immune response. Also called immunoglobulins, antibodies are secreted proteins produced by immune cells that are designed to recognize a wide range of foreign pathogens. After a bacterium, virus, or other pathogen enters the bloodstream, antibodies target antigens [proteins, fat molecules, and other pieces of the pathogen] that are specific to that foreign invader. These antibodies then alert the immune system to the presence of the invaders and attract lethal 'effector' immune cells to the site of infection.

For the last hundred years, immunologists have firmly held that the role of antibodies was solely to recognize pathogens and signal the immune system to make an immune response. The conventional wisdom was that the dirty work of killing the pathogens was to be left to other parts of the immune system. Now, Scripps has demonstrated that antibodies also have the ability to kill bacteria. This suggests that rather than simply recognizing foreign antigens and then activating other parts of the immune system to the site of infection, the antibodies may further enhance the immune response by directly killing some of the bacteria themselves. Antibodies appear to make ozone, which they detected through its chemical signature, which no other known molecule has. Never before has ozone been detected in biology.

It has been known that all antibodies have the ability to produce hydrogen peroxide, but they need to first have available a molecule known as 'singlet' oxygen—another highly reactive oxygen species—to use as a substrate. Singlet oxygen is an energetically charged form of oxygen that forms spontaneously during normal metabolic processes. Phagocytes like neutrophils produce singlet oxygen and are the most likely source of the substrate for antibody production of hydrogen peroxide. Antibodies attract neutophils to the site of an infection. Once there, the neutrophils will engulf and destroy bacteria and other pathogens by blasting them with singlet oxygen and other oxidative molecules. The antibodies combine singlet oxygen with water to produce hydrogen peroxide, producing ozone as a side product.

Another interesting finding is that the antibodies carry the reaction through an unusual intermediate chemical species of dihydrogen trioxide, a reduced form of ozone. Dihydrogen trioxide has also never before been observed in biological systems.

The research article, "Evidence for Antibody-Catalyzed Ozone Formation in Bacterial Killing" is authored by Paul Wentworth, Jr., Richard A. Lerner, et. al. and appears in the November 18, 2002 "Science Express," the advanced publication edition of the journal Science. The article will appear in *Science*.

CUBAN OZONE THERAPY SPECIALISTS - PUBLISHED PAPERS

Ozone Research Center, Cuba
Centro de Investigaciones del Ozono

Cuban studies supplied with permission by Dr. Thomas Moreira

OZONE THERAPY

Preclinical Studies:

- Ajamieh H., Merino N., Candelario-Jali E., Menéndez S., Martínez G., Re L., Giutiani A. and León O.S. "Similar protective effect of ischemic and ozone oxidative preconditionings in liver ischaemia/reperfusion injury", **Pharmacological Research**, 2002 (in press).
- Borrego A., Zamora Z y Hernández F. "Efecto del aceite de girasol ozonizado de uso oral sobre la respuesta de anticuerpos específica a la vacuna cubana anti Hepatitis B", **Revista Mexicana de Ciencias Farmacéuticas**, 2002 (in press).
- Zamora Z., Reyes N., González Y. y Moleiro J. "Efecto del aceite de teobroma ozonizado y la dexametasona sobre la reacción de anafilaxia cutánea pasiva", **Revista Mexicana de Ciencias Farmacéuticas**, 2002 (in press).
- Zamora Z., Schulz S. y Menéndez S. "Evaluación del efecto del ozono aplicado intraperitonealmente en un modelo de peritonitis en ratas", **Revista Mexicana de Ciencias Farmacéuticas**, 2001 (in press).
- Al-Dalain S.M., Martínez G., Candelario-Jalil E., Menéndez S., Re L., Giuliani A. and León O.S. "Ozone treatment reduces markers of oxidative and endothelial damage in an experimental diabetes model in rats", **Pharmaceutical Research**, 44(5):391-396 (2001).
- Candelario-Jalil E., Mohammed-Al-Dalain S., León O.S., Menéndez S., Pérez G., Merino N., Sam S. and Ajamieh H.H. "Oxidative preconditioning affords protection against carbon tetrachloride-induced glycogen depletion and oxidative stress in rats", **J. Appl. Toxicol.**, 21:297-301 (2001).
- Remigio A., Zullyt Z., González Y. y Moleiro J. "Estudio del efecto mutagénico del Ozono administrado por insuflación rectal en roedores", **Revista CENIC Ciencias Biológicas**, 32(1):59-62 (2001).
- Peralta C. Xaus C., Bartrons R., León O.S., Gelpi E. and Roselló-Catafau J. "Effect of ozone treatment on reactive oxygen species and adenosine production during hepatic ischemia-reperfusion", **Free Rad. Res.**, 33:595-605 (2000).
- Barber E., Menéndez S., León O.S. et al. "Prevention of renal injury after induction of ozone tolerance in rats submitted to warm ischaemia", **Mediators of Inflammation**, 8:37-41 (1999).
- Peralta C., León O.S., Xaus C. et al. "Protective effect of ozone treatment on the injury associated with hepatic ischemia-reperfusion: antioxidant-prooxidant balance", **Free Rad. Res.**, 31:191-196 (1999).
- Rojas R.L., Martínez C.L., Turrent J. y Menéndez S. "Administración de ozono y reacción anafilática in vivo del cobayo", **Revista CENIC Ciencias Biológicas**, 30(2):119-120, 1999.
- Barber E., Menéndez S., Barber M. O., Merino N., Calunga J.L. "Estudio renal funcional y morfológico en riñones de ratas pretratadas con ozono y sometidas a isquemia caliente", **Revista CENIC Ciencias Biológicas**, 29(3):178-181 (1998).
- González A., Basabe E., Merino N., Menéndez S., Gómez M., Capote A. "Evaluación histopatológica del oído interno de curieles sometidos a toxicidad por estreptomicina y tratados con ozono", **Revista CENIC Ciencias Biológicas**, 29(3):189-191 (1998).

- León O.S., Menéndez S., Merino N. et al. "Ozone oxidative preconditioning: a protection against cellular damage by free radicals", **Mediators of Inflammation**, 7:289-294 (1998).
- León O.S., Menéndez S., Merino N., López R., Castillo R., Sam S., Pérez L., Cruz E., Jouseph F. y Fernández A. "Influencia del precondicionamiento oxidativo con ozono sobre los niveles de calcio", **Revista CENIC Ciencias Biológicas**, 29(3):134-136 (1998).
- Lezcano I., Garcia G., Martínez G., Molerio J., Zamora Z., Fernández C., González A. y Castañeda D. "Efectividad de la manteca de cacao ozonizada para el tratamiento de la candidiasis vaginal", **Revista CENIC Ciencias Biológicas**, 29(3):206-208 (1998).
- Lezcano I., J. Molerio, M. Gómez M., Contreras R., Roura G. y Díaz W. "Actividad in vitro del OLEOZON frente a agentes etiológicos de infecciones de la piel", **Revista CENIC Ciencias Biológicas**, 29(3):209-212 (1998).
- Remigio A., González Y., Zamora Z., Fonseca G. y Molerio J. "Evaluación genotóxica del OLEOZON mediante los ensayos de micronúcleos en médula ósea y sangre periférica de ratón", **Revista CENIC Ciencias Biológicas**, 29(3):200-202 (1998).
- Rodríguez Y., Menéndez S., Bello J.L., Matos E., Espinosa A., Turrent J., Pimienta L., Ramos S. y Otero C. "Actividad antitumoral del ozono", **Revista CENIC Ciencias Biológicas**, 29(3):196-199 (1998).
- Turrent J., Alfonso C., Ancheta O., Menéndez S., Casacó A., Carballo A., Rodríguez S. "Estudio estructural y ultraestructural de la histología del recto en ratones previamente tratados con ozono mediante insuflación rectal", **Revista CENIC Ciencias Biológicas**, 29(3):174-177 (1998).
- Turrent J., Legrá G., Menéndez S. y Luis M.C. "Ozonoterapia en la enfermedad de Sutton (periadenitis mucosa necrótica recurrente)", **Revista CENIC Ciencias Biológicas**, 29(3):157-160 (1998).
- Zamora Z.,. Turrent J., Menéndez S. y Carballo A. "Anafilaxia cutánea pasiva en ratas usando suero de ratones previamente tratados con ozono mediante insuflación rectal", **Revista CENIC Ciencias Biológicas**, 29(3):125-127 (1998).
- Acevedo F., González J., Moleiro J. et al. "Ensayo de toxicidad dérmica de 120 días del aceite ozonizado, OLEOZON, en ratas Cenp. SPRD", **Avances en Biotecnología Moderna**, 4, T-4 (1997).
- Martínez G., León O.S., Rodríguez C., Merino N. et al. "Estudio de la toxicidad aguda dérmica del aceite ozonizado OLEOZON® en ratas", **Revista CENIC Ciencias Biológicas**, 28(1):35-47 (1997).
- Hernández F. y Menéndez S. "Efecto de la ozonoterapia intramuscular sobre el metabolismo de conejos normocolesterolémica", **Revista de Investigaciones Biomédicas**, 9:40-47 (1990).
- Arruzazabala M.L., Noa M., Menéndez S. y Gómez M. "Efecto del ozono sobre los metabolitos del ácido araquidónico en pulmón aislado de curiel", **Revista CENIC Ciencias Biológicas**, 20(1-2-3) 8-11 (1989).
- Fernández S.I., Quinzan C., Menéndez S., Gómez M. y Acosta P. "Estudio del efecto del ozono intratesticular sobre la espermatogénesis", **Revista CENIC Ciencias Biológicas**, 20(1-2-3-):16-20, 1989.
- Friman M., Walker D., Eng L., Menéndez S. y Gómez M. "Efecto del ozono en las células endoteliales circulantes en ratas", **Revista CENIC Ciencias Biológicas**, 20(1-2-3):25-28 (1989).
- Hernández F., Noa M., Menéndez S. y Gómez M. "Estudio del metabolismo de los lípidos de animales de experimentación sometidos a ozonoterapia", Revista CENIC Ciencias Biológicas, 20(1-2-3):48-53 (1989).
- Noa M., Hernández F., Herrera M., Menéndez S., Capote A. y Aguilar C. "Observaciones morfológicas en ratas tratadas con ozono por vía intramuscular", **Revista CENIC Ciencias Biológicas**, 20(1-2-3):20-23 (1989).

- Rodríguez M.D, Menéndez S., Gómez M., Ancheta O. et al. "Efecto del suero ozonizado sobre embriones en cultivo", **Revista CENIC Ciencias Biológicas**, 20(1-2-3):32-37 (1989).
- Rodríguez M.D., Menéndez S., Gómez M. et al. "Estudio teratogénico del agua ozonizada", **Revista CENIC Ciencias Biológicas**, 20(1-2-3):11-14 (1989).
- Rodríguez M.D., Menéndez S., Gómez M. y García H. "Estudio teratogénico del ozono administrado por insuflación rectal a ratas wistar", **Revista CENIC Ciencias Biológicas**, 20(1-2-3):28-32 (1989).

Clinical Studies:
- Alvarez I. y Hernández F. "Valores de referencia de la enzima Glutation S Transferasa eritrocitaria en una muestra poblacional", **Revista CENIC Ciencias Biológicas**, 30(1):3-6 (1999).
- Corcho I., Hernández F., Yánez L. y Reyes T. "Estudio in vitro del efecto del ozono sobre la expresión de linfocitos T y la función fagocítica en sujetos sanos", **Revista CENIC Ciencias Biológicas**, 30(1):24-26 (1999).
- Alvarez I., Hernández F. y Rosales M. "La GST eritrocitaria y su relación con la ozonoterapia endovenosa", **Revista CENIC Ciencias Biológicas**, 29(3):128-133 (1998).
- Basabe E., Bell L., Menéndez S., Bell R. y Núñez J.A. "Perfil hormonal de niños con discapacidad auditiva tratados con ozonoterapia", **Revista CENIC Ciencias Biológicas**, 29(3):153-156 (1998).
- Basabe E., Borroto V., Bell L., Menéndez S., López C., Alarcón M.A. "Respuesta mediante efecto Doppler del tronco vértebro-basilar de pacientes con síndrome cócleo-vestibular incompletos tratados con ozonoterapia y acupuntura", **Revista CENIC Ciencias Biológicas**, 29(3):185-188 (1998).
- Basabe E., Miranda M., Rodríguez G., Menéndez S., Sheshukova N., Soto A., Carballo M.C., García X. y Alarcón M.A. "Resultados psicopedagógicos en niños con discapacidad auditiva, después de 3 y 5 años del tratamiento de ozonoterapia", **Revista CENIC Ciencias Biológicas**, 29(3):169-173 (1998).
- Borrego L., Borrego LL., Díaz E., Menéndez S., Borrego L.R. y. Borrego R. A. "Ozono más cobaltoterapia en pacientes con adenocarcinoma prostático", **Revista CENIC Ciencias Biológicas**, 29(3):137-140 (1998).
- Corcho I., Hernández F., Reyes N. et al. "Cambios del sistema inmune en procesos inflamatorios durante la aplicación de la ozonoterapia", **Revista CENIC Ciencias Biológicas**, 29(3):203-205 (1998).
- Fernández J.I., Turrent J., Colmenero M.J. y Menéndez S. "La ozonoterapia en pacientes con neuroangiopatía diabética", **Revista CENIC Ciencias Biológicas**, 29(3):165-168 (1998).
- Menéndez S., Fernández J., Turrent J. y Colmenero M. "La ozonoterapia en pacientes con neuroangiopatía diabética", **Revista CENIC Ciencias Biológicas**, 29(3):165-168 (1998).
- Rodríguez M.M., García J.R., Menéndez S., Devesa E. y Valverde S. "Ozonoterapia en la enfermedad cerebro-vascular isquémica", **Revista CENIC Ciencias Biológicas**, 29(3):145-148 (1998).
- Rodríguez M.M., Menéndez S., García J.R., Devesa E. y Cámbara A. "Ozonoterapia en el tratamiento de los síndromes Parkisonianos del anciano", **Revista CENIC Ciencias Biológicas**, 29(3):149-152 (1998).
- Rodríguez M.M., Menéndez S., García J.R., Devesa E. y., González R. "Ozonoterapia en el tratamiento de la demencia senil", **Revista CENIC Ciencias Biológicas**, 29(3):141-144 (1998).
- Turrent J. y Menéndez S. "Ozonoterapia en el asma bronquial: bases terapéuticas para su aplicación", **Revista CENIC Ciencias Biológicas**, 29(3):161-164 (1998).
- Turrent J., Legrá G., Menéndez S. y Luis M.C. "Ozonoterapia en la enfermedad de Sutton (periadenitis mucosa necrótica recurrente)", **Revista CENIC Ciencias Biológicas**, 29(3):157-160 (1998).

- Hernández, F., Menéndez S. and Wong, R. "Decrease of blood cholesterol and stimulation of antioxidative response in cardiopathy patients treated with endovenous Ozone Therapy", **Free. Radic. Biol. Med.**, 19(1):115-119 (1995).
- Hernández F. "La ozonoterapia y la peroxidación de los lípidos. Relaciones y efectos en la aterosclerosis", **Revista CENIC Ciencias Biológicas**, 24(1-2-3):25-29 (1993).
- Romero A., Blanco R., Menéndez S., Gómez M., y Ley J. "Ateroesclerosis obliterante y ozonoterapia. Administración por diferentes vías", **Angiología**, 45(5):177-179 (1993).
- Romero A., Menéndez S., Gómez M., y Ley J. "La ozonoterapia en los estadios avanzados de la arterosclerosis obliterante", **Angiología**, 45(4):146-148 (1993).
- Ugarte C., Wong R., Gómez M. y Menéndez S. "Utilidad del ozono en la angioplastia transluminal percutánea", **Revista Imagen**, 4 (1992).
- Menéndez S., Peláez O., Gómez M., Copello M. et al. "Aplicación de la ozonoterapia en la retinosis pigmentaria", **Revista Cubana de Oftalmología**, 3(1):35-39 (1990).
- Aguilar E., Torres M.A., Ramos J.M., Oztolaza A., Gómez M., Menéndez S., García R., Guza L.A., Vargas M., Verdecia M. y Lezcano G. "Recuperación de la inmunosupresión humoral en un quemado crítico por ozonoterapia. Presentación de un caso", **Revista CENIC Ciencias Biológicas**, 20(1-2-3):106-110 (1989).
- Behar R., García C.E., Sardiñas J., Menéndez S., Lemagne C. y Alvarez C. "Tratamiento de la úlcera gastroduodenal con ozono", **Revista CENIC Ciencias Biológicas**, 20(1-2-3):59-61 (1989).
- Calvo H., Menéndez S., Gómez M., Lacasa A. y Molerio J. "Experiencias preliminares en la utilización del ozono en pacientes de terapia intensiva del Hospital "Carlos J. Finlay", **Revista CENIC Ciencias Biológicas**, 20(1-2-3):128-135 (1989).
- Ceballos A., Balmaseda R., Wong R., Menéndez S. y Gómez M. "Tratamiento de la osteoartritis con ozono" , **Revista CENIC Ciencias Biológicas**, 20(1-2-3):151-153 (1989).
- De las Cajigas T., Bastard V., Menéndez S., Gómez M. y Eng L. "El aceite ozonizado en infecciones de la piel y su aplicación en el consultorio del médico de la familia", **Revista CENIC Ciencias Biológicas**, 20(1-2-3):81-84 (1989).
- Delgado J., Wong R., Menéndez S. y Gómez M. "Tratamiento con ozono del Herpes Zoster", **Revista CENIC Ciencias Biológicas**, 20(1-2-3):160-161 (1989).
- García R., Menéndez S., Gómez M., Cuza L.A., Ramos J., Sanfiel A., Díaz W., Verdecia M., Vargas M., Lezcano G., Torres M.A. y Enríquez E. "El ozono como coadyuvante en el tratamiento de un paciente quemado crítico" , **Revista CENIC Ciencias Biológicas**, 20(1-2-3):111-115 (1989).
- García R., Ruíz E., Ramos R., Verdecia M., Cuza L.A., Vargas M., Lezcano G., Torres M.A., Menéndez S., Gómez M. y Enríquez E. "Estudio electrocardiográfico en grandes quemados tratados por autohemoterapia con ozono", **Revista CENIC Ciencias Biológicas**, 20(1-2-3):104-106 (1989).
- Hernández F., Menéndez S. y Eng L. "Efecto de la ozonoterapia intravascular sobre el sistema de la Glutatión peroxidasa", **Revista CENIC Ciencias Biológicas**, 20(1-2-3):37-40 (1989).
- Menéndez F., Díaz G. y Menéndez S. "Ozonoterapia en la artritis reumatoidea", **Revista CENIC Ciencias Biológicas**, 20(1-2-3):144-151 (1989).
- Menéndez S., Peláez O., Gómez M., Copello M., Mendoza M. y Díaz W. "Aplicación de la ozonoterapia en la retinosis pigmentaria", **Revista CENIC Ciencias Biológicas**, 20(1-2-3):84-90 (1989).
- Menéndez S., Peláez O., Gómez M., Copello M., Mendoza M. y Díaz W. "La ozonoterapia en el campo de la oftalmología", **Revista CENIC Ciencias Biológicas**, 20(1-2-3):91-93 (1989).
- Quiñones M., Menéndez S., Gómez M. et al. "Ozonoterapia en el tratamiento de las úlceras de miembros inferiores causadas por insuficiencia venosa crónica", **Revista CENIC Ciencias Biológicas**, 20(1-2-3):76-81 (1989).

- Rabell S., Menéndez A., Alonso P.L., Ruibal A. et al. "La terapia con ozono y la prevención de la sepsis en el enfermo crítico", **Revista CENIC Ciencias Biológicas**, 20(1-2-3):124-127 (1989).
- Ramos J., Torres M., Aguilar E., Ostolaza A., Gómez M., Menéndez S., García R., Cuza L.A., Vargas M., Verdecia M. y Lezcano G. "Estudio inmunológico de 25 pacientes grandes quemados tratados con ozono", **Revista CENIC Ciencias Biológicas**, 20(1-2-3):116-120 (1989).
- Rodríguez B.R., Menéndez S., Quesada X., Vecino C. y Herrera F. "Utilización de la ozonoterapia en el tratamiento de las hiperlipidemias", **Revista CENIC Ciencias Biológicas**, 20(1-2-3):153-156 (1989).
- Romero A., Menéndez S., Gómez M. y Ley P. "La ozonoterapia en la claudicación intermitente de evolución desfavorable", **Revista Cubana de Cirugía**, 28(6):543-548 (1989).
- Romero A., Menéndez S., Gómez M., Díaz W. y Carballo A. "La ozonoterapia en la aterosclerosis obliterante", **Revista CENIC Ciencias Biológicas**, 20(1-2-3):70-76 (1989).
- Santiesteban R., Menéndez S., Gómez M. y Francisco M. "Ozonoterapia en la atrofia óptica", **Revista CENIC Ciencia Biológicas**, 20(1-2-3):93-99 (1989).
- Sardiñas J., Behar R., Garcia C.E., Menéndez S. et al. "Tratamiento de la giardiasis recidivante con ozono", **Revista CENIC Ciencias Biológicas**, 20(1-2-3):61-64 (1989).
- Torres M.A., Agular E., Ramos J., Guza A., Oztolazaba L.A., Verdecia M., Vargas M., Lezcano G., García R., Menéndez S., Gómez M. y Enríquez E. "Coagulación intravascular diseminada y ozonoterapia. Criterios de evaluación", **Revista CENIC Ciencias Biológicas**, 20(1-2-3):120-124 (1989).
- Triana I., Menéndez S., Peláez O. et al. "La ozonoterapia en el campo de la oftalmología", **Revista Cubana de Oftalmología**, 2(3):168-172 (1989).
- Velasco N., Menéndez S., Fernández J. et al. "Valor de la ozonoterapia en el tratamiento del pie diabético neuroinfeccioso", **Revista CENIC Ciencias Biológicas**, 20(1-2-3):64 - 70(1989).
- Wong R., Ceballos A., Menéndez S. y Gómez M. "Ozonoterapia analgésica", **Revista CENIC Ciencias Biológicas**, 20(1-2-3):139-144 (1989).
- Wong R., Rivero R., Menéndez S. y Gómez M. "Ozonoterapia en la esteatosis hepática", **Revista CENIC Ciencias Biológicas**, 20(1-2-3):157-159 (1989).
- Wong R., Soler A., Torrientes D., Noriega A., Menéndez S. y Gómez M. "Ozonoterapia y estudio gasométrico", **Revista CENIC Ciencias Biológicas**, 20(1-2-3):136-139 (1989).
- Ruíz A., León R., Menéndez S. y Villalonga J. "Cimetida vs agua ozonizada en el tratamiento de la úlcera gastroduodenal. Estudio preliminar", **Revista Cubana de Medicina**, 27(3):7-13 (1988).

OZONIZED SUBSTANCES APPLIED TO OZONE THERAPY

- **Clinics:**
- Falcón L., Daniel R., Menéndez S., Landa N. y Moya S. "Solución para la epidermofitosis de los pies en integrantes de las FAR", **Rev. Cubana de Medicina Militar**, 29(2):98-102 (2000).
- Falcón L., Menéndez S., Daniel R., Garbayo E., Moya S., Abreu M. "Aceite ozonizado en Dermatología. Experiencia de 9 años", **Revista CENIC Ciencias Biológicas**, 29(3):192-195 (1998).
- Hernández F. y Menéndez S. "Aspectos bioquímicos en el uso de aceite ozonizado para el tratamiento de la giardiasis. Estudio preliminar", **Revista CENIC Ciencias Biológicas**, 28(1):3-6 (1997).
- Cruz O., Menéndez S., Reyes O. y Díaz W. "Aplicación de la ozonoterapia en el tratamiento de conductos radiculares infectados", **Revista Cubana de Estomatología**, 31(2):47-51 (1994).

- Mena L., Menéndez S. y Omechevarría E. "Efectos del ozono en el tratamiento de la gingivoestomatitis herpética aguda", **Revista Cubana de Estomatología**, 31(1):14-17, (1994).
- Grillo R., Falcón L. y Menéndez S. "Tratamiento de herpes simple genital con aceite ozonizado. Estudio Preliminar", **Revista Cubana de Medicina Militar**, 4(1) (1990).
- De las Cajigas A., Pérez A., Menéndez S. y Gómez M. "Estudio mutagénico del ozono por autohemoterapia", **Revista CENIC Ciencias Biológicas**, 20(1-2-3):41-44 (1989).
- Wong R., Delgado J., Menéndez S. y Gómez M. "Ozonoterapia y úlcera lingual", **Revista CENIC Ciencias Biológicas**, 20(1-2-3):103-104 (1989).

Microbiology:
- Sechi L.A., Lezcano I., Nuñez N., Espino M., Dupre I., Pinna A., Molicotti P., Fadda G., Zanetti S. "Antibacterial activity of ozonized sunflower oil (Oleozon)", **J Appl. Microbiology**, 90(2):279-284 (2001).
- Lezcano I., Núñez N. Espino M. and Gómez M. "Antibacterial activity of ozonized sunflower oil, Oleozon, against *Staphylococcus aureus* and *Staphylococcus epidermis*", **Ozone Sci. & Eng.**, 22(2):207-214 (2000).

Chemistry:
- Díaz M., Lezcano I., Molerio J. and Hernández F. "Spectroscopic characterization of ozonides with biological activity", **Ozone Sci. & Eng.**, 23(1):35-40 (2001).
- Menéndez S., Falcón L. et al. "Ozonized sunflower oil in the treatment of tinea pedis", **Mycoses**, 44 (2001).

Toxicology:
- Jardines D., Ledea O., Zamora Z. "Triglicéridos insaturados ozonizados como precursores de ácidos dicarboxílicos urinarios de ratas Wistar", **Revista CENIC Ciencias Químicas**, 32(2):65-69 (2001).
- Jardines D., Zamora Z., Correa T., Rosado A. y Moleiro J. "Perfil de ácidos orgánicos urinarios en ratas tratadas con oleozon por vía oral", **Revista CENIC Ciencias Químicas**, 29(2):79-84 (1998).
- Fernández S.I., Quinsan C., Menéndez S. y Gómez M. "Evaluación mutagénica del aceite ozonizado administrado intragástricamente", **Revista CENIC Ciencias Biológicas**, 20(1-2-3):14-16 (1989).
- Fernández S.I., Quinzan C., Menéndez S. y Gómez M. "Estudio en animales de experimentación de posibles efectos teratogénicos y mutagénicos por vía intraperitoneal e intramuscular", **Revista CENIC Ciencias Biológicas**, 20(1-2-3):45-47 (1989).
- Gell A., Pérez O., Lastre M. et al. "Ozonoterapia en gerbils infectados experimentalmente con *Giardia lamblia*", **Revista CENIC Ciencias Biológicas**, 20(1-2-3):55-58 (1989).
- Noa M., Hernández F., Herrera M., Menéndez S., Capote A. y Aguilar C. "Estudio histológico de vías digestivas de ratones tratados con aceite ozonizado", **Revista CENIC Ciencias Biológicas**, 20(1-2-3):23-24 (1989).

Your comments and suggestions are well received. Contact to: ozono@infomed.sld.cu
(c) copyright 1999-2002 Ozone Research Center.

Reading, Viewing, & Listening

Ed McCabe Publications

Oxygen Therapies,

A New Way of Approaching Disease

O₃ VS. AIDS

Flood Your Body With Oxygen,

Therapy for Our Polluted World

Ed McCabe Audios

Introduction to Oxygen Therapies

Breaking the Ozone Silence

Ed McCabe Videos

Introduction to Oxygen Therapies

Ed McCabe on Medical Ozone

All Ed McCabe publications are available from
The Family Health News
9845 NE 2nd Ave.
Miami Shores, FL 33138
Tel: 305-759-9500
Tel: 305-759-8710

Ed's Suggested Background Music and Visuals
Wingmakers
'First Source'

Other Publications

The Use of Ozone in Medicine
Renate Viebahn-Haensler
From: The Family News
9845 NE 2nd Ave.
Miami Shores, FL 33138
Tel: 305-759-8710
Tel: 305-759-9500

NEW! *Ozone, A Medical Breakthrough?*
(Film and Video)
and
Ozone & The Politics of Medicine (Film and Video)
Geoff Rogers
Threshold Films
#141-1857 West 4th. Avenue
Vancouver, B.C. V6J 1M4, Canada
Tel: 800-216-4403

The Peroxide Story
George Borell
ECHO 300 South 4th Street
Delano, MN, 55328

For More Information About Low Level Laser Therapy
2035, Inc.
Email 2035@pobox.com
Tel: (877) 862-5669, Fax (650) 649-2642
Web http://www.wowapipublishing.com
(Mention this book and receive a discount!)

Hormone Heresy
What Women Must Know About Their Hormones
"Mothers Don't Let Your Daughters Get Cancer"
Sherrill Sellman, Get Well International
P. O. Box 690416, Tulsa, OK 74169
Info@ssellman.com
www.ssellman.com
Tel: 877-215-1721

Modern Foods, The Sabotage of Earths Food Supply
David Casper, MA & Thomas Stone, ND, CN
Center Point Press
PMB 143, 12463 Rancho Bernardo Road, San Diego, CA 92128
Tel: 858-513-4372

Fats that Heal Fats that Kill
To get your free tape and purchase Udo's books, tapes, and for more
information on how to eat. Tel: 1-800-446-2110. Udo's book is also
available through health food stores and bookstores. Website:
www.fatsthatheal.com

Udo Erasmus
Alive Books, 7436 Fraser Park Drive, Burnaby, BC Canada V5J 5B9
Tel: 604-435-1919

Cook Right 4 Your Type
Dr. Peter J.'Adamo
G.P.Putnam's Sons

Your Body's Many Cries for Water
ABC of Allergies Asthma and Lupus
F. Batmanghelidj, M.D.
(Available through book stores)

Dr Charles Farr
Complete set of medical peroxide references, treatises, workbooks
Available from:

IOMA
The Oxidative Medicine Assoc.,
P. O. Box 891954, Oklahoma City, OK 73189
Tel: 405 634 1310

𝕭𝖗𝖊𝖆𝖙𝖍 𝖔𝖋 𝕲𝖔𝖉 𝕾𝖔𝖈𝖎𝖊𝖙𝖞-

The Breath of God Society is a private grass roots gathering of like-minded individuals interested in oxygen and related issues.

Society members encourage:

1. The proper oxygenation and cleansing and repair and care of our bodies.

2. The proper oxygenation and cleansing and repair and care of our pets, our farm animals, and our crops.

3. The proper oxygenation and cleansing and repair and care of our environment.

Membership in the Breath of God Society also entitles members to a free subscription to the society e-newsletter:

INHALE!

The World of Oxygen Therapies

INHALE! Is the Breath of God Society oxy-newsletter, an e-mail publication that typically informs members about various oxygen related subjects, news, information, tips, and announcements. We stick to topics that interest you, and we keep the sends short and to the point.

Ed McCabe is the Chief Editor of INHALE! Subscribing to this newsletter is the way to get *on his mailing list*. Submissions are encouraged, and membership in the society and subscriptions to the newsletter are open to all breathers.

To join the Breath of God Society, go to the Breath of God Society website **www.breathofgodsociety.org** and click on 'subscribe'.

The INHALE! Newsletter is not sent on a regular basis; only when we have news for you that we believe you will wish to see. Although we might occasionally carry brief ad links to help defray mailing costs, unlike most newsletters, this is not an advertising vehicle. Please likewise be assured that your e-mail address is private, and used solely by your private society.

For 500+ searchable references, articles, and the latest information on all the Oxygen Therapy issues and equipment, surf over to Mr. Oxygen,[TM] Ed McCabe's[©] home website:

www.oxygenhealth.com

~ May the Blessings Be ~